THE ROUTLEDGE COMPANION
TO MEDIA AND TOURISM

The Routledge Companion to Media and Tourism provides a comprehensive overview of the research into the convergence of media and tourism and specifically investigates the concept of mediatized tourism.

This *Companion* offers a holistic look at the relationship between media and tourism by drawing from a global range of contributions by scholars from disciplines across the humanities and social sciences. The book is divided into five parts, covering diverse aspects of mediatization of tourism including place and space, representation, cultural production, and transmedia. It features a comprehensive theoretical introduction and an afterword by leading scholars in this emerging field, delving into the ways in which different forms of media content and consumption converge, and the consequential effects on tourism and tourists.

The collection is an invaluable resource for students and scholars of tourism studies, cultural studies, and media and communication, as well as those with a particular interest in mediatization, convergence culture, and contemporary culture.

Maria Månsson is a Senior Lecturer at the Department of Strategic Communication at Lund University, Sweden. She has a PhD in Service Studies from Lund University that focused on *mediatized tourism*. Her research deals with media's influence on tourism and tourists' performances at destinations, place marketing, and place branding with a particular focus on popular culture.

Annæ Buchmann has a PhD in Social Science and is a Lecturer in Events, Tourism, and Leisure at the University of Newcastle, Australia. Annæ has published widely in film/media tourism and horse tourism considering multispecies perspectives. Her interdisciplinary research looks at long-term strategies for creating sustainable communities and organisations in urban and regional destinations.

Cecilia Cassinger obtained her PhD in Marketing in 2010. She is an Associate Professor at the Department of Strategic Communication, Lund University, Sweden. Her research concerns the consequences and transformative potential of place brand communication and how communication is strategically used to mitigate conflicts among users in places, such as cities, regions, and nations.

Lena Eskilsson is a Senior Lecturer in Human Geography working at the Department of Service Management and Service Studies at Lund University, Sweden. She has a specialisation in place development and marketing with finalised research projects focusing on the bridging of the film and tourism sectors, as well as the influence of media on tourism and tourist behaviour at destinations.

'Through high-level theorizing, truly interdisciplinary lenses and illustrative case studies, this companion succeeds in merging media studies and tourism perspectives in critical ways that allow for new understandings of changing media landscapes and emerging tourist behaviors. The diversity of media channels, media contents, locations and forms of tourism discussed is impressive and genuinely engaging. A milestone for research on mediatized tourism!'

Ulrike Gretzel, Senior Fellow, Annenberg School for Communication and Journalism, University of Southern California.

'This book is set to become a must-have for those who are interested in understanding the interplay of the traditional and new media with the tourism industry. The thoroughly researched chapters deal with various theories, methodologies, countries, media channels and aspects of tourism including travel planning, decision making, experiences and marketing'.

Eli Avraham, Communication Department, University of Haifa, Israel

THE ROUTLEDGE COMPANION TO MEDIA AND TOURISM

Edited by Maria Månsson, Annæ Buchmann, Cecilia Cassinger, and Lena Eskilsson

Routledge
Taylor & Francis Group

LONDON AND NEW YORK

First published 2021
by Routledge
2 Park Square, Milton Park, Abingdon, Oxon OX14 4RN

and by Routledge
52 Vanderbilt Avenue, New York, NY 10017

Routledge is an imprint of the Taylor & Francis Group, an informa business

British Library Cataloguing-in-Publication Data
A catalogue record for this book is available from the British Library

Library of Congress Cataloging-in-Publication Data
Names: Månsson, Maria, editor. | Buchmann, Annæ, editor. | Cassinger, Cecilia, editor. | Eskilsson, Lena, editor.
Title: The Routledge companion to media and tourism / edited by Maria Månsson, Annæ Buchmann, Cecilia Cassinger and Lena Eskilsson.
Description: London; New York: Routledge, 2021. |
Includes bibliographical references and index. Identifiers: LCCN 2020002519 |
ISBN 9781138366282 (hardback) | ISBN 9780429430398 (ebook)
Subjects: LCSH: Tourism—Social aspects. | Mass media. | Popular culture.
Classification: LCC G155.A1 R6836 2020 | DDC 306.4/819—dc23
LC record available at https://lccn.loc.gov/2020002519

ISBN: 978-1-138-36628-2 (hbk)
ISBN: 978-0-367-49688-3 (pbk)
ISBN: 978-0-429-43039-8 (ebk)

Typeset in Bembo
by codeMantra

CONTENTS

Contents

FIGURES

TABLES

CONTRIBUTORS

Hazel Andrews (Dr) is a Reader in Tourism, Culture & Society at LJMU. Hazel is interested in issues of identity, selfhood, the body, and existential anthropology. Hazel is the author/editor of seven books that engage with these and other areas of research interest including *Liminal Landscapes: Travel, Experience and Spaces In-Between* (2012), *Tourism and Violence* (2014), *Tourism Ethnographies* (2018), and the monograph *The British on Holiday* (2011), which was featured in the Royal Anthropological Institute's *Reviewer meets Reviewed* series at the British Museum in London (2016). Hazel has given numerous guest lectures and conference presentations, including keynote speeches, organised international conferences, and symposia, and is chair of the RAI's Tourism Committee.

Alyaa Anter is an Assistant Professor at Mass Communication College at Ajman University, UAE. She received a PhD from Radio & TV department, Mass Communication faculty at, Cairo University, 2013. Her research interest centres on digital media, tourism communication, innovation in media, image of countries, and political communication. She participated in many international conferences.

Sue Beeton, Professor, is a travel and tourism researcher and writer, President of the Travel and Tourism Research Association, and Foundation Chair of the College of Eminent Professors at William Angliss Institute in Australia. For over a quarter of a century, Prof. Beeton has conducted tourism-based research into community development, film-induced tourism and pop culture, and nature-based tourism. As well as producing numerous academic papers, book chapters, and reports, Prof. Beeton has published a range of research-based books, including *Ecotourism: a practical guide for rural communities, Community Development Through Tourism* and *Tourism and the Moving Image*, as well as two editions of the acclaimed monograph, *Film-Induced Tourism.*

Pat Brereton is a Professor in the School of Communications at Dublin City University, Ireland. He has written extensively on all aspects of environmental media since his PhD was published as *Hollywood Utopia: Ecology in Contemporary American Cinema* (2005), while his latest book – *Environmental Literacy: New Digital Audiences* (Routledge 2019) – reiterates the need for environmentalism to permeate a broad range of disciplines. Furthermore, he has written extensively on Irish and new media and has always been interested in the growing links between film tourism and environmental engagement. More recently, he has worked on a number of audience research projects.

Annæ Buchmann (Dr) is a Lecturer in events, tourism, and leisure at the University of Newcastle, Australia. Dr. Buchmann has published widely in film/media tourism and has been a member of the 'International Tourism and Media' – network since its inception and presented at all its conferences. Annæ also researches horse tourism and recreation, and uses her Environmental Science background when addressing aspects of sustainability and resilience/emergency preparedness including disaster management. Her research is disseminated in engaging presentations and publications including peer-reviewed journal articles, industry newsletters, and newspaper articles, and Annæ is esteemed as an expert reviewer and member of advisory boards.

Susan Carson (Dr) is a Professor in the School of Communication, Creative Industries Faculty at Queensland University of Technology, Brisbane, Australia. In 2017, she co-edited *Performing Cultural Tourism: Communities, Tourists and Creative Practices* (Routledge) with Dr Mark Pennings. Dr Carson has published widely on cultural tourism in Australia and more recently on the relationship between tourism and digital media as well as Indigenous tourism. She is a Program Director in the School of Communication and has published on the scholarship of creative practice as research and postgraduate pedagogies.

Cecilia Cassinger obtained her PhD in marketing, 2010. She is an Associate Professor at the Department of Strategic Communication at Lund University, Sweden. Her research concerns the consequences and transformative potential of place brand communication and how communication is strategically used to mitigate conflicts among users in places, such as cities, regions, and nations. She has published articles in the academic journals such as *Place Branding and Public Diplomacy*, *Place Management and Development*, *International Journal of Tourism Cities*, and *European Journal of Cultural Studies*.

Maria Criselda G. Badilla is an Associate Professor of the Asian Institute of Tourism at the University of the Philippines, Diliman, Quezon City. She has obtained her Doctor of Philosophy in Communication from the same university. Her research interests are new media, tourism marketing, destination branding, and image formation.

Michael S. Daubs is a Senior Lecturer and the Programme Director for Media Studies at Victoria University of Wellington, New Zealand. His research interrogates and critiques the power and promise of mobile and ubiquitous media through investigations of apps and code, digital labour, and the mediatization of social movements. He is the co-editor (with Vincent R. Manzerolle) of *Mobile and Ubiquitous Media: Critical and Internal Perspectives* (Peter Lang, 2017).

Jörgen Eksell is an Assistant Professor at the Department of Strategic Communication at Lund University. He holds a PhD in Service Studies from Lund University, 2013. His research interests lie in place branding, hospitality studies, and strategic communication. The last few years he has participated in several multidisciplinary research projects focusing on contemporary urban challenges such as perception of safety, antitourism, and overtourism.

Lena Eskilsson is a Senior Lecturer in Human Geography working at the Department of Service Management and Service Studies at Lund University, Sweden. She has a specialisation in place development and marketing and has during the last years worked with research projects focusing on the bridging of the film and tourism sectors as well as the influence of media on tourism and tourist behaviour at destinations. Currently she is working on a multi-disciplinary project on sustainable urban tourism development.

José Fernández-Cavia, PhD, is an Associate Professor in Advertising and Public Relations at the Universitat Pompeu Fabra (Barcelona, Spain). He's the author of more than ten books and thirty journal articles on communication issues. He's also a founding member of the research group 'Communication, Advertising and Society' (CAS) and leads the research project ITOURIST: 'The tourist on the Web: informational habits and destination choice,' funded by the Spanish government. His main research interests lie in communication, advertising, and place branding. Currently he serves as Head of Department at the Department of Communication, UPF.

Natàlia Ferrer-Roca (Dr) is an Adjunct Professor at the Department of Organization, Business Management and Communication at the University of Girona, Catalonia. She has a PhD in Media Studies from Victoria University of Wellington (New Zealand), MA in Communications Policy (Distinction) from Westminster University (London), and a degree in Journalism from the Autonomous University of Barcelona. Her research is inter-disciplinary, connecting political economy and cultural economy of media, media policy, cultural industries, and destination branding. Moreover, she is Associate Director (Research) of The Place Brand Observer (http://placebrandobserver.com). Her research has been published in the *Journal of Media Business Studies, Place Branding and Public Diplomacy, Media Industries,* and *Studies in Australasian Cinema.*

Jennifer Frost (Dr) is an Associate Professor in the Department of Management, Sport and Tourism at La Trobe University, Australia. Her research interests include travel narratives, tourism and well-being, and the role of tourism and events in rural regeneration. She is a member of the organising committee for the International Tourism and Media network, which has staged 8 biennial conferences since 2004.

Warwick Frost (Dr) is an Associate Professor in the Department of Management, Sport and Tourism at La Trobe University, Australia. His research interests are broadly in the interplay of tourism, heritage, and the media. He is the convenor of the International Tourism and Media network, which has staged 8 biennial conferences since 2004. He is particularly interested in the shared cultural heritage of the nineteenth-century Pacific Goldfields of Australia, New Zealand, and California.

Francesc Fusté-Forné is a Researcher at the Blanquerna-Ramon Llull University School of Communication and International Relations (Barcelona, Catalonia), where he earned a PhD on Communication Studies. Earlier, he earned a PhD on Tourism, Law and Business at the University of Girona (Girona, Catalonia), where he is currently Associate Professor. His research is focused on the study of food. He is undertaking research on the role of gastronomy with regard to mass media and as a driver of social changes. Also, he studies culinary heritages from a geographical and tourist perspective.

Fani Galatsopoulou (Dr) is a Senior Teaching Fellow and Researcher at the School of Journalism and Mass Communications at the Aristotle University of Thessaloniki, Greece.

She is an education specialist in the fields of lifelong continuing education and adult professional training in tourism. She has 24 years of teaching experience in graduate and postgraduate level. Her main research interests lie in the fields of formal and informal education and vocational training, digital media, and communication in tourism and travel. She has participated in several research projects, conferences, and conventions about digital media, travel journalism, and sustainable tourism.

Ana Oliveira Garner (Dr) is a media and communications scholar, and documentarist, who has been teaching audiovisual and communications since 2009. In her PhD thesis, she analysed how people use social media to tell personal stories via Micro Visual Narratives and created an app that aimed at reimagining social media. Her main interests include Audiovisual; Critical Analysis of Media; Social Media; Digital Narratives; and Cultural Studies. A passionate traveller and world culture explorer, she has lived in Brazil, the USA, Spain, Hong Kong, and the UK, and runs a blog about cultural differences and travel tips: abordodomundo.com

Nicolai J. Graakjær (b. 1972, Danish) is a Professor of Music and Sound in Market Communication in the Department of Communication and Psychology at the University of Aalborg, Denmark. His research interests span from musicology, sound studies, market communication, media studies to social psychology. His publications include *Analyzing Music in Advertising* (Routledge, 2015), *Sound and Genre in Film and Television* (MedieKultur, 2010, Co-editor), *Music in Advertising: Commercial Sounds in Media Communication and other Settings* (Aalborg University Press, 2009, Co-editor) as well as papers in, e.g., *Critical Discourse Studies, Visual Communication, European Journal of Marketing,* and *Popular Music and Society.*

Rasmus Grøn (b. 1972, Danish) is an Associate Professor in the Department of Communication and Psychology at Aalborg University. His research interests and publications cover the fields of human geography, aesthetics, experience theory, critical design, and media sociology.

Szilvia Gyimóthy is an Associate Professor in Tourism Marketing at Copenhagen Business School. Her research is focused on how tourism is shaping places and place-making practices in the wake of global mobility and mediatized travel. Szilvia is an interdisciplinary scholar, bridging across the fields of market communication and branding, tourism geography, and consumer culture studies. Her research projects explored the commodification of rural places and regions along popular consumer culture trends, including the Nordic terroir, adventure sports, Bollywood films, and contemporary pilgrimage. More recently, Szilvia has been working with conceptualising sharing economy phenomena, and the marketisation of social relationships in communitarian businesses.

Tim Hannigan is a Midlands4Cities/AHRC-funded PhD candidate at the University of Leicester, UK, working on ethical issues in contemporary British travel writing. He is also an experienced guidebook writer, and author of several narrative history books, including *Murder in the Hindu Kush* (The History Press 2011), *Raffles and the British Invasion of Java* (Monsoon 2012), and *A Brief History of Indonesia* (Tuttle 2015). His research has been published in the *Journal of Commonwealth Literature, Journeys, Studies in Travel Writing,* and *Terrae Incognitae*, and he contributed the chapter on Robert Macfarlane to *The Handbook of British Travel Writing* (De Gruyter, 2019).

Kyungjae Jang is an Associate Professor in the Department of Integrated Global Studies, School of Integrated Arts and Sciences at Hiroshima University. He holds a PhD and MA in tourism studies from Hokkaido University, and a BA from Korea University. Dr. Jang has conducted participatory research on transnational Japanese contents tourism in the USA, Tunisia, Korea, and Taiwan. His previous publications include *Contents Tourism in Japan* (with Philip Seaton, Takayoshi Yamamura, and Akiko Sugawa-Shimada), and "Between Soft Power and Propaganda: The Korean Military Drama Descendants of the Sun," *Journal of War & Culture Studies* (2019).

André Jansson is a Professor of Media and Communication Studies at Karlstad University, Sweden. He is also the Director of the Geomedia Research Group at the same university. His work gravitates around questions of mediatization, space/place, and power relations in culture and society. His most recent books are *Transmedia Work: Privilege and Precariousness in Digital Modernity* (Routledge, 2019, with Karin Fast) and *Mediatization and Mobile Lives: A Critical Approach* (Routledge, 2018). He has published articles on media and tourism in several journals, including *Annals of Tourism Research, Tourist Studies,* and *European Journal of Communication.*

Philipp Dominik Keidl is a Postdoctoral Fellow in the DFG-Graduiertenkolleg "Configurations of Film" at Goethe University, Frankfurt. Philipp's most recent publications concentrate on non-fiction fan practices, transmedia, material culture, and moving image heritage. He holds a PhD in Film and Moving Image Studies from Concordia University in Montreal, and an MA in Preservation and Presentation of the Moving Image from the University of Amsterdam. Before joining Goethe University, Philipp taught film studies at the University of Northern British Columbia and was a Research Fellow at Tallinn University.

Clio Kenterelidou (Dr) Economist and Communications Specialist (MA, PhD) has significant academic and professional experience in Public and Political Communication, communications strategy, participation-development-public relations, strategic planning. She has participated in international research initiatives and has produced research papers in journals and refereed conference proceedings, authored chapters in books, and co-edited one collective volume: all in the field of Communication, Journalism, and the Media. She is an Expert of the European Commission for Research and Innovation, for Communication, the Hellenic General Secretariat of Research and Technology, and for the European Communication Monitor (EUPRERA & European Association of Communication Directors).

Kathleen M. Kuehn is a Senior Lecturer in Media Studies at Victoria University of Wellington. Her research focuses on the intersection of digital media, cultural labour, and consumer culture. She is the author of *The Post-Snowden Era: Mass Surveillance and Privacy in New Zealand* (BWB Texts, 2016) and is currently researching the cultural labour of women in the craft beer industry.

Giulia Lavarone (PhD in Film Studies, University of Padua, 2010) is a Research Fellow at the University of Padua (Italy). She has authored several scholarly articles on films by Jacques Rivette, Agnès Varda, Matteo Garrone, Quentin Tarantino, and many others, focusing on intertextuality, the relationship between cinema and other arts, cinema and the city, film and landscape. In 2014–2018, she undertook postdoctoral research in Padua within two interdisciplinary projects on film-induced tourism, publishing scholarly articles, and the book

Cinema, media e turismo (Padova University Press). She teaches in Padua and in Milan (Italy) at the Università Cattolica del Sacro Cuore.

Sara Leckner holds a PhD in Media Technology from the Royal Institute of Technology in Sweden. She is currently working as an Associate Professor at the Department of Computer Science and Media Technology at Malmö University, Sweden. Her research focuses on new media devices and services, multiple channel publishing, media perception and utilisation, and media theory. Currently she is a member of the Fair Data project focusing on business solutions and fair data handling of commercial user data.

Antonio Loriguillo-López is a Post-Doctoral Fellow at the Communication Sciences Department in Universitat Jaume I de Castelló de la Plana (Spain). He holds a PhD in Communication Sciences by Universitat Jaume I, where he has taught Visual Narrative and Video Games Analysis. His main research interests lie in the fields of anime and post-classical narration in contemporary film and television. His research has been published in journals such as *Quarterly Review of Film and Video* or *International Journal on Media Management*.

Yunuen Ysela Mandujano-Salazar is an Associate Professor at the Autonomous University of Ciudad Juarez (Mexico). She holds a Doctorate in Social Sciences and a Master in Studies of Asia and Africa Specialty Japan. Her research interests focus on contemporary Japanese culture and society, particularly on domestic media discourses. Her recent publications include 'Media idols and the regime of truth about national identity in post-3.11 Japan' in the *Routledge Handbook of Japanese Media* edited by Darling-Wolf (2018) and the article 'Exploring the construction of adulthood and gender identity among single childfree people in Mexico and Japan' (2019).

Maria Månsson is a Senior Lecturer at the Department of Strategic Communication at Lund University, Sweden. She has a PhD in Service studies from Lund University that focused on *Mediatized tourism*. Her research deals with media's influence on tourism and tourists performances at destinations, place marketing, and place branding with a particular focus on popular culture. She is currently involved in research projects dealing with different urban challenges such as safety and risk perceptions in regard to travelling and issues related to overtourism. She is a member of the organisation of *the International tourism and media* network and has organised the biannual conference.

Pere Masip is a Professor in the Blanquerna School of Communication and International Relations at Ramon Llull University (Barcelona, Catalonia). He holds a PhD in Communication. His main research interests are digital journalism, and the impact of technology on journalistic and communication practices. He has participated in several national and international projects. He is currently coordinating a research project funded by the Spanish Ministry of Economy and Competitiveness entitled Active audiences and agenda-setting in the Digital Public Sphere. He has published articles in journals such as *The International Journal of Press and Politics*, *Journalism Studies*, *Digital Journalism*, *International Communication Gazette*, *American Behavioral Scientist*, and *Journalism Practice*.

Aine Mc Adam is a PhD candidate at the Department of Sociology at Maynooth University, Ireland. She is the recipient of the prestigious John and Pat Hume Scholarship and is currently working towards the completion of her interdisciplinary thesis. As a tutor over the last number of years in sociology, Aine is acutely aware of the needs of undergraduate students and has taken courses in Tourism and Media studies. She currently also works as an expert research advisor for an Irish national children's charity.

Katarina Miličević, PhD, is a CEO and Co-founder of *thinktourism*, the innovative startup supporting tourism planners and travellers with the creative and data-driven insights and solutions. Her professional expertise is in strategic tourism planning, market research, destination marketing, and branding, featured in a number of scientific and professional publications. She is an Assistant Professor at the School of Economics and Business at the University of Ljubljana (Slovenia) and a Senior Lecturer at the Zagreb School of Economics and Management (Croatia). She is also a President of HI CROATIA (Croatian Youth Hostel Association), which is a member of Hostelling International.

Sabrina Mittermeier (University of Augsburg) completed her dissertation, a cultural history of the Disneyland theme parks at LMU Munich in 2018; it is slated to be published with Intellect Books in 2020. She is the co-editor of *The Routledge Companion to Star Trek* (2021), a forthcoming essay collection on *Star Trek: Discovery* (Liverpool UP 2020), as well as the volume *Here You Leave Today – Time and Temporality in Theme Park*s (Wehrhahn 2017) and has further published on diverse topics of American popular culture. She is working on a postdoc project on LGBT Public History in the US and West Germany.

Christian Hviid Mortensen, PhD, is a former curator of media heritage at the Media Museum in Denmark (2007–2018). Currently Christian is a Postdoc at the IT University of Copenhagen, where he researches the dynamics of hybrid heritage experiences spanning the digital and physical realms. His research interest is the dynamics between media, culture, heritage, and memory. Christian is also on the editorial board of *MedieKultur: Journal of Media and Communication Research*.

Gianna Moscardo is a Professor in the College of Business, Law and Governance at James Cook University. Her research interests include customer and tourist experience design, heritage interpretation, sustainability communication, and new approaches to tourism planning and development for sustainability. Dr. Moscardo is the current Chair of the Building Excellence in Sustainable Tourism Education Network (BEST EN), an international organisation committed to the creation and dissemination of knowledge to support the sustainable tourism education, research, and practice.

Apoorva Nanjangud (1992) is a PhD candidate on the ERC-funded project, 'Worlds of Imagination' at the Department of Arts and Culture Studies at Erasmus University Rotterdam. Under the supervision of Prof. Stijn Reijnders, her project looks into contemporary Bollywood tourism practices and examines the transnational tourism flows generated thereof.

Jan Henrik Nilsson holds a PhD in social and economic geography from Lund University, Sweden. At present, he is an Associate Professor in economic geography at the Department of Service Management and Service Studies at Lund University. His main research interest lies within the geographies of mobility, transport, tourism, and hospitality, and their relationship to environmental sustainability and climate change. He has published a number of international journal articles and book chapters on tourism and hospitality, transport geography, urban geography, and the historical geography of tourism and hospitality. He is currently working on a multi-disciplinary project on sustainable urban tourism development.

Adriaan Odendaal is a designer and web-developer from South Africa, currently working for a social entrepreneurship startup from Johannesburg. He completed his undergraduate in Visual Studies and Film Studies at Stellenbosch University and the University of Cape Town and recently graduated with his MA in Media Arts Cultures from Aalborg University as part of an Erasmus Mundus Joint Master Degree. His focus is on critical design, digital culture, game studies, and software studies. He has worked on game design, as well as critical and speculative design projects such as the *Sound Souvenirs* along with co-author Karla Zavala.

Carl Magnus Olsson is an Associate Professor at the Department of Computer Science and Media Technology at Malmö University in Sweden. His research interests lie in experiential computing and the challenges for designing consumer technologies and software products. He has spent time as guest researcher at Northern Illinois University, and has a PhD from the University of Limerick, Ireland. He has been a project manager for several research projects related to data-driven innovation that improves public services.

Barbara Pavlakovič is a Teaching Assistant at the University of Maribor, Faculty of Tourism, Slovenia. Her working and research fields cover various aspects of tourism, communication and safety in tourism. She graduated at Faculty of Social Sciences at the University of Ljubljana in communication studies, developed her public relation skills in several organisations as PR practitioner, and continued her career in academia. Currently, she is a PhD student at Faculty of Organizational Sciences at the University of Maribor.

John Pearce is a Lecturer and Research Officer in the College of Business Law and Governance at James Cook University. He holds undergraduate degrees in marketing and psychology and has spent several years involved in tourism industry research and practice. He is currently a graduate student at Southern Cross University working on storytelling and its role in destination marketing. He is passionate about sustainability, and his current research interests include pro-environmental behaviour, using novel methodologies, mindfulness, and experimental social psychology.

Mark Pennings (Dr) is a Senior Lecturer in Visual Arts at Queensland University of Technology, Brisbane, Australia. In 2017, he co-edited *Performing Cultural Tourism: Communities, Tourists and Creative Practices* (Routledge) with Associate Professor Susan Carson. His research focuses on contemporary art and culture, and art and the experience economy in the context of tourism. Recent papers include 'Potential affordances of public art in public parks: Central Park and the High Line' in *Proceedings of the Institution of Civil Engineers – Urban Design and Planning (2017)* and '*Art museums and the global tourist: Experience Centers in experiencescapes*' *Athens Journal of Tourism* (2015).

Les Roberts is a Senior Lecturer in Cultural and Media Studies at the University of Liverpool. The core focus of his work is situated within the interdisciplinary fields of spatial anthropology and spatial humanities. He is the author/editor of a number of books that engage with these and other areas of research interest, including *Deep Mapping* (2016); *Locating the Moving Image: New Approaches to Film and Place* (2014); *Liminal Landscapes: Travel, Experience* *and Spaces In-Between* (2012); *Film, Mobility and Urban Space* (2012); and *Mapping Cultures: Place, Practice, Performance* (2012). His latest book, *Spatial Anthropology: Excursions in Liminal Space*, was published by Rowman & Littlefield in 2018.

Sofia Sampaio is a Senior Researcher at the Centre for Research in Anthropology (CRIA) of the Instituto Universitário de Lisboa (ISCTE-IUL) and a Visiting Fellow at the School of Social Sciences (CPDOC) of the Fundação Getúlio Vargas in Brazil. She holds a PhD in Cultural Studies from the University of Lisbon and an MA in Anglo-American Studies from the University of Porto. Her research focuses on moving images and tourism-related visual practices, combining film studies, archival work, and anthropological approaches to tourism and cinema.

Philip Seaton is a Professor in the Institute for Japan Studies at Tokyo University of Foreign Studies. His research concentrates in two interlinked areas: war memories and contents tourism. He analyzes representations of history within media and tourist sites, and their relationship to collective memory. He is the author/editor of five books, including *Contents Tourism in Japan* (2017, Cambria Press, co-authored with Takayoshi Yamamura, Akiko Sugawa-Shimada, and Kyungjae Jang) and *Contents Tourism and Pop Culture Fandom* (in press, Channel View Publications, co-edited with Takayoshi Yamamura), and has guest edited special editions of *Japan Forum* and *Journal of War & Culture Studies*. His website is www.philipseaton.net

Tom Sintobin is an Assistant Professor of Cultural Studies at Radboud University, Nijmegen, the Netherlands. His research interests include Dutch literature from the 19th and 20th century, Dutch and Belgian culture around 1900 and Cultures of Tourism. He wrote a monograph on the Belgian author Stijn Streuvels (*Wie schaft er op de woorden? Vijf keer Streuvels lezen* (KANTL, 2005)) and recently co-edited the volume *Gender, Companionship, and Travel. Discourses in Pre-modern and Modern Travel Literature* (Routledge, 2018).

Božo Skoko, PhD, is a Professor at the Faculty of Political Science of the University of Zagreb where he is Head of the Public Relations postgraduate study. The areas of his scientific research include communication, international relations, national identity, and image and branding destination. He is a long-time strategic communications consultant and co-founder of Millenium promocija, the leading Croatian public relations agency. He is a former journalist and editor at Croatian Radio television (HRT). He has published seven books and over seventy scientific papers on public relations, the media and managing the identity and image of Croatia.

Åsa Thelander is an Associate Professor at the Department of Strategic Communication at Lund University. Her main research focus is in the field of visual communication. It includes studies of communication strategies involving visuals. It has resulted in publications focused on advertising, Instagram, and art events. The research interest also includes studies of the visual practices i.e. taking and making photographs in different contexts and media. Aside the interest in visual communication, Åsa has experience from and published articles on qualitative methods, particularly visual methods.

Anke Tonnaer is an Assistant Professor of Anthropology and Development Studies at Radboud University, Nijmegen, the Netherlands. Her research interests include indigenous tourism, heritage, and practices of representation, with a regional focus on the Pacific, particularly Indigenous Australia. Amongst others, she has written on the place of memory in tourism (2016) and the role of wilderness (forthcoming, 2019). Other current work addresses 'new wilderness' projects in Europe, including the return of the wolf (forthcoming).

Andreja Trdina is an Assistant Professor at the University of Maribor, Faculty of Tourism, Slovenia. She has a PhD in Media Studies and in her research focuses on popular culture, media, and class and distinction with special regard to contemporary material/consumer culture. She is currently dealing also with research on mediatization of tourism, travel as social and cultural practice, and politics of mobility and belonging. She has participated in various research projects and among others published articles in *Javnost-The Public, Slavic Review,* and *Comedy Studies.*

Maja Turnšek is an Assistant Professor and Vice-Dean for Research at the University of Maribor, Faculty of Tourism, Slovenia. Her background is in media and communication studies. She lectures on communication, psychology, and marketing in tourism. Her main research interests cover political economy of new media, sharing economy, travel-related platform work, storytelling and humour in tourism, and experience economy.

Anne Marit Waade is a Professor in Global Media Industries at Aarhus University. Her main research interests include Nordic noir, location studies, aesthetics in television drama, crime series and travel series, creative industry, the export of Danish television series, place branding, location placement, and screen tourism. Her publications include titles such as 'Just follow the trail of blood': Nordic Noir tourism and screened landscapes (2020), *Locating Nordic Noir* (2017), 'Melancholy in Nordic Noir' (2017), 'Local Colour in German and Danish Television Drama' (2015), 'When Public Service Drama Travels' (2016), and *Wallanderland* (2013).

Kim Williams (Dr) is a Lecturer in the Faculty of Higher Education at William Angliss Institute, Melbourne, Australia. Kim lectures in event management and human resources. Her research background is diverse but tends to focus on human resources issues, with a prime emphasis on professional development and training. She is also interested in event management, gastronomy, fashion, and wine tourism. Kim has published in a variety of journals such as *Journal of Heritage Tourism, Managing Leisure, Journal of Teaching in Travel & Tourism,* and *The Australian Journal of Teacher Education,* and has contributed to a number of research books.

Rebecca Williams (Dr) is a Senior Lecturer in Communication, Culture and Media Studies at the University of South Wales. She is the author of *Post-object Fandom: Television, Identity and Self-Narrative* (2015, Bloomsbury) and editor of *Torchwood Declassified* (2013, I.B. Tauris) and *Transitions, Endings, and Resurrections in Fandom* (in press, University of Iowa Press). She is currently writing *Theme Park Fandom: Distinction, Immersion & Participatory Culture* for University of Amsterdam Press.

Nicholas Wise is a Reader in International Urban Change in the Faculty of Business and Law at Liverpool John Moores University. He has published broadly in the areas of tourism studies focusing on place, destination image, and regeneration linked to his background in human geography. His recent tourism research appears in *Tourism Management Perspectives, Cities,* and *Journal of Community Psychology.* He has edited/co-edited several special issues of journals and he has co-edited eight books, including *Sport, Events, Tourism and Regeneration;*

Urban Transformations: Geographies of Renewal and Creative Change; *Urban Renewal, Community and Participation: Theory, Policy and Practice*; and *Events, Places and Societies*.

Jia Xie (Dr) is a Postdoctoral Research Fellow in the School of Tourism Management at Sun Yat-sen University, China. She earned her doctorate in sociology at the University of Glasgow. Her primary research interest is backpacker tourism, sociology of tourism, and new media studies.

Takayoshi Yamamura (PhD) is a Professor of Center for Advanced Tourism Studies at Hokkaido University, and he holds a PhD in urban engineering from the University of Tokyo. He is one of the pioneers of 'Contents Tourism' and 'Anime Induced Tourism' studies in Japan, and he has served the Chair of several governmental advisory boards such as the Meeting of International Tourism Promotion through Animation Contents of The Japan Tourism Agency, etc. His main English work includes *Contents Tourism and Pop Culture Fandom* (co-edited with P. Seaton, 2020), *Contents Tourism in Japan* (with P. Seaton, A. Sugawa-Shimada and K. Jang, 2017).

Karla Zavala is a designer and project manager from Peru. She completed her undergraduate in Communication for Social Development at the University of Lima and recently graduated with her MA in Media Arts Cultures from Aalborg University as part of an Erasmus Mundus Joint Master. She has over 10 years of design and communications experience, having worked as a communications project manager for the Peruvian Ministry of Education and for the UN. As a researcher she has a particular interest in digital aesthetics, critical design, software studies, and digital culture.

Malin Zillinger earned her doctorate in human geography and has ever since been interested in questions related to tourism, time, and space. Working as a Researcher at the Department of Service Management and Service Studies at Lund University, she has been involved in tourism research projects on mobility, information search behaviour, guiding, and innovation. Besides such research fields, Zillinger has been interested in method development since the days when she wrote her thesis. Apart from research, Zillinger is teaching on tourism and methods, and contributing to deepening the dialogue between academia and industry.

ACKNOWLEDGEMENT

The book project is sprung from the related 7th International Tourism and Media (ITAM) conference in 2016, hosted by Lund University at Campus Helsingborg Sweden, where attendees from different disciplines presented and discussed media and tourism from a range of perspectives.

1

INTRODUCTION

In the juncture of media convergence and tourism – towards a research agenda

Maria Månsson, Cecilia Cassinger, Lena Eskilsson, and Annæ Buchmann

Introduction

In the world of media convergence, every important story gets told, every brand gets sold, and every consumer gets courted across multiple platforms.

(Jenkins, 2006, p. 3).

This chapter presents an overview of the extant research on media and tourism focusing particularly on how contemporary tourism is mediatized and presents the framework and organisation of this book.

The Routledge Companion to Media and Tourism is the first of its kind in offering an in-depth examination of the breadth and scope of the intersecting fields of media and tourism research.

The book deals with the production and consumption of tourism as it relates to and is transformed by media convergence. Media convergence is applied here in order to explore and contextualise changes in media and their consequences for tourism consumption. Convergence affects the relationship between existing technologies, industries, markets, genres, and audiences (Jenkins, 2004), and by extension the tourism system and its markets. The close interaction between different media products blur sector-related boundaries, which have become fluid. Moreover, media convergence implies the collapse of media texts and technological platforms into cultural performances like tourism (Jansson & Falkheimer, 2006; Jenkins, 2006). The intertwining of media content and processes of production and consumption has had and continues to have a profound effect on tourists, the tourism system, and its related industries. At its simplest, media changes the meaning of tourism in time and space by prolonging the consumption experience and journey. Travel starts long before arriving at the destination with online bookings, interactive maps through which one can experience local neighbourhoods, travelblogs, Instagram accounts, and Facebook groups through which connections are established. In addition, popular books, television series, films, and computer games attract unprecedented number of visitors and affect expectations of destinations. The tourism landscape is also increasingly fragmented, and tourists have more power to co-create and steer their journeys and experiences. This, in turn, creates an incredibly rich research landscape.

Our concern in editing this book is to add to the already existing body of research on the role of popular culture and media in tourism by considering the wider implications of media practices, logics, and conditions for contemporary tourism. In order to achieve this, we would like to shift the attention from the effect of media to a focus on the embeddedness of media in tourism experiences and practices and how media communicatively constructs a socio-cultural reality (cf. Couldry & Hepp, 2013). This book will explore and expand this notion. The aim of the volume is thus to provide a comprehensive account of state-of-the-art research that advance a specialised research agenda on the junction of media convergence and tourism. The volume brings together a large number of contributions across disciplines and geographies that shed light on tourism as a highly mediatized social and cultural process. These original contributions provide insights into multiple perspectives on the media and tourism relationship, whilst critically discussing the leading views in their disciplinary area. In summary, the following chapters illuminate the production and consumption of tourism as it relates to and is transformed by media convergence.

Tracing current trends in media and tourism

Tourism has long been intertwined with media. For example, Urry (1990) suggested that the tourist gaze could be influenced by non-tourist activities, including film and television, and Butler (1990) specifically stating that films play a vital role in influencing the travel preference. Since then a relatively wide and fragmented tradition has emerged around the relation between various types of media and tourism. This is timely since 'media deconstruct previous understandings of tourism with de-differentiation and dematerialization, renegotiate the reality of tourism (ontology), introduce a new way of seeing (epistemology), and provide new solutions (methodology) and instruments (methods) for today's tourism research' (He, Wu, & Li, 2019, n.p.). The rapid digitalisation of our modern society 'has played a crucial role in the construction of new notions of place' (Leotta, 2016, n.p.) and has profound consequences for the experience of everyday life and travel decision-making processes.

Yet, despite the growing interest for the interrelationship between media and tourism, the previous research is scattered and published in different outlets located in different disciplines. Several books have sought to lay the groundwork in this regard. The 2003 volume *Visual Culture and Tourism*, edited by Lubbren, demonstrated that tourism and visual culture have a long-standing history of mutual entanglement and was the first known work to specifically discuss these two aspects together. *The Media and the Tourist Imagination. Converging Cultures* (2005) edited by Crouch, Jackson, and Thompson examines the way media re-imagines travel and tourism and how tourism practices are affected or altered by the media. *Mediating the Tourist Experience. From Brochures to Virtual Encounters* (first published in 2013 and then re-published 2016) edited by Scarles and Lester also demonstrated how nowadays tourists play an active role in co-creation through the mediation of their own tourist experience as well as the experiences of other tourists. In addition, *The Routledge Handbook of Cultural Tourism* (2013) edited by Smith and Richards looked at the tourist experience itself and the various ways in which tourists seek to be more actively and interactively engaged. Several contributions consider demographic changes in engaging with and using media in the tourism experience. In further detail, *The Routledge Handbook of Popular Culture and Tourism* (2018) edited by Lundberg and Ziakas specifically addresses popular cultural production and consumption trends, analysing their consequences for tourism, spatial strategies, and destination competitiveness. In an important attempt to theoretically consolidate the field Månsson (2011) advocated the concept of mediatized tourism. However, current research remains fragmented and too often focuses on mere descriptions or singular case studies with little strategic relevance for similar areas, such as literary and heritage tourism. Moreover, the emphasis is typically still on a Western perspective with only few exceptions (e.g., Iwashita, 2003; Kim, 2012; Kim, Long, & Robinson, 2009;). What is needed is a broader understanding of how tourism is mediatized by various actors and of the drivers and impacts of diverse forms of media and tourism.

Mediatized tourism – setting the scene

The space of media is transdisciplinary, and it has been suggested that it is no longer relevant to talk about individual media (e.g. television, film, social media, and so on) (Couldry & Hepp, 2013). Therefore, as an overarching perspective, this book is guided by the concept of media convergence and by extension mediatization. The term mediatization directs the focus to multidisciplinary studies concerned with broader societal and cultural processes and changes but also studies of the influence of particular media on particular fields and domains (Ekström, Fornäs, Jansson, & Jerslev, 2016, p. 1100). Here, we combine this approach with the concept of 'convergence' (Jenkins, 2006) to move distinctively beyond ideas such as media representations and the notion of the passive tourist.

Mediatization emerged as a concept in media and communication studies in the past decade to emphasise the transformative aspects of the contemporary media landscape (Couldry, 2008). The concept is used to analyse critically the interrelation between changes in media and communications on the one hand, and changes in culture and society on the other (Couldry & Hepp, 2013; Hepp & Krotz, 2014). Furthermore, mediatization emphasises the intensified and changing importance of media and how it spreads to other parts of our culture and society (Hjarvard, 2008, 2013, 2016). According to Kaun and Fast (2014, p. 12), 'mediatisation encompasses all processes of change that are media induced or that are related to a change in the media landscape over time'. Thus, it is a useful concept when it is used as a frame for understanding the relationship between media and cultural change (Hepp, 2009). It is in this sense that mediatization becomes interesting for the interrelationship of tourism and media. Schultz (2004) identifies four processes related to mediatization: extension, substitution, amalgamation, and accommodation. The first process focuses on how media extend human communication in place and time. The second process, substitution, stands for media's role in replacing social activities that formerly had to take place face-to-face. For example, tourists can now write about and post photos of places on social media like Facebook and TripAdvisor whilst travelling; before, it was not until tourists came home that they could show and tell about their travel experiences. Amalgamation represents the insight that 'media activities not only extend and (partly) substitute non-media activities; they also merge and mingle with one another' (Schultz, 2004, p. 88). Hence, media activities intermingle with other kinds of activities. The fourth process is accommodation, this process highlights media's influence on sectors outside the media sphere – for example, politicians' adaptation to the language of media when presenting themselves (Schultz, 2004). To begin with, mediatization was mainly addressed within the political field, now many other areas such as religion, culture and everyday life is in focus (Kaun & Fast, 2014). It is therefore also of relevance for tourism since media and tourism has always been intertwined. Mediatization can help to capture these intensified processes (cf. Andersson, 2017; Jansson, 2002; Månsson, 2011).

Jansson (2015, p. 82) argues that 'mediatization is a historical meta-process whereby a variety of social realms, in organizational settings as well as everyday life, become increasingly adapted to and dependent upon media technologies and institutions'. By applying mediatization, it is possible to address the transformative processes in everyday life that are influenced by an intertwining of media, culture, and society (cf. Jansson, 2015). From a media perspective, Jansson (2002) linked mediatization in tourism to the embeddedness of tourist gaze in the consumption of media images. Similarly, Jensen and Waade (2009) highlight mediatization, whilst exploring the interrelationship of media and tourism. They argue that media do not just change tourist's performances when they are visiting a destination, as media have a profound impact on all social interactions and ways of communication. Månsson (2011, 2015) argues from the perspective of a tourism scholar for the use of new media perspectives such as mediatization and convergence to understand the intertwining of media and tourism. Scarles and Lester (2013, 2016) build on this call for further studies to explore the interrelationship of media and tourism when they brought together a global spectrum of researchers. Their aim was to show the complexity of the processes in the relationship between tourism and media, in order to highlight the plurality that is involved in mediating tourist behaviours and destinations. Other researchers addressing similar issues are, for example, Reijnders (2011); Gyimóthy, Lundberg, Lindström, Lexhagen, and Larson (2015); and Gyimóthy (2018). Nevertheless, there is still a scarcity of research addressing media and tourism by the processes associated with mediatization (cf. Cohen & Cohen, 2012). The concept of mediatization is not straightforward and has been criticised for providing a linear account of media's influence on society (e.g. Corner, 2018). Here, we contend that mediatization is a useful concept to capture how different types of media shapes and frames tourists' experiences and practices.

This Companion is thus guided by notions of media convergence and mediatization and brings together scholars from different disciplinary backgrounds within the social sciences whilst addressing media and tourism from a range of perspectives. The contributions are highlighted in the next section that shows the different parts of the book and the included chapters.

Organisation of the book

This edited volume is composed of 44 chapters written by 62 authors. The contributions are organised according to five themes: part I) critical and conceptual entrance points to the field, part II) mediatized places and spaces, part III) the circle of representation, part IV) tourists as media producers, and part V) transmedia tourism. The themes serve as different ways of capturing and theorise the mediatization of tourism and traces its manifestations across various sites and platforms. The contributions are grouped based on topicality and relationship with the different themes, rendering a blending of contrasting or complementary accounts.

Part I – critical and conceptual entrance points to the field

The first part of the book deals with critical and conceptual entrance points into the field of media and tourism and presents a framework against which the other chapters in the book may be read. Part I opens with two invited contributions (Chapters 2 and 3) written by two leading scholars in media and communication studies, André Jansson professor at Karlstad University, Sweden, and Anne-Marit Waade professor at Aarhus University, Denmark. Both contributors deal with the research frontiers of the field and offer a meta-perspective on the embeddedness of media in tourism. In the subsequent Chapter 4, Sabrina Mittermeier argues that the theme park is a particularly interesting site at which to elaborate on the convergence of media and tourism due to its dual status as a highly mediatized place and tourist destination. Chapter 5, by Kathleen M. Kuehn and Michael S. Daubs, explores issues of tourism and media convergence in relation to redefinitions of value around labour and cultural space in Google 'Trekker' programme, a crowdsourcing initiative to capture landscape imagery for Google Maps. In Chapter 6, Maja Turnšek, Andreja Trdina, and Barbara Pavlakovič examine the process of 'Othering' in the promotion of Melania Trump's hometown as a tourist destination. Chapter 7, by Karla Zavala and Adriaan Odendaal, considers alternative ways to improve the preservation of culture heritage online through post-digital archival practices. Giulia Lavarone's contribution – Chapter 8 – is concerned with the cinematic tourist experience through the lens of film theory. Lavarone understands mediatization as a socio-spatial concept and applies it to critically consider the relocation of cinema. In a related vein, Sofia Sampaio extends cinema tourism studies by means of a cultural materialist perspective. In Chapter 9, she demonstrates how cinema tourism can benefit from the consideration of the intertwined material and meaning-making capabilities of both cinema and tourism.

Part II – mediatized places and spaces

The second part of the book, mediatized places and spaces, analyses mediatization from different spatial perspectives. Chapter 10 by Nicolai Graakjær and Rasmus Grøn explores how auditory aspects of televised football can promote football tourism. The chapter analyses how in-stadium experiences with focus on the role played by the spectator sounds is mediated through TV transmissions. In Chapter 11, Yunuen Ysela Mandujano-Salazar discusses place-related media activities from the Japanese idol group Arashi and the development of a special segment of domestic tourism. Chapter 12 by Maria Criselda G. Badilla analyses through official websites and social media sites how destination management organisations use different channels to create a single destination image of South-East Asia. In Chapter 13, Natália Ferrer-Roca analyses the role of feature films in branding and marketing the destination New Zealand. The chapter explores the relationships between feature filmmaking and destination marketing, economic development, and country reputation. In Chapter 14, José Fernández-Cavia discusses the role played by official destination websites in the process of building and disseminating destination brands and attractions. Chapter 15 by Apoorva Nanjangud examines the emerging travel representations and the narrative structure in Bollywood cinema by analysing three popular Bollywood blockbusters. In Chapter 16, Francesc Fusté-Forné and Pere Masip investigate how legacy media builds a food-based storytelling. The chapter analyses representations of food and tourism in Spanish newspapers. The last chapter in Part II, Chapter 17, by Hazel Andrews and Les Roberts explores questions around the symbolic role of the 'other' and 'stranger' in narratives of places and mobility using the case of the HBO drama *True Detective*.

Part III – circle of representation

The third part in the Companion is focused on the circle of representation. This is a concept discussed by, for example, Jenkins (2003) and Urry (1990). It is seen as a rather passive view where tourists are passive media consumers who only search for images seen before whilst travelling. However, Urry and Larsen (2011) address the performative aspects involved in the mediatized gaze. Mediatization therefore challenges the passive view on tourists as this concept highlights tourists' agency in creating images (Månsson, 2011). The contributions in this part seek to challenge the notion of the passive tourist. The construction of images is examined from a tourist as well as an organisational perspective.

In Chapter 18, Cecilia Cassinger and Åsa Thelander employ the concept of gaze as a theoretical lens through which to examine the role of visual social media in co-creating images of destinations. Their study shows how different gazes discipline locals' imaginations of a city with a bad reputation. Chapter 19 by Pat Brereton explores the representation of Ireland through the global franchises *Star Wars* and *Game of Thrones*. The aim is to see whether the franchises can help to promote green eco-tourism to Ireland. The same cases are also addressed in Part IV (Chapter 27) but then from a fan tourism perspective. In Chapter 20, Alyaa Anter investigates the role of the National Geographic Abu Dhabi channel as a cultural outlet in framing touristic destinations in the UAE. Philip Seaton and Sue Beeton discuss the rewriting of history in Chapter 21 and how new values are assigned to heritage tourist sites associated with popular culture by using a Japanese and an Australian case. In Chapter 22, Božo Skoko and Katarina Miličević explore the interdisciplinary field of public relations and film as a part of the destination branding process, and destination image as an outcome of destination branding process by using Croatian film-induced tourism as a case. Looking at a more traditional medium, guidebooks, Tim Hannigan finds in Chapter 23 that guidebooks, despite being intended to facilitate leisure travel, continue to contain traces of anti-tourism, probably sustained by reliance on readers' own self-construction. In Chapter 24 by Warwick Frost and Jennifer Frost film tourism is also addressed by focusing on Pavlova westerns and its impact on the cultural landscape of New Zeeland. Jörgen Eksell and Maria Månsson problematize in Chapter 25 the performance of place-making practices by tourists as they engage with music, using the city of Liverpool as a case. Finally, in Chapter 26, Tom Sintobin and Anke Tonnaer address the circulation of tourism media products by exploring filmic representations, posted on social media, by tourists and semi-professional travellers of their visits to 'primitive' places.

Part IV – tourists as media producers

The fourth part of this book examines tourists as transformative producers that extend key works by media studies scholars including Jenkins (1992, 2006), Maorimoto (2018), and Langley and Zubernis (2017), as well as by tourism scholars like Buchmann, Moore, and Fisher (2010), Buchmann (2014) and Campos, Mendes, do Valle, and Scott (2017). All authors highlight the active role tourists take in creating meaningful and transformative products and experiences.

In Chapter 27, Áine Mc Adam investigates the nexus and disparity between the media, tourism, and cultural heritage in Ireland. She shows how tourism and heritage are interconnected and that media persists in selling the 'gaze'. Following on, Christian Hviid Mortensen demonstrates in Chapter 28 how cultural memory can become mediatized and cherished even outside its country of origin with an effective narrative for ascribing value to objects. Rebecca Williams shows the transmediality of *Hannibal's* Florence read through the lens of the participatory culture that surrounds the series, rather than the multiple iterations that exist in different texts and forms of media in Chapter 29. Similarly, Antonio Loriguillo-López analyses how anime location seekers reflect a profile of active users, expert locators of verified information, and distributors of cores of new knowledge on anime for other fans in Chapter 30. Following on from this, Philipp Dominik Keidl introduces in Chapter 31 the phenomenon of fan-run museums and shows how locations that promise encounters of other fans and their works in their local environments also mobilise them. Kyungjae Jang and Takayoshi Yamamura demonstrate in Chapter 32 that social media plays a role in strengthening trivial associations with transnational mediatization of sites by creative fandom. Ana

Oliveira Garner in Chapter 33 analyses how tourism is further submitting to the logic of social media whilst also implying an active role by tourists in the construction of the visual discourses about a place. In Chapter 34, Annæ Buchmann highlights the constant changes and mediatization of the Sherlock Holmes fandom, and its particularly transformative culture. Chapter 35, by Andreja Trdina, Barbara Pavlakovič, and Maja Turnšek, extends the concept of cultural intimacy and examines the role cultural capital plays in structuring this sensibility among fans.

Part V – transmedia tourism

The final part of the Companion focuses on transmedia tourism. Transmedia storytelling is something that is associated with Jenkins (2006). Furthermore, transmedia is also something that is connected to and part of the understanding of mediatization (Jansson, 2018). Transmedia tourism is therefore something that is highly relevant to address in order to enhance the understanding of media and tourism. Transmedia storytelling is central to understand how tourism discourses and practices are shaped as they are systematically dispersed across multiple digital platforms.

In Chapter 36, John Pearce and Gianna Moscardo argue that stories and storytelling are a core link between media and tourism by developing a storytelling framework for tourism. In Chapter 37, Nicholas Wise argues that we are moving beyond traditional forms of destination image generation, where Destination Marketing Organisations (DMO) portray a destination, to user-generated images. His contribution demonstrates the tensions that arise in differences between DMO's and tourists' representation of destinations. Chapter 38 by Sara Leckner and Carl Magnus Olsson is closely linked to Chapter 39 by Lena Eskilsson, Maria Månsson, Jan Henrik Nilsson, and Malin Zillinger since both chapters address tourist information in a digital age. Leckner and Olsson explore the challenges and opportunities that arise in transferring value when a physical tourist information bureau is closing down to become purely digital. Eskilsson et al. analyse tourists' information search behaviour and their use of different information channels from a transmedia tourism perspective by studying German tourists' travelling to Sweden. In Chapter 40, Fani Galatsopoulou and Clio Kenterelidou explore transmedia storytelling in an attempt to contribute to the conceptualisation and definition of nautical tourism in a sustainable and anthropocentric way. Kim Williams, in Chapter 41, examines how transmedia narratives can influence tourism, whilst also providing a framework to understand the syntheses between cultural heritage institutions, their visitors, and the fashion sector. Chapter 42, by Susan Carson and Mark Pennings, deals with the impact of smartphone use and the way such technologies facilitate new interactions between tourists and heritage sites. Jia Xie, Chapter 43, explores the impact of smartphone on Chinese backpackers' mobility pattern. She concludes that backpackers are empowered by the advanced technology whilst at the same time run the risk of being distracted from exploring the destination.

Final thoughts

The Companion ends with an afterword written by Associate Professor Szilvia Gyimóthy, a leading Danish tourism scholar. In her closing commentary, she comments on the socio-spatial (re/dis)ordering power of entwined tourism and media practices, and the conceptualization of participatory placemaking. Based on the contributions to this Companion, she highlights three processes of participatory placemaking: place-wrecking, place-assembling, and place-enhancing. These concepts and processes are highly interesting within inter-disciplinary studies, as illustrated in this Companion, and provide important insights for practitioners.

In closing, we editors contend that the connection between media convergence and tourism needs further in-depth exploration and theorisation. Given that we live in a society where different types of media (broadly defined) are used as resources to organise and make sense of daily life, the extant research on how media shapes tourism experiences and practices is surprisingly limited. The contributions in this book pave the way for deepening our understanding of the relationship of media and tourism. They highlight that the relationship is reciprocal by demonstrating that tourists are not passive actors, but co-creators of contents and meanings, which influence our understanding of time and space. In these contributions, we see possibilities to shift the focus of mediatized tourism from a more or less predetermined process by the logics of the media industry to one that is powered by tourists' practices.

References

Andersson, M. (2017). Mediatization from below. In O. Driessens, G. Bolin, A. Hepp, & S. Hjar-vard (Eds.), *Dynamics of mediatization. Institutional change and everyday transformations in a digital age* (pp. 35–56). Cham, Switzerland: Springer International Publishing.

Buchmann, A. (2014). Not all that glitters is gold: Three case studies of film tourism in New Zealand. In K. Dashper (Ed.), *Rural tourism: An International Perspective* (pp. 361–374). Newcastle Upon Tyne, UK: Cambridge Scholars Publishing.

Buchmann, A., Moore, K., & Fisher, D. (2010). Experiencing film tourism. Authenticity and fellow-ship. *Annals of Tourism Research, 37*(1), 229–248.

Butler, R. W. (1990). The influence of the media in shaping international tourist patterns. *Tourism Recreation Research, 15*, 46–55.

Campos, A. C., Mendes, J., do Valle, P. O., & Scott, N. (2017). Co-creating animal-based tourist ex-periences: Attention, involvement and memorability. *Tourism Management, 63*, 100–114.

Cohen, E., & Cohen, S. A. (2012). Current sociological theories and issues in tourism. *Annals of Tour-ism Research, 39*(4), 2177–2202.

Corner, J. (2018). 'Mediatization': Media theory's word of the decade. *Media Theory, 2*(2), 79–90.

Couldry, N. (2008). Mediatization or mediation? Alternative understandings of the emergent space of digital storytelling. *New Media and Society, 10*(3), 373–391.

Couldry, N., & Hepp, A. (2013). Conceptualizing mediatization: Contexts, traditions, arguments. *Communication Theory, 23*, 191–202.

Crouch, D., Jackson, R., & Thompson, F. (Eds.). (2005). *The media and the tourist imagination. Converg-ing cultures.* Abingdon, UK and New York, NY: Routledge.

Crouch, D., & Lubbren, N. (Eds.). (2003). *Visual culture and tourism.* Oxford, UK: Berg Publishers.

Ekström, M., Fornäs, J., Jansson, A., & Jerslev, A. (2016). Three tasks for mediatization research: Con-tributions to an open agenda. *Media, Culture & Society, 38*(7), 1090–1108.

Gyimóthy, S. (2018). The Indianization of Switzerland: Destination transformations in the wake of Bollywood films. In C. Lundberg & V. Ziakas (Eds.), *The Routledge handbook on popular culture and tourism* (pp. 376–388). Abingdon, UK: Routledge.

Gyimóthy, S., Lundberg, C., Lindström, K., Lexhagen, M., & Larson, M. (2015). Popculture tourism: A research manifesto. In D. Chambers & T. Rakic (Eds.), *Tourism research frontiers: Beyond the bound-aries of knowledge* (pp. 13–26). Bingley, UK: Emerald.

He, Z., Wu, L., & Li, X. (2019). Mediatized tourism: A mediatic turn of research paradigm? Research Paper, TTRA Conference June 25–27 2019 Melbourne, Australia.

Hepp, A. (2009). Differentiation: Mediatization and cultural change. In K. Lundby (Ed.), *Mediatiza-tion. Concepts, changes, consequences* (pp. 19–38). New York, NY: Peter Lang Publishing.

Hepp, A., & Krotz, F. (2014). Mediatized worlds – Understanding everyday mediatization. In A. Hepp & F. Krotz (Eds.), *Mediatized worlds. Culture and society in a media age* (pp. 1–15). Basingstoke, UK: Palgrave Macmillan.

Hjarvard, S. (2008). The mediatization of society. A theory of the media as agents of social and cultural change. *Nordicom Review, 29*(2), 105–134.

Hjarvard, S. (2013). *The mediatization of culture and society.* Abingdon, UK and Oxon, UK: Routledge.

Hjarvard, S. (Ed.) (2016). *Medialisering: mediernes rolle i social og kultural forandring.* Köpenhamn: Hans Reitzel.

Iwashita, C. (2003). Media construction of Britain as a destination for Japanese tourists: Social constructionism and tourism. *Tourism and Hospitality Research, 4*(4), 331–340.

Jansson, A. (2002). Spatial Phantasmagoria. The mediatization of tourism experience. *European Journal of Communication, 17*(4), 429–443.

Jansson, A. (2015). Interveillance: A new culture of recognition and mediatization. *Media and Communication, 3*(3), 81–90.

Jansson, A. (2018). *Mediatization and mobile lives: A critical approach.* London, UK: Routledge.

Jansson, A., & Falkheimer, J. (2006). Towards a geography of communication. In J. Falkheimer & A. Jansson (Eds.), *Geographies of communication. The spatial turn in media studies* (pp. 9–25). Göteborg, Sweden: Nordicom.

Jenkins, H. (1992). *Textual poachers. Television fans and participatory culture.* London: Routledge.

Jenkins, H. (2004). The cultural logic of media convergence. *International Journal of Cultural Studies, 7*(1), 33–43.

Jenkins, H. (2006). *Convergence culture. Where old and new media collide.* New York & London, UK: New York University Press.

Jenkins, O. (2003). Photography and travel brochures: The circle of representation. *Tourism Geographies, 5*(3), 305–329.

Jensen, J. L., & Waade, A. M. (2008). *Medier og turisme.* Århus: Academica.

Kaun, A., & Fast. K. (2014). *Mediatization of culture and everyday life.* Huddinge, Sweden: Södertörns högskola, Karlstads universitet.

Kim, S. (2012). The relationships of on-site film-tourism experiences, satisfaction, and behavioural intentions: From the film-tourism perspective. *Journal of Travel & Tourism Marketing, 29*(5), 472–484.

Kim, S., Long, P., & Robinson, M. (2009). Small screen, big tourism: The role of popular Korean television dramas in South Korean tourism. *Tourism Geographies, 11*(3), 308–333.

Langley, T., & Zubernis, L. (2017). *Supernatural psychology: Roads less traveled.* New York, NY: Sterling.

Leotta, A. (2016). Navigating movie (M)apps: Film locations, tourism and digital mapping tools. *Media Culture, 19*(3), n.p.

Lundberg, C., & Ziakas, V. (Eds.). (2018), *The Routledge handbook on popular culture and tourism.* Abingdon, UK: Routledge.

Månsson, M. (2011). Mediatized tourism. *Annals of Tourism Research, 38*(4), 1634–1652.

Månsson, M. (2015). *Mediatized tourism: The convergence of media and tourism performances* (Dissertation). Lund, Sweden: Media-Tryck, Lund University.

Morimoto, L. (2018). The 'Totoro Meme' and the politics of transfandom pleasure. *East Asian Journal of Popular Culture, 4*(1), 77–92.

Reijnders, S. (2011). *Places of the imagination: Media, tourism, culture.* Farnham, UK: Ashgate Publishing.

Scarles, C., & Lester, J. (2013). *Mediating the tourist experience.* Farnham, UK: Ashgate Publishing Group.

Scarles, C., & Lester, J. (Eds.). (2016). *Mediating the tourist experience. From brochures to virtual encounters.* London, UK: Routledge.

Schultz, W. (2004). Reconstructing mediatization as an analytical concept. *European Journal of Communication, 19*(1), 87–101.

Smith, M., & Richards, G. (Eds.) (2013). *Routledge handbook of cultural tourism.* London, UK: Routledge.

Urry, J. (1990). *The tourist gaze. Leisure and travel in contemporary societies.* London, UK, Thousand Oaks, CA & New Delhi, India: Sage.

Urry, J., & Larsen, J. (2011). *The tourist gaze 3.0.* London, UK: Sage.

PART I

Critical and conceptual entrance points to the field

2

INVITED CONTRIBUTION – THE JANUS FACE OF TRANSMEDIA TOURISM

Towards a logistical turn in media and tourism studies

André Jansson

Introduction

Tourists are people in need of particular types of information. When Thomas Cook & Son introduced the first version of their famous European timetable, its title was *Cook's Continental Time Tables & Tourist's Handbook*. Since then, the publication has come out on a regular basis with only minor interruptions (e.g., during the Second World War), and is today called the *European Rail Timetable*. Its endurance points to the long-standing need among tourists to gain more than phantasmagorical representations of foreign places and people. Tourists also need reliable facts and instructions, even concrete handbooks, for organizing their mobilities and going about various forms of interaction when away from home. Besides timetables, such things as maps, guidebooks, phrasebooks, and various travel documents (passports, visa, tickets, etc.) belong to the standard logistical media equipment among tourists. Most of which have been around since the early days of modern tourism, in one shape or another.

Strangely enough, however, research on media and tourism is a relatively recent occurrence, arising more than a century after the above-mentioned logistical innovations of Thomas Cook & Son. Overviewing the field (as it may now be called), one also discovers that the wording 'media and tourism' mainly refers to the *representational* aspects of media, especially the role of popular culture and marketing in shaping tourism destinations and practices, rather than to the role of media technologies in coordinating people's activities in time and space. In its most unmitigated fashion, 'media and tourism' refers to *media-induced* tourism, that is, leisure trips organized around a particular media format, film or celebrity associated with a particular place. While such media tourism constitutes the centre of the field – as reflected also in the composition of this volume – questions of *logistical media* (Peters, 2008) remain more peripheral.

There are several reasons to this bias, especially the deep-seated tendency in the wider areas of both human geography and media studies to prioritize the cultural products of mass media industries over technological and material matters. At the same time, there are signs of change. Questions of tourism logistics seep into the field of media and tourism studies, largely due to the rise of *transmedia* as a dominant mode of cultural circulation. As I will argue in this short essay, this is a much-needed development, which (if it continues) should have a foundational impact on how we think about the relations between media and tourism. A logistical turn would above all shed light over media's role in the often-troubled coming together of, or collision between, tourism phantasmagoria and real-life matters.

In the forthcoming sections, I will elaborate this argument via three principal statements concerning transmedia and tourism:

1   Transmedia is the dominant mode of cultural circulation today.
2   Transmedia tourism is an expanding social condition.
3   Transmedia tourism extends processes of de-differentiation.

By way of conclusion, I will reflect briefly upon the prospects of a logistical turn in media and tourism studies.

Transmedia is the dominant mode of cultural circulation today

The term transmedia is above all associated with research on participatory culture, notably the interactions between fan cultures and popular media franchises like Teenage Mutant Ninja Turtles (Kinder, 1991), The Matrix (Jenkins, 2006), and Transformers (Fast, 2012). Transmedia is then taken as the underpinning of a particular type of storytelling, through which media industries and ordinary users jointly produce a symbolic universe. Fans and other cultural agents may intervene (within certain limits) in the unfolding of a story and contribute in other ways to the circulation of meaning. This means that the emphasis is placed on the recreation, remixing, and recontextualization of texts and their meanings. According to this theoretical legacy, transmedia storytelling is largely a means of strengthening audience engagement and brand identities (e.g., Fast & Örnebring, 2017; Freeman, 2015). There are also studies pointing to how the same ideas can be applied in relation to, for instance, social mobilization and activism (e.g., Ramasubramanian, 2016; Zimmerman, 2016).

With the rise of social media and mobile platforms (notably, the smartphone), however, we need to think about transmedia beyond the narrow confines of popular media representations. Today, it is not only certain brands and narratives that circulate in an open-ended, transitory manner. Most information does. This is not to assert that, say, travel itineraries, train tickets, and boarding passes issued to the ordinary traveller are intended to be 'remixed' or part of a 'world-building' effort. Yet, such pieces of information are affected by the fundamental logics of transmedia as a *mode of cultural circulation*. They may be accessed through a variety of digital entry points (e.g., an email account handled via multiple devices) and passed on, or shared, in different formats (text-file, pdf, screen-shot, printed paper, etc.) through different channels (email, social media, text messages, etc.) to oneself and others. As original texts are adapted and shared, they may enter into broader processes of social communication – among family, friends, fellow travellers, and so forth – and eventually acquire new meanings. The same thing goes for most other forms of mediated communication today. As such, transmedia should be contrasted from, and has taken over after, mass media as the dominant mode of cultural circulation in affluent modern societies (see Fast & Jansson, 2019).

The broadened understanding of transmedia that I suggest here (see also Jansson & Fast, 2018) implies that we are now dealing with a phenomenon with more far-ranging social consequences than mere textual co-creation. Taken as a general mode of cultural circulation, transmedia refers to how social practices, in addition to texts *per se*, are moulded by and negotiated through different platforms and devices, and interweave with various forms of offline communication. We may call this complex development *transmediatization*. As a consequence, transmedia cannot be set apart from everyday life but saturates most of its parts. Transmedia also affects social domains beyond 'the everyday' and weave them together with more ordinary spaces of communication. The transmediatization of tourism is a case in point.

Transmedia tourism is an expanding social condition

Just like transmedia studies, the field of media and tourism studies has been largely preoccupied with popular cultural themes and representations. As noted above, 'film tourism', 'celebrity tourism', and 'television tourism' have been given much attention, especially during the last two decades. There is also a more direct connection between the fields. The production of theme-parks and tourist destinations around popular film and tv narratives and characters can be seen as an example of how tourism development blends with culture-industrial world-building (e.g., Couldry, 2002; Månsson, 2011; Reijnders, 2011). Thus, tourism often occurs within carefully scripted, half-virtual/half-real, 'transmediascapes', to which tourists also contribute with their semiotic, enunciative, and textual productivity (cf. Fiske, 1992). This is well captured in several of the chapters in this volume, such as in Keidl's analysis of the fan-created Star Wars-related Film Figures Exhibition in Monchengladbach, Germany, and Loriguillo-Lopez's work on anime-induced 'scene hunting tourism' in Japan, both chapters highlighting the cultural and material bottom-up processes involved in much media tourism today.

If we apply the broader understanding of transmedia, however, many more questions emerge. The relations between transmedia and tourism then concern not just the narrative worlds of popular culture but tourism as an increasingly media-reliant social condition. In a previous study, Karin Fast and I applied this approach onto the world of work and labour, arguing that the transmedia mode of cultural circulation conditions anything from how work tasks are carried out and work-places designed to how the boundaries between work and leisure are drawn and what kinds of jobs are actually available to people, and how (e.g., within the expanding 'gig-economy') (Fast & Jansson, 2019). We thus defined transmedia work not as a particular category of work but as a *social condition* affecting the world of work in multiple, situated ways. At the same time, this social condition is marked by certain common denominators, such as extended peer-to-peer visibility and corporate surveillance. If we transfer this understanding to the realm of tourism, we can define transmedia tourism as *a social condition saturated by (1) online sharing practices that (2) make tourism and tourists visible to other transmedia users and (3) enable real-time feedback on circulated content, which in turn facilitates (4) automated processes of surveillance and, ultimately, (5) the commodification of social practices and relations at large* (cf. Fast & Jansson, 2019, p. 6).

A crucial point here is that the social normalization of transmedia affects more than tourism practices *per se*. Besides leading to, for instance, heightened expectations as to the circulation of tourist images via social media, transmedia affects how and to what extent tourist practices, outlooks, and cultural products reach into everyday life. Two obvious consequences are worth mentioning. First, people in general are to an increasing extent made part of the *tourism of others*, especially through social media where images, comments, 'likes', and so forth are continuously circulated. Research shows that most tourists stay in touch with friends and family while travelling using a variety of online channels (e.g., Lo, McKercher, Lo, Cheung, & Law, 2011; Munar & Jacobsen, 2014). This means that the tourism world and its particular (oftentimes extra-ordinary) qualities and norms become inseparable from the ordinary lifeworld.

Second, and related to the first point, transmedia contributes to making *tourism planning* a perpetual, everyday process. With 'smart tourism technologies', notably mobile apps and online systems for booking, rating, reviewing, and comparing destinations, itineraries, and modes of travel, the practice of online tourism planning is increasingly part of everyday life (Huang et al., 2017; Xiang, Wang, O'Leary, & Fesenmaier, 2015). Attractive destinations, hotels, tours, hosts, etc. may be bookmarked for future purposes and shared and reworked together with prospective travel Companions regardless of place and time. One individual may thus entertain several different travel-plans at the same time. Some of these may eventually get realized but oftentimes these prospective journeys remain on the virtual drawing table.

Altogether, the phantasmagorical and expressive aspects of 'virtual tourism' are increasingly entwined with more instrumental and logistical travel procedures and preparations. As we will see next, this development is symptomatic of a broader social process of de-differentiation that transmedia plays into.

Transmedia tourism extends processes of de-differentiation

The question of de-differentiation is closely bound up with theories of post-tourism and, in a broader sense, postmodernity that flourished in the 1980s and 1990s. While modern society was marked by differentiation, meaning that various spheres of activity developed their own institutional logics, norms, regulations, etc., the postmodern condition epitomized the gradual breakdown of such distinctions (e.g., Featherstone, 1991; Lash, 1990). As Lash and Urry argued in 1994, electronic media and the pervasiveness of visual consumption throughout culture was a key force behind this development, ultimately problematizing the boundaries between representations and reality.

> Since what we increasingly consume are signs or images, so there is no simple 'reality' separate from such modes of representation. What is consumed in tourism are visual signs and sometimes simulacrum; and this is what is consumed when we are supposedly not acting as tourists at all.
>
> *(Ibid., p. 272)*

Lash and Urry spoke of neither the Internet nor the mobile phone, since these technologies had not yet made their way into everyday life. Still, the explosion of images that global telecommunications had brought along, and which turned ordinary consumers into armchair travellers (see also Feifer, 1985), was enough to start questioning the boundaries between the everyday and the extraordinary, between home and away, between real places and simulations. These observations eventually led the authors to speak about the 'end of tourism' and the corresponding rise of the 'cool, self-conscious and role-distanced' post-tourist (Lash & Urry, 1994, p. 276). De-differentiation meant that tourism was increasingly everywhere and anywhere, and that consumers had to develop reflexive strategies to maintain social and cultural distinctions.

The 'end of tourism' thesis is problematic in many ways, especially since people are obviously more and more keen on spending money on organized leisure travel. Its implications have been problematized by several scholars, including Urry himself (for a discussion, see Haldrup & Larsen, 2009, pp. 21–26). Still, as given by my characterization of transmedia tourism above, we are now dealing with a social condition that accentuates many of the trends identified in the debates on postmodernity and the de-differentiated 'semiotic society' (Lash, 1990). This concerns especially the dilemma of ubiquitous transmedia connectivity through smartphones and other devices, which implies not only that touristic impulses saturate everyday life to an ever-increasing extent, but also that it gets increasingly difficult to shelter the liminal space of tourism from unwanted intrusions (see Fan, Buhalis, & Lin, 2019). Transmedia circulation means that, for example, information pertaining to the world of work may end up on the same platforms, and in the same flows, as leisure-oriented communication. This fusion of social worlds is further accentuated by the fact that social media and e-mail accounts are often used as general entry-points to online tourism services, ranging from restaurant reservations and traffic information to ticketing and Wi-Fi access, which in turn (often by default) sparks extended flows of customized online advertising.

This is to say that transmedia tourism brings a new flavour to de-differentiation. The most distinct, and novel, quality of transmedia tourism, I argue, is not that tourist images and simulations saturate and shape everyday realities. It is rather the opposite. The world of mediated tourism phantasmagoria is perforated by hands-on plans and procedures that previously were not part of the mediated tourism-world. Such real-life matters are typically of a logistical nature, pertaining *both* to the organization of tourism and to other realms of social life, such as, managing one's work and dealing with family logistics. This can be seen as a form of ongoing tourism disenchantment whereby it gets more and more difficult for the individual tourist to escape the psychological pressures of ordinary obligations and expectations. At the same time, however, transmedia makes it easier to turn touristic dreams into reality (at least for some people). This is the Janus Face of transmedia tourism: the logistical annexation of tourism phantasmagoria as a blessing and a curse.

A logistical turn?

I have in this chapter argued that studies of media and tourism should account for the continuously widened implications of transmedia. We cannot isolate the significance of transmedia to popular brands or narratives, such as in film tourism, but should see transmedia as a mode of circulation that also comprises various kinds of logistical information that was previously sealed off from the more phantasmagorical realms of tourism media. Of course, we should not exaggerate this condition; after all, the classical tourist map, to take an example, has always been both a representational, possibly phantasmagorical, *and* logistical device. One could even argue that the physical map is phantasmagorical precisely *because* it is also logistical, giving tourists an idea of how to navigate and what to eventually see with their own eyes. The distinction was never clear-cut. Still, the current condition, where the same digital devices and platforms are used for accessing and managing an open-ended spectrum of contents, brings these realms even closer together. In this volume, a related example is provided by Kuehn and Daubs in their chapter on Google trekker as a 'ubiquitous mapping project' through which ordinary users around the world help recording and charting various 'off the grid' spaces. The process of de-differentiation is not only extended; it also takes on a new shape since logistical elements are by default brought into the touristic transmediascape – and vice versa.

Accordingly, the expansion of transmedia tourism, as described here, requests scholars of media and tourism to look closer into matters of travel planning, navigation, coordination, and evaluation – things that are commonly thought of as alien to the liminal spaces of tourism and rarely associated with 'media tourism'. Furthermore, I argue, a vital turn to such issues should be accompanied by an ambition to resist over-simplified divisions between 'logistical' and 'representational' affordances of media, as well as between 'the rational' and 'the phantasmagorical' in tourism. A logistical turn, as I envision it, means that logistical dimensions of media and tourism are studied and theorized in relation to the *overall picture* of modern travel and leisure. From such a perspective, the importance of an all-embracing, holistic research Companion like this can hardly be overstated. While it is still too early to declare that a logistical turn is actually taking place, the renewed interest in seemingly mundane or old-fashioned things like 'tourist information' and 'tourist bureaus' and their migration to digital platforms – as discussed in the chapters by Leckner et al. and Eskilsson et al. – points to an increasingly important area of study. Not least, we should critically acknowledge and analyse how the latest digital innovations in tourism logistics, ultimately the 'transmediatization of everything', affect different groups in society. Whose travel dreams and whose mobilities are being lubricated and whose are not?

References

Couldry, N. (2002). *The place of media power: Pilgrims and witnesses of the media age*. London, UK: Routledge.

Fan, D. X., Buhalis, D., & Lin, B. (2019). A tourist typology of online and face-to-face social contact: Destination immersion and tourism encapsulation/decapsulation. *Annals of Tourism Research, 78*, 102757.

Fast, K. (2012). *More than meets the eye: Transmedial entertainment as a site of pleasure, resistance and exploitation* (Doctoral dissertation, Karlstad University, Sweden).

Fast, K., & Jansson, A. (2019). *Transmedia work: Privilege and precariousness in digital modernity*. London, UK: Routledge.

Fast, K., & Örnebring, H. (2017). Transmedia world-building: The shadow (1931–present) and transformers (1984–present). *International Journal of Cultural Studies, 20*(6), 636–652.

Featherstone, M. (1991). *Consumer culture and postmodernism*. London, UK: Sage.

Feifer, M. (1985). *Going places*. London, UK: Macmillan.

Fiske, J. (1992). The cultural economy of fandom. In L. A. Lewis (Ed.), *The adoring audience: Fan culture and popular media* (pp. 30–49). London, UK: Routledge.

Freeman, M. (2015). Up, up and across: Superman, the Second World War and the historical development of transmedia storytelling. *Historical Journal of Film, Radio and Television, 35*(2), 215–239.

Haldrup, M., & Larsen, J. (2009). *Tourism, performance and the everyday*. London, UK: Routledge.

Huang, C. D., Goo, J., Nam, K., & Yoo, C. W. (2017). Smart tourism technologies in travel planning: The role of exploration and exploitation. *Information & Management, 54*(6), 757–770.

Jansson, A., & Fast, K. (2018). Transmedia identities: From fan cultures to liquid lives. In M. Freeman & R. R. Gambarato (Eds.), *The Routledge Companion to transmedia studies* (pp. 340–349). London, UK: Routledge.

Jenkins, H. (2006). *Convergence culture: Where old and new media collide*. New York: New York University Press.

Kinder, M. (1991). *Playing with power in movies, television, and video games: From muppet babies to teenage mutant ninja turtles*. Berkeley: University of California Press.

Lash, S. (1990). Learning from Leipzig—or politics in the semiotic society. *Theory, Culture & Society, 7*(4), 145–158.

Lash, S., & Urry, J. (1994). *Economies of signs and space*. London, UK: Sage.

Lo, I. S., McKercher, B., Lo, A., Cheung, C., & Law R. (2011). Tourism and online photography. *Tourism Management, 32*(4), 725–731.

Månsson, M. (2011). Mediatized tourism. *Annals of Tourism Research, 38*(4), 1634–1652.

Munar, A. M., & Jacobsen, J. K. S. (2014). Motivations for sharing tourism experiences through social media. *Tourism Management, 43*, 46–54.

Peters, J. D. (2008). Strange sympathies: Horizons of German and American media theory. In F. Kelleter & D. Stein (Eds.), *American studies as media studies* (pp. 3–23). Heidelberg, Germany: Universitätsverlag.

Ramasubramanian, S. (2016). Racial/ethnic identity, community-oriented media initiatives, and transmedia storytelling, *The Information Society, 32*(5), 333–342.

Reijnders, S. (2011). *Places of the imagination: Media, tourism, culture.* London, UK: Ashgate.

Xiang, Z., Wang, D., O'Leary, J. T., & Fesenmaier, D. R. (2015). Adapting to the Internet: Trends in travelers' use of the web for trip planning. *Journal of Travel Research, 54*(4), 511–527.

Zimmerman, A. (2016). Transmedia testimonio: Examining undocumented youth's political activism in the digital age. *International Journal of Communication, 10*(21), n.p.

3

INVITED CONTRIBUTION – MIND THE GAP

Interdisciplinary approaches to media and tourism

Anne Marit Waade

Introduction

The fact that both tourism and media have become crucial parts of contemporary consumer culture has given rise to a range of academic studies of the relationship between media and tourism in disciplines such as culture studies, visual studies, tourism studies, media studies, business studies, urban planning, place branding, consumer culture studies, communication technology studies, geography, and anthropology. In particular, John Urry's groundbreaking work on the tourist gaze has influenced and inspired research in the field since its first edition was published in 1990. A second edition came out in 2001, and a third edition (co-authored with Jonas Larsen) in 2011. Both tourism studies and media studies have undergone significant expansion and changes throughout the last couple of decades, partly because of various technological, transport, and market-driven developments.

In this chapter, I will map the interdisciplinary approaches to media and tourism as cultural phenomena and fields of research. My primary focus will be on media studies and tourism studies, although I am well aware that the field also attracts attention from other disciplines. In general, the relationship between media and tourism has drawn extensive attention in tourism studies, whereas so far, media and communication studies have only rarely addressed this relationship (Jansson & Falkheimer, 2006). Both tourism studies and media studies are interdisciplinary fields that combine theories, methods, and perspectives from the social sciences and humanities. However, generally speaking, tourism studies is more closely related to business, management, and marketing studies, whereas media studies, significantly, embeds its disciplinary approaches in the humanities and combines understandings of narratives, aesthetics, and genres with audience studies, media production, and media industry, based on social sciences methods. Communication studies, which, in most contexts, differs from media studies, is mainly informed by the social sciences and often includes journalism studies. These various disciplinary traditions and foundations will influence the way that the relationship between media and tourism is approached, and whether the tourist, the tourism industry, the screen industry (film, television computer games), or new media technologies (e.g. GPS-based locative media, social media, mobile media, augmented reality/AR, virtual reality/VR technologies, or algorithm-based production and distribution) primarily informs the main objective and focus of a particular study. This book, *The Routledge Companion to Media and Tourism*, illustrates this interdisciplinary approach, in which the contributors primarily base their work on one discipline or another, but at the same time reach out, bridge, and combine the various disciplinary terms, methods, and perspectives. This book bridges the gaps between the different disciplinary approaches and develops synergies between the fields. To 'mind the gap' between media and tourism studies, I hope that in the future, more media studies scholars will contribute to these fields and develop media-specific approaches to understanding the relationships between media and tourism, and how this strong connection influences new modes of media production; travel journalism; marketing; audience and user patterns; digital, online, and social media technologies; and locative media features that target tourists. Tourism has become a significant industry and culture, and thus in various ways drives media production, distribution, and media technology development.

Framework for mapping disciplinary approaches to media and tourism research

The relationship between media and tourism is characterised by its interdisciplinary approach, which draws on disciplines such as tourism studies, media studies, geography, and anthropology, in which the different disciplines typically emphasise one of the following agents and aspects.

Tourists: How tourists use media to prepare for their trips, navigate and photograph while they are travelling, and curate and share memories when they return home.

Tourism destinations: How media drive destinations to promote themselves, and to provide visitors with practical navigation and information tools, as well as to deliver operational management tools and strategic planning tools.

Media production and journalism: How tourism influences new modes of cross-sector collaboration, generates new sources of funding (e.g. location placement in screen productions), and influences screen productions and genres (e.g. travel series, tourist drama series) and printed media and journalism formats (e.g. travel journalism and travel magazines).

Media technologies: How tourism influences the development of locative media, social and mobile media features, and new user and fandom patterns.

If we take my field of research – screen tourism – as an example, we clearly see that many scholars study screen tourists, their motivations for, and experiences of walking in the footsteps of their favourite characters and television drama series (e.g. Buchmann, Moore & Fisher, 2010; Connell, 2012; Kim, 2012; Lee, 2012; Leotta, 2011; Reijnders, 2011; Roesch, 2009; Sjöholm, 2010), and others focus on the screen tourism destinations, and how film and television influence tourism (e.g. Beeton, 2016; Månsson & Eskilsson, 2013). In contrast, if we take the screen industry as a point of departure and examine studies of how tourism influences journalism and screen productions, including their narratives and aesthetics, the academic contributions are more limited, and the tourism aspect is subordinated (e.g. Hanusch & Fürsich, 2014; Roberts, 2016; Turnbull & McCutcheon, 2017; Waade, 2016; Wheatley, 2016). In connection with this, the study of how tourism influences the development of new location-based technologies, including screen tourism apps and interactive maps, is a new and emerging field of study (Leotta, 2016; Lexhagen, Larson, & Lundberg, 2013; van Es & Reijnders, 2018). It is likely that we will see more of this kind of research related to media and tourism in the years to come.

In my previous work with Kim Toft Hansen, we proposed the location study model as a framework for empirical investigations of the relationship between places and television drama series, including screen tourism (Hansen & Waade, 2017). If we apply our off-screen location model to map the primary disciplines that are involved when we look at the relationships between places, media, and tourism, we may differentiate among (a) media-specific, (b) tourism-specific, (c) geography-specific, and (d) policy-specific approaches (Figure 3.1).

The *media-specific approaches* to studying media and tourism encompass studies informed by media theories (e.g. mediatisation, transmedia) or an understanding of how the various media (print media, film, television, internet, social media, mobile media, and locative media) have specific affordances, features, and conditions when it comes to production, content, and use (cf. the sections in this volume). This approach also includes media industry studies that focus on media policy, how media production and cross-sector collaboration are prompted by tourism (Ibrus, 2019), or the role that screen agencies, media clusters, media intermediaries, travel journalism, media platforms, and online services play when it comes to developing tourism. From a media-specific point of view, screen tourists are considered audiences, users, or fans, and screen tourism a mode of long tail marketing of certain media franchises including films and television series (Grey, 2010; Hills, 2002; Jenkins, 2006). Additionally, the field of smart tourism emphasises how the different technologies (location-based media, social media, mobile media) variously reinforce, influence, and change media and tourism practices (Femenia-Serra & Neuhofer, 2018).

If we turn to the *tourism-specific approaches* to studying media and tourism, the primary focus is on the tourists, how media influences their motivation and experiences and on developing and marketing the tourism destination. As already stated, DMO (Destination Marketing Organisations) studies and tourism studies more generally have significantly shed light on the relationship between media and tourism (Gyimóthy, 2015, 2018a; Lundberg, Lexhagen, & Mattsson, 2012; Reijnders, 2011). In this context, it is worth mentioning that tourism marketing is a well-established field of research in tourism studies, with its own academic journals, conferences, and networks. Since tourism as a cultural phenomenon and creative economy is related to many other disciplines, the tourism-specific approaches to studying media and tourism could also include elements from other disciplines, such as heritage studies (e.g. McAdam in this volume),

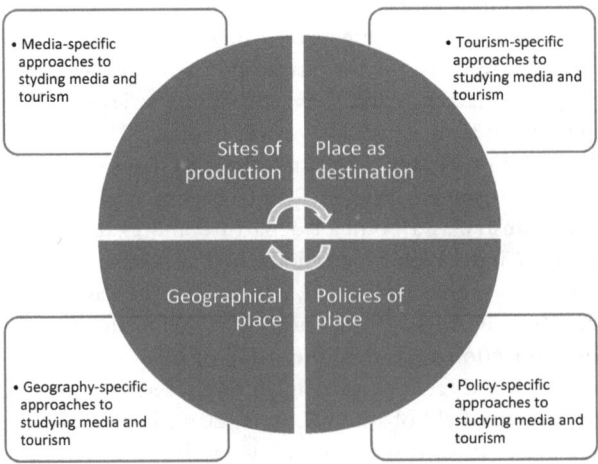

Figure 3.1 Interdisciplinary framework for studying media and tourism inspired by the off-screen location model in *Locating Nordic Noir*

Source: Hansen & Waade, 2017.

musicology, food studies (e.g. Williams and Fusté-Forné in this volume), sports studies (e.g. Graakær in this volume), or folklore studies, to mention just a few. Furthermore, and as a distinct category, *policy approaches* to studying media and tourism are mainly informed by the political sciences, including business and administration studies, and consider the place not only as a site of media production or as a tourist destination but as a political and administrative organisation that encompasses various sectors and political obligations. The two previous categories may overlap, for example, media policy focuses on media production and regulation, and destination management addresses how to brand and develop a destination. Although tourism-specific approaches encompass strategies for, and studies of how to attract tourists, policy approaches have a broader scope concerning how a place (nation, region, city) can capitalise on media and tourism. For example, the tax incentive is a political strategy for attracting screen productions, and through this, to possibly develop screen tourism. Finally, the *geography-specific approach* is related to physical, geographical settings and landscapes, and how media and tourism represent, stage, and influence the actual places (Hansen & Waade, 2017; Joyce, 2019; Saunders, 2019). The relationship between media and tourism is a well-established research focus in geography studies (e.g. Edensor, 2002), in the same way as geography is one of the cornerstones of tourism studies. The geography-specific approach emphasises the character of the places themselves, their physical, historical, social, political, and cultural features, and how these features correspond to, contrast with, or combine ideas and imagery in mediatized tourism (Månsson, 2011; Saunders, 2019).

Even though media studies and tourism studies are distinct disciplines with different academic training, theories, methods, and mindsets embedded in different institutional faculties and departments, it is important to mention that media and tourism has emerged as an interdisciplinary field of research with its own institutional units and communities, for example, the international academic conference that focuses on the relationship between media and tourism, *International Tourism and Media Conference* (ITAM), which occurs every second year, and with which several of the contributors to this volume are affiliated. There also exists an informal research network that focuses on the relationship between popular media and tourism, and from which the *Popculture Tourism Research Manifesto* originates (Gyimóthy, Lundberg, Lindström, Lexhagen & Larson, 2015), whose network recently released *The Routledge Handbook of Popular Culture and Tourism* (Lundberg & Ziakas, 2018). However, both groups consist mainly of tourism scholars, and both groups focus mainly on single case studies that present tourism as influenced by popular music, films, and television series (e.g. vampire tourism, Bollywood tourism etc.). To my knowledge, at this stage, there are no similar networks and conferences based in media and communication studies contexts.

Concluding remarks: Knowledge gaps and future challenges for media and tourism studies

Research that addresses the relationship between media and tourism evolves and changes rapidly, following the development of media technologies, consumer culture, and tourism practices. Thus, it may be difficult to identify general knowledge gaps and future challenges. Still, it is worth giving them a thought. Therefore, this edited volume is a welcome contribution to the field. If the overarching ambition is to develop media and tourism as distinct and interdisciplinary fields of research, I think the following gaps and challenges need to be addressed in future works and initiatives:

- We need to strengthen meta-disciplinary studies and develop *interdisciplinary, conceptual, and methodological frameworks* for understanding and approaching the relationship between media and tourism;
- We need to go beyond single case studies, and *initiate comparative and large-scale research studies* at the international level with a special focus on 'blind spots' in current research;
- We need to strengthen the *media-specific approaches* to media and tourism research, including (a) how distinct mobile, social, and location-based technologies influence tourism practices; (b) how tourism affects media-industry practices, such as location placement (a certain kind of product placement) as a business model, long tail marketing, the role of fan tourism in transmedia franchises; and (c) how different genres, formats, narratives and aesthetics make a difference when it comes to media-influenced tourism;
- Since media and tourism are characterised as cross-sector collaborations, we need to conduct *empirical studies of how these cross-sector collaborations and co-creation work,* for example, when it comes to (a) the implications of different professional and disciplinary languages, values and mindsets; (b) new types of partnerships and collaborative practices throughout the media and tourism industries (Ibrus, 2019) and private–public partnerships; (c) spillover effects, including knowledge, economy, and societal spillover (cf. Allen, Grimes, & Kerr, 2013; Fleming, 2015); (d) develop frameworks for cross-sector stakeholder analysis; and (e) link studies of media and tourism (including media-induced tourism) to culture tourism and the entertainment industry as broader fields and concepts;
- To supplement to the foregoing, it is particularly interesting to study industry events and markets that present, celebrate, and manifest media and tourism as emerging and cross-sector fields of practice, for example, screen tourism industry conferences such as *Seen on Screen* (Film London), *Mixed Reality* (Ystad municipality and Film Skåne), and screen industry fairs such as the international film commission's conference *Cineposion* (AFCI), and the MIPCOM and MIPTV screen fairs in Cannes, which emphasise the distribution of screen content throughout markets and countries;
- We need to develop research on the relationship between media and tourism that goes beyond the usual suspects, such as (a) screen tourism, (b) fiction, and (c) the Western world, and the Anglophone media conglomerates such as HBO, Netflix, and Disney (the EU-funded research project, *Worlds of Imagination* is a good example hereof);
- We welcome research that takes into account new modes of innovative engagement and practice-based research in the field, for example, how researchers may interact with and innovate in media and tourism collaborations, as in the Danish research project, Innocoast, funded by the Danish Innovation foundation (also see Waade, 2020);
- Finally, and in extension of the foregoing ideas, we welcome new cross-sector and collaborative training of professionals in the field, whereby media professionals and tourism professionals work together and develop a common vocabulary, mindset, and tools.

In the years to come, the relationship between media and tourism will continue to emerge, and it will continue to be combined with other cultural markers such as food, music, festivals, sport, and celebrities, and thus challenge research and studies that encompass and contextualise both the production aspects (media and tourism industries) and consumer and tourism practices. I already look forward to seeing how new interdisciplinary collaborations and initiatives emerge in the research fields of media and tourism, and from this ground-breaking volume.

References

Allen, C., Grimes, A., & Kerr, S. (2013). *Value and culture: An economic framework.* NZ, Cultural Policy Research Programme, online report. Retrieved from https://mch.govt.nz/sites/default/files/Value%20and%20Culture%20An%20Economic%20Framework%20Aug%202013%20pdf%20%28D-0500475%29.PDF

Beeton, S. (2016). *Film-induced tourism* (2nd Ed.). Bristol, UK: Channel View.

Buchmann, A., Moore, K., & Fisher, D. (2010). Experiencing film tourism: Authenticity and fellowship. *Annals of Tourism Research, 37*(1), 229–248

Connell, J. (2012). Film tourism – Evolution, progress and prospects. *Tourism Management, 33*, 1007–1029.

Edensor, T. (2002). *National identity, popular culture and everyday life.* Oxford, UK: Berg Publishers.

Femenia-Serra, F., & Neuhofer, B. (2018). Smart tourism experiences: Conceptualisation, key dimensions and research agenda. *Investigaciones Regionales – Journal of Regional Research, 42*, 129–150.

Fleming, T. (2015). *Cultural and creative spillovers in Europe: Report on a preliminary evidence review.* Arts Council England's model for measuring spill-over effects in the creative sector in Europe. Tom Fleming Creative Consultancy. Retrieved from https://www.artscouncil.org.uk/sites/default/files/Cultural_creative_spillovers_in_Europe_full_report.pdf

Hanusch, F., & Fürsich E. (2014). *Travel journalism: Exploring production, impact and culture.* Basingstoke, UK: Palgrave Macmillan.

Grey, J. (2010). *Show sold separately: Promos, spoilers, and other media paratexts.* New York: New York University Press.

Gyimóthy, S. (2015). Bollywood-in-the-Alps: Popular culture place-making in tourism. In A. Lorentzen, K. Topsøe Larsen, & L. Schrøder (Eds.), *Spatial dynamics in the experience economy* (pp. 158–174). New York, NY: Routledge.

Gyimóthy, S. (2018a). Transformations in destination texture: Curry and Bollywood romance in the Swiss Alps. *Tourist Studies, 18*(3), 292–314.

Gyimóthy, S. (2018b). The Indianization of Switzerland. In C. Lundberg & V. Ziakas (Eds.), *Routledge handbook on popular culture and tourism* (pp. 376–388). London, UK & New York, NY: Routledge.

Gyimóthy, S., Lundberg, C., Lindström, K. N., Lexhagen, M., & Larson, M. (2015). Popculture tourism: A research manifesto. In D. Chambers & T. Rakic (Eds.), *Tourism research frontiers: Beyond the boundaries of knowledge* (pp. 13–26). Bingley, UK: Emerald.

Hansen, K. T., & Waade, A. M. (2017). *Locating Nordic Noir. From beck to the bridge.* Basingstoke, UK: Palgrave Macmillan.

Hills, M. (2002). *Fan culture.* London, UK & New York, NY: Routledge.

Ibrus, I. (2019). *Emergence of cross-innovation systems – Audiovisual industries, co-innovating with education, health care and tourism.* Croydon, UK: Emerald Publishing.

Jansson, A., & Falkheimer, J. (2006). Towards a geography of communication. In J. Falkheimer & A. Jansson (Eds.), *Geographies of communication. The spatial turn in media studies* (pp. 9–25). Göteborg, Sweden: Nordicom.

Jenkins, H. (2006). *Convergence culture: Where old and new media collide.* New York & London, UK: New York University Press.

Joyce, S. (2019). Experiencing northern Ireland as Game of Thrones destination. In F. Barber, H. Hansson, & S. Dybris McQuaid (Eds.), *Ireland and the north* (pp. 95–118). Oxford, UK: Peter Lang.

Kim, S. (2012). The relationships of on-site film-tourism experiences, satisfaction, and behavioural intentions: From the film-tourism perspective. *Journal of Travel and Tourism Marketing, 29*(5), 472–484.

Lee, C. (2012). 'Have magic, will travel': Tourism and Harry Potter's United (Magical) Kingdom. *Tourist Studies, 12*, 52–69.

Leotta, A. (2011). *Touring the screen: Tourism and New Zealand film geographies.* Bristol, UK: Intellect.

Leotta, A. (2016). Navigating movie (m)apps: Film locations, tourism and digital mapping tools. *M/C Journal, 19*(3).

Lexhagen, M., Larson, M., & Lundberg, C. (2013). The virtual fan(g) community: Social media and pop culture tourism. In A. M. Munar, S. Gyimóthy, & L. Cai (Eds.), *Tourism social media: Transformations in identity, community and culture* (pp. 133–158). Bingley, UK: Emerald.

Lundberg, C., Lexhagen, M., & Mattsson, S. (2012). *Twication: The twilight saga travel experience.* Östersund, Sweden: Jengel Förlag AB.

Lundberg, C., & Ziakas, V. (2018). *Routledge handbook on popular culture and tourism.* London, UK & New York, NY: Routledge.

Månsson, M. (2011). Mediatized tourism. *Annals of Tourism Research, 38*(4), 1634–1652.

Månsson, M., & Eskilsson, L. (2013). *Euroscreen: The attraction of screen destinations. Baseline report assessing best practice*. Rzeszów, Poland: Pracownia Pomysłów.

Reijnders, S. (2011). *Places of the imagination: Media, tourism, culture*. Farnham, UK: Ashgate Publishing.

Roberts, L. (2016). Landscapes in the frame: Exploring the hinterlands of the British procedural drama. *New Review of Film and Television Studies, 14*(3), 364–385.

Roesch, S. (2009). *The experiences of film location tourists*. Bristol, UK: Channel View Publications.

Saunders, R. A. (2019). Geopolitical television at the (b)order: Liminality, global politics, and world-building in *The Bridge*. *Social & Cultural Geography, 20*(7), 981–1003.

Sjöholm, C. (2010). Murder walks in Ystad. In: B.T. Knudsen & A.M. Waade (Eds.), *Re-investing authenticity: Tourism, place and emotions* (pp. 154–169). Bristol, UK: Channel View Publications.

Turnbull, S., & McCutcheon, M. (2017). Investigating miss fischer: The value of a television crime drama. *Media International Australia, 164*(1), 56–70.

Urry, J., & Larsen, J. (2011). *The tourist gaze 3.0 – Leisure and travel in contemporary societies*. London, UK & Los Angeles, CA: Sage Publications.

van Es, N., & Reijnders, S. (2018). Making sense of capital crime cities: Getting underneath the urban facade on crime-detective fiction tours. *European Journal of Cultural Studies, 21*(4), 502–520.

Waade, A. M. (2016). Nordic noir tourism and television landscapes: In the footsteps of *Kurt Wallander* and *Saga Norén*. *Scandinavica, 55*(1), 41–65.

Waade, A. M. (2020, forthcoming). Screening the west coast: Developing *new Nordic Noir* tourism in Denmark. In S. Reijnders, L. Bolderman, A. Waysdorf, & N. van Es (Eds.), *Locating imagination. Popular culture, tourism and belonging*. London, UK & New York, NY: Routledge.

Wheatley, H. (2016). *Spectacular television. Exploring televisual pleasure*. London, UK & New York, NY: I. B. Tauris. *Worlds of imagination (2019)*. Retrieved from: https://www.worldsofimagination.eu.

4

THEME PARKS – WHERE MEDIA AND TOURISM CONVERGE

Sabrina Mittermeier

Introduction

Henry Jenkins (2006) has famously defined convergence as 'the flow of content across multiple media platforms, the cooperation between multiple media industries, and the migratory behaviour of media audiences who will go almost anywhere in search of the kinds of entertainment experiences they want' (p. 2). Theme parks are a key example of the convergence culture he describes, as they are usually owned and operated by multi-media conglomerates and are at the forefront of what has become known as 'mediatized tourism' (Månsson, 2011), motivating fans of media products to travel in search of what they perceive as authentic experiences of the fictional story worlds underlying them. Even though studying mediatized tourism without considering the theme park, I would argue, is almost impossible, the parks are still vastly understudied, despite existing scholarship coming from a variety of disciplines. This is exemplified by the fact that existing definitions of the form, i.e. that by Kagelmann (1993) or Clavé (2007) approach the theme park from an industry standpoint, solely as a tourist destination – which serves well their own background of tourism studies, yet leaves a lot to be desired for those in media and cultural studies, which is the disciplinary approach I would like to take for this chapter.

To make up for this lack in scholarship, I define the theme park from the perspective of cultural studies and based in part on Jenkins' work, calling it 'a participatory medium that relies on strategies of theming to entertain an audience within a transnational consumer culture' (Mittermeier, 2017, p. 6). I will use this definition here to discuss the intricate relationship between tourism and media in the theme park; touching upon the theme park as a medium; the theme park as a transnational tourist destination; its participatory nature, made possible by a multimedia environment; and its role within consumer culture. Even though the form is an American invention (the original Disneyland that opened in 1955 is usually credited as the first theme park), it has since spread all over the world, and extended far beyond the Walt Disney Company's achievements; in recent years, the Chinese market has become particularly important. Themed entertainment has also long since expanded beyond the confines of these parks, whether to the resorts in Las Vegas, lavish shopping malls, cruise ships, or even heritage attractions: a phenomenon famously described as 'Disneyization' by Bryman (2004). Yet, this chapter wants to discuss theme parks in their original form, and to do so, I will use examples from Disney's theme parks around the world, as well as the Wizarding World of Harry Potter sections of Universal's theme parks, as these are among the most-visited destinations every year and are particularly suited to discuss the role of media, since they are also among the most technologically sophisticated theme parks worldwide.

The theme park as an immersive medium

First, I will discuss the theme park as a medium. Bolter and Grusin (2000) have pointed towards the 'hypermediacy' of theme parks (p. 170) and have used them as one case in which their concept of 'remediation' is most apparent in today's media landscape. Remediation means that one medium is transformed into another, such as turning a film into a theme park attraction, a practice that Disneyland first pioneered. Because of this, studies of Disneyland, or theme parks in general, have often focused on the park experience as being 'like walking into a movie' (Hine, 1986, p. 151). Theme parks as well as other immersive environments (such as themed restaurants, hotels, or shopping malls) are designed using the strategies of theming, which Lukas (2007) has defined as 'the use of an overarching theme... to create a holistic and integrated spatial organization of a consumer venue' (p. 1). Mitrasinovic (2006) further distinguishes between 'literal thematization' (detailed reproduction) and 'interpretive thematization' (the evocative and metaphoric use of colours, fabrics, and the like) (pp. 154–155). To theme a space, designers not only use architecture but also music and other sounds, smells, costumes for its employees, as well as food and merchandising articles. Disney calls the practice of designing and building theme parks 'Imagineering', a term coined by Walt Disney himself, meaning the 'blending of creative imagination with technical know-how' (cited in Sklar, 2010, pp. 10–11). Theming, then, is only one part of this larger process that also encompasses several other technologies to tell a story; in addition to film and music, this can include audio-animatronics to simulate humans or animals, other kinds of interactive information technology; different kinds of vehicles for actual or simulated transport; and many more such tools, often state of the art. The purpose of all of these is the immersion of the visitor into the venue and the story or information it tries to convey. Immersion means being completely submerged into something, as Murray (1997) has argued,

> [we] seek the same feeling from a psychologically immersive experience that we do from a plunge in the ocean or swimming pool: the sensation of being surrounded by a completely other reality, as different as water is from air, that takes over all of our attention, our whole perceptual apparatus.
>
> *(pp. 98–99)*

Immersion thus implies 'a transition, a 'passage' from one realm to another, from the immediate physical reality of tangible objects and direct sensory data to *somewhere else*' (Huhtamo, 1995, p. 159; original emphasis). For the visitor to experience such immersion, they have to apply suspension of disbelief, the willingness to ignore the elements of a theme park (or other themed space) that might destroy the illusion of another reality, and theme park designers thus usually place great importance on removing any such interferences and guide the visitor's point of view (Lonsway, 2009, p. 125).

Joe Rohde, one of Disney's most prolific Imagineers, has coined another term that helps us to understand the practice of theme park design and makes clear just how much storytelling is at the core of this experience: 'narrative placemaking'. Rohde (2007) defines narrative placemaking as 'the building of ideas into physical objects' and explains the intricacies of telling stories through three-dimensional space rather than the more linear, two-dimensional one of a film: '...the guests' real physical bodies are all moving inside the imaginary narrative space. Guests make choices as to how to travel through the space or where to look. ... [so,] linear storytelling doesn't read. The place itself, in every detail, must reiterate the core ideas that drive the story'. In a well-designed themed space, the visitor is directly involved in the narrative: 'In spoken narrative, and in theatrical narrative, the audience is passive, but in narrative placemaking, the audience is a part of the visual environment. They are given roles within the narrative'. This shows then that the theme park is in and of itself a participatory storytelling medium, and as Waysdorf and Reijnders (2018) argue, it provides 'a spatial and embodied connection to a narrative world' (p. 185), further putting focus on the active role of the visitor. Freitag (2017) further has argued that since theme parks incorporate a number of media within themselves, they consequently become 'composite' or 'hybrid media' (p. 706).

So, while remediation is only one part of how the theme park creates immersion, it is one that has always been incredibly important for its marketing: part of the enormous initial and continuous success of Disneyland has been the synergy it creates between the theme park and other Disney products. One example of this is Disneyland's park icon, Sleeping Beauty Castle. When the park opened in 1955, the animated film *Sleeping Beauty* had not yet been released (it would not be until 1959), and the home of the titular princess did in fact not even look much like the castle found at the park – but its name promoted the film before it came out, and once it did, a walk-through exhibit showcasing its artwork was incorporated inside the building, drawing guests that had now become fans of the film. This has become a standard practice for theme parks that are not owned and/or operated by Disney as well. In more recent years, this remediation process has also been reversed, as theme park attractions have been turned into other media, such as films, or comics – Disney's *Pirates of the Caribbean* (2003–2017) film series is still the biggest testament to this trend. A ride originally built under Walt Disney's supervision but opened only after his death in 1967, it quickly became one of the most popular attractions at Disneyland – so much so that when the Magic Kingdom theme park in Florida opened without it in 1971, guest demand for it forced the company to rectify this mistake. Both Tokyo Disneyland and Disneyland Paris feature a version of it, so it had garnered an international audience familiar with it when the first film, *Pirates of the Caribbean: The Curse of the Black Pearl* (2003) opened in cinemas. Its film's success in both domestic and international markets was so massive that Disney then began to add characters and story elements from the film to the rides, reversing the remediation process once more. Shanghai Disneyland, which opened in 2016, features a whole themed area called Treasure Cove that is based on the films and its attraction Pirates of the Caribbean: Battle for the Sunken Treasure is completely based on the films and hardly shares similarities with the 'original' theme park rides anymore, bringing the process full circle.

Freitag (2017) has built on Bolter and Grusin's work and introduced the term 'intermedial transposition' for 'the cases in which a particular movie is translated into the medium of the theme park' (p. 710), which can also extend to entire themed areas, such as the above-described Treasure Cove. One other notable example of this is The Wizarding World of Harry Potter in the Universal Studios theme parks in Orlando, FL and Hollywood, CA in the U.S., and Osaka, Japan. Such themed areas, including rides, shows, shops, and restaurants all themed to the same film (or in the case, whole franchise of books and films), massively rely on the recognition value of the product they are based on. This is especially true for rides, as they usually greatly compress storylines into a few key scenes and characters, making it rather confusing to those visitors not familiar with the intellectual property (IP) to understand what is going on. For instance, the Harry Potter and the Forbidden Journey ride (found at all three theme park locations) is specifically set during the events of the fourth film, *Harry Potter and the Goblet of Fire* (2005), as hinted at in the queue area that leads visitors through the Hogwarts castle and grounds, but then also incorporates elements from other films, such as the spiders living inside the Forbidden Forest that originally make an appearance in *Harry Potter and the Chamber of Secrets* (2002). As Aronstein (2012) has argued, such attractions 'are effective only because the … narratives on which they are based are omnipresent; without the intertextual interpretive context provided by the films, these rides would be meaningless – a series of disconnected images' (p. 67). The whole Wizarding World follows the visual iconography established by the *Harry Potter* films, also outside the rides. For instance, the Diagon Alley area found in Universal Studios Florida is a direct recreation of the sets for the film, and the merchandise sold at any of the locations, such as the famed Bertie Botts' Every Flavour Beans also uses the art design created for the films, and score composed for them is used throughout the area.

All of this makes possible a high level of immersion for fans of the franchise, as they experience it as an 'authentic adaptation of the *Harry Potter* story-world, a place *Harry Potter* fans can … experience the story-world in an embodied manner' (Waysford & Reijnders, 2018, p. 174). This also points towards the role of authenticity in these spaces – one of the most discussed aspects of theme parks academically is that of whether or not they provide any authentic experience at all. Older scholarship usually reads them critically as fake and inauthentic, particularly when they attempt to recreate historical periods, but more recent studies have engaged with it more in terms of an 'authenticity of experience' (Bolter & Grusin, 2000, p. 172). It gives the audience agency in that they have to bring their own emotional investment in the narrative space to fully experience it. As studies on immersive media have shown, previous knowledge, or at least an interest in the story, is vital to its success (Hofer & Wirth, 2008, p. 167). Universal thus wants the fans of this franchise to fill their theme parks, and as there does not seem to be an end to its popularity, the acquisition of this IP was obviously a smart move – it establishes their parks as tourist destinations specifically for this group of fans. Theme parks thus have become spaces of fandom, where visitors engage in cosplay and feel a sense of community, similar to that of fan conventions (Waysdorf & Reijnders, 2018, p. 184), and travelling to them for many 'represents a realization of an earlier imaginary journey' (Reijnders, 2016, p. 673) that fans have taken when reading/watching/playing the media they are based on. As argued above, theme parks have always been banking on this imaginary journey, establishing them as part of what Karpovich (2010) hails as 'film-induced tourism' (p. 10). As mentioned above, Månsson (2011) has built on this term, moving away from simply connecting this phenomenon to film fandom, and describes it as 'mediatized tourism', encompassing the constant convergence of media products in tourist practices and again grant the individual tourist more agency in her experience.

Theme parks and participatory culture

The concept of mediatized tourism also marks theme parks as part of the previously discussed concept of convergence culture as it not only is based on consumers 'mak[ing] connections among dispersed media content' (Jenkins, 2006, p. 3) but also thrives on the convergence of several media and technologies. Technology is at the core of the theme park experience, and many attractions use cutting edge or even state-of-the-art technology for their storytelling – the Harry Potter and the Forbidden Journey ride, for instance, is a mixed media ride that uses audio-animatronics, 3D film, and KUKA industrial robotics. The Ollivander's show and interactive wands are another great example of how technological advancements contribute to making the experience at the Wizarding World as immersive as possible. In Harry Potter lore, Ollivander is a wand maker, and children from the (British) magical community are brought to him to let their wand choose them (not vice versa) at age 11, when they are old enough to attend the Hogwarts School of Witchcraft and Wizardry. At Universal Studios, this is transformed into an attraction: a group of guests is brought into the wand store, and then one guest, usually a child, is chosen by the actor portraying Ollivander to go through the ritual. However, all guests can purchase wands after, thus turning shopping into an 'immersive, imaginative act' (Waysdorf & Reijnders, 2018, p. 183) – they are quite literally buying into the story world. There are two kinds of wands available: replicas from those created for the films and associated with a specific character, adding yet another element of cosplay to the experience, and interactive wands. These latter wands have a uniform design but use infrared sensors to communicate with specific locations around the themed area. Each wand is sold with a map that shows locations throughout the Wizarding World where spells can be performed, and which spell movements are required to perform them. When the visitor visits these locations and performs the spell as indicated, there is a reaction to it – usually found in window displays around the shops in Hogsmeade and Diagon Alley, such as books hovering or creatures moving. The use of these interactive media thus makes another level of embodiment possible, adding another layer of immersion into the story world of the *Harry Potter* films and books, as it draws in visitors via a form of role-play.

Such interactive games are becoming more and more prevalent in theme parks. Disney debuted the Sorcerers of the Magic Kingdom interactive game at the Magic Kingdom in Walt Disney World in 2012 that is essentially a more sophisticated version of the wands at Universal and operates via the use of a collectible card game (tapping into another common fandom interest). They have also recently introduced *Play Disney Parks*, an app that is mainly geared at keeping guests entertained while waiting in line but also adds other gamification elements to the experience at all of their American theme parks. It includes trivia, themed music playlists and other mini games, that lets users collect achievements at attractions, letting the visitor interact with the park on another, virtual level. With this, Disney is reacting to the success of other mobile games such as the augmented reality-based *Pokémon Go* that many visitors also played in their parks and extends the visitor experience to a virtual level – an experience that should appeal to millennial and even younger audiences in particular.

The use of personal mobile devices in their parks, however, does not end there. In 2013, Disney first began to roll out MyMagic+ at Walt Disney World, a sophisticated system geared at improving guest flow and vacation planning. Every guest creates an account at the company's online platform called MyDisneyExperience, that is also accessible via a dedicated app. The system lets guests book their hotel, theme park entrance, dining reservations, as well as several other services geared towards improving their experience, such as Fastpass+, granting faster access to certain rides at a previously picked time slot, as well as Memory Maker/Photopass, that includes professional photographs taken while at the parks and hotels, such as with character performers or on rides. The system is tied to the so-called Magic Bands, wearables (either as a wristband or keychain) that use radio-frequency identification near-field communication technology to gain park entrance, activate Fastpass, connect pictures taken, open hotel room doors, and charge food and other purchases to a credit card connected to the account. When scanned at any location, Magic Bands present the Cast Member (Disney's term for its employees) with the guest's name and potentially, other added information, such as birthday celebrations. All of this does not only smooth the guest experience and flow, it also makes possible for a more immersive experience, such as being 'magically' greeted by name by both cast members and characters such as Mickey Mouse, or receiving surprises based on their birthday or honeymoon celebration. Since guests can also make changes and check any of their plans on the go using the smartphone app, it actively connects their personal devices to the theme park experience. Such technological advances turn the theme park into an even more mediatized environment than before and also makes it depend much more heavily than previous on active consumer participation – another key element of convergence culture. This 'participatory culture' (Jenkins, 2006, p. 3) is also what I have highlighted in my definition of theme parks, as outlined at the beginning of this chapter.

Another aspect where participatory culture has become important, not only for how visitors interact with the space of the theme park, but also for how theme parks market themselves, is social media. Social media engagement is becoming more and more important for tourist destinations, and theme parks are no exception. The aforementioned Memory Maker/Photopass packages are offered in different versions in every Disney theme park around the world, and other theme parks often provide similar programs. Via the use of apps, visitors can view and download the photos that have been taken of them almost instantly and thus make them available via social media, such as Facebook, Twitter, or most importantly for photos, Instagram. Disney has increasingly tried to engage with bloggers and other social media personalities, as one recent case shows: many 'influencers,' as well as regular guests, had begun taking posed pictures and selfies in front of a purple wall in the Magic Kingdom theme park's Tomorrowland section, earning it its own hashtag on Instagram (#purplewall). Disney reacted by redesigning said wall, making it more visually interesting and even creating a signature Purple Wall Slush beverage that was sold in the area for a while (Clark, 2018).

Because practices of riding and restaurant reviews have become popular to showcase via video channels, YouTube also plays a central role in the engagement with the theme parks, particularly in travel blogs. Generally, many theme park visitors (especially those visiting Disney's resorts, but also Universal and others) have begun to put a lot of effort into planning their stays at these tourist destinations – a phenomenon that the above mentioned My Disney Experience system reacts to but also makes even more necessary than before. Thus, reviews spread via social media are an important resource for these visitors, who use online fora such as Disboards, countless blogs and vlogs, as well as other social media channels to both consume and create them.

The theme park and transnational consumer culture

The Disney theme parks naturally also run their own channels to promote the parks and provide information. This, I would argue, also fosters transnational engagement of their parks. Not only does the company-run Disney Parks Blog cover all their theme parks around the world (in addition to the American locations, these include resorts in Paris, Tokyo, Hong Kong, and Shanghai), fan-run travel blogs have also increasingly started to write about the more far-off destinations, providing trip reports and news. Social media makes such transnational connections among fans a lot easier than before, and has also eased travel planning abroad that increasingly lies in the hands of the individual consumer rather than travel agencies.

In addition to more extensively cross-promoting their international parks to a wider audience, Disney is also weaving intertextual references between some of their attractions around the world. The Imagineers working on the TokyoDisney Sea theme park introduced the backstory of S.E.A., the Society of the Explorers and Adventurers, and created original characters that are part of the narrative of such rides as the local version of the Tower of Terror.

Ever since, characters and locations connected to this same backstory have been added across Disney parks around the world, including Hong Kong Disneyland's Mystic Manor, the Miss Adventure Falls ride in Typhoon Lagoon (part of Walt Disney World), the Magic Kingdom's Skipper Canteen restaurant, and even parts of the children's activities areas on Disney Cruise Line's Disney Magic cruise ship. This narrative is likely only apparent to those theme park visitors that are more immersed in the Disney parks' fandom, usually repeat visitors that also have the means to visit several of their resorts around the world. As Månsson (2011) has argued, 'one way for producers to use media convergence is intertextuality: media products' interrelationship through either hidden or open reference' (p. 1637), and Disney is using this practice to engage their fans even more in their story worlds. Such intertextuality between attractions and across parks strengthens narrative coherence and thus immersion and also relies on visitor's active participation to decipher these cross-references, marking it as yet another case of how theme parks are part of convergence culture.

This practice also provides yet another impetus for fans of Disney theme parks to travel to each of them, since all of them not only offer unique attractions not to be found elsewhere but also increasingly cater to people that are specifically fans of their attractions that stand alone from any other remediation processes, i.e. are not adaptations of films or other media. There is an existing and ever-growing fandom of the Disney parks in and of itself, and that extends to particular attractions, often those that have existed the longest, adding an element of nostalgia. For instance, the Haunted Mansion ride, that in different versions can be found in most Disneyland theme parks, has such a cult following that Disney has begun to produce a large number of merchandise articles on it, and it frequently becomes a source for cosplay of its fans. In the Asian theme parks, Disney the Duffy Bear and friends, a series of original characters created especially for the parks, have also garnered a massive audience willing to spend a lot of money on its products.

Merchandise is only one visible marker of theme parks as consumerist spaces. One key characteristic of theme parks over their antecedents, the amusement parks, is that they operate on all-inclusive pricing system – you usually only pay one entrance fee at the gate, and then all rides and entertainment are included in these. The cost of theme parks tickets, particularly those operated by Disney and Universal, is steep. While the amount per day drops the longer you stay, prices easily average 100 US\$ for a one-day ticket to enter each of these parks. Yet, food and beverages are also not yet included in this price, so even if one foregoes the purchase of souvenirs, the entrance fee does not cover the daily cost – and theme park food is anything but cheap. This makes clear just how much these tourist destinations are not accessible to just anyone – which also distinguishes theme parks from other forms of mediatized tourism, such as movie site tours in public spaces. Theme parks like these are usually only affordable by the middle and upper-classes, and as I have argued elsewhere (Mittermeier, 2017), Disney in particular has always specifically targeted this segment of the population.

Conclusions

Theme parks, even if only accessible to specific audience, are an excellent example to discuss how tourism and media relate to each other. They are not only media themselves but also highly mediatized spaces, and in many ways, the original destinations of mediatized tourism. As venues operated by multimedia conglomerates, they are also at the heart of the capitalist consumer culture permeating almost all of contemporary tourism. They aim to provide their visitors with a highly immersive experience that is often dependent on the remediation of other media such as film, and as such, they have become not only tourist destinations specifically catering to fans, they have also sparked their own fandoms and give people a space to perform their fandom individually and with others. To make immersion possible, they also rely on an array of different media and technology, and as interactive experiences become more and more feasible and popular, now also increasingly incorporate visitors' personal mobile devices. As tourist destinations, they are also relying on the same technology, involving the visitors in planning their stays, and providing sophisticated systems to do so. Additionally, they extensively rely on visitors to create promotion for their parks via social media. Arguably, then, the agency of their guests is more important for theme parks than ever before, cementing their status as both media and tourist destinations in today's participatory culture. Theme parks thus not only are part of convergence culture, they are its epitome – a true convergence of media and tourism.

References

Aronstein, S. (2012). Pilgrimage and medieval narrative structures in Disney's park. In S. Aronstein & T. Pugh (Eds.), *The Disney middle ages: A fairy-tale and fantasy past* (pp. 57–76). New York, NY: Palgrave Macmillan.

Bolter, J. D., & Grusin, R. (2000). *Remediation. Understanding new media.* Cambridge, MA: MIT Press.

Bryman, A. (2004). *The Disneyization of society.* London, UK: Sage.

Clark, C. (2018). The popular purple wall at Disney world just got its own signature slushie — and your Instagram will never be the same. *Business Insider.* Retrieved from: https://www.businessinsider.com/purple-wall-slushie-at-disney-world-magic-kingdom-2018-5?IR=T.

Clavé, S. A. (2007). *The global theme park Industry.* Oxfordshire, UK: CABI.

Freitag, F. (2017). Like walking into a movie: Intermedial relations between theme parks and movies. *Journal of Popular Culture, 50,* 704–722.

Hine, T. (1986). *Populuxe.* New York, NY: Knopf.

Hofer, M., & Wirth, W. (2008). Präsenzerleben – Eine medienpsychologische Modellierung. *Montage A/V, 17*(2), 159–175.

Huhtamo, E. (1995). Encapsulated bodies in motion. Simulators and the quest for total immersion. In S. Penny (Ed.), *Critical issues in electronic media* (pp. 159–186). Albany: State University of New York Press.

Jenkins, H. (2006). *Convergence culture: Where old and new media collide.* New York: New York University Press.

Kagelmann, H. J. (1993). Themenparks. In H. J. Kagelmann & H. Hahn (Eds.), *Tourismuspsychologie and tourismussoziologie* (pp. 407–415). München, Germany: Quintessenz.

Karpovich, A. I. (2010). Theoretical approaches to film-motivated tourism. *Tourism and Hospitality Planning & Development, 7*(1), 7–20.

Lonsway, B. (2009). *Making leisure work: Architecture and the experience economy.* New York, NY: Routledge.

Lukas, S. A. (2007). The themed space: Locating culture, nation, and self. In S. A. Lukas (Ed.), *The themed space: Locating culture, nation, and self* (pp. 1–22.). Plymouth, UK: Lexington Books.

Månsson, M. (2011). Mediatized tourism. *Annals of Tourism Research, 38,* 1634–1652.

Mitrasinovic, M. (2006). *Total landscape, theme parks, public space.* Farnham, UK: Ashgate.

Mittermeier, S. (2017). *Middle-class kingdoms. A cultural history of Disneyland and its variations, 1955–2016* (Doctoral dissertation, LMU, Munich, Germany [forthcoming with Intellect in 2020]).

Murray, J. (1997). *Hamlet on the holodeck: The future of narrative in cyberspace.* New York, NY: Free Press.

Reijnders, S. (2016). Stories that move: Fiction, imagination, tourism. *European Journal of Cultural Studies, 19,* 672–689.

Rohde, J. (2007). From myth to mountain: Insights into virtual placemaking. *ACM SIGGRAPH Computer Graphics, 41*(3), 1–es.

Sklar, M. (2010). Introduction: Imagineering: A history that's no mystery. In The imagineers (Eds.), *Walt Disney imagineering. A behind the dreams look at making MORE magic real* (pp. 10–11). New York, NY: Disney Editions.

Waysdorf, A., & Reijnders, S. (2018). Immersion, authenticity and the theme park as social space: Experiencing the wizarding world of Harry Potter. *International Journal of Cultural Studies, 2,* 173–188.

5

CINEMATIC TOURISM IN A TIME OF MEDIA CONVERGENCE

A spatial framework

Giulia Lavarone

Introduction

This chapter will focus on cinematic tourism from a cultural perspective and the disciplinary background of media and, in particular, film studies. It will discuss some of the research perspectives which deploy film theory to understand the cinematic tourist experience with the aim of building interdisciplinary knowledge. The term 'cinematic tourism' (Tzanelli, 2007) can refer either to an actual or to a virtual tourist experience; therefore, it will be used here to extend 'film tourism'. Adopting spatial categories, later described, one section will focus on *tourist space in films*. It will synthesize and discuss a handful of relevant contributions which, in the second half of the 2000s, tried to explore the virtual tourist experience and the *pre-mediation* – in Grusin's (2010) terms, of the actual tourist experience, recurring to film theories elaborated from 1970s onward. The subsequent section will focus instead on *films in tourist space* and will suggest an under-developed line of inquiry, involving contemporary film theories in the examination of the actual tourist experience. In particular, an understanding of cinematic tourism in the framework of *relocation of cinema* will be presented, in relation to the more general questioning of the role of cinema in the times of media convergence.

This conceptual entrance point to the field of cinematic tourism, which relies on film theory, is proposed here due to the consideration that film studies are largely recognized to have provided a limited contribution to the exploration of film tourism, and further research is often required, as stated in the first section. Especially when film tourism was being established as a research field, film and media studies were solicited to contribute in the exploration of the 'effects' of media products on tourism, for example, by identifying film elements that are more likely to induce tourism, a question often asked in a business perspective. To understand cinematic tourism within the theoretical framework of *mediatization* also denotes embracing the conceptual shift from exploring the 'effects' of media, 'conceived of as separate from society and culture' (Hjarvard, 2008, p. 105), on tourism, to questioning the intrinsic changes led by mediatization to the tourist experience itself. This has been developing over the years through the intensifying interdisciplinary research on film tourism in a cultural perspective, pushing media and film studies to adopt different points of view.

The term mediatization refers to *the process whereby society to an increasing degree is submitted to, or becomes dependent on, the media and their logic* (Hjarvard, 2008, p. 113). It has been used 'to describe media's increasing social and cultural relevance since the emergence of mass media (print, cinema, radio, television)' (Couldry & Hepp, 2013, p. 197). When looking at contemporary times, it mostly deals with the increasing role played by interactive media in our daily lives. The concept of 'mediatized tourism' has been understood in relation to the relevance of interactive media in producing and sharing content, thus highlighting tourists' agency (Månsson, 2011). This perspective will not be discussed in this chapter, which mostly deals with 'traditional' media like cinema and television, even if questioned in the frame of contemporary media convergence (Jenkins, 2006). While acknowledging the latter as a distinctive feature of contemporary times, the choice of privileging cinema might be sustained by the observation that it still works as a sort of 'grand master' (Fanchi, 2012) in the media environment, both in the production of transmedia narratives, where film usually is still given the most important content (Zecca, 2012), and in the creation of user-generated content, where film persists as a sort of ideal 'horizon' (Fanchi, 2012). An 'idea of cinema' seems to survive in contemporary times, even in the dispersed forms assumed by what was previously a unitary experience (Casetti, 2015). This suggests that some acquisitions of film theory, including those elaborated prior to the times of convergence, may still be relevant to understand a deeply changed scenario, and its relations with tourism.

Film tourism and film studies

Film tourism has been explored since the late 1990s in various disciplinary fields, such as tourism economics and management, cultural geography, visual anthropology, sociology, cultural studies, media studies, fan studies. Beeton (2010, developed in Beeton, 2015) and Connell (2012) both point out a progression in the development of film tourism inquiry, from a prevalent business and marketing perspective to an interdisciplinary 'higher inquiry' (Beeton, 2010, p. 4), or 'alternative discourse' (Connell, 2012, p. 1008), involving media studies and social sciences. When it comes to film studies, tourism researchers have often pointed out that the contribution of this discipline seemed to be still limited, making frequent calls for cross-disciplinarity (e.g. Beeton, 2010, 2015; Connell, 2012; Croy & Heitmann, 2011). Film scholars exploring the field also stressed this point (e.g. Leotta, 2011; Mazierska & Walton, 2006).

The limited interest in film tourism within film studies, at least initially, might be due to an under-estimation of this research field's potential to stimulate deep theoretical questionings, beyond serving economics and marketing perspectives. Actually, research within film studies was solicited by initial questions posited by tourism scholars, regarding, for example, the identification of 'the elements that make a film "create" film-induced tourism' (Beeton, 2005, also quoted in Leotta, 2011, p. 8). In Italy, for example, Provenzano (2007) tried a classification of the diegetic functions of landscape in film, relating it with their propensity to induce tourism. Nevertheless, in the 'boom period in film tourism research' (Connell, 2012), broader interdisciplinary knowledge was pursued beyond this perspective, for example, with the aim of exploring the cinematic tourist experience.

In her progress review aimed at 'critically evaluat[ing] film tourism as a subject of cross-disciplinary academic study', Connell (2012) points out – within film studies – two research areas which should be particularly relevant for cross-disciplinary work on film tourism. The first is related to film viewing, analysed from psychoanalytic, cognitive, and neurological perspectives. The second focuses on space, involving the relationship between film and landscape. Within this scope, Connell identifies two main lines of inquiry. The first deals with the experience of virtual travel inherent in film and television viewing. The second deals with practical and conceptual issues deriving from the practice of *displacement*, i.e. when films or TV programmes are shot in locations which differ from the fictional setting. Some of these issues, concerning the experience of film viewing (especially in its traditional psychoanalytical understanding) and that of virtual travel, as well as the focus on the spatial dimension, will play a significant part in this chapter.

It must be said that in the last few years, interdisciplinary approaches between media and tourist studies have been gradually developing. An example is Beeton's (2015) recent work where she analyses film tourism relying on audience theories, including it among audience activities after film/TV viewing and relating it to different audience orientations before, during and after viewing. Beeton's study explicitly refers to the innovative methodological approach introduced by Kim (2012), who had explored how different dimensions of audience involvement with TV dramas influence on-site film tourism experiences.

A spatial framework

Acknowledging the key role played by spatial issues in film research and in the humanities at large over the last decades, and also their relevance in the specific field of film tourism research highlighted by Connell (2012), the following sections will be organized borrowing a conceptual framework which concerns the relationships between films and space. Not all of the contents within the chapter will expressly deal with spatial issues and thus necessarily adhere to this framework, yet it will actually prove its effectiveness for organizing and understanding most of them. This framework, also mentioned by Connell (2012), lays on the distinction traced by Mark Shiel (2001) between *space in films* and *films in space*:

> Cinema is a peculiarly spatial form of culture, of course, because (of all cultural forms) cinema operates and is best understood in terms of the organization of space: both *space in films* – the space of the shot; the space of the narrative setting; the geographical relationship of various settings in sequence in a film; the mapping of a lived environment on film; and *films in space* – the shaping of lived urban spaces by cinema as a cultural practice; the spatial organization of its industry at the levels of production, distribution, and exhibition; the role of cinema in globalization.
>
> *(p. 5)*

The first category, that of *(tourist) space in films*, concerns virtual voyages accomplished through film viewing, with film space thus becoming a (virtual) tourist space, also potentially *pre-mediating*, in Grusin's terms (2010), an actual tourist experience. The second category, that of *films in (tourist) space*, will be used in relation to the space of film exhibition or, if we adjust Shiel's quoted definition to the contemporary scenario of the *relocation of cinema* (Casetti, 2012), in relation to the numerous spaces where new and varied interactions between films and their audiences occur, tourist space included.

In both sections, we deal with the *mediatization of the tourist space*. It is thus possible to refer to Henri Lefebvre's (1974/1991) triadic model of the social production of space, invoked by Jansson (2013) in his conceptualization of mediatization as a sociospatial concept. As regards cinematic tourism, in fact, we could say that tourist space is *mediatized* in the sense that it is virtually explored through media (in Lefebvre's terms, *conceived space*), or that it is physically invaded by media devices (*perceived space*). Finally, that in many cases, media have assigned it a touristic use and led to the establishment of new rituals, such as those accomplished by cinematic pilgrims (*lived space*).

Tourist space in films: virtual and pre-mediated travels

The exploration on the relationship between film viewing and travelling has found a fertile ground in postmodern attraction both with images taking the place of 'reality' and with the tourist considered as a key figure of contemporary age. This relationship has been largely examined in two senses, deeply intertwined: film viewing as virtual travel and film viewing as *pre-mediating* subsequent corporeal travel. The first meaning has been widely explored both in tourist studies and in film theory. An example are two works of paramount importance in the respective fields: John Urry's research on the tourist gaze (Urry & Larsen, 2011, first edition 1990), and Giuliana Bruno's spatial understanding of cinema as a 'means of transportation' for the spectator involved in a haptic, and not merely optic, experience (Bruno, 2002). However, we should note that within the contemporary scenario of media convergence, virtual cinematic travels are accomplished across various media, for example when 'cinematic viewers (…) surf the web for virtual consumption of the potential tourist destinations' suggested by the film (Tzanelli, 2007, p. 16). In Urry's work, further to the idea of film viewing as virtual travel introduced in relation to Feifer's (1985) description of the 'post-tourist', the question of actual travel is central. Starting from the observation that 'the tourist gaze is increasingly media-mediated' (Urry & Larsen, 2011, p. 116), leading to the introduction of the term 'mediatized gaze' (p. 20), Urry discusses actual tourist flows induced by the media.

Precisely because of its mediatization, tourist gaze can be questioned through media and film theory. Jackson (2005) does exactly that, starting from the analysis of her own tourist experience in Los Angeles, a spatial exploration which had been *pre-mediated* by film and TV viewing. Among other theoretical perspectives (such as the mass culture debate, Althusserian Marxism, and postmodernism), Jackson selects the Metzian theory of the film gaze as the most appropriate to analyse her experience and reconsider the concept of tourist gaze. In borrowing elements from the notion of film gaze as discussed by film theorist Christian Metz, in fact, Jackson aims at instilling psychoanalytical factors in the traditionally more sociological understanding of the tourist gaze. It is well known that film theories of the 1970s described the experience of film viewing through a comparison with the pre-Oedipal phase of the 'mirror stage' and have identified visual pleasures related to narcissism, voyeurism, exhibitionism, and fetishism. Jackson's proposal is to read the tourist's 'desire to look' through analogue subconscious elements, which would overtake the 'critical distance' from the object of gaze, usually ascribed to the tourist gaze in socio-cultural terms. As she clearly states, tensions persist in collapsing elements from these distant theories, and we could add that criticisms may arise from the choice of theories focused on gaze, contrasted by more recent approaches stressing the role of other senses, like Bruno's (2002) aforementioned research. This is yet an interesting attempt of paving the way for the production of truly interdisciplinary knowledge through the adoption of an 'integrated conceptual approach'.

Not by chance, Jackson's (2005) essay belongs to one of the two interesting collective projects on media and tourism, published at the mid-2000s, which assembled contributions by authors from the two disciplines and were expressly conceived with the aim of building an interdisciplinary body of knowledge. Acknowledging the mediatization of the tourist experience, the two works share a search for new interdisciplinary methodologies and concepts in order to analyse virtual travel. In some cases, these perspectives are fully developed while in others they are just hinted at as suggestions for future research. In the 2006 special issue of *Tourist Studies* on *Tourism and the moving image*, the editors Mazierska and Walton argue that 'there are no touristic films or television programmes as such, but only ways to see them as a product of or a stimulus to the tourist gaze' (p. 9), and thus any media text can be analysed from a tourist perspective. They envisage new interdisciplinary branches of media studies and tourist studies, which could be called 'tourist media studies'.

A conceptual exploration relying on this interdisciplinary nexus is led in the volume *The media & the tourist imagination* (Crouch, Jackson, & Thompson, 2005), in which some contributions, like the one by Jackson (2005) previously discussed, reveal a strong interest for media and film theory. The aim of this collective work was to design an analytical tool expressly created to explore the relationship between media and tourism. The key concept stemmed from the project is that of 'tourist imagination', moulded on 'melodramatic imagination' discussed by Peter Brooks in 1995. 'Tourist imagination' is defined as 'the imaginative investment involved in the crossing of certain virtual boundaries within the media or actual boundaries within the physical process of tourism' (p. 2). What is interesting to notice is that, according to the editors who follow an intuition included in the essay by Fish (2005), a clash is likely to occur in films and TV programmes between the tourist imagination and the most typical narrative forms, usually based on conflict. In studying texts which, by the way, do not depict tourist activities, such as British TV rural dramas, Fish argues that media producers construct their audiences as 'televisual tourists', looking for escape. To read these media texts as touristic texts helps the interpreter understand some narrative and formal choices (for example, in their representation of rurality). Yet when 'tourist forms' meet 'media forms', as Fish' analysis shows, 'discursive structures of the media can (…) provide their own formal constraints to the tourist imagination'. The need for dramatic conflict, stated by narratological theory and evident to the media practitioners interviewed by Fish, shakes the tourist idyll and thus finally, 'viewers have alternative choices' for their interpretation (Crouch et al., 2005, p. 6). To analyse the virtual cinematic tourist experience would thus mean to explore contrasts between tourist imagination and media forms.

Tools provided by film theory, semiotics especially, are in place for exploring film viewing intended as a form of mediatized tourist experience. We could mention Leotta's (2011) study of film and tourism in New Zealand, which uses the aforementioned concept of 'tourist imagination' to explore how the chosen films 'construct viewers as textual tourists' (p. 3). Drawing on structuralist semiotics, he focuses his attention on the 'textually constructed avatar of the spectator' (p. 4) to analyse how each film's narrative, stylistic features and in particular its construction of space stimulate 'tourist imagination'. In *the Lord of the Rings*, for example, he identifies elements such as the inclination to an aesthetics of sublime, with aerial plans minimizing the scale of the characters in front of landscape, the construction of an omnipotent gaze, the narrative based on travel, and many others. This work can be also mentioned as an example of the influence of the concept of 'tourist imagination', often applied in subsequent research.

Films in tourist space: the relocation of cinema

With reference to the spatial framework adopted, the previous section questioned tourist space in films, or *conceived space* in Lefebvre's (1974/1991) terms, even if spatial issues where not always expressly concerned. Now, the focus on space will be even greater when questioning some changes, led by mediatization, to the actual tourist space. These changes are linked to the presence of films both in terms of *perceived* and of *lived* space.

If we talk about contemporary tourism, cinematic and not, it is evident that within tourist space, different forms of encounters with media and films occur, through the pervasive presence of screens and audiovisual displays, which modify our *perceived* space. It is thus possible to notice that while their mediatising presence in tourist sites is a widespread research subject in tourist studies (in relation to film tourism, see, for example, Edensor, 2005), this has been more rarely observed through film theory. The film tourist experience is still under-explored as a film viewing experience in itself, which can be understood, following a brilliant yet under-developed intuition by Gaudiosi (2012), within the wider film theory debate on the *relocation of cinema*. Hence, 'the process in which a media experience is reactivated and re-purposed elsewhere in respect to the place it was formed, with alternate devices and in new environments' (Casetti, 2012). Cinema is somehow relocated during the tourist experience, when films are watched on tablets, smartphones, screens placed on a bus, VR viewers, and so on, even split in fragments (and thus losing one of the traditional features of cinema, i.e. uninterrupted watching). This may happen either on organized movie tours or independent visits. For some tourists, this might be the first contact ever with the film, while in other cases, already known films may be interpreted in new ways, consistent or not with interpretations suggested by tourist marketing. This experience is to be read within the frame of technological convergence, with contents flowing through different devices.

Interestingly, this experience may also be discussed in other terms which highlight the persistence, in this exploded and fragmented 'galaxy' of viewing experiences (Casetti, 2015), of an 'idea of cinema', intended as the 'form of an experience' which still survives. According to this theory, the 'idea of cinema' would invite to understand as 'cinema', the most diverse situations born of *relocation*,

> such as watching a film at home, on a journey, in a waiting room, on a DVD player or on a computer, and chatting about what we are watching on a social network, after having downloaded it from the internet.
>
> *(Casetti, 2012)*

Casetti (2015) lists the elements which 'provid[e] continuity' in the 'endurance of a certain type of experience':

> In addition to the particular way of seeing images offered by cinema, and certain recurring subjects, an important role is played by environmental factors (the fact of being in front of a screen), cultural factors (the memory of what cinema was), linguistic factors (the presence of a 'cinematic language', whatever that term is taken to mean), and broadly psychological ones (the need for cinema, to use an old term coined by Edgar Morin). It is on the basis of a multiplicity of elements that we recognize our experiences as 'cinematic' and thereby give cinema another chance to live.
>
> *(p. 6)*

What I would like to suggest here is that the film tourist experience can be read within the framework of the *relocation of cinema* in two senses. On one hand, the migration of films to different devices and environments; on the other hand, the re-appearance of this persisting 'idea of cinema', through a sort of imaginative restoration of previous film viewing experiences, which can only be accomplished in a unique place like the shooting location.

Nostalgia, identified as a key element in the film and TV tourist experience (Kim, 2019), may involve aspects of the film/TV series but also memories of the experience itself of film or TV viewing. Cunningham (2008), who discusses film pilgrimages in relation to theories of cinephilia, itself strongly linked with nostalgia (Elsaesser, 2005), states that 'cinephilic pilgrims' try to 'reify (that is, to ground within the real) an inherently ephemeral experience of the past, while simultaneously utilizing real spaces as 'portals' through which to once again access, personally experience, and even occupy, the past' (p. 126). The temporal dimension comes out here, accessed again through a spatial key. The past experience(s) are that of previous film viewing(s), understood by Cunningham relying on Siegfrid Kracauer's theories, as not repeatable. Precisely in this hiatus, in this impossible repetition, lies the persistence and strength of that 'idea of cinema', which in a somehow enigmatic yet suggestive phrase, Casetti (2012) says can be 'measured against an *almost* that wants to seem like a *nearly completely*, even though it tends to be not at all'. While films can be watched repeatedly through new devices during the tourist experience, what cannot be repeated, yet persists as a sort of ideal horizon, are previous film viewing experiences. The past experiences of film viewing must be intended here, as Cunningham explicitly does, regardless of their occurrence in a movie theatre.

My hypothesis is that these experiences may be imaginatively restored through the ritual performed in a unique and 'auratic' place (see Buchmann, Moore, & Fisher, 2010, p. 241; on TV, Couldry, 2000, p. 81), the one where shooting occurred or in other cases, the one where the film is set. Only this place can serve as an 'access-point' to the media world (Couldry, 2000), as it enjoys the status of 'place of imagination' (Reijnders, 2011). It is evident that we are now dealing with a *mediatized tourist space* in another sense, that of a tourist space rendered as such by media, and of which media have established new uses, like rituals specific to film and TV pilgrimages – in Lefebvre's (1974/1991) terms, with *lived space*.

Conclusions

The theoretical framework of *mediatization* suggests a shift from studying the 'effects' of media texts on tourism to questioning 'the wider consequences of media's embedding in everyday life' (Couldry & Hepp, 2013, p. 195), tourist practices included. Considering the specific field of cinematic tourism, this chapter suggests an entrance point which consists in exploring the cinematic tourist experience through the lens of film theory. This point of view is proposed in response to the numerous calls for cross-disciplinarity addressed to media and film scholars over the years.

Acknowledging the relevance of the relationship between film and space in film tourism research, a spatial framework is introduced in the second section to explore the cinematic tourist experience. On one hand, relying on Shiel's (2001) understanding of cinema as a 'peculiarly spatial form of culture', his influential distinction between *space in films* and *films in space* is borrowed in relation to tourist space. On the other hand, following Jansson's (2013) suggestion to read *mediatization* as a sociospatial concept recurring to Lefebvre's (1974/1991) model on the social production of space, an understanding in terms of *conceived, perceived,* and *lived* space is put forward.

The conceptual viewpoint claiming the role of film theory in understanding cinematic tourism is adopted in the third section through synthesizing and discussing some relevant perspectives emerged in the mid-2000s, within contributions expressly aimed at carrying out interdisciplinary work. In this section, the cinematic tourist experience is examined as a virtual tourist experience accomplished through film viewing, or, in a deeply intertwined meaning, as an actual tourist experience, influenced by constructions of the tourist gaze provided by media. Both of these meanings have been accounted for through readings which rely on classical film theories from the 1970s on (for example, film gaze, narratology, and semiotics). *Tourist space in films* has thus been discussed, understanding *mediatization of tourist space* in terms of *conceived space*.

In the fourth section of the chapter, new lines of inquiry are suggested, relying on recent film theories, especially when dealing with the concept of the *relocation of cinema*. In this section, the actual cinematic tourist experience and *films in tourist space* are questioned. On one hand, the watching of films during the tourist experience through various devices and in new environments is discussed. The presence of screens and audio-visual displays in tourist sites, largely explored in tourist studies, are examined within a framework borrowed from film theory. On the other hand, the tourist experience is said to entail memories of previous film or TV viewing experiences, whose imaginative restoration is pursued through rituals accomplished by cinematic pilgrims within the tourist space. The latter has thus been intended as *mediatized* both in terms of *perceived* and of *lived space*, thinking either of the physical presence of devices or of new uses assigned to the space itself.

My mainly spatial understanding of cinematic tourism has led to invoke theories, like that of the *relocation of cinema*, which, while arising from the contemporary context of media convergence, affirm an ideal persistence of what has been the *cinematic experience*, even in the dispersed forms it assumes nowadays. Recognizing this invites us to deploy film theory, also in its classical configurations, to understand cinematic tourism even in this deeply changing scenario.

References

Beeton, S. (2005). *Film-induced tourism*. Clevedon, UK: Channel View.

Beeton, S. (2010). The advance of film tourism. *Tourism and Hospitality Planning & Development,* 7(1), 1–6.

Beeton, S. (2015). *Travel, tourism and the moving image*. Retrieved from: https://www.amazon.it/dp/B01MRCB35R/ref=dp-kindle-redirect?_encoding=UTF8&btkr=1.

Bruno, G. (2002). *Atlas of emotion. Journeys in art, architecture and film*. New York, NY: Verso.

Buchmann, A., Moore, K., & Fisher, D. (2010). Experiencing film tourism. Authenticity & fellowship. *Annals of Tourism Research, 37*(1), 229–248.

Casetti, F. (2012). The relocation of cinema. *Necsus-European Journal of Media Studies, 1*(2). Retrieved from: https://necsus-ejms.org/the-relocation-of-cinema/.

Casetti, F. (2015). *The Lumière galaxy. Seven key words for the cinema to come*. New York, NY: Columbia University Press.

Connell, J. (2012). Film tourism. Evolution, progress and prospects. *Tourism Management, 33*(5), 1007–1029.

Couldry, N. (2000). *The place of media power. Pilgrims and witnesses of the media age*. London, UK & New York, NY: Routledge.

Couldry, N., & Hepp, A. (2013). Conceptualizing mediatization: Contexts, traditions, arguments. *Communication Theory, 23*, 191–202.

Crouch, D., Jackson, R., & Thompson, F. (2005). Introduction: The media and the tourist imagination. In D. Crouch, R. Jackson, & F. Thompson (Eds.), *The media and the tourist imagination. Converging cultures* (pp. 1–13). London, UK: Routledge.

Croy, W.G., & Heitmann, S. (2011). Tourism and film. In P. Robinson, S. Heitmann, & P. Dieke (Eds.), *Research themes for tourism* (pp. 188–204). Wallingford, UK: CABI.

Cunningham, D. (2008). 'It's all there, it's no dream': *Vertigo* and the redemptive pleasure of the cinephilic pilgrimage. *Screen, 49*(2), 123–141.

Edensor, T. (2005). Mediating William Wallace. Audio-visual technologies in tourism. In D. Crouch, R. Jackson, & F. Thompson (Eds.), *The media and the tourist imagination. Converging cultures* (pp. 105–118). London, UK: Routledge.

Elsaesser, T. (2005). Cinephilia or the uses of disenchantment. In M. De Valck & M. Hagener (Eds.), *Cinephilia. Movies, love and memory* (pp. 27–43). Amsterdam, The Netherlands: Amsterdam University Press.

Fanchi, M. (2012). *Cinema-Grand Master*. Il film e la sua esperienza nell'epoca della convergenza. In F. Zecca (Ed.), *Il cinema della convergenza. Industria, racconto, pubblico* (pp. 193–204). Milano & Udine, Italy: Mimesis.

Feifer, M. (1985). *Going places: the ways of the tourist from imperial Rome to the present day*. London, UK: Macmillan.

Fish, R. (2005). Mobile viewers: Media producers and the televisual tourist. In D. Crouch, R. Jackson, & F. Thompson (Eds.), *The media and the tourist imagination. Converging cultures* (pp. 119–134). London, UK: Routledge.

Gaudiosi, M. (2012). Proiezioni cartografiche: il cinema tra geografie transmediali e spazi urbani. In F. Zecca (Ed.), *Il cinema della convergenza. Industria, racconto, pubblico* (pp. 179–192). Milano & Udine, Italy: Mimesis.

Grusin, R. (2010). *Premediation. Affect and mediality after 9/11*. Basingstoke, UK: Palgrave Macmillan.

Hjarvard, S. (2008). The mediatization of society. A theory of the media as agents of social and cultural change. *Nordicom Review, 29*(2), 105–134.

Jackson, R. (2005). Converging cultures; converging gazes; contextualizing perspectives. In D. Crouch, R. Jackson & F. Thompson (Eds.), *The media and the tourist imagination. Converging cultures* (pp. 183–197). London, UK: Routledge.

Jansson, A. (2013). Mediatization and social space: Reconstructing mediatization for the transmedia age. *Communication Theory, 23*, 279–296.

Jenkins, H. (2006). *Convergence culture. Where old and new media collide*. New York: New York University Press.

Kim, S. (Sangkyun) (2012). Audience involvement and film tourism experiences: Emotional places, emotional experiences. *Tourism Management, 33*(2), 387–396.

Kim, S. (Seongseop), Kim, S., (Sangkyun) & Petrick, J. F. (2019). The effect of film nostalgia on involvement, familiarity and behavioral intentions. *Journal of Travel Research, 58*(2), 283–297.

Lefebvre, H. (1974/1991). *The production of space*. Oxford, UK & New York, NY: Blackwell.

Leotta, A. (2011). *Touring the screen. Tourism and New Zealand film geographies*, Bristol, UK: Intellect.

Månsson, M. (2011). Mediatized tourism. *Annals of Tourism Research, 38*(4), 1634–1652.

Mazierska, E., &Walton, J. K. (2006). Tourism and the moving image. Introduction. *Tourist Studies, 6*(1), 5–11.

Provenzano, R. (Ed.) (2007). *Al cinema con la valigia. I film di viaggio e il cineturismo*. Milano, Italy: Franco Angeli.

Reijnders, S. (2011). *Places of the imagination. Media, tourism, culture*. Farnham, UK: Ashgate.

Shiel, M. (2001). Cinema and the city in history and theory. In M. Shiel & T. Fitzmaurice (Eds.), *Cinema and the city. Film and urban societies in a global context* (pp. 1–18). Oxford, UK: Blackwell.

Tzanelli, R. (2007). *The cinematic tourist. Explorations in globalization, culture and resistance*. New York, NY & Abingdon, UK, Routledge.

Urry, J., & Larsen, J. (2011). *The tourist gaze 3.0*. Los Angeles, CA & London, UK & New Delhi, India & Singapore & Washington, DC: Sage.

Zecca, F. (2012). *Cinema reloaded*. Dalla convergenza dei media alla narrazione transmediale. In F. Zecca (Ed.), *Il cinema della convergenza. Industria, racconto, pubblico* (pp. 9–37). Milano & Udine, Italy: Mimesis.

6

WHAT DO MELANIA TRUMP TOURISM AND DRACULA TOURISM HAVE IN COMMON? 'OTHERING' IN THE WESTERN MEDIA DISCOURSE

Maja Turnšek, Andreja Trdina and Barbara Pavlakovič

Introduction

At the time of the presidential campaign of Donald Trump and especially on the days following his election for president of the United States of America a small part of the news-frenzy about the surprising result benefited also the Slovene public relations. Slovene tourism received an extremely high proportion of reporting. Numerous Western news networks reported that Slovenia is now hoping for a tourism boost since this is the birthplace of the new First Lady Melania Trump. Sevnica, the small municipality in southeast of Slovenia, where tourism is almost non-existent, is now said to be hoping for an influx of American tourists, who wish to visit the First Lady's (humble) origins (e.g. Bradley, 2016; Fenwick, 2016; Fox News, 2016; Morris, 2016).

The aim of this paper is to address these hopes by analysing the phenomenon of emerging Melania Trump tourism through the prism of the 'classical' theoretical thought on the process of 'Othering' in Western discourses (Said, 1977). We take the cases of Dracula tourism and Melania Trump tourism as two distinct examples of processes of 'Othering' in the complex interrelation of media and tourism (we build on the work of Hovi (2014) on Dracula tourism in Transylvania, Romania). We find that there are major similarities between the emerging Slovene tourism in response to Melania Trump and tourism in Romania in response to Bram Stoker's (1897) fiction novel Dracula. Specifically, by juxtaposing Melania tourism in Slovenia to Dracula tourism in Romania, we reflect here on the specifics of the process of 'Othering' in the Western media and the ambivalent local tourism dissent towards what they perceive as skewed representation on the one hand and an opportunity for increased media attention and tourism growth on the other.

We have divided the chapter in three sections. In the literature review, we first set the theoretical backbone of this research: the process of 'Othering' in mediatized tourism. In the second section, we discuss Hovi's (2014) work on Dracula tourism and identify the characteristics that are similar to emerging Melania Trump tourism. Finally, in the third section, we present the results ascertained with two research methods: discourse analysis of the Western media representations of Melania Trump hometown Sevnica and in-depth interviews with local Melania tourism stakeholders. The interviewees are representatives of (a) Slovenian Tourist Board; (b) Public Institute for Culture, Sport, Tourism and Youth Activities Sevnica; and (c) Sevnica Municipality. The interviews were focused on their perception of media representation of Sevnica and on their strategy on how to dissent the media representations or potentially use the media coverage for promotion of the destination. The contribution ends with concluding thoughts on the process of 'Othering' in the Western media and the dissenting, yet ambivalent local tourism reactions.

'Othering' in mediatized tourism

'Othering' can be defined as the process of constructing and maintaining a dichotomy between 'Us', as marked by a particular (Western) identity, and the Other(s). The conceptualisation rests on postcolonial theories of Orientalism whereby the Orient is showed to be 'the Other' of the West as its contrasting image, idea, personality, and experience (Said, 1977). In the colonial discourse, the Other is a subject of difference that is 'almost the same, but not quite', since the discourse is constructed around ambivalence and difference between 'Us' versus 'the Other' (Bhabha, 1984). Haldrup and Larsen (2010, p. 79) claim that the effects of Orientalism are material and embodied, as through acts of representation Orientalism establishes the very difference between the East and the West and thus transforms the former into 'materials' for the imagination of the latter. In addition, they argue that orientalist techniques are not only crucial in making 'other' spaces 'controllable' and 'manageable' but also making these spaces and the cultures, people, objects belonging to them 'consumable'.

Salazar (2012) identifies the discourses of the past: orientalism, colonialism, and imperialism as the fertile ground for nostalgic and romantic tourism dreams. It has been repeatedly demonstrated how 'Othering' in hegemonic discourses of 'Orientalism' in tourism industry still frames tourism performances (Haldrup & Larsen, 2010) and imaginations (Bryce, 2007). In todays' media saturated society, the construction of Otherness has been accomplished through media texts in particular as these are omnipresent and provide constant flow of images about places but also in tourism and especially in the intersection of both. Britton (1979) critiques the tourism marketing for failing to represent the 'Third World' destinations as real places but instead represents them as 'paradise', 'unspoiled', 'sensuous'. Echtner and Prasad (2003) identify three recurring images in tourism marketing brochures representing different 'Third World' countries: the myth of the unchanged, the myth of the unrestrained, and the myth of the uncivilised. They show how the representations surrounding these myths replicate colonial forms of discourse, emphasising binaries between the First and Third Worlds and maintaining broader geopolitical power structures. Similarly, Bryce (2007) suggests that 'Othering' discourse is deployed in British promotional brochures on the destinations of Turkey and Egypt. Henderson and Weisgrau (2007) show how guidebooks about India mirror the accounts of 19th-century British colonial tourists. Coming to similar conclusions, Diekmann and Hannam (2012) analysed Western movie representations of the slums of India and the walking tours around the slums.

While most of the research was done on the Western tourism/media discourses of 'the Other', Bandyopadhyay and Morais (2005) provide an analysis of how Indian government's media campaigns resist to the representations of India in American tourism media. Their findings revealed that the two representations are different in ways that reflect the colonial nature of international tourism and the postcolonial stage of India's nationalism. They stress the importance of local dissonance as an important tool for understanding the conflicting ideological forces within the processes of Othering and its counter-reactions. The focus of this chapter is thus not only to illustrate the process of 'Othering' by the West but to also delve deeper into the process of dissonance with the 'Othering' from the perspective of those who are represented as the 'Other'. And the seemingly unlikely Companions of the Dracula and Melania Trump both represent two distinct examples of processes of 'Othering' in the field of media and tourism.

Juxtaposing Melania tourism with Dracula tourism: The Eastern European 'Other'

While the classical literature on the Othering focuses on the colonial past and the Western discourses of the so-called 'Third World' countries, this chapter applies the 'Othering' framework within Europe itself. Specifically, we follow the line of thought that argues that the Balkans has emerged as a cultural and religious 'Other' to 'Europe proper' since the Byzantine or Ottoman rule (Bakic-Hayden & Hayden, 1992). Additionally, Bakic-Hayden and Hayden (1992, p. 4) claim that in this century, an ideological 'Other' – communism, has replaced the geographical/cultural 'Other' of the Orient. The symbolic geography of eastern inferiority, however, remains.[1] This older symbolic geography of Eastern Europe as the 'Other' was reinforced and strengthened later by 'Othering' discourses on communism (particularly within the cold war and in the context of an ideological and political geography of the democratic, capitalist west versus the totalitarian, communist east). There is an overlap or continuity in the nature and the logics of the rhetoric and images used to represent that dichotomy no matter how different the historical processes were (ibid., pp. 3–4).

Hovi (2014) researched these demarcating borders in Dracula tourism in Transylvania. Dracula tourism was spurred by popular fictive work *Dracula* written by Bram Stokers in 1897, subsequent numerous adaptations and other works borrowing the same character. By comparing Hovi's (2014) descriptions of Dracula tourism in Transylvania to the newly developing Melania tourism in Slovenia, we can identify important similarities. First, and most obvious, in both cases, we can talk about tourism induced by a collective imaginary about a real existent person, that has been transformed and intertwined with imagination and burdened with ideological discourse about the Eastern European as the 'Other'. In the Dracula case, this is imaginary created and built around the historic character of Vlad Dracula or Vlad the Impaler. In the Melania tourism, this is collective imaginary created and built around the current First Lady of United States of America. Second, in both cases, we talk about tourism related to birthplaces of the two real-life persons, traversing from the smallest destinations to larger and larger areas, eventually covering whole countries. Third, both persons were born in areas that are part of Eastern Europe, bearing the collective Western imaginary of ex-socialist and communist regimes and the exotic Eastern Europe 'Other'. While Slovenia by its own understanding positions itself within Central Europe, at the time Melania was born, it was still part of Socialist Republic of Yugoslavia and is nowadays still often perceived as part of the Eastern Europe. Thus, both Romania and Slovenia become part of the Western collective imaginary of the ex-communist Eastern Europe.

Fourth, tourism in the two cases has less to do with the real-life persons than it has with their collective imaginary, what we will term here as characters, pointing out the imagined, collectively constructed understandings. This is very clear in the case of Dracula since his vampire characteristics are obvious fiction, taking only clues from the real-life person myths, intertwined with reality.[2] From this reality, the fictional monster that was built by Bram Stoker is still today an important and reanimated element in collective imaginary and horror stories. We of course cannot talk about fiction to such extent in the case of Melania tourism, but it is important to point out that Melania tourism is spurred by collective, mostly media-driven construction and representation of Melania Trump. To some extent, one fuels the other, similarly as Vlad the Impaler fuelled imagination of Bram Stoker, but at the end it is still a constructed media representation that fuels Melania tourism and not the real-life person.

Fifth, and most important, in both cases, the constructed imaginary of the character and the travel destination does not stem from the local culture but is created by the West. Hovi (2014) claims that although there are many cases within the tourism industry where the local tourist industry has to negotiate between outside expectations and local cultural values, Dracula tourism is unique. It is a combination of a known historical figure with a fictional character that derives completely from outside the history and culture of the original historical figure. Although many historical and mythical figures have been absorbed into Western popular culture and related tourism like Robin Hood, William Wallace, King Arthur, or many characters of the Wild West, this has all been done, Hovi argues, more or less within the same Anglo-American culture and on the culture's own terms.

> The combination of the historical character and the character from popular culture has often been done with the interest and understanding of the culture where they came from. In the case of Dracula tourism, the character of Vlad the Impaler has been 'forcefully' attached to the Western imaginary vampire Dracula without any input from Romanian culture.
>
> *(ibid., pp. 16–17)*

And this fifth similarity is the starting point of our research and the basis for our research question: what is the constructed media imaginary of Melania Trump tourism in the Western media and how do the local tourism stakeholders react to it? This research question leads us to the process of 'Othering' in the complex interrelation between media and tourism.

Western media representations of Melania's hometown

We build here on the cultural studies paradigm, both in media and tourism research (Crouch, 2009). Since the operating of specific discursive structures in the media provide 'formal constraints to the tourist imagination' (Crouch, Jackson, & Thompson, 2005, p. 6), we analyse media content as imaginative resources in constructing the image of Melania Trump and Slovenia. Specifically, we follow works that focus on the issues of media discourse and representation of the 'Other' and at the same time read tourism as socially constructed and negotiable imaginaries. Method used here is qualitative discourse analysis of media representations of Melania tourism in the Western media that were read by the local stakeholders. According to discourse theory (Åkerstrøm Andersen, 2003), the world comes into being through the statement as event. We apply deconstructive reading of media representations in order to understand how collective imaginaries and corresponding subject positions are constructed. Our aim is to expose dissimulating characteristics of discourses and to chart real material effects they have on the structuring of social relations (Howarth, 2000, p. 136), also the destination marketing attempts by the Slovene stakeholders.

The sample for the analysis was composed of ten USA and six British online media pieces, out of which four included video material next to written text and photography. We have selected the media content that was collected as part of their media monitoring and shared with us by the Slovene and Sevnica destination marketing stakeholders whom we interviewed (see below). The reason for this was to make sure that the content we analysed was the content that was known and responded to by the local stakeholders. Since the focus was on Western media reporting, we focused only on reports in English, which were almost exclusively from USA and UK media houses and for the sake of more reliable comparison we thus excluded the small number of contents in English from other countries.

In the analysis, we identified key nodal points as privileged discursive points that attempt to fix meaning: points that organise a discourse on Melania tourism as central reference points. The media representations of Melania Trump hometown revolved around two main discursive points of 'Othering': (a) the 'humble' communist origin and (b) 'exaggerated' Melania tourism boost.

The 'humble' communist origin

First, Western media write especially about Sevnica as a rather poor and isolated area with the intention of pointing out the humble origin of the future First Lady, always connected to the ex-communist past. Such representation is often accompanied by pictures of a typical ex-communist housing block in which Melania Trump and her family lived in the first couple of years. We argue that by reiterating the discursive division of Western 'reason and modernity' vs. eastern 'stasis and passivity' what is offered to readers in Western media is the occupation of specific subject position[3] from which to 'know' Sevnica as the less developed ex-communist industrial town.

Exaggerated tourism boost

Second, the reports of the expectations of the 'tourism boost' that are supposed to be felt in Slovenia and Sevnica especially were largely exaggerated in the media, with titles such as 'Slovenia's tourism industry could win big after Trump victory' (Fox News, 2016). In some cases, the news about the 'future tourism boost' were presented with partly hidden sarcasm: the journalists interviewed local workers in Sevnica, asked questions that people politely answered, but the overall news report was meant to highlight the current state of non-existent tourism in Sevnica and the lack of strategy around the idea of the emerging tourism boom. For example, numbers for 2014 were low since there were 76,535 tourist arrivals from USA to Slovenia and only 649 total arrivals to Sevnica (Statistični urad RS, n.d.). In other cases, the potential future tourism 'boom' was represented from the viewpoint of 'the less developed East' being a beneficiary of the intervention of the West: the intervention being the future 'flocks' of Western tourists. This was combined with news on the local acceptance and celebration of Melania Trump becoming the first lady with titles such as 'Melania Trump's hometown is partying after election victory' (Frej, 2016). Such news served as an interesting small piece of information on how the different, ex-communist culture is celebrating Donald Trump's victory – framing the East as the beneficiary of the West. In tourism representations of Sevnica, Westerns subjects were invited to uncover Slovenia which has been in this way made available for the Western production of knowledge and fascination about it, for example, in the piece titled 'Guide to Melania Trump's homeland of Slovenia' (Dwyer, 2016). Such discourse relies on two related 'Othering' assumptions: that Slovenia is still 'hidden' and 'mysterious' and that this invites intervention and exploration by the rational Western subjects.

Local reactions to the media representations

In course of the research, we have performed four in-depth interviews with representatives of local and national organisations, which are responsible for the field of tourism in Sevnica and Slovenia. The interviewees are representatives of (a) Slovenian Tourist Board; (b) Public Institute for Culture, Sport, Tourism and Youth Activities Sevnica; and (c) Sevnica Municipality. The interviews were focused on their perception of media representation of Sevnica and on their strategy on how to dissent the media representations or potentially use the media coverage for promotion of the destination. The results of the analysis and comparison with the Dracula case (Hovi, 2014) show specific steps in the response to Western representations. These are (a) dissent against what they find inauthentic and skewed representations, (b) recognition of economic interests in destination marketing and development, and finally (c) attempts at reconciliation of economic interests and media coverage of the destination.

Dissent against Western media representations

The destination management representatives from Melania Trump hometown very much recognised the Western media attempts at representing Sevnica and Slovenia as the 'Eastern European Other'. Interviewee from Sevnica Municipality for example said: 'Journalists were interested in exposing the history of Melania's family and presenting her hometown as industrial, underdeveloped and under socialist rule. They believed that there is no cultural or social life in Sevnica, that we have nothing to show'. Another interviewee from Sevnica Municipality adds: 'They used pictures of industrial chimneys and cloudy sky'. Furthermore, they did not like the fact that some journalists wanted to film local dressmakers at their work to present Melania's humble working class origins, trying to represent the current destination's industrial and living situation via pictures that would imply that it is the same as it was 30 years ago, thus presenting a more 'backwards' image than it really is. Additionally, one of the interviewees emphasised that the American media were predispositioned with the relations between Americans and Russians and would present the destination from this over simplistic dualism.

Recognition of own economic interests

As the first press coverage was focused on political aspects of Melania, interviewed stakeholders were not keen to expose themselves to media. Their strategy was not to take any political side and not to speak about politics. This diplomacy regarding Melania's name derives firstly from being careful about the political connotation of Donald Trump and secondly from legal restrictions Melania herself imposed with the help of her legal representative, who is responsible to guard over any inappropriate action towards her name taken in Slovenia.

Although they expressed their reserve towards using Melania Trump in their promotional activities, they started to consider her influence on Slovenia tourism promotion and tourism economy. Slovenian tourism board calculated that the media coverage itself in the 'peak' three months would cost over 100.000 €, and the Sevnica municipality estimates that this number could reach over million euro. They furthermore noticed that it was easier to receive media coverage, for example, when organising cultural events. Additionally, they recognised own economic interests in potentially higher recognisability of Slovenia at the USA market. The growth index of USA tourist arrivals to Slovenia from 2016 to 2017 was 129. Statistical data show that in 2017 there were 116,263 tourist arrivals from USA to Slovenia and 1,457 total arrivals to Sevnica, which has doubled its numbers since 2014 (Statistični urad RS, n.d.). Yet, they are cautious by interpreting this data in too simplistic way since the number has grown steadily over the last eight years and Melania media coverage is claimed to be just one of the reasons for the growth. However, Sevnica Municipality representative stated that this spurred interest also in tourists from other parts of the world: 'German RV tourists became regular guests in Sevnica, which happened only after Melania media coverage'. As argued by Lisec and Potočnik Topler (2016), who served as PR consultants to the Sevnica tourism stakeholders, the media attention has been a big stress for the Sevnica community at the beginning but turned out to be a significant potential for the local tourism industry. After the initial surprise and dissent against the skewed media coverage, the local destination management started to perceive it as an opportunity: new tourism products were introduced, such as guided tours, souvenirs, and local food in relation to Melania Trump. According to interviewees, the highest demand was for a locally made cake 'Melania'. Sevnica municipality is especially proud of their newly introduced souvenir line 'First lady'. In addition, local tourism agency started to promote Sevnica tours in nearby spas and this cooperation brought positive results in terms of higher numbers of Sevnica visitors. In Sevnica itself, tourism infrastructure is improving, restaurants and the local Castle museum are reporting higher visits, and the only hotel in the municipality has been renovated and reopened.

Attempts at changing media representations

Local stakeholders worked hard on refocusing the journalists' interests away from politics, personal life of Melania Trump, and the communist past. Their aim was to direct them towards tourism, high living standard and modern, globally interconnected (thus not 'communist') industry in Sevnica. Municipality and Public Institute for Culture, Sport, Tourism, and Youth activities formed a communication strategy that included promotion of tourism. Municipality representative: 'Each journalist received an information file, full of data about the destination. We emphasised the Sevnica castle, mountain Lisca, gastronomic delights, sustainable tourism, river Sava, coexistence with nature, recreation opportunities, cycling, farming, and so on'. They launched Sevnica tourism web site in English, where there was detailed information about destination attractions, events, and other offers. They also took journalists on tours around Sevnica, accompanied by English-speaking local tour guide. Sevnica Municipality interviewee:

> It was very important to us, that the journalists walked around the town so they could feel the local vibe. Journalists that spent at least couple of hours in our town reacted differently than those, who just flew in for public statement and then directly out.

This is supported by public institute representative statement: 'We changed the image of the destination only by explaining and by showing the journalists what we have. This was the only way'.

Another emphasis of the destination marketing attempts was to emphasise globally recognised industry from Sevnica. When journalist heard that products from Sevnica factories are used in Kempinski hotels or by BMW, their interest was raised and as Municipality representative noted: 'This was the way to get their attention, and once we realised it, we used it as a communication strategy'. Lastly, municipality was always very proud to highlight accessible public health system, public education, general infrastructure, optical network, clean and tidy environment, beautiful views, diverse landscape, and most of all safety, since there are not many risks while staying in Sevnica. Mayor of Sevnica constantly used 'Google Earth' web page in order to direct journalists to the most interesting points in Sevnica and surroundings. However, the main manner of changing the destination image was to present the already existing positive features of the destination, which did have an effect in the eyes of the local management since the journalists started focusing less on Melanias' previous life, but started to ask about the current local way of life, tourism and alike. As the Sevnica Municipality representative optimistically claimed: 'And through these actions the myth of Eastern Europe was debunked'.

Conclusions

We understand tourism and media as interrelated and 'enmeshed within multiple flows, events and other spaces rather than as detached sites replete with depthless signs' (Edensor, 2005, p. 115). As the case of Melania Trump tourism shows, similarly as to Dracula tourism, the discourses of Western media 'Othering' are met with local dissent. However, both cases also show a more nuanced picture of economic interests intertwining with media coverage and the local attempts at trying to change the contested media representation. Rather, what we are witnessing is a rare account of its existence mostly in imagination, of envisioning it as a Western intervention that would 'aid the East'. As such the case of Melania tourism is an example of blurring of reality and fiction – in the sense of currently existing only (or mostly) in media and local stakeholder's collective imaginaries. Furthermore, a fiction that is built around dissenting discourses of the West and the East and that is a result of these conflicting discourses. The study showed that local tourism management and marketing stakeholders, similarly as the tourism stakeholders in Romania in reaction to Dracula tourism, recognise and dissent to Western media representations. On the other hand, local stakeholders have been quick to replace their outrage by recognition of own economic interest in the heightened media attention. Consequently, local destination management attempts were made towards developing new tourism products. This was combined by strategically planned destination marketing actions towards reshaping simplistic media representations and showing the broader picture of tourism attractions at the destination.

Notes

1 See Todorova (2009) for an account of how the Western image of the Balkans came into being and how in the same manner as the Orient the Balkans was constructed as the Other of Europe grounded in Saidian (1977); opposition/dichotomy East-West.
2 Such as the story of Count Vlad Dracula supposedly impaling some twenty thousand men, women and children to build "the forest of the impaled" in order to scare of the Turkish army (Hovi, 2014, pp. 39–40).
3 When an individual or the collective is offered a particular position in the discourse from which they can speak and act in a meaningful way (Åkerstrøm Andersen, 2003).

References

Åkerstrøm Andersen, N. (2003). *Discursive analytical strategies: Understanding Foucault, Koselleck, Laclau, Luhmann.* Bristol, UK: The Policy Press.

Bakic-Hayden, M., & Hayden, R. M. (1992). Orientalist variations on the theme "Balkans": Symbolic geography in recent Yugoslav cultural politics. *Slavic Review, 51,* 1–15.

Bandyopadhyay, R., & Morais, D. (2005). Representative dissonance: India's self and western image. *Annals of Tourism Research, 32,* 1006–1021.

Bhabha, H. (1984). Of mimicry and man: The ambivalence of colonial discourse. *October, 28*, 125–133.

Bradley, M. (2016, November 27). Hometown pride: Melania Trump's hometown is hoping for a Tourism Boom. *NBC News.* Retrieved from: http://www.nbcnews.com/nightly-news/video/hometown-pride-melania-trump-s-hometown-is-hoping-for-a-tourism-boom-818612803865.

Britton, R. A. (1979). The image of the Third World in tourism marketing. *Annals of Tourism Research, 6*, 318–329.

Bryce, D. (2007). Repacking orientalism: Discourses on Egypt and Turkey in British outbound tourism. *Tourist Studies, 7*, 165–191.

Crouch, D. (2009). The diverse dynamics of cultural studies and tourism. In T. Jamal & M. Robinson (Eds.), *The SAGE handbook of tourism* (pp. 82–97). London, UK: Sage.

Crouch, D., Jackson, R., & Thompson, F. (2005). Introduction: The media and the tourist imagination. In D. Crouch, R. Jackson, & F. Thompson (Eds.), *The media and the tourist imagination: Converging cultures* (pp. 1–13). London, UK & New York, NY: Routledge.

Diekmann, A., & Hannam, K. (2012). Touristic mobilities in India's slum spaces. *Annals of Tourism Research, 39*, 1315–1336.

Dwyer, C. (2016, November 17). Guide to Melania Trump's homeland of Slovenia. *CNN Travel.* Retrieved from: https://edition.cnn.com/travel/article/slovenia-travel-guide/index.html.

Echtner, C. M., & Prasad, P. (2003). The context of third world tourism marketing. *Annals of Tourism Research, 30*, 660–682.

Edensor, T. (2005). Mediating William Wallace: Audio-visual technologies in tourism. In C. David, R. Jackson, & F. Thompson (Eds.), *The media and the tourist Imagination: Converging cultures* (pp. 105–118). London, UK & New York, NY: Routledge.

Fenwick Elliot, A. (2016, November 9). Rolling hills and a 900-year-old castle: Inside Melania Trump's tiny Slovenian hometown as it gears up for a major boost in tourism. *Daily Mail.* Retrieved from: http://www.dailymail.co.uk/travel/travel_news/article-3920196/Melanias-Slovenian-hometown-eyes-Trump-win-boon-tourism.html.

Fox News (2016, November 20). Slovenia's tourism industry could win big after Trump victory. *FoxNews.com.* Retrieved from http://www.foxnews.com/travel/2016/11/10/slovenias-tourism-industry-could-win-big-after-trump-victory.html.

Frej, W. (2016, November 09). Melania Trump's hometown in Slovenia is partying after election victory. *Huffington Post.* Retrieved from: https://www.huffpost.com/entry/melania-trump-hometown-slovenia-partying_n_582338f3e4b0e80b02ce3400.

Haldrup, M., & Larsen, J. (2010). *Tourism, performance and the everyday: Consuming the orient.* New York, NY: Routledge.

Henderson, C. E., & Weisgrau, M. K. (2007). *Raj rhapsodies: Tourism, heritage and the seduction of history.* Aldershot, UK: Ashgate Publishing, Ltd.

Hovi, T. (2014). *Heritage through Fiction: Dracula Tourism in Romania* (Doctoral disertation, University of Turku: The Finnish Doctoral Programme for Russian and East European Studies). Retrieved from: http://doria32-kk.lib.helsinki.fi/bitstream/handle/10024/98458/AnnalesB387Hovi.pdf?sequence=2.

Howarth, D. (2000). *Discourse.* Buckingham, UK: Open University Press.

Lisec, A., & Potočnik Topler, J. (2016, September). Sevnica in the middle of the 2016 American presidential campaign. In M. R. Djuricic, I. Cirovic, & N. Milutinovic (Eds.), *Proceedings of the 3rd international Conference Higher education in function of development of tourism in Serbia and Western Balkans* (pp. 135–142). Symposium conducted at Business and Technical College of Vocational Studies, Užice, Serbia.

Morris, H. (2016, November 11). Will Slovenia be the unlikely beneficiary of Trump's election triumph? *The Telegraph.* Retrieved from: http://www.telegraph.co.uk/travel/destinations/europe/slovenia/articles/sevnica-slovenia-birthplace-of-melania-trump-hoping-to-capitalise-on-president-trump/.

Said, E. W. (1977). *Orientalism.* London, UK: Penguin.

Salazar, N. B. (2012). Tourism imaginaries: A conceptual approach. *Annals of Tourism Research, 39*, 863–882.

Statistični urad RS (Statistical Office of Slovenia) (n.d.). *Turizem (Tourism).* Retrieved from: https://www.stat.si/StatWeb/Field/Index/24.

Stoker, B. (1897). *Dracula.* London, UK: Archibald Constable and Company.

Todorova, M. (2009). *Imagining the Balkans.* New York, NY: Oxford University Press.

7

CONFRONTING THE GAZE, GRIPPING THE VIRTUAL

A cultural materialist perspective on cinema-tourism studies[1]

Sofia Sampaio

Introduction

There has been a growing interest in the intersections of tourism and cinema. This trend is not surprising, given the importance accorded to vision and visuality in tourism (e.g. Adler, 1989). More recent perspectives (e.g. Coleman & Crang, 2002) have been concerned with the embodied and performative rather than contemplative nature of tourist experiences, underlining the active involvement of other senses. Yet, sight has largely retained its primacy in tourism studies. More importantly, this primacy has been crucial to the development of what I have loosely been calling 'cinema-tourism studies'.

Coined in the course of my work in the field, this umbrella term has allowed me to address different, but overlapping, research on film and tourism.[2] I have identified three major strands. Firstly, there is research conducted within media and film studies, which regards cinema as an inherently travel technology (e.g. Bruno, 2002; Corbin, 2014) and film spectatorship as a form of 'virtual tourism' (e.g. Gibson, 2006; Strain, 2003). Some of this work has analysed the language of cinema to demonstrate how films deliver complex 'cinematic journeys' that are not necessarily of a touristic nature (e.g. Eleftheriotis, 2010). Another influential strand has been 'film-induced tourism' (e.g. Beeton, 2005), which has developed within applied tourism studies. Its main aim is to analyse the potential of moving images to enhance tourism either by influencing spectators on their destination choices or by attracting them to the places where the images were shot or post-produced. Though committed to interdisciplinary knowledge, these studies are ultimately rooted in tourism management, destination image planning, and marketing (e.g. Hudson & Ritchie, 2006). Finally, a more social-sciences-oriented strand has sought to understand 'cinematic tourism' through the eyes of its social actors, such as tourists, locals, spectators, grassroots organisations, tourism developers, and operators (e.g. Tzanelli, 2007). Research within this strand is sensitive to the role that 'imagination' and agency plays in tourism (e.g. Crouch, Jackson, & Thompson, 2005).

My research has acknowledged all three strands but holds greater affinities with film studies and the social sciences. Working from a cross-disciplinary perspective that is strongly indebted to cultural studies and visual anthropology, I have researched the relationship between tourism and visuality within the domain of film in both 'explicitly' and 'non-explicitly touristic media' (Mazierska & Walton, 2006, p. 10).[3] In the cultural studies tradition, I have kept an eye on the intertwined material and meaning-making capabilities of both cinema and tourism, stressing not only their embeddedness in a variety of social practices but also the manner in which the products and texts thus produced are subject to the audience's responses and appropriations (Sampaio, 2014).

The chapter analyses two key concepts – the tourist gaze and virtual tourism – to critically assess their overwhelming impact on the field. Both concepts can be traced back to Urry's (2002) theoretical work on the gaze, even if their meanings and implications have moved well beyond that frame. Despite evincing a slight interest in moving images, the tourist gaze has been deployed as a methodological tool in film analysis; virtual tourism, on the other hand, has remained bound up with general platitudes and abstractions mostly heralding the arrival of a postmodern, dedifferentiated, and post-touristic era. In any event, both concepts have failed to engage with the particulars of visuality in tourism contexts and their relationship with broader and increasingly mediatized tourism contexts. In the concluding section, I draw attention to media archaeology, more-than-representational theories and the concept of mediatization as examples of recent developments in media research that can offer valuable contributions to the field.

Confronting the gaze

The tourist gaze, a concept firmly embedded in sociological theories that foreground the occularcentric qualities of (Western) modernity, has become the major doorway to cinema-tourism studies. It has also become, as I wish to demonstrate, something of a methodological tool in film analysis. The major reference when discussing this concept is Urry's eponymous work (2002, first published in 1990). The corollary of older theoretical trends that had been suggesting links between tourism, vision, and modernity (e.g. MacCannell, 1999, first published in 1976), Urry's work not only defined tourists as collectors of signs and gazes but went one step further to consider tourism as a matter of consumption of visual experiences (Urry, 2002, pp. 3, 42, 44). Drawing on Michel Foucault's panopticon model (Foucault's 'medical gaze' is the main reference), Urry (2002) introduced the term 'tourist gaze' to describe an all-encompassing perception device sustained by professional experts, which organises intercultural encounters in travel contexts along the lines of pleasure, difference, and the 'out-of-the-ordinary' (p. 145). The keyword here is 'frame'. For Urry (2002), the typical tourist experience consists in seeing 'named scenes through a frame, such as the hotel window, the car windscreen or the window of the coach' (pp. 90–91). It is what gets inside of this frame, not what stays outside of it, that defines the touristic.

The tourist gaze has attracted much criticism. It has been pointed out that the gaze is too static, passive, and ascetic, thus failing to capture the full range of active, pleasurable, and dynamic activities that make up the tourist experience, notably outside European contexts (Perkins & Thorns, 2001). The gaze marginalises the body, underestimates the role played by other senses (Crouch & Desforges, 2003; Veijola & Jokinen, 1994), and downplays the agency of tourists (MacCannell, 2001). Even if allowing for a range of thematic options – romantic, collective, spectatorial, reverential, environmental, mediatized (Urry, 2002, pp. 149–151) – the tourist gaze remains bound to a deterministic grid that enables tourists to engage with the extraordinary but not with the unexpected (MacCannell, 2001). Shortly after the book's publication, Urry (1992) acknowledged that the concept had not been sufficiently elaborated but confirmed its main contours, which have remained intact in subsequent revised editions.[4]

For the purposes of this chapter, I wish to emphasise two aspects. Firstly, Urry (2002) attributes a central role to vision (e.g. pp. 13, 117, 145, 146) but devotes scarce attention to visuality. Indeed, the tourist gaze fails to engage with how people *look* and what they actually *see* in tourism contexts. The visual particulars of the tourist gaze are only sketchily provided; when they *are* considered (e.g. Crawshaw & Urry, 1997), it is only to be recaptured by a closed and self-reproducing 'hermeneutic circle' (Urry, 2002, p. 129). In fact, because the tourist gaze transforms sights into signs – tourists are likened to semioticians who decode what they see as 'typical', 'authentic', 'traditional', and 'real' – the visual objects of the gaze are ultimately irrelevant. As Urry (2002) put it: 'We do not literally "see" things. Particularly as tourists we see objects constituted as signs' (p. 117). That is, while conceived of as a visual device that encourages tourists to look, the tourist gaze ultimately prevents tourists from seeing. Like Barthes' guidebook, the gaze is, therefore, 'an agent of blindness' (Barthes, 1993, p. 76). For Urry (2002), the visual may be 'the organising sense within the typical tourist experience' (p. 146), but its presence and significance as a bodily sense is rather disappointing.

Secondly, Urry is not concerned with the workings and implications of the gaze in moving images. His analysis is grounded in photography, not cinema. The model for the tourist gaze is the static photographic frame. Film, however, represents the attempt (through movement) 'to overcome the limits of the traditional picture and its frame' (Gunning, 2006, p. 34). Except for a brief mention of the role the railway played in the development of a 'more mobilised gaze' or 'tourist glance' (Urry, 2002, p. 153), which elicits a fleeting comparison with TV and film, Urry relegates film to a pre-touristic moment of travel anticipation (p. 3). Unlike photography, described as a 'new mode of visual experience' (p. 125), no special attention is paid to cinema's role in producing an innovative form of mobile perception, in constructing, that is, 'a new form of observer' (Gunning, 2006, p. 34).

And yet, cinema-tourism scholars have made ample use of this concept, not only as a theoretical framework but also as a methodological tool. The former aspect comes naturally, as most studies of the cinematic gaze draw on the same theories of modernity that have inspired the tourist gaze (e.g. Denzin, 1995; Friedberg, 1993). That the tourist gaze has become an established tool in film analysis, however, merits closer inspection. Different authors have drawn on Metz (e.g. 1986) and psychoanalysis to explain how the tourist gaze is built into a film and then activated during the viewing experience. The overlap between the tourist and the cinematic gazes[5] is typically explained as a result of the spectator's identification either with the camera ('primary identification') or with the main character ('secondary identification') (Strain, 2003, pp. 20–21). Accordingly, the tourist gaze is built through objective shots, which give an impression of unmediated 'looking', and subjective shots, which take on a character's point of view, being especially effective as 'a point of entry for spectators' when that character is a traveller and the protagonist (Strain, 2003, p. 143). Because of these cinematic devices, shots (whether objective or subjective) taken from a hotel window, a car windscreen, the window of a coach, bus, or train are immediately recognised as 'touristic'.[6] The tourist and the cinematic gazes are thus difficult to tell apart. They are also confused with the act of looking, despite the fact that their origins lie with Lacanian psychoanalytic theory, where the gaze is 'something missing from perception' and therefore cannot be confused with the eye (Saper, 1991, pp. 36, 43).[7]

The regular and rather straightforward application of the tourist gaze to film analysis – which produces a kind of short circuit between theory and methodology – has left its mark on the field, introducing some major biases. Most studies tend to focus on films that bear an obvious relation to tourism (cf. Mazierska & Walton, 2006, p. 8), being thus positioned, from the very beginning, inside the orderly scope of the tourist gaze. Moreover, they give precedence to fiction films and the classical narrative mode (cf. Sampaio, 2014), in relation to which formalist theories of spectatorship have most fully developed. Finally, the stories that most studies prioritise feature protagonists that are invariably tourists – usually personified by white European or American citizens travelling abroad alone, as a couple or in a group. All these thematic and structural choices invite a tourist-gaze-oriented theoretical approach and film analysis. Though adopting different focuses and serving different disciplinary ends, most cinema-tourism studies have thus been busy analysing similar films (or even the same films)[8] through a tourist gaze lens that cannot but confirm methodologically its theoretical premises, in a vicious interpretive circle that adds little to our understanding of film-mediated visuality in tourism contexts.

Gripping the virtual

The concepts of 'tourist gaze' and 'virtual tourism' are closely connected. Whether overtly stated or not, the former concept has authorised scholars to implicate moving images in the shift, allegedly taking place in contemporary societies, 'from a corporeal tourism to a virtual tourism' (Gibson, 2006, p. 157). What this shift implies is that films cease to be regarded as mere vehicles of pre-touristic promotional practices to themselves become a touristic practice.

Again, film and media scholars have turned to theories of spectatorship to explain this process: what the screen shows and the spectators see is akin to 'a tourist spectacle', so that 'in watching a film, the audience virtually travels' (Gibson, 2006, p. 163). This notion builds (rather ambiguously) on two different propositions: (1) certain films are (designed to be) consumed as tourist attractions (e.g. the Merchant Ivory films deliver 'cultural tourism' to American audiences – Gibson, 2006, p. 160); (2) because cinema is part of a vaster mobile visual culture, the movie-going experience is itself a tourist experience. In any of these versions, what seems to count is the film-as-travel-spectacle (and cinema as a spectacle-generating device), rather than the spectator, who merges with the images and disappears.[9]

The notion that virtual tourism is predominantly a visual experience permeates Urry's work. The superimposition of the different frames encountered in travel (such as the hotel window and the car windscreen) with the frames of audio-visual technologies like television – expressively described as 'a frame at the flick of a switch' (Urry, 2002, p. 91) – is what ultimately supports Urry's claim that we live in post-touristic times. This idea takes on rather simple contours: the tourist gaze has become so ubiquitous (on TV and the internet, in cinema and advertising – and, we could add, in more recent mobile display devices like tablets and smart phones) that we no longer need to move physically away from home to become tourists. Ours is an age of 'dedifferentiation' between seeing and travelling, ordinary life and vacationing; even when we are not 'really' *being* tourists, we are likely to *be* 'virtual' tourists. In fact, being a tourist seems to be the all-embracing condition of post-modernity: we are all tourists now (cf. Bauman, 1998; Urry, 2002, p. 74).

As powerful and enrapturing as this account may be, it nevertheless draws on a series of sweeping generalisations that are difficult to either prove or disprove. A less abstract understanding of virtual tourism has foregrounded the kinaesthetic dimension of film to explain the transformation of the tourist gaze into a 'cinematic-tourist glance' or 'cinematic-travel glance' (Gibson, 2006, p. 161). This line of inquiry takes the tourist gaze as 'an experiential structure' grounded in mobility, rather than a 'disembodied eye' (Strain, 2003, p. 2), to insert it in a long tradition of 'perception technologies' (Strain, 2003, p. 73) that, since the 19th century, have sought to endow travel representations with greater realism. Over the years, the quest for realism has resulted in the development and appropriation of a range of immersive technologies, such as the diorama, the panorama, stereoscopy, photography, cinema, and a wealth of hybrid recreational and educational practices like the Hale's tours (Rabinovitz, 2006), the *Maréorama* (Barbosa, 2017), and the travel lecture (Ruoff, 2006). All these devices and attractions offered visual, mobile, and often multisensorial supplements to the representations of travel that permeated late 19th-and early 20th-century popular culture. This continues to be the case with more recent immersion-driven technologies, such as travel games (Strain, 2003), IMAX (Acland, 1998), and virtual reality (VR) devices that combine head-mounted displays and computer-generated moving images to promote, deliver, or enhance different kinds of touristic experiences (Leotta & Ross, 2018).

Far from being an invention of our contemporary ('post-modern') times, the vicarious or 'armchair tourist' has been around for a long time, both as a historical entity and a fantasy. The longstanding presence of this figure suggests that imagination and image making have played a major part in tourism's social and cultural development. Clearly, the concept of virtual tourism touches on important issues. Nevertheless, it has also served as a stand in for empirical work, preventing a better grasp of vision and visuality in concrete tourism contexts. An overly emphasis on the virtual has too often detracted attention from the array of social and material forms and practices that continue to underpin both tourism (including in its mediatized forms) and cinema experiences. That is to say, cinematic tourism and film-induced tourism are not just about watching films and then going on a trip. Similarly, virtual tourism is not just about gaining access to travel experiences by proxy through visual and immersive technologies. A lot more is happening. Tourism has convincingly been described as 'a productive system that fuses discourse, materiality and practice' (Franklin & Crang, 2001, p. 17). The same ought to be said of moving images, which rely on practices that are just as material as doing tourism.

Finally, regardless of the claims of social theorists and VR promoters that we no longer need to travel to become travellers, most people seem to be more interested in travelling than in watching travel images on a screen. In fact, most people seem to be interested in doing *both*, often at the same time. How moving images are being used, in a context of multi-media convergence (Månsson, 2011), to supplement, enhance, and convey people's experiences in tourism is an area of enquiry that requires further attention and more suitable approaches.

Conclusions

In this chapter, I discussed how the concepts of tourist gaze and virtual tourism have had distorting effects in cinema-tourism studies. Well established in tourism studies, the tourist gaze has become a methodological tool. When applied directly to film analysis, it is easily confused with looking, thus countering its Lacanian origins. The gaze also tends to conflate touristic practices of seeing with scopic regimes of power (cf. Mitchell, 2002), making it difficult, if not impossible, to conceive of any other uses and understandings of tourism-related images and visuality outside of its totalising purview. These shortcomings are reflected in the corpora of films, which tend to be limited in number and kind. Similarly, the concept of virtual tourism has stimulated work of a speculative rather than empirical nature, encouraging dematerialised and ahistorical approaches to tourism that disconnect it from the realm of material affordances and (changing) social practices, often to merely (even if unwittingly) reproduce the industry's self-promotional images. Materiality, nevertheless, underpins and, indeed, suffuses all kinds of touristic activities, including those that are mediatized or occularcentric. It cannot be ignored.

But how to proceed from here? Is it possible to do cinema-tourism studies without succumbing to the overpowering influences of the tourist gaze and virtual tourism? Urry devoted little attention to cinema and the moving images. His theoretical engagement with tourist visuality drew largely on photography, which is also why the application of his work to other media is likely to fall short. Moreover, much of the work carried out so far has fallen prey to an effects-oriented approach to media, which reduces tourism-related images to semiotic markers of tourist sites (MacCannell, 1999), generators of pre-touristic anticipation (Urry, 2002), or a 'pull factor' of tourism destinations (Riley & Van Doren, 1992). Research under these influences also shows a bias for content- and representation-based analysis and seems to assume that tourists and movie-goers are passive consumers of all-powerful culture industries (cf. Tzanelli, 2007).

Outside the field of tourism-cinema studies, most of these approaches have been called into question. In media studies, there has been a shift away from discourse and representation and towards 'non-representational' or, in David Morley's finer expression, 'more-than-representational' theories (Krajina, Moores, & Morley, 2014, p. 694). Similarly, media archaeology – an ambitious research programme that draws on visual studies, film history, and media studies to understand the changing nature of cinema in contemporary societies – has pursued a line of enquiry that acknowledges cinema's historical embeddedness 'in other media practices, other technologies [and] other social uses' (Elsaesser, 2016, p. 19). Finally, the work that has been developed within the sociology of media around the concept of 'mediatization' (Couldry & Hepp, 2013; Hjarvard, 2008) has come up with a series of theoretical and methodological tools that promise to grapple, in a more satisfactory manner than Urry's thoughts on post-tourism, the vexing question of the 'consequences' of the media in contemporary societies. In all these fields, the media has come to be viewed as an integral part of contemporary societies rather than a separate world that impacts on other areas only sporadically; single-media analyses have fallen out of favour, given the need to account for the interconnections between different media, and theorisation has been carried out alongside empirical work (of a social, cultural, or historical kind) – not instead of it.

There are plenty of ways in which cinema-tourism studies can benefit from these approaches. Among other challenges, the field needs to overcome its 'unconscious multidisciplinarity' (Karpovich, 2010, p. 10) and build more solid connections with cutting-edge knowledge. It also needs to engage more critically and creatively with empirical work, turning away from the current pandemic of case studies that merely repeat well-established (if not necessarily truthful) conclusions (cf. Beeton, 2011; Connell, 2012). Finally, it is important to consider a wider range of research objects, such as hitherto neglected archival moving images and marginal touristic and filmmaking practices, which will necessarily require a renewed toolkit of theories, concepts, and methods. In view of recent and on-going societal developments, we can aver that the interest in the links between cinema and tourism is likely to continue to grow in upcoming years. The field's ability to produce scientifically sound and socially relevant research in the future will largely depend on its commitment to undertake this kind of integrated and rather ambitious theory-object-methodology renovation.

Notes

1 I thank the editors for their comments and support during the review process. Research for this chapter was funded by Fundação para a Ciência e a Tecnologia (FCT), through grant IF/00313/2013.
2 The expression 'cinema-tourism studies' avoids the submission of one of the elements of the dyad to the other, offering an alternative to the more limited, albeit more widely used, 'film' or 'cinematic' tourism. The chapter adopts this expression throughout.

3 Respectively, in promotional tourism films (Sampaio, 2017) and mainstream narrative films (Sampaio, 2014).

4 Urry addressed most of these critiques in Chapter 8 of the second edition (2002, pp. 145–156). Co-authored with Jonas Larsen, the third edition (2011) incorporated new material and widened its analytical scope.

5 One of the best accounts of the "cinematic gaze" is Mulvey's essay (1989, first published in 1975), a feminist critique of classical Hollywood cinema. Feminist critics of the tourist gaze have been adroit in blending these two gazes (e.g. Pritchard & Morgan, 2000).

6 This is clear in Merchant Ivory's film *A Room with a View* (1985), when Lucy Honeychurch (Helena Bonham-Carter), the archetypal Georgian tourist (cf. Sampaio, 2012), opens the window of her pension room with a view overlooking the Piazza della Signoria, in Florence.

7 See the anecdote reported by Lacan in Seminar XI and retold by Dash & Cater (2015, p. 275), who propose an inspiring application of the Lacanian gaze to tourism.

8 Consisting mostly of blockbusters produced and distributed under the aegis of "global Hollywood" (Miller, Govil, McMurria, Maxwell, & Wang, 2005). Examples include: *The Lord of the Rings*; *Captain Corelli's Mandolin*; *Harry Potter*; *The Da Vinci Code*; *Gladiator*; *The Beach*; *Braveheart*, among others.

9 But see Tzanelli (2007, pp. 3–5), who attempts to bridge the gap between the (disembodied) eye and the gazing body.

References

Acland, C. (1998). Imax technology and the tourist gaze. *Cultural Studies, 12*(3), 429–445.

Adler, J. (1989). Origins of sightseeing. *Annals of Tourism Research, 16*(1), 7–29.

Barbosa, S. H. (2017). Beyond the visual: Panoramatic attractions in the 1900 World's Fair. *Visual Studies, 32*(4), 359–370.

Barthes, R. (1993). *Mythologies*. London, UK: Jonathan Cape.

Bauman, Z. (1998). *Globalization: The human consequences*. Cambridge, UK: Polity Press.

Beeton, S. (2005). *Film-induced tourism*. Clevedon, UK, Buffalo, NY & Toronto, Canada: Channel View Publications.

Beeton, S. (2011). Tourism and the moving image – Incidental tourism promotion. *Tourism Recreation Research, 36*(1), 49–56.

Bruno, G. (2002). *Atlas of emotion: Journeys in art, architecture, and film*. London, UK: Verso.

Coleman, S., & Crang, M. (2002). *Tourism: Between place and performance*. New York, NY & Oxford, UK: Berghahn.

Connell, J. (2012). Film tourism: Evolution, progress and prospects. *Tourism Management, 33*(5), 1007–1029.

Corbin, A. (2014). Travelling through cinema space: The film spectator as tourist. *Continuum: Journal of Media and Cultural Studies, 28*(3), 314–329.

Couldry, N., & Hepp, A. (2013). Conceptualizing mediatization: Contexts, traditions, arguments. *Communication Theory, 23*, 191–202.

Crawshaw, C., & Urry, J. (1997). Tourism and the photographic eye. In C. Rojek & J. Urry (Eds.), *Touring cultures: Transformations of travel and theory* (pp. 176–195). London, UK & New York, NY: Routledge.

Crouch, D., & Desforges, L. (2003). The sensuous in the tourist encounter. Introduction: The power of the body in tourist studies. *Tourist Studies, 3*(1), 5–22.

Crouch, D., Jackson, R., & Thompson, F. (2005). *The media and the tourist imagination: Converging cultures*. London, UK & New York, NY: Routledge.

Dash, G., & Cater, C. (2015). Gazing awry: Reconsidering the tourist gaze and natural tourism through a Lacanian-Marxist theoretical framework. *Tourist Studies, 15*(3), 267–282.

Denzin, N. K. (1995). *The Cinematic society: The voyeur's gaze*. London, UK: Sage.

Eleftheriotis, D. (2010). *Cinematic journeys: Film and movement*. Edinburgh, UK: Edinburgh University Press.

Elsaesser, T. (2016). Media archaeology as symptom. *New Review of Film and Television Studies, 14*(2), 181–215.

Franklin, A., & Crang, M. (2001). The trouble with tourism and travel theory? *Tourist Studies, 1*(1): 5–22.

Friedberg, A. (1993). *Window shopping: Cinema and the postmodern*. Berkeley & Los Angeles: University of California Press.

Gibson, S. (2006). A seat with a view: Tourism, (im)mobility and the cinematic-travel glance. *Tourist Studies, 6*(2), 157–178.

Gunning, T. (2006). 'The whole world within reach': Travel images without borders. In J. Ruoff (Ed.), *Virtual voyages: Cinema and travel* (pp. 25–41). Durham, NC & London, UK: Duke University Press.

Hjarvard, S. (2008). The mediatization of society: A theory of the media as agents of social and cultural change, *Nordicom Review, 29*(2), 105–134.

Hudson, S., & Ritchie J. R. B. (2006). Promoting destinations via film tourism: An empirical identification of supporting marketing initiatives. *Journal of Travel Research, 44*, 387–396.

Karpovich, A. I. (2010). Theoretical approaches to film-motivated tourism. *Tourism and Hospitality Planning & Development, 7*(1), 7–20.

Krajina, Z., Moores, S., & Morley, D. (2014). Non-media-centric media studies: A crossgenerational conversation. *European Journal of Cultural Studies, 17*(6), 682–700.

Leotta, A., & Ross, M. (2018). Touring the 'world picture': Virtual reality and the tourist gaze. *Studies in Documentary Film, 12*(2), 150–162.

MacCannell, D. (1999). *The tourist: A new theory of the leisure class.* Berkeley: University of California Press.

MacCannell, D. (2001). Tourist agency. *Tourist Studies, 1*(1), 23–37.

Månsson, M. (2011). Mediatized tourism. *Annals of Tourism Research, 38*(4), 1634–1652.

Mazierska, E., & Walton, J. K. (2006). Tourism and the moving image. *Tourist Studies, 6*(1), 5–11.

Metz, C. (1986). *The imaginary signifier: Psychoanalysis and the cinema.* Bloomington & Indianapolis: Indiana University Press.

Miller, T., Govil, N., McMurria, J., Maxwell, R., & Wang, T. (2005). *Global Hollywood 2.* London, UK: BFI.

Mitchell, W. J. T. (2002). Showing seeing: A critique of visual culture. *Journal of Visual Culture, 1*(2), 165–181.

Mulvey, L. (1989). Visual pleasure and narrative cinema. In L. Mulvey (Ed.), *Visual and other pleasures* (pp. 14–38). Houndmills, Basingstoke, UK & London, UK: Macmillan.

Perkins, H. C., & Thorns, D. C. (2001). Gazing or performing? Reflections on Urry's tourist gaze in the context of contemporary experience in the antipodes, *International Sociology, 16*(2), 185–204.

Pritchard, A., & Morgan, N. J. (2000). Privileging the male gaze: Gendered tourism landscapes, *Annals of Tourism Research, 27*(4), 884–905.

Rabinovitz, L. (2006). From Hale's tours to star tours: Virtual voyages, travel ride films and the delirium of the hyper-real. In J. Ruoff (Ed.), *Virtual voyages: Cinema and travel* (pp. 42–60). Durham, NC & London, UK: Duke University Press.

Riley, R., & Van Doren, C. S. (1992). Movies as tourism promotion: A 'pull' factor in a 'push' location. *Tourism Management, 13*(3), 267–274.

Ruoff, J. (2006). Show and tell: The 16mm travel lecture film. In J. Ruoff (Ed.), *Virtual voyages: Cinema and travel* (pp. 217–237). Durham, NC & London, UK: Duke University Press.

Sampaio, S. (2012). 'I wish something would happen to you, my friend': Tourism and liberalism in E. M. Forster's Italian novels'. *Textual Practice, 26*(5), 895–920.

Sampaio, S. (2014). Watching narratives of travel-as-transformation in *The Beach* and *The Motorcycle Diaries. Journal of Tourism and Cultural Change, 12*(2), 184–199.

Sampaio, S. (2017). Tourism, gender and consumer culture in late and post-authoritarian Portugal. *Tourist Studies, 17*(2), 200–217.

Saper, C. (1991). A nervous theory: The troubling gaze of psychoanalysis in media studies. *Diacritics, 21*(4), 32–52.

Strain, E. (2003). *Public places, private journeys: Ethnography, entertainment, and the tourist gaze.* New Brunswick, Canada: Rutgers University Press.

Tzanelli, R. (2007). *The cinematic tourist: Explorations in globalization, culture and resistance.* London, UK & New York, NY: Routledge.

Urry, J. (1992). The tourist gaze 'revisited.' *American Behavioral Scientist, 36*(2), 172–186.

Urry, J. (2002). *The tourist gaze* (2nd Ed.). London, UK, Thousand Oaks, CA & New Delhi, India: Sage.

Urry, J., & Larsen, J. (2011). *The tourist gaze 3.0.* London, UK, Thousand Oaks, CA & New Delhi, India: Sage.

Veijola, S., & Jokinen, E. (1994). The body in tourism. *Theory Culture & Society, 11*(3), 125–151.

8

PROMOTING CULTURAL HERITAGE IN A POST-DIGITAL CONTEXT

A speculative future for the online archive

Adriaan Odendaal and Karla Zavala

Introduction

Many memory institutions, such as museums, retain much of their cultural authority as geographically specific destinations, with their prerogative to preserve cultural heritage in the collective memory of the public enshrined by physically exhibited artworks and artifacts. These institutions have come to increasingly rely on digital technology to make their collections available to online publics in attempts to boost preservation efforts and promote cultural agendas (Berry, 2016, p. 103; Petrelli, Marshall, O'Brien, McEntaggart, & Gwitt, 2017, p. 281). The popularization of the term 'memory institutions' in the early 1990s, as a collective noun for museums, archives, and libraries, was itself 'linked to the new possibilities opened up by the advent of the Internet [...] that could facilitate seamless access to collection information' (Robinson, 2012, p. 415).

However, the evolution of internet technology and the accompanying digitization of culture inherently undermine the authority of these institutions to archive, store, and preserve cultural memory. The communicational infrastructure of the World Wide Web allows internet users to skirt institutional gatekeepers and access digitized cultural heritage more directly. As Blom (2016) writes in *Memory in Motion: Archives, Technology, and the Social*, digitization makes 'memory materials more or less instantaneously available to anyone, anywhere in the world', presenting a radical democratization of cultural memory (p. 13). Yet, former curator at the Solomon R. Guggenheim Museum, Jon Ippolito, argues that memory institutions, 'no matter how digitized, remain hamstrung by their own history as centralized repositories' (2014, p. 79). And while memory institutions such as Tate, MoMA, or the Smithsonian have come to respond in innovative ways to a Web 2.0 user-centred digital paradigm, now a new paradigm of ubiquitous computing promises even more fundamental changes to how we navigate digital culture, how digital heritage enters and remains in our collective cultural memory, and how memory institutions fit within wider mediatized cultural and economic contexts.

There has been an observed shift to what some call 'ubiquitous computing' – the 'sociocultural and technical thrust to integrate and/or embed computing pervasively, to have information processing thoroughly integrated with or embedded into everyday objects and activities' (Ekman, 2016, p. 5). Browser-based experiences of the web are increasingly supplanted by that of hidden interfaces. Internet-of-Things objects that upend expectations of how and where personal computers are accessible, while augmented reality (AR) super-imposes Web-based experiences onto physical spaces. Ambient computational environments change location-based experiences according to personal browsing habits while wearable technology reenvisages the standardized interface of the internet. And then there are increasingly smart assistants that allow users to ostensibly 'browse' the Web through a number of seamless and screenless experiences (Ekman, 2016; Manwaring & Clarke, 2015). It is a veritable post-digital condition. The user-experiences of the internet is shifting away from authoritative websites actively accessed, or searched for, by the user. Cultural content becomes increasingly digital while the digital inversely becomes more embedded and invisible as it is delivered directly to users – even as embodied experiences – through smart objects and digitally encoded environments.

The subject of this chapter is to discuss digital memory, cultural preservation, and souvenirs using post-digital critique and critical design theory as theoretical framework. As a particular case in point, we employ the *Sound Souvenirs,* a research-led and speculative design project presented as a post-digital archive experience and exhibited by the authors and their collaborators[1] at the 2017 *New Interfaces for Musical Expression* (NIME) conference in Copenhagen. The project is placed in the junction of ubiquitous computing and mediatized tourism where contemporary memory institutions find themselves.

We argue that a consequent analysis of the *Sound Souvenirs* project offers a unique interdisciplinary entry-point into the larger discussion of mediatized tourism presented throughout this book. Although the scenario addressed by this project might seem somewhat speculative, far-removed from the more immediate issues memory institutions face as part of the tourism industry, exploring and experimenting with the 'preferable futures' of cultural preservation might not only present potential solutions to future problems memory institutions might face – but might perhaps also offer interpolated insights into the present.

Critical design and preferable futures

Critical design presents a particularly powerful practice-based research approach to explore questions of digital preservation and post-digital archival mnemonics. This approach was first introduced by Dunne (2005) in his book *Hertzian Tales: Electronic Products, Aesthetic Experience and Critical Design* as a way to challenge narrow assumptions about the role everyday products perform in life by exploring new technologies in a way that can generate a critical narrative about the technological future. Dunne and Raby (2013) argue that critical design is indeed 'a critical medium for exploring the implications of new developments in science and technology' (p. vi). Therefore, this approach provides an optimal practice to reflect on the possibilities and affordances of mediatized tourism in relation to the preservation and promotion of cultural heritage in a ubiquitous computing context. It should be noted that the theoretical value of this exercise is not in finding concrete solutions to persisting problems but to pose 'difficult questions' that can allow the exploration of useful possibilities given that the method provides a space to imagine, as the authors propose, 'how things could be' (Dunne & Raby, 2013, p. 12). Importantly, critical design theory consists of a 'design spectrum' that includes practices such as 'design fiction', 'future design', 'anti-design', 'radical design', and for our purposes: 'speculative design', through which potential futures are investigated.

Dunne and Raby refer to a diagram used by Stuart Candy to illustrate different kinds of potential futures (Dunne & Raby, 2013, p. 3). They divide these 'futures' into three 'cones': probable, plausible, and possible. The first cone of the probable future is where non-speculative design solves problems within an existing system, whereas the cone of plausible futures offers space for 'planning and foresight, the space of what could happen' (Dunne & Raby, 2013, p. 3). Hereby, even though a 'design solution' can have a fictional nature and hence only experimental application in the real world, critical design opens up new perspectives for solving actual problems. Additionally, a final cone is intersected between the probable and the plausible: preferable futures. This intersection is the area in which the *Sound Souvenirs* project was designed.

Within a speculative future, the *Sound Souvenirs* project addresses the mediatization of cultural preservation in a post-digital context, deploying consumer objects from cultural tourism as critical design objects. Thus, instead of focusing our efforts on the digital channels of institutions, such as websites and social media, our speculative design project changed the focus to more experimental, and theoretically substantiated, mnemonic strategies for a 'post-digital archives' set within a larger cultural and economic context. This was done by leveraging the role of the public user as activator of digital archival memory. For in a prescient future of post-digital ubiquitous computing, users are increasingly being placed at the centre of the technological thrust towards a more intuitive, semantic, contextually-aware, and personally embedded Web.

Søndergaard (2009), who supervised the *Sound Souvenirs* project at Aalborg University, writes of his own early digital archive experiments: 'The important thing is to do something different, something unexpected' as a digital archive experience 'might define an entirely new level of cultural and social production' (p. 28). Thus, this chapter does not merely offer an investigation into a pressing problem for memory institutions and digital cultural preservation but also offers a method for thinking beyond the immediate problems and common solutions. This is something that is needed if institutions are not only to survive but also thrive in an increasingly mediatized context. The speculative design process was thus started by asking the question: What if we can design a cultural preservation strategy that is centred on the user's mnemonic role, instead of on the institution's archive, taking into account the effects a shifting post-digital technological context might have?

The *Sound Souvenirs*

The *Sound Souvenirs* project was exhibited at the 2017 NIME Conference in Copenhagen, a gathering of researchers, musicians, and media art practitioners focused on inventive and experimental musical interface design, technologies, and performances (NIME, 2017, n.p.). With the conference being a time-based event, we departed from the prerogatives of the conference itself and created a fictional institution called the *Memorial to Forgotten Sounds*, tasked with preserving cultural content important to the posterity of the conference. Consequently, we populated our experimental memory institution's online archive with sound art pieces extracted from digitally recorded performances from past NIME conferences. Sound art in this sense is understood as a media art practice that regards recorded sound as a form of representation inextricably linked to technological mediation, dating back from the early 20th-century inventions of futurist Luigi Russolo who built artistic machines to replicate the noises of the industrial age.

Integral to our speculate design was not the archive itself but rather its mnemonic strategy – how that which is stored in the archive is recalled or remembered, and thus preserved in cultural memory. We created *Sound Souvenirs* for this purpose: design objects that had the compact semblance of a souvenir, laser-cut from luminous translucent acrylic to

Figure 8.1   Visitors to the *Sound Souvenir* pop-up 'memory institution' at NIME (2017)
Source: Luis Bracamontes.

resemble five distinct soundwaves. These souvenirs could function as tangible mnemonic activators of their corresponding online archival sound file with visitors using an AR application to access and experience this digitally archived cultural heritage of the conference (Figure 8.1). Central to this archive was that visitors could receive their *Sound Souvenir* to take home, both as mnemonic device and touristic memento. Our installation consequently served as something of a tourist stop for 2017 NIME conference goers: a pop-up stand for a fictional memory institution vying for their attention in the exhibition space, offering souvenirs not only to attract visitors – but ultimately to fulfil our institution's goal of keeping the conference's cultural heritage in the public's collective memory through post-digital archival mnemonics.

Storage vs digital memory

The archive for our fictional memory institution was created as an online repository reminiscent, in aesthetics and user-interface design, of early internet databases. In this, it served as a discrete site with no social media presence, backlinks pointing towards it, user-interactivity or any other mechanism to promote access to it. It served merely as institutional 'online storage' to digitally preserve the conference's cultural heritage.

The possibility of such online archives' ability to recall that which it stored was already called into question early in the history of the internet. Pioneering internet scholar Howard Caygill (1999) wrote that the technological basis of the Web makes any archive appear 'as an effect of the links made possible by the technological work of memory rather than a given (and carefully policed) store of information' (1999, p. 2). This idea was expanded, Ernst (2013) who argues that the idea of the archive in internet communication moves the archive toward an economy of circulation (2013, p. 99). With this he contended that 'so-called cyberspace is not primarily about memory as cultural record but rather about a performative form of memory as communication' (Ernst, 2013, p. 99). For Ernst (2013), there is no such thing as 'storage' on the internet as any archive and its cultural content becomes a function of constantly circulating 'memory'. The media and other files comprising the *Memorial to Forgotten Sounds* online archive are thus merely dormant data, informational bits, stored on a localizable physical server. As an online archive, it only ever exists if and when these files are recalled by a user on a networked computer and compiled into a navigable website containing now accessible media files. The online archive thus only exists when it is *remembered* at behest of a user. The internet, Ernst argues, doesn't have an organized memory. 'If there is memory'. writes Ernst, 'it operates as a radical constructivism: always just situationally built, with no enduring storage' (2013, p. 138).

Notwithstanding, an online archive creates the illusion of permanence as the communicational computation involved in recalling and reconstructing a website in a browser is largely obscured and happens almost instantaneous. As Chun (2016) emphatically argues in *Ubiquitous Computing: Computing, Complexity, and Culture*: memory is not something that remains, if memory appears to be static or stable it is simply because it is constantly regenerated (p. 168). The distinction between digital memory and storage underscores a fundamental misconception memory institutions have about the mediatization of cultural memory in a post-digital context. The conflation of memory and storage creates a misleading ethos that can easily lead to a neglect of the continuous compounding socio-cultural forces that makes our cultural memory seem stable and permanent (2016, pp. 168–169). In this online regime of memory, it is through the user's embodied action that an online archive is retrieved from server storage and continuously 'situationally built'. Therefore, a post-digital context fundamentally displaces the cultural authority for preservation from institutions unto the users. Berry (2016) contends that 'Computation therefore threatens to *de-archive* the archive, disintermediating the memory institutions and undermining the curatorial functions associated with archives' (2016, p. 107). As a result, Berry argues that a post-digital context demands a new social ontology circumscribing new ways of interacting and exploring digitized archives (Berry 2016, p. 106).

Framed by ideas of the post-digital condition, our critical design appropriation of souvenirs objects as archival mnemonic strategy aimed to leverage individual user memory as means of generating 'digital memory' and thus activating our otherwise obscured online archive and its forgotten cultural content. This premise is based on the idea that souvenirs are social 'touchstones of memory', able to bring 'the past into the present and [make] past experience live' (Morgan & Pritchard, 2005, p. 37). As Gordon (1986) significantly wrote in *The Souvenir: Messenger of the Extraordinary*, 'as an actual object', the souvenir 'concretizes or makes tangible what was otherwise only an intangible state' (1986, p. 1). The metaphorical transition from impermanent (memory) to tangible state (souvenir) that Gordon refers to is carried out by the 'physical presence [that] helps locate, define and freeze in time a fleeting, transitory experience' (Gordon, 1986, p. 1). This allows us to use the souvenir as an overcoded design object in a post-digital context where the experience of the web becomes embedded in the experience of everyday objects.

Post-digital archive experience

As material objects that can trigger memories, souvenirs move beyond the domain of touristic commodities and can facilitate memory institutions such as museums in preserving cultural heritage through creating acquirable, *memorable,* representative facsimiles of their collection. However, through the critical design of the *Sound Souvenirs*, we challenge the assumptions that such souvenirs correspond solely to material culture (paintings, statues, sculptures, architectural landmarks). We hold that the same role can be deployed for the immaterial bits of culture in a post-digital context.

McGugan and Petichakis (2009) write about souvenirs using Wenger's conception of reification: 'the process of giving form to our experience by producing objects that congeal this experience into *thingness*' (1998, p. 58). They argue that the process of reification itself situates the souvenir as a 'mediating artefact', hinting at the potential role souvenirs can play in preservation strategies of online archives in an era of ubiquitous computing (McGugan & Petichakis, 2009, p. 235). As the online archive's digital composition is totally assembled out of code, by giving a tangible shape to the archive's content (reification), a souvenir can act as a medium to aid the user's memory and bridge the transition from intangible memory to tangible activator of that memory. By selecting a souvenir as mnemonic strategy for the NIME digital archive, we thus aimed to leverage the individual's role in recalling the archival content, while on the other hand responding to the digitality of the archive and its digitally immaterial artifacts to be 'remembered'.

To address the question of how to meaningfully compile the seemingly immaterial digital files as tangible souvenirs, we turn to the post-digital concept of 'neomateriality' (Paul, 2015). This phenomenological approach to the digital focuses on the materiality of an object conceived through digital technologies. Neomateriality 'describes an objecthood that incorporates networked digital technologies, and embeds, processes, and reflects back the data of humans and the environment, or reveals its own coded materiality and the way in which digital processes see our world' (Paul, 2015, p. 1). Paul further writes that this concept is contextual to a post-digital scenario, which implies objects being 'conceptually and practically shaped by the Internet and digital processes yet often manifest in material form' (Paul, 2015, p. 1). By designing the *Sound Souvenirs* as such objects of neomateriality, they could function as reified artifacts of the 'digitally immaterial' soundwaves preserved by the *Memorial to Forgotten Sound* online archive.

Paul (2015) comments that 'artistic practice engaging with conditions of neomateriality often highlight this condition by turning code and abstraction into the material framework of an object' (2015, p. 2). In the process of designing the *Sound Souvenirs*, we thus necessarily highlighted the digital condition of the selected sounds. The digital representation of the sound pieces comprised its mediation through audio editing software – its 'immateriality' as binary code compiled and graphically represented. Through converting the digital representation of the soundwave into vector graphics, and drawing on 8-bit aesthetics to visually highlight its compressed digitality, we wanted to emphasize the skemorphic digitality of this neomaterial object. The result of the mediation process was a *Sound Souvenir* in the form of a materialized soundwave, laser-cut from luminous acrylic, connected to a rounded base (Figure 8.2). These were then engraved with their respective online archival URLs and packaged for distribution as materialized 'sound files' (with our souvenir naming convention including the '.wav' extensions of the actual files), focusing attention on the post-digital condition of the *Sound Souvenirs*.

As part of our design premise the *Sound Souvenirs* could be regarded as Tangible User Interfaces (TUI). The term describes the coupling of '[p]hysical representations (e.g. spatially manipulable physical objects) with digital representations (e.g. vector graphics and audio files), yielding user interfaces that are computationally mediated but not generally identifiable as "computers" per se' (Ullmer & Ishii, 2000, p. 916)). Thus, the *Sound Souvenirs* become 'playback devices' for the digital files in our online archive. The name *Sound Souvenir* is itself taken from Bijsterveld and Van Dijk who use the term to describe memories of music as including 'the sensory experiences of having listened to particular recordings and interacted and tinkered materially with the devices that play them' (2009, p. 11). The materiality of the soundwaves represented by the souvenirs thus not only materially occupy physical space but also serves as cues to recall the digital content associated with the souvenir through activation by the AR app in the pop-up installation, as well as through the URL engraved in the souvenir.

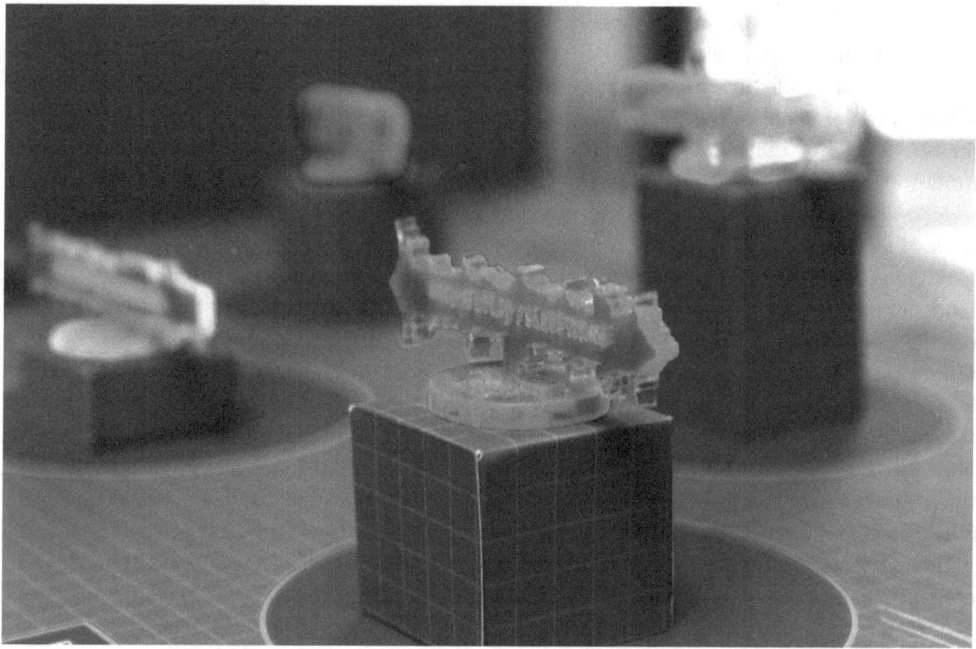

Figure 8.2   The material *Sound Souvenirs* on display at NIME (2017)
Source: Luis Bracamontes.

This experiment implies that the *Sound Souvenir* as TUI plays the role of mnemonic interface for a digital archive that leaves aside the web browser. This creates a post-digital experience of the memory institution's archive, with our 'memory object' taking advantage of ubiquitous computing technology and becoming a post-digital material cue to recall cultural memory. Petrelli et al. (2017) challenge museum practices in an age of increasing ubiquitous computing in a similar experiment of his own, writing that '[b]y combining principles from ubiquitous computing and tangible interaction, it is possible to close the gap currently existing between the exhibition floor and the online services and design visitor experiences that take both aspects into account' (p. 281–282). This convergence underscored our speculative experiment – as our archival content is only available to the public online, even if the institutional context is location-based. It is this convergence that also allows us to leverage memory institution's place within a context of mediatized tourism.

Mediatized cultural tourism and the post-digital archive

To deploy souvenirs as central mnemonic strategy for our archive, it was necessary to not only look at the material, socio-cultural, and psychological functions of souvenirs, but to also consider their dimension as distinctly economical artefacts integral to memory institutions' situation in the tourism industry. Souvenirs are often conceived of as commercial object produced for, and procured as part of, the experience of visiting a museum. As such, museum gift shops have become 'increasingly marketed as attractions in their own right', an expected part of any museum experience (Larkin 2016, p. 109). Yet, even within this commodifying context, souvenirs retain their function as memory objects. Procuring a souvenir becomes an important way to concretize 'memorable tourism experience', those 'tourism experiences that are positively remembered and recalled after the events have occurred' (Sthapit, Coudounaris, & Björk, 2018, p. 631).

It is this facet of souvenirs as embodiments of 'memorable tourism experiences' that enabled us to conceive of the *Sound Souvenirs* as more than just objects for archival mnemonics. We aimed to use the souvenir as a 'memory object' that conflates post-digital archival mnemonics and the location-based situatedness of memory institutions within the tourism industry. And inversely, we could appropriate commodifiable souvenirs as more than just a facet of a memory institution's commercial operations.

Given the affordance of critical and speculative design, we were in the privileged position of working outside of the financial and logistical concerns of museums. Yet, this also allowed us to critically exploit underlying cultural dimensions of mediatized tourism. During its installation at the NIME conference, the *Sound Souvenirs* were consequently distributed free off cost to visitors of our pop-up institution. However, visitors were required to make a metaphorical transaction before being able to procure one. The transaction entailed what we called 'paying with memory', which effectively meant that visitors had to go through the process of recalling the online archive before being able to take a souvenir of the specific archival sound file home. The action of activating the archive was facilitated by a tablet made available to visitors through which they could visit the online archive through an AR app that responded to the QR encoded exhibition environment the souvenirs were set in (Figure 8.1).

The use of AR or other digital interactivity has been increasingly implemented in museums since the early 1990s. These digital interactions are often designed in order to enhance the experience of visitors through animating otherwise static objects or allowing additional information or media content to be accessed (Petrelli et al., 2017, p. 282). Yet, the AR mediation of the souvenir acquisition served to enhance the 'memorable experience' of the post-digital archive within the context of our memory institution, while also coupling it with the functions of our archival mnemonics. As Petrelli et al. (2017) write of his own experiments in mediatized souvenirs:

> While the most common experience is to buy a souvenir after the [museum] visit, there is evidence to suggest that the souvenir should instead be an integral part of the visit itself, it should be constructed during the visit in such a way that it becomes the embodiment of the personal experience.
>
> *(p. 283)*

Through this process, the *Sound Souvenirs* subverts its own role as mere object for consumption, as in its consumption it enacts our post-digital mnemonic strategy for cultural preservation. It serves both as procurable object to reify a 'memorable tourism experience', as well as a 'memory object' for the cultural memory kept in 'situationally built' circulation by the post-digital archive itself.

Conclusions

The interplay between the digitality and objecthood of the *Sound Souvenir* at the centre of our fictional memory institution's archive prompted significant discussions among visitors about our installation at NIME. Of particular interest was the potential a project such as this could have when implemented in earnest at existing memory institution gift shops – especially given the novelty digitally over-coded souvenirs present. Throughout the process, however, we never explicitly addressed the question of feasibility. Our chosen design methodology was, after all, primarily meant to prompt discussion about digital heritage preservation in a post-digital context. Our consequent critical and speculative design approach allowed us to create a 'what if' scenario responding to institutional reluctance to forego traditional cultural authority and adapt to new technocultural paradigms. Through this, we explored the impending liquidity of digital content in a post-digital world which renders the cultural archive as we know it invisible, yet somehow contextually omnipresent. In that sense, we argued, the materiality of the interfaces mediating cultural archives have the potential to be contextual cues to access archives of the future as well, mediated through new interfaces that are yet to come. Yet, this also gives interpolatable insights into the present. For already, what gets preserved online is not that which gets digitally stored at the behest of memory institutions but, in an ostensive extension of Jenkins' conversion culture, that which is remembered and interacted with by digital media users in the public. In an era where digital technology permeates all aspects of culture, the traditional archive tasked with preserving culture becomes subject to a new regime of public memory and with that, a new regime of diffused control.

The particular use of the souvenir as a critical design object allowed us to leverage institutional situation within an increasingly mediatized tourism industry. It is in the theoretical gap between post-digital critique and mediatized tourism that we could thus explore ways memory institutions can improve their digital archival practices, critically deploying touristic consumer objects as a way to concretize the roles experience and memory can play in a post-digital context. Implicit in this kind of approaches, however, is also a critical distance. Part of a speculative design approach is to use the medium of technology self-reflexively, taking into critical consideration that with the solutions technological development present, there are always residual problems. Technological development never presents a catch-all solution. This framed the overarching discussion of the project, aimed not at predicting the future, or offering practical responses to it, but instead trying to provoke radical futures that can lead the reader to imagine and reflect on current and emerging technocultural contexts. In this, critical design also offers a disciplinarily experimental entry-point for re-thinking other facets of mediatized tourism and their larger cultural implications, and possible applications.

Note

1 *Sound Souvenirs* was a project developed during the EMJMD Media Arts Cultures program at Aalborg University along with Luis Bracamontes and Sultana Ismet Jerin. Among their contributions, Bracamontes was largely responsible for designing the augmented reality experience, while Jerin made an invaluable contribution with research regarding tangible user interfaces.

References

Berry, D. M. (2016). The post-archival constellation: The archive under the technical conditions of computational media. In I. Blom, T. Lundemo, & E. Røssaak (Eds.), *Memory in motion: Archives, technology, and the social* (pp. 103–129). Amsterdam, The Netherlands: Amsterdam University Press.

Bijsterveld, K., & van Dijk, J. (2009). Chapter three: The preservation paradox in digital audio. In K. Bijsterveld & J. van Dijk (Eds.), *Sound Souvenirs audio technologies, memory and cultural practices* (pp. 11–25). Amsterdam, The Netherlands: Amsterdam University Press.

Blom, I. (2016). Introduction: Rethinking social memory: Archives, technology, and the social. In I. Blom, T. Lundemo, & E. Røssaak (Eds.), *Memory in motion: Archives, technology, and the social* (pp. 11–41). Amsterdam, The Netherlands: Amsterdam University Press.

Caygill, H. (1999). Meno and the Internet: Between memory and the archive. *History of the Human Sciences, 12*(2), 1–11. doi: 10.1177/09526959922120216.

Chun, W. H. K. (2016). Ubiquitous memory: I do not remember, we do not forget. In U. Ekman, J. D. Bolter, L. Díaz, M. Søndergaard, & M. Engberg (Eds.), *Ubiquitous computing, complexity, and culture* (pp. 159–173). New York, NY: Routledge.

Ekman U. (2016). Introduction: Complex ubiquity-effects. In U. Ekman, J. D. Bolter, L. Díaz, M. Søndergaard, & M. Engberg (Eds.), *Ubiquitous computing, complexity, and culture* (pp. 1–20). New York, NY: Routledge.

Ernst, W. (2013). Discontinuities: Does the archive become metaphorical in multimedia space? In J. Parikka (Ed.), *Digital memory and the archive* (pp. 113–140). Minneapolis: University of Minnesota Press.

Dunne, A. (2005). *Hertzian tales. Electronic products, aesthetic experience, and critical design*. Cambridge: MA: MIT Press.

Dunne, A., & Raby, F. (2013). *Speculative everything : Design, fiction, and social dreaming*. Cambridge, MA: MIT Press.

Gordon, B. (1986). The Souvenir: Messenger of the extraordinary. *Journal of Popular Culture, 20*(3), 135–146.

Ippolito, J. (2014). Death by institution. In R. Rinehart & J. Ippolito (Eds.), *Re-collection: Art, new media, and social memory* (pp. 75–87). Cambridge, MA: MIT Press.

Larkin, J. (2016). 'All museums will become department stores': The development and implications of retailing at museums and heritage sites. *Archaeology International, 19*, 109–121. doi: 10.5334/ai.1917.

Manwaring, K., & Clarke, R. (2015). Surfing the third wave of computing: A framework for research into eObjects. *Computer Law & Security Review, 31*(5), 586–603. doi: 10.1016/j.clsr.2015.07.001.

McGugan, S., & Petichakis, C. (2009). Messengers of the extraordinary: The role of souvenirs in academic development. *International Journal for Academic Development, 14*(3), 233–235.

Morgan, N., & Pritchard, A. (2005). On souvenirs and metonymy: Narratives of memory, metaphor and materiality. *Tourist Studies, 5*(1), 29–53.

NIME (2017). *NIME – The International Conference on New Interfaces for Musical Expression*. Retrieved from: http://www.nime.org/.

Paul, C. (2015). From immateriality to neomateriality: Art and the conditions of digital materiality. In *Proceedings of the 21st International Symposium on Electronic Art* (pp. 1–4). Retrieved from: http://isea2015.org/proceeding/submissions/ISEA2015_submission_154.pdf.

Petrelli, D., Marshall, M. T., O'Brien, S., McEntaggart, P., & Gwitt, I. (2017). Tangible data souvenirs as a bridge between a physical museum visit and online digital experience. *Personal and Ubiquitous Computing, 21*(2), 281–295. doi: 10.1007/s00779-016-0993-x.

Robinson, H. (2012). Remembering things differently: Museums, libraries and archives as memory institutions and the implications for convergence. *Museum Management and Curatorship, 27*(4), 413–429. doi: 10.1080/09647775.2012.720188.

Søndergaard, M. (2009). The digital archive experience. In M. Jacobsen & M. Søndergaard (Eds.), *Re_Action: The digital archive experience* (pp. 25–48). Aalborg, Denmark: Aalborg University Press.

Sthapit, E., Coudounaris, D. N., Björk. P. (2018). The memorable souvenir-shopping experience: Antecedents and outcomes. *Leisure Studies, 37*(5), 628–643.

Ullmer, B., & Ishii, H. (2000). Emerging frameworks for tangible user interfaces. *IBM Systems Journal, 39*(3), 915–931.

Wenger, E. (1998). *Communities of practice: Learning, meaning, and identity*. Cambridge, UK: Cambridge University Press.

9

PHYSICAL DIGITAL LABOUR AND THE COMMODITISATION OF CULTURAL SITES

Mediatising tourism through social mapping

Kathleen M. Kuehn and Michael S. Daubs

Introduction

This chapter explores issues of social mapping, labour, and the valorisation of space as they function in service of expanding Google's digital empire through a case study of the Google 'Trekker' programme. Google Trekker is a crowdsourcing initiative designed to capture hard-to-reach landscape imagery for Google Maps. Volunteers carry a 42.5lb/19kg 'trekker pack' outfitted with 15 5-megapixel cameras that, once activated, take a photo every 2.5 seconds to record 360-degree panoramic views of remote or unique places inaccessible to Google's vehicle-based capturing equipment (e.g., Street View car), which captures footage from a camera attached to the top of a moving vehicle for integration into Google Maps' Street View function.

The chapter applies a combination of digital media and communication theory and critical human geography to the study of media and tourism. It considers how Google relies upon digital technologies and their users to socially construct space and place according to a market logic. Drawing from Marxian approaches to digital labour (e.g., Fuchs, 2014; Terranova, 2000) and space (e.g., Farman, 2010, 2014; Lefebvre, 1991), we examine how the Trekker project redefines traditional boundaries between the material and immaterial, physical and digital, public and private. By extending Marxist theories of digital labour to the production of the tourist experience, our analysis of Google Trekker's user-generated mapping project reveals one of many invisible layers behind the mediatization of contemporary tourism. We unpack how the Trekker program problematises existing 'digital labour' critiques by speaking to its *physical* elements to show how physical demands and related assumptions about class-based modes of production are largely absent from contemporary conversations about informational work. In the process, we demonstrate how Google Trekker constitutes a convergent, novel form of 'physical digital labour' whereby digital content is voluntarily produced via manual processes. Through this discussion, we consider how Google Maps and the Trekker project reconceptualise and commoditise space and place, including significant cultural and sometimes sacred sites, as consumable, *mediatized* tourist destinations.

The concept of mediatization is one that critically analyses 'the interrelation between changes in media and communications on the one hand, and changes in culture and society on the other' (Couldry & Hepp, 2013, p. 197). In developing the concept, Andreas Hepp

(2009) refers to how the 'moulding' forces of media can 'exert a certain 'pressure' on the way we communicate' (p. 47). Friedrich Krotz (2008) views mediatization as a 'metaprocess', i.e., one of a number of long-term processes such as globalisation and industrialisation that increasingly influence most aspects of not just our daily experiences but also social and cultural development (p. 23). Krotz (2007, p. 27) further argues that metaprocesses are conceptual constructs through which people try to make the world 'manageable'. When images collected through the Google Trekker project are incorporated into Google Maps, they render physical space and distant places in ways that makes them understandable and manageable by making them *consumable*. Trekker's embodiment of two intertwining metaprocesses—mediatization and commodification—reinforces Google Maps as a modern, digital representation of the ideological nature of maps and mapmaking. Google thus serves as a primary example of 'how corporations that control the flows of information and the infrastructure behind those flows now wield powerful global control' (Farman, 2010, p. 877).

Maps, data, and the Google 'empire'

We begin by challenging the common-sense assumption that maps are objective representations of reality. Lefebvre (1991) argues that space is not a product or thing—i.e., something that exists *a priori;* to assume otherwise is to assume that space is a given, rather than produced and practised. Instead, space is social, a 'set of relations between things (objects and products)' (p. 83). Maps are designed to serve specific functions, e.g., to resolve spatial problems or orient users in specific ways or for specific purposes. They reflect a specific aspect of 'reality' while erasing or misrepresenting others. The inherent bias of a map is therefore a product of its functionality. A cartographer makes decisions about what details to include, exclude, falsify, or exaggerate in order to make the map effective in communicating space for the user; these ends are often politically and ideologically motivated (Farman, 2010, p. 872). Representations of space and place thus construct and stabilise geographical knowledge, which has always been closely associated with power (Graham, 2014).

The history of cartography demonstrates how the production of maps has been politically and ideologically motivated. Map design has long served colonial expansion as Edney (1997) argued: 'Imperialism and mapmaking intersect in the most basic manner. Both are fundamentally concerned with territory and knowledge…The empire exists because it can be mapped, the meaning of empire is inscribed into each map' (cited in Farman, 2010, p. 870). The Mercator Projection map, for example, functioned not just for the purposes of nautical navigation but to also reinforce Europe's colonial domination by placing the continent as the central landmass from which all nautical activity radiated outward. The 1979 *McArthur's Universal Corrective Map of the World* challenges such Anglo-constructed norms of spatial orientation by reversing the southern and northern hemispheres to present an 'upside down' view of the world.

As these two examples illustrate, maps are 'unstable signifiers, heavily imbued with the cultural perspectives of the society that created them' (Farman, 2010, p. 874); they are not merely representations of space but also representations of the cultures and contexts responsible for their production. To think of maps as objective thus obfuscates the reality of their production, the contexts and cultural spaces they valorise, and who gets a say in what spaces are ultimately produced. This obfuscation is particularly true for computerised, digital systems. Robin Mansell (2012) argues that our reliance on such systems masks

the values and motivations of those who are designing the system. When these developments are seen as the outcomes of a complex self-organizing system, the assumption is that this is simply the result of an optimizing evolutionary (a 'natural selection' of the fittest) process.

(p. 116)

Mansell's (2012) point underscores the utopian promise of 'neogeography', the creation of spatial data and personalised mapping by non-professionals (Graham, 2010). Enabled by 'Web 2.0' interactive online technologies, neogeography aims to challenge many of these existing knowledge and value regimes. User-driven digital mapping opens up new possibilities for representing space, as 'everyday' people augment the material world with digital content to construct new geographies of knowledge with locally relevant information, particularly for underrepresented and disadvantaged places. In practice, however, global digital divides continue to cast uneven 'data shadows' over the world's economic peripheries (Graham, 2014).

Graham's (2014) empirical documentation of locally relevant and locally produced content indexed by Google Maps, Wikipedia, and other key information platforms demonstrates the continued dominance of North American and Western Europe in geographical knowledge production. Comparatively more geographical information is written about the Global North, while linguistic barriers also render user-generated information inaccessible to entire populations. In addition, the 'digital division of labour' is also significantly skewed; Africa, Asia, and South America, for instance, report significantly fewer content producers per capita than Western nations, suggesting inequalities of local voice and participation. These uneven data shadows and divisions of labour matter because '[a] lot of people and places are both literally and figuratively left off the map' (p. 114). These absences 'influence what we know and what we can know about the world', since the voice and representation of some people and places remain visible and more dominant than others (p. 114).

As 'virtual palimpsests' of place (Graham, 2010), neogeography intensifies mapping's spatial (as opposed to temporal) bias. For Innis (2007), space-biased media that communicate information over vast distances tend to favour the centralisation of power within societies. They also have a higher information capacity than time-biased media, which provide people with an 'artificially extended and verifiable memory' (p. 30). Space-biased media make it easier to copy and further disseminate information, which transforms information into a commodity. Information becomes 'mechanised', i.e., stored, processed, and disseminated via a complex system of technologies and tools (Innis, 1991, p. 190). Because space-biased media favour centralised structures, a select few are able to control these informational tools, resulting in what Innis (2007) referred to as a 'monopoly of knowledge'; those with authority over information tools have the ability to shape worldviews because they are in a position to define what representations of reality are 'legitimate'.

Thus, as Sharma points out: 'Civilizations that emphasise space over time tend to be imperial powers, involved in the conquering of space at the expense of the maintenance of culture over time...by all such determinations, global capital depends on spatially biased cultures' (cited by Farman, 2014, p. 84). This dynamic can be seen in the Google Trekker project. In its current form, the programme calls upon tourism boards, non-profits, universities, research organisations, or other third parties with access to unique 'off the grid' spaces and places to 'help map the world' ('Be the Next Trekker', 2015). Recorded locations integrated into Google Maps include sites like Mount Everest, the Grand Canyon, the Galapagos Islands, and Mount Fuji. The New South Wales (NSW) National Parks and Wildlife Service partnered with Trekker to record hiking trails and bushwalking routes for prospective tourists to explore, plan, and/or decide whether to visit from the comfort of their own homes ('Google Street View Trekker in NSW', 2014). The Trekker program has also photographed the interior and exteriors of global cultural sites including the Eiffel Tower, Burj Khalifa, and Taj Mahal.

These public–private partnerships highlight one way by which Google transforms nature into sites of consumption. No longer appreciated simply for their beauty, environmental uniqueness, or historical importance, these sites are presented as consumable destination experiences via revenue-generating tourism, naturalising the capitalist ideologies of both Google and tourist organisations. When a user clicks on a particular site, its commercial value is immediately evident: Trekker's user-generated imagery is integrated with Google Maps' consumer reviews, Google Flights, popular visit times, and recommendations of 'similar searches'. Space and place are, first and foremost, rearticulated as consumption opportunities. Google relies upon the presumed neutrality of computer-generated maps and the immediacy and presumed objectiveness of aerial and panoramic photographs obscure this social construction and commodification of space. The combination of cartography, photography, and 'trust' in Google reinforces the idea that Google presents space 'as it is', masking its ideological construction.

Though Google's global, deterritorialised form is fundamentally different from traditional notions of imperial power, its dominance over the digital economy is witnessed by its near monopoly over search engines (and thus access to information), web traffic, location services, ad revenue, and—via Maps—the production, representation, and valorisation of space. As Trekker enhances the value of Google's Street View maps, it enables Google to compete more aggressively in the map data market, which it now dominates with over one billion unique monthly users (compared to Apple Maps' 55 million). This translates to 67% of the navigation app market, far exceeding its closest competitors Waze (12%) and Apple (11%) (Panko, 2018). Google Maps' expansive integration into other websites, mobile applications, and platforms only further extends its dominance. Mapping data holds extensive commercial value for third parties in the form of targeted marketing and other forms of ubiquitous surveillance, often at the expense of user privacy (Andrejevic, 2007).

A discussion about the production of space and place in relation to mapmaking requires that we not only 'interrogate whose space we are talking about' but also whose labour goes into its production (Farman, 2014, p. 85). We now turn towards how Google's attempt to relate space and global capital (i.e., its spatial conquest) requires the participation of volunteer 'digital labourers' (i.e., users and public institutions) to help generate the content that enables the firm to establish a 'monopoly of knowledge'.

Space, maps, and physical digital labour

Google Trekker's user-generated production of non-tangible digital content is a form of value-producing labour. Programmes like Trekker function on the democratic promise that participatory digital media invites amateurs and professionals alike to circumvent corporate control of knowledge by actively participating in its creation. Terms like 'produsage' (Bruns, 2008) or 'prosumption' (Tapscott & Williams, 2006) speak to the collapse in traditional distinctions between 'producers', 'consumers', and 'users' in the contemporary digital economy. Critical digital media scholarship has argued that 'immaterial labour' captures Internet users into a set of social relations not unlike those Marx observed from the factory floor. For example, the free, voluntary labour of producers generates immense value for corporate firms, harnessed towards the ends of capital accumulation, often at the expense of user privacy (Andrejevic, 2007). This perspective has been challenged by other scholars (e.g., Baym & Burnett, 2009; Hesmondalgh, 2010) who insist the appropriation of immaterial labour is not exploitative in the same sense or argue that such activities empower users who seek remit in non-monetary forms of capital (e.g., social, cultural, human).

Fuchs (2014) explicates the politics of free labour in the digital economy by distinguishing work from labour, arguing 'digital *labour*' is a form of 'alienated digital work'. Conversely, 'digital *work*'

> makes use of the body, mind or machines or [some] combination…as an instrument of work in order to organize nature, resources extracted from nature, or culture and human experience, in such a way that digital media are produced and used…Digital work includes all activities that create use-values that are objectified in digital media technologies, contents and products generated by applying digital media.
>
> *(p. 352)*

By definition, digital labourers create content but do not own or control the means, tools, conditions, or results of production. It is thus the state of alienation from one's work that constitutes digital production as labour—alienation from the 'work itself, from instruments and objects of labour and from the products of labour' (p. 351). Fuchs (2014) also calls for the demystification of the material/immaterial split around digital labour, arguing that global ICT products and infrastructure are not immaterial but bound to a wide range of material forces and varying social relations of production. Describing the 'international division of digital labour (IDDL)', Fuchs points to the range of work and labour conditions obfuscated in the celebrations of unalienated work in the network society on one end, and post-material, post-industrial labour critiques on the other. In the digital economy, productive forces have evolved towards informational, rather than industrial or material forces; however, as Fuchs notes, 'The informational productive forces do not eliminate, but sublate (*aufheben*) other productive forces…In order for informational products to exist, a lot of physical production is needed. This includes agricultural production, mining and industrial production' (p. 10). Other examples include the slave-like conditions of mineral extraction in Africa; the industrialised labour of ICT manufacturing and assembly in China; call centre service workers in India; Silicon Valley's software engineers and the unpaid 'produsage' of digital media users/prosumers. In other words, what we consider to be 'informational goods' (like Google Maps) still require a wide range of labour—manual and digital, material and immaterial.

The Google Trekker project merges the digital and material in distinct ways. The highly physical nature of production becomes the responsibility of the Trekker participant who must carry (and care for) the Google Trekker backpack through remote environments and/or precarious places and lengthy distances. For example, staff from London's Canal and River Trust walked over 100 miles in a month to record some of the country's most famous waterways. Here, one can see the convergence of manual or physical labour and digital labour, what we refer to here as 'physical digital labour'; those carrying the Trekker pack are undergoing extensive physical exertion in order to generate digital media and content—the imagery and locative data used to populate Google's Street View maps. The Trekkers' physical labour produces a digital output for which they are not paid using tools they do not own and, thus, labour from which they are ultimately alienated.

The labour involved requires little skill or mastery of technical knowledge. The Trekker Pack's automated photo-taking features, for instance, free Trekkers from locating the 'perfect' shot. The user can also control the cameras from his or her Android phone (Google has an exclusive agreement with Android). While the machine does not necessarily control the worker, the worker is subject to the machine as the production of images 'point toward disembodiment, the dislocation of the subject, and objectivity' (Farman, 2010 p. 875). This automation also serves to obfuscate the production process and alienate the digital worker, i.e., the Trekker whose physicality enables the shot.

As cartographic science came to dominate the art of mapmaking, notions of authorship lost status; whereas hand-drawn maps were once associated with an individual creator or artist, mass produced and computer-generated maps

> are more commonly associated with the machinery that produces them than the person or organisation capturing or compiling them. This association between machine and product distances maps like Google Earth from a sense of subjectivity and instead emphasises the objective nature of photographic representations of Earth.
>
> *(Farman, 2010, p. 875)*

The presumed neutrality of Trekker's computer-generated maps renders falsified representations or political, ideological motivations invisible because their aerial, panoramic, or multi-dimensional qualities seemingly re-present the 'real'. Simultaneously, as labourer work in service of the machine, the production (and consumption) of images obfuscates the manual and digital labour invested in all Google maps.

We argue Trekker's function is to expand the commercial, experiential, and visual consumption of space, and reproduce Google's dominance and visibility in the global market. Trekker maps, like all maps, are constructed within particular social relations of production and within a particular political economy that reinforces Google's imperial dominance over the production and control of space—processes also inextricably tied to the valorisation of space.

The relationship between mapmaking and imperial power's concern with 'territory and knowledge' (Edney, 1997) plays out in the way Google uses its existing dominance to harness the productive labour of its users in the production of Trekker-view maps. Although physical digital labour is integral to the Google Trekker project, the programme is primarily reserved for non-profit organisations like researchers, community trusts, tourism boards, and universities—many of which are public and drastically under-funded institutions—that have ready access and intricate 'local' or specialised knowledge of exclusive, remote, or unique locations. In other words, Google enables the construction of 'knowledge' by granting exclusive spatial production to those with a vested interest in a specific re-presentation of that space (and thus the production of specific knowledges).

Partnerships with tourism boards particularly emphasise the project's commercial underpinnings as even nature or 'works' such as temples and monuments become 'beauty spots', sites where 'ravenous consumption picks over the last remnants of nature and of the past' (Lefebvre, 1991, p. 84). Local tourism boards and non-profits have been incentivised to develop panoramic interiorisations of important or unique cultural sites, making digital replications of global spaces virtually accessible for both commercial and cultural imperatives. Users and public institutions feel like they are a part of something by mapping places special to them, but in doing so assist in positioning Google as the *de facto* authority on what value we assign to space.

The Google Trekker project folds public institutions into a public–private partnership model that engages the labour of those institutions in the process; in doing so, it reinforces a wider set of ideological and political economic arrangements resonant with a neoliberal ethos of self-made entrepreneurship. For example, the Tourism Board in Victoria, Australia, partnered with Trekker in 2014 to map some of the region's most popular tourist attractions

(Roper, 2014). As part of the city's 'Play Melbourne' campaign, the public–private partnership followed its 2013 'Remote Control Tourists' project that allowed potential visitors to experience the city in real time by directing four camera-and-microphone outfitted 'tourists' to various locations they wanted to see—and consume—via Facebook and Twitter requests. The final Remote Control Tourist Google Map of Melbourne made available for public use at the campaign's end now features an online interactive city guide that only partially represents the restaurants, bars, cafes, and attractions originally crowdsourced from participants.

Google Trekker's lack of objectivity and political/ideological motivations are even more visible when looking at its Street View world map (see www.google.com/streetview/understand/), which outlines the geographic spaces the programmer has thus far mapped. That which is deemed 'mappable' (included) becomes just as relevant as that which is not (excluded). Notably, much of the developing world remains unmapped, alongside places (Northern Canada, Antarctica) where little commerce or tourism occurs (i.e., spaces with little commercial or consumer value). This mapping resonates with the larger Google Maps project that privileges the interiors and floor plans of businesses, universities, tourist destinations, malls, airports, and other commercial venues: that is, sites and spaces of consumption. Of course, some places refuse to be mapped, demonstrating that Google's spatial conquest has not gone uncontested. In 2011, officials in Bangalore denied Google's attempt to map the city with Street View car on the grounds of 'security concerns'; Google eventually convinced the Indian government to allow Trekker to photograph the country's cultural monuments (Anwer, 2014).

For Farman (2010), Google's desire is 'to map out a new territory: the digital empire' whereby intellectual property trades as the new currency (p. 876). The relationship with Google folds non-profit and tourist organisations into public–private partnerships that raise important questions about the extent to which these new representations of space and place are public or private goods. Despite the volunteer labour of users/participants, Google maintains all rights over collected imagery. The public organisation simply gains 'exposure' (or digital access, like everyone else) in exchange for performing physical digital labour. As such, organisations become dependent upon Google to access the 'intellectual property' they produced.

Commoditising cultural and spiritual sites

The re-presentation of cultural sites as tourist destinations seems an inevitable development, partly because they were always intended to act as such. However, much like the way natural sites are converted into spaces of and for consumption, Trekker reconceptualises symbolic and sacred sites in a similar way: positioned as products and destinations for consumption, they're separated from their original cultural contexts and spiritual and cultural meanings.

The Taj Mahal exemplifies this dynamic at play. As described by cultural geographer Tim Edensor (1998), '[t]he world's most famous tomb' was built on the banks of the River Jamuna in Agra, India as a mausoleum for Mumtaz Mahal, wife of Mughal emperor Shah Jahan, who died in childbirth in 1631 (p. 1). As a tomb or memorial site, visits to the Taj Mahal embody 'Thanatourism, or Dark Tourism, travel associated with death' (Seaton, 2002, p. 73), further imbuing the site with deep cultural and spiritual significance. These 'romantic' notions that allow the story to persist are arguably misrepresentations, however, as the Taj Mahal likely 'intended to symbolize [Shah Jahan's] own glory' (Begley, 1979, p. 10). Even so, the Taj Mahal's spiritual significance remains evident in its design and 'perhaps comes closer to evoking heavenly visions than any other work of Islamic architecture' (p. 12). The building contains a large number of Koran inscriptions that effectively constitute 'a thematically unified inscriptional program, analogous in its cognitive significance to the iconographic programs of religious monuments decorated with images' (p. 13). In short, whether considered a monument to the dead or representation of the Shah's power and glory, the Taj Mahal is a deeply sacred space.

However, Google Maps' presentation of this site is nearly devoid of such significance. Street View's brief textual interlude does include a passing mention of it being 'built over 350 years ago by a powerful ruler in honor of his beloved wife' ('Taj Mahal'). However, the rest of this introduction is dedicated to framing Google as the best way for 'the whole world' to learn about and 'experience' the Taj Mahal, even if they may never have the chance to get there:

> From the comfort of your computer, tablet or phone, take a 360-degree journey around the Taj Mahal as if you were walking around the building itself. Enjoy postcard views or discover rarely-seen perspectives; get up close to the hand-carved details and explore expansive landscapes.
>
> *(Google maps, Taj Mahal, 2018)*

Google rhetorically offers an objective and authentic experience while simultaneously asserting itself as the best—if not only—institution to offer that experience. The suggestion is that the Taj Mahal is best experienced through the mediatized tourist experience Google Maps offers — a confidence that stems from Google's perceived superiority, i.e., their monopoly of knowledge. However, the Taj Mahal is reduced to a series of images to be consumed: postcard views, hand-carved details, and landscapes. De-contextualised from its origin and local environs, this presentation separates (alienates) the Taj Mahal from its vast range of cultural, historical, or spiritual meanings.

The Taj Mahal's exploitation is not a new phenomenon. As Edensor (1998) notes, various entities have commodified the monument (or at least its name): e.g., Donald Trump's former Atlantic City casino, a five-star Bombay hotel, 'a blues singer and countless Indian restaurants throughout the world' (p. 1). The difference between these examples and Google Trekker is that none of the former stake any claim in presenting an 'objective' representation. Google, on the other hand, does make such a claim and, with it, reduces the Taj Mahal in its mediatization as a consumable 'virtual journey' stripped of cultural significance yet valued for its tourist appeal.

Conclusions

Local tourism boards and non-profits need to develop ways of monetising important or unique cultural sites in an era of systematic defunding. Trekker provides 'free' tools for struggling agencies to generate, re-brand, promote, and mediatise 'new' tourist experiences. However, by extending Marxist theories of digital labour to the production of the tourist experience, our analysis reveals the many invisible layers behind contemporary tourism's mediatization. Until recently, scholarship has generally overlooked the material *and* physical labour invested in digital content production. As our case study shows, digital physical labour lies at the centre of Google's contemporary reign over monopolies of knowledge and expansion of empire. Here, this expansion is in pursuit of commercial rather than public/government interests, which works to reinforce Google's dominant role within the global information economy. With a focus on mediating and mediatising sites of tourism and consumption, Google Trekker exemplifies the ideological nature of maps and mapmaking. It mediatizes tourism by rendering nature, locations, and monuments as spaces to be consumed, while masking that process by relying on the immediacy and perceived objectivity of computer-generated cartography and photographs to present itself as a 'real', disembodied representation of the world. In doing so, Google not only shapes physical movement, resource allocation (time, energy, capital), and cultural engagement in seemingly objective, non-ideological terms, while also shaping the ways labour, space, and place are engaged, experienced, and ultimately valued.

References

Andrejevic, M. (2007). *iSpy: Surveillance and power in the interactive era*. Lawrence: University of Kansas.

Anwer, J. (2014 February 23). Meet the Google trekker photographing Indian monuments. *The Times of India*. Retrieved March 9, 2019 from: http://timesofindia.indiatimes.com/tech/tech-news/hardware/Meet-the-Google-trekker-photographing-Indian-monuments/articleshow/30896468.cms.

Baym, N. K., & Burnett, R. (2009). Amateur experts International fan labour in Swedish independent music. *International Journal of Cultural Studies, 12*(5), 433–449.

"Be the Next Trekker" (2015). *Google Maps*. Retrieved September 29, 2015 from: http://www.google.co.nz/maps/about/partners/streetview/trekker/

Begley, W.E. (1979). The myth of the Taj Mahal and a new theory of its symbolic meaning. *The Art Bulletin, 61*(1), 7–37.

Bruns, A. (2008). *Blogs, Wikipedia, Second Life, and Beyond: From Production to Produsage*. New York, NY: Peter Lang Publishing.

Couldry, N., & Hepp, A. (2013). Conceptualizing mediatization: Contexts, traditions, arguments. *Communication Theory, 23*(3), 191–202.

Edensor, T. (1998). *Tourists at the Taj: Performance and meaning at a symbolic site*. New York: Routledge.

Edney, M. (1997). *Mapping an empire: The geographical construction of British India, 1765-1843*. Chicago: University of Chicago Press.

Farman, J. (2010). Mapping the digital empire: Google Earth and the process of postmodern cartography. *New Media & Society, 12*(6), 869–888.

Farman, J. (2014). Map interfaces and the production of locative media space. In R. Wilken & G. Goggin (Eds.), *Locative media* (pp. 83–93). Hoboken, NJ: Taylor and Francis.

Fuchs, C. (2014). *Digital labour and Karl Marx*. New York, NY: Taylor & Francis.

Graham, M. (2010). Neogeography and the palimpsests of place: Web 2.0 and the construction of a virtual earth. *Journal of Economic and Social Geography, 101*(4), 422–436.

Graham, M. (2014). Internet geographies: Data shadows and digital divisions of labour. In M. Graham & W. H. Dutton (Eds.), *Society and the Internet: How networks of information and communications are changing our lives* (pp. 99–116). Oxford, UK: Oxford University Press.

Hepp, A. (2009). Researching 'mediatized worlds': Non-mediacentric media and communication research as a challenge. In N. Carpentier, I. T. Trivundža, P. Pruulmann-Vengerfeldt, E. Sundin, T. Olsson, R. Kilborn, ..., & B. Cammaerts (Eds.), *Media and communication studies interventions and intersections* (pp. 37–48). Tartu, Estonia: Tartu University Press.

Hesmondhalgh, D. (2010). User-generated content, free labour and the cultural industries. *Ephemera: Theory & Politics in Organization, 10*(3/4), 267–284.

Innis, H. (1991 (1951)). *The bias of communication*. Toronto: University of Toronto Press.

Innis, H. (2007). *Empire and communications*. Toronto: Dundern Press.

Krotz, F. (2007). *Mediatisierung: Fallstudien zum Wandel von Kommunikation*. Wiesbaden, Germany: VS Verlag.

Krotz, F. (2008). Media connectivity: Concepts, conditions, and consequences. In A. Hepp, F. Krotz, S. Moores, & C. Winter (Eds.), *Connectivity, networks, and flows: Conceptualising contemporary communications* (pp. 13–31). Cresskill, NJ: Hampton Press.

Lefebvre, H. (1991). *The production of space* (D. Nicholson-Smith, Trans.). Cambridge, MA: Blackwell.

Mansell, R. (2012). *Imagining the internet: Communication, innovation and governanace*. London, UK: Oxford University Press.

Panko, R. (2018, 10 July). *The popularity of Google maps: Trends in navigation apps in 2018*. Retrieved from: https://themanifest.com/app-development/popularity-google-maps-trends-navigation-apps-2018

Roper, P. (2014, 19 August 2014). Tourism Victoria straps on Google's panoramic backpack camera to map popular destinations. Retrieved from: https://www.marketingmag.com.au/news-c/tourism-victoria-straps-on-googles-panoramic-backpack-camera-to-map-popular-destinations/.

Seaton, A. V. (2002). Thanatourism's final frontiers? Visits to cemeteries, churchyards and funerary sites as sacred and secular pilgrimage. *Tourism Recreation Research, 27*(2), 73–82.

Tapscott, D., & Williams, A. D. (2008). *Wikinomics: How mass collaboration changes everything*. New York: Penguin.

Terranova, T. (2000). Free labor: Producing culture for the digital economy. *Social Text, 18*(2), 33–58.

PART II

Mediatized places and spaces

10

FOOTBALL TOURISM AND THE SOUNDS OF TELEVISED MATCHES

Nicolai J. Graakjær and Rasmus Grøn

Introduction

In Scandinavia – representing the setting of the authors – live television from matches in the English top tier began in 1969 (Dahlén, 2008). During the 1970s and 1980s, the televised production was markedly refined, and the coverage obtained a prominent position through designated weekly programs on national, monopolised public service stations. Arguably, these programs have helped form a widespread awareness of English football and specific clubs. Since then, the TV coverage of the English (and other major) football league has proliferated domestically as well as globally, and media and football have formed a symbiotic partnership (Scherer & Rowe, 2014), where TV transmissions of football are not only a prerequisite for the leagues' business model, but a presumably important catalyst for the 'delocalisation' of sporting identification and loyalties (Scherer & Rowe, 2014). Thus, the scene has been set for *football tourism*, where the attendance of the spectacular of a professional football game is the sole or main reason for the journey. Based on the premise that media representations of events and destinations can create specific preconceptions and influence the inclination to physically experience similar events at these destinations (Jansson & Falkheimer, 2006; Urry & Larsen, 2011), this chapter will explore how auditory aspects of televised football can be considered to promote football tourism. The topic will be approached according to the authors' research backgrounds in the fields of media aesthetics and spatial experience design.

First, we will outline the current impact of football tourism in an English context, and its attachments to the football game as event as well as the attraction of the stadium as place. Second, a case study will explore how the 'in-stadium experience' (Rowe, 2014, p. 754) is mediated in TV transmissions, focusing specifically on the spectator sounds as a vital component for this experience. Hereafter, the mediation will be related to the live experience, and we discuss whether football tourism is reflecting and contributing to an impoverishment of the desired in-stadium experience, as this experience is becoming increasingly commodified and mediatized.

The rise of football tourism

Football tourism is a specific example of sports tourism, a field of growing significance and attention within research as well as policy (see e.g. Hinch & Higham, 2005). Heather Gibson provides a basic definition of sport tourism as '…a leisure-based travel that takes individuals temporarily outside of their home communities to play, watch physical activities or venerate attractions associated with these activities' (Gibson, 1998, p. 49). Hereby, Gibson distinguishes between *active sport tourism* (experience of participation), *event sport tourism* (experience of spectating), and *nostalgia sport tourism* (experience of veneration) (see also Houlihan & Malcolm, 2016; Kurtzman, 2005). Football tourism pertains primarily to the field of event sport tourism (Deery, Jago, & Fredline, 2004), and like most other forms of event tourism (ibid.), media coverage – here televised matches in particular – plays a pivotal role for the spread and continuance of football tourism.

Football has developed into a global media phenomenon exemplified most significantly by the English Premier League (EPL). In 2015, EPL was estimated to reach a global television audience of 730 million people across 185 countries (Rudkin & Scharma, 2017). Although notoriously difficult to determine due to a multiplicity of influencing factors, the virtually unlimited media access does not seem to have had a negative effect on the physical attendance to football games – quite the opposite (EPFL n. d.). However, it is not only the television coverage that is transcending borders, as the attendance to football games has also become a transnational phenomenon. In 2014, 800.000 foreign tourists visited England to watch games in the EPL (indicating a 50.000 rise since 2010), spending an estimated 684 million £ in the process, which was significantly more than the average tourist in Britain (Rudkin & Scharma, 2017; Visit Britain, 2015). Moreover, football tourism has positively influenced the temporal distribution of tourism, as it is relatively evenly spread throughout the year, mirroring the season of EPL spanning a period of ten months.

'Being there' and the importance of place

However, not only the event, but also *the place* of the event must be considered as a motive for football tourism. This place can of course be the city, or country, where the game is played, and many cities employ the presence of a famous football club as part of their destination branding (Gammelsaeter, 2017; Proctor, Dunne & Flanagan, 2018). However, 40% of interviewed football tourists in Britain state 'watching football' as the sole purpose of their generally brief stays (2.5 days in average; Visit Britain, 2015), and research has not shown any causal relation between the amount of football tourism and the general tourist appeal of the places in question. Instead, as implied in Gibson's (1998) category of Nostalgia Sport Tourism, the main place of interest and 'veneration' for the football tourist is the venue itself: the football stadium. A football club's stadium is a topoi permeated with historical aura for the club's fans, who often have a 'topophilic relationship' (Bale, 1993, p. 64) to the stadium as a nodal point of collective emotions and memories. At stadia, the nostalgia is often enforced by statues, memorials, museums, and guided stadium tours (Gammon & Fear, 2005). Concurrently, many stadia have in recent years evolved from utilitarian containers for sport events (and their spectators) into tourist attractions (Gammon & Fear, 2005; Giniesta, 2017). This is partly due to worldwide, strategic reconstructions of stadia into catalysts for tourist and destination development since the turn of the millennium (King, 2010; Spirou, 2010). These stadia share a strong focus on service, comfort, and security in order to attract affluent customers – not least tourists. However, the tourist attraction of the football stadium it also closely related to the pervasive mediation of football, as the global media exposure makes these venues more commonly known. Thus, football stadia are also *mediatized places,* whose auratic appeal for tourists, including many 'topophilic' fans, is solely based on mediated impressions and experiences. In the forthcoming analysis, however, the stadium will not be studied as an attraction in itself but viewed as the significant context for the experience of the game as a tourist event.

The appeal of the physical in-stadium experience of attending a football game is commonly linked to the state of 'being there' (Weed, 2010, p. 105) as a witness to the unpredictable drama in its full kinaesthetic and emotional intensity (Schäfer & Roose, 2010), which subsequently can provide the 'witness' with a substantial social capital. Ideally, the spectator participates in a strongly atmospheric and social experience (Edensor, 2015), 'liminally' (Turner, 1988) demarcated from the mundane everyday life, in strong co-presence and rhythmic interaction with the fellow spectators. How, then, does the televised version of a football game offer spectators in front of the screen a sense of the atmosphere of 'being there', and how does this offer compare to actual attendance? In exploring this question, we will focus on the *aesthetics of sounds* of televised football with a particular interest for the spectators' role. Attending a football game is a multisensorial intense social experience. Not only visually (looking at and reflecting oneself in the surrounding crowd) and tactilely (by the close proximity of co-spectators), but first and foremost, auditorily, as the sound of the crowd is an indispensable condition for the fullness of the game as an atmospheric, social, and sensory event: 'Without sound, the stadium is empty' (Gaffney & Bale, 2004, p. 31; see also Charleston, 2008; Nishio, Larke, Heerde & Melnyk, 2016). The importance of spectator sound also goes for the televised presentation of football, where 'the crowd's roars, chants and cheers are crucial to the experience and atmosphere of the game' (Lury, 2005, p. 82; see also Kennedy & Hills, 2009, p. 58). However, so far, studies on televised football have focused on other aspects of program aesthetics – that is, the visuals and the verbal dimensions of the commentary – and the appeal of the spectators' sounds have not been specified.

Sounding out televised football

The following analysis is based on a case – the EPL match between Tottenham Hotspur and Leicester City played on the 29th of October, 2016, and shown on Discovery Networks Denmark's channel '6'eren'. The case has been selected because the authors were able to attend and make observations at the stadium as well as to receive a copy of the televised match for subsequent textual analysis. The analysis focuses first on the sounds from the televised match which is then compared to the sounds from the actual venue. We suggest that the case can be considered prototypical and hence paradigmatic (Flyvbjerg, 2006) of televised matches from the EPL, as it represents a so-called *clean feed* produced by Premier League Production and offered by Sky Sports and BT Sport to '6'eren' among other channels throughout the world (Milne, 2016). Thus, apart from the local commentary, all other audio-visual elements are offered globally. This already hints at the importance of spectator sounds, as they are an integral, universal part of the soundscape at the venue. This part can be further fleshed out by analysing, first, the placement of the spectator sounds during the program, and, second, their functions, that is, what they can be considered to *do* (Rodman, 2010, p. 81) in terms of the appeal of televised football.

From the perspective of their *placement*, there is no part or sequence of the televised broadcast that do not include spectator sounds during the match. The sounds appear uninterruptedly, and there are no auditory lacunae or pauses. By comparison, commentary, although also a continuing feature of the broadcast, is filled with (mostly short-lasting) pauses. Moreover, the spectator sounds vary with respect to their salience and forcefulness. Basic sounds include scattered whistling, shouting, and clapping, which primarily manifest the spectators' immediate reactions and attitudes toward developments during the match. Often, these sounds are substituted or overpowered by even more pronounced and coordinated sound events when groups of spectators unite in prolonged rhythmic clapping and melodic chanting.

With regard to the spectator sounds' relation to the visuals, they are diegetic in the sense that they originate from 'the "story world" of the narrative text' (Rodman, 2010, p. 53). Moreover, whereas the match is represented from numerous specific points of view, the sounds of the match are characterised by one, unspecific point of audition. The latter might be termed a 'zone of audition' (inspired by Chion, 1994, p. 90) to indicate not only the omnidirectional nature of sound perception but also the composite quality of the soundscape compared to the specificity and mono-directionality of the individual visual points of view.

The composite organisation of the venue sounds reflects the simultaneous operation of multiple microphones throughout the venue (Wittek, 2013, n. p.). The parallel employment of multiple microphones offers a composite soundscape that do not exactly correspond to the sounds the viewer would be able to hear if he/she was actually positioned as suggested by the camera. Performer sounds picked up from the so-called shotgun microphones (Wittek, 2013) placed close to the field are, for example, not perceivable from most points of views offered by the camera. Also, the soundscape is not affected by neither the frequent cuts from one camera position to another (representing a wide range of angles and distances) nor the occasional interruption of live visuals when sequences of replay in slow motion are shown shorn of the sounds that used to accompany them.

The functions of spectator sounds include two overlapping dimensions. First, the constant presence indicates that the sounds help establish a sense of *continuity* and granted that the achievement and maintenance of continuity is 'a fundamental aspect of the management of liveness' (Scannell, 2014, p. 158), spectator sounds play an important role in this respect as well. Thus, spectator sounds can be considered as 'lived' or acted out from the perspective of the attendees and, in turn, as a constituent of 'liveness' from the perspective of the television viewer: 'The liveness of broadcast coverage is the key to its impact, since it offers the real sense of access to an event in its moment-by-moment unfolding' (Scannell, 1996, p. 84).

Second, the sounds evoke dimensions of *atmosphere*. The sounds generally indicate that the stadium is almost filled with enthusiastic spectators (the attendance at the case match was 31.868 equalling 97% of the stadium capacity), which in turn indicates to television viewers that the event is attractive, important, and well worth of the engagement (cf. Gaffney & Bale, 2004 above). The chanting furthermore adds to these impressions as it highlights a sense of community and common identity among (sections of the) spectators and hence a sense of history and tradition. For example, the chant 'Oh when the Spurs (go marching in)', that has long functioned as the signature anthem for Tottenham emerges on several occasions during the game. The power and significance of the spectators' sounds are indeed illustrated right before the kick-off, where one minute of silence is held to honour fallen soldiers as part of the annual *Remembrance Sunday*. The absence of spectator sound during this short period of time presents a contrast and an 'ear-opener' to the sound's magnitude the moment it returns. Furthermore, as a specific aspect of providing a sense of engagement and atmosphere, the spectator sounds function as a sort of an *auditory seismograph* that constantly monitors developments of the game. To television viewers, the regular deployment of visual slow motion means that the spectators' sounds remain the primary connection to the venue. For example, approximately half way through the first half of the case match, the visuals follow the home team creating a chance accompanied by an increase in spectator shouts and cheering. After the ball flies close past the goal of the opponents, the spectators forcefully clap and cheer. However, seconds later, as the visuals run a slow-motion replay of the chance, the spectators stop clapping and begin to boo and shout in negative terms. Here, the spectators are reacting to the opponents' goalkeeper trying to 'steal time' – an incident not shown on television but seen at the venue.

While the functions of spectator sounds are thus pivotal to the experience of football on television, there are of course some significant differences compared to actually attending the match at the venue. Basically, the transmission cannot transmit the corporeal and atmospheric experience of your own physical presence, and, not less important, 'the sheer physical presence, in every way, of the crowd' (Scannell, 2014, p. 174; see above). In terms of the crowd's auditory presence, i.e. spectator sounds, the possible verbal aspects of chanting emerge more clearly at the venue. Second, television viewers are not positioned to appreciate the 'territorial' and cultural dimension of spectator sounds, as different groups of supporters (home and away supporters, respectively) continuously combat to dominate the soundscape and to be able to define, defend, and defy social identities (Hognestad, 2012). Third, and related thereto, television viewers are not able to appreciate with much nuance (as the televisual use of multiple microphones tends to level out the sound quality) the venue as an example of 'aural architecture' that allows for a specific quality of soundscape to emerge. Fourth, and arguably most significantly, television viewers are detached in a distanced spectator position, as they are not able to feel immersed and participate in the production of sound (rhythms and chants). That being said, the analysis has indicated that the mediated spectator sounds not only play an important role for the experience of televised football. Also, due to their suggestion of atmosphere and 'liveness', they can be considered to promote an impression and expectation of the atmosphere related to the particular club and venue.

Discussion

Obviously, (the sounds of) televised football does not represent the only media contribution to the mediation of football or the potential enticement of football tourism.

First, the issue of football tourists' motivations is 'monumentally troublesome' (Smith & Stewart, 2007, p. 175), as these motivations are influenced by a multi-variable complex of cultural, social, psychological, economic, and contextual factors. Furthermore, motivation varies according to the individual's fan identity, i.e. the degree of attachment to the 'sport object'. Here, Smith and Stewart points to the variables of attraction (the degree of pleasure and interest in fan activities), centrality (the activity's importance to the person's life style), and identification (the degree of association between the sport object and the person's self-concept) (Smith & Stewart, 2007, p. 167ff; see also Giulianotti, 2002). However, we suggest that among the factors relevant to the potential promotional impact on potential tourists, the televised spectator sounds offer a distinct contribution.

Second, televised football has indeed become but one part of an extensive and multi-facetted landscape of media consumption that includes fan sites, blogs, chatrooms, social media platforms, etc., positing mediated sports in a 'multimodal, multichannel, and multi-platform environment' (Hutchins & Rowe, 2012, p. 2). The development of digital media grants the user a more participatory role in commenting, organising, distributing, and producing media content (Månsson, 2011, p. 1639). For example, Tottenham Hotspurs' official Danish fan website (www.spursk.dk) hosts, among other things, a range of chatrooms for fans to create consumer-to-consumer networks in order to practice 'exchange of know-how' and 'co-production of experiences' (Månsson, 2011, p. 1639). Furthermore, some fans turn into producers by uploading videos where they comment on their teams' performances and/ or display live footage from matches. While this footage is mostly of a rather raw quality, some fans make citizen sport journalism that approximates the technical and aesthetic level of professional productions (see e.g. Tottenham Fan Chris Cowlin, 2013; Månsson, 2011, p. 1645). However, regardless of the importance of new, social media in influencing the experience of football and the practices of football fandom, televised football matches must still be considered central: 'The TV screen is not going away and will remain at the centre of the networked media sporting environment for some time yet' (Boyle, 2014, p. 750; see also Scherer & Rowe, 2014). Today, the digitalisation of TV consumption has compromised the hegemony of the TV screen in favour of other devices. However, this development from 'broadcast to broadband' (Hutchins & Rowe, 2012, p. 5) has only enforced the pole position of the televised football match as media product, which must still be considered as the main construction of 'the football game' as media experience and cultural phenomenon.

Presently, football is entwined in the pervasive mediatization of late modern society (Månsson, 2011), interrelating live and mediated football experiences in a multitude of ways. Here, research points to a profound impact of mediatization, as the spectators' approach and expectations of the in-stadium experience are increasingly shaped by the conventions of mediation (Hutchins & Rowe, 2012; Turner, 2013). This is further enforced by the new stadium designs (see above), where the focus on service and security leads to a 'sanitisation' (Turner, 2013, p. 86) of football grounds that impoverishes the somatic and social qualities of the in-stadium experience and encourages a more passive 'armchair fan culture' of attendance mimicking the distanced comfort of the TV lounge's 'cinematic consumer experience' (Turner, 2013, p. 88).

Consequently, the imperative of 'being there' are currently being modified by new popular 'middle regions' between the couch and the stadium such a sport bars and public live screening events, where broadcasted games are displayed on big screens in social environments (Weed, 2007, 2010). These venues can be regarded as (less expensive) alternatives to football tourism as they facilitate 'imagined journeys' by simulating 'the pleasures and benefits of such physical proximity' (Weed, 2010, p. 105). But they have also augmented the scope of football tourism, as it has become common place during big tournaments to invite fans without tickets to indulge in festival zones (Frew & McGillivray, 2008), where live football games are screened in staged environments of festivity and consumption. And while these events represent a mediatization of football tourism, they also allow for more performative and socially aroused expressions of fandom than the 'sanitised' stadia.

However, considering the rising numbers of football tourists, this tendency should be conceived as an extension of the football tourism that does not substitute but supplements the in-stadium experience. Furthermore, while the critique seems reasonable with respect to the visual mediation of the match at the venue of newly (re)constructed stadia, it neglects the role played by the sounds at the venue. Televised visuals arguably offer viewers a highly privileged access (e.g., close ups and slow-motion replays) to almost every relevant detail of the match, whereas the experience of actually being at the venue imply limited visual access to match details. Modern stadia, however, increasingly include big video screens to remedy this limitation (Whannel, 2014), thereby enforcing the 'armchair culture', as 'the rhetoric of mediatisation embedded in such devices as the instant replay, the simulcast and the close up, are now constitutive of the live event itself' (Turner, 2013, p. 88). By contrast, the experiences of the sounds at the venue offer something else and hereunder a significant add-on to the experience of the sounds of the televised match. This is not only because spectators are immersed in a social situation and are potentially co-producers of sounds and atmospheres, which is what gatherings at sport bars and public live screening events can likewise accommodate.

Spectators can also engage in auditory rivalries (between groups of spectators) while being corporeally immersed in the 'aural architecture' of the stadium and thus being able to 'listen to the place' neither of which is obtainable at sport bars and public live screening events. However, while co-production of atmospheres is surely a potential of being there, an atmospheric draining of English football stadiums, sometimes to the point of 'anaesthesia' (Gaffney & Bale, 2004, p. 25), has long been widely observed (see also Charleston, 2008; Turner, 2013). The football tourist potentially contributes to this development by embodying a lurking outsider detached from the cultural context in which the game is embedded. Thus, in the words of a devoted supporter of the football club Chelsea: '[T]here are more tourists, it's an experience and they're there to capture it on their iPad rather than interact' (Gibson, 2013). The tourist spectator attends and (hopefully) revels in the spectacular but does not necessarily and easily contribute to the co-creation of the experience. Football tourists in the form of (global) flaneurs do not easily blend and interact with (local) football supporters (Giulianotti, 2002). Following a well-known and paradoxical logic of tourism, the mere prevalence of football tourists potentially impoverishes the experience that was likely to has been the original and prime aim of attending.

Conclusions

In this chapter, we have identified football tourism as a significant and growing type of sports tourism, which is stimulated and influenced by the intertwined relationship between tourism and media. Hereafter, we focused specifically on the sonic dimension of the football event by comparing the live and TV experience of a football game from the EPL. From the perspective of the theme of mediatized places and spaces, our study revealed an ambiguous relationship between mediatization and place. First, there is no denying that the bodily, situated experience at the venue is something quite different from watching the match from the armchair. Actual attendance offers tourists the opportunity to co-produce atmospheres and obtain a sonic impression of the particular, spectacular venue. Second, the case analysis supported our initial hypothesis on the media representation as a potential catalyst of live attendance (i.e. football tourism), as the (production of the) spectator sounds of the televised match offer an impression and expectation of the in-stadium experience, as it provides a sense of atmosphere and 'liveness' by simulating the immediacy of 'being there'. However, the pervasive – most significantly, visually based – mediatization of the 'football event' contributes to an approximation of the in-stadium experience to the media construction of 'the football game' and its affiliated codes of behaviour. Thus, the attendance of football games, that are being played in increasingly sterile, restrictive stadium environments, is in danger of being drained of its unique atmospheric qualities and brought to resemble the comfortable, but passive 'armchair position' of the TV consumer. Paradoxically, the football tourist potentially partakes in this development as an affluent outsider and flaneur.

References

Bale, J. (1993). *Sport, space, and the city*. London, UK: Routledge.
Boyle, R. (2014). Television sport in the age of screens and content. *Television & New Media, 15*(8), 746–751.
Charleston, S. (2008). Determinants of home atmosphere in English soccer: A committed supporter perspective. *Journal of Sport Behavior, 31*(4), 312–328.
Chion, M. (1994). *Audio-vision. Sound on screen*. New York, NY: Columbia University Press.
Dahlén, P. (2008). *Sport och medier. En introduktion* [Sport and media. An introduction]. Kristiansand, Norway: IJ-forlaget.
Deery, M., Jago, L., & Fredline, L. (2004) Sport tourism or event tourism: Are they one and the same? *Journal of Sport & Tourism, 9*(3), 235–245.

Edensor, T. (2015). Producing atmospheres at the match: Fan cultures, commercialization and mood management in English soccer. *Emotion, Space and Society, 15*, 82–89.

EPFL. (n.d.). Average stadium utilization of professional soccer matches of The Premier League in England from 2010 to 2017. In *Statista – The Statistics Portal*. Retrieved 1 November 2018 from: https://www.statista.com/statistics/799770/the-premier-league-stadium-utilization-england/.

Flyvbjerg, B. (2006). Five misunderstandings about case-study research. *Qualitative Inquiry, 12*(2), 219–245.

Frew. M., & McGillivray, D. (2008). Exploring hyper-experiences: Performing the fan at Germany 2006. *Journal of Sport & Tourism, 13*(3), 181–198.

Gaffney, C., & Bale, J. (2004). Sensing the stadium. In P. Vertinsky & J. Bale (Eds.), *Sites of sport: Space, place, experience* (pp. 25–39). London, UK: Routledge.

Gammelsaeter, H. (2017). Media visibility and place reputation: Does sport make a difference? *Journal of Place Management and Development, 10*(3), 288–298.

Gammon, S., & Fear, V. (2005) Stadia tours and the power of backstage. *Journal of Sport Tourism, 10*(4), 243–252.

Gibson, H. (1998). Sport tourism: A critical analysis of research. *Sport Management Review, 1*, 45–76.

Gibson, O. (2013, 16 November). Atmosphere and fans' role in Premier League games becoming a concern. *The Guardian*. Retrieved November 1, 2018, from: https://www.theguardian.com/football/2013/nov/16/premier-league-fans-atmosphere-concern.

Giniesta, X. (2017). The business of stadia: Maximizing the use of Spanish soccer venues. *Tourism and Hospitality Research, 17*(4), 411–423.

Giulianotti, R. (2002). Supporters, followers, fans, and flaneurs: A taxonomy of spectator identities in football. *Journal of Sport and Social Issues, 26*(1), 25–46.

Hinch, T., & Higham, J. (2005). Sport, tourism and authenticity. *European Sport Management Quarterly, 5*(3), 243–256.

Houlihan, B., & Malcolm, D. (2016). *Sport and society* (3rd Ed.) Los Angeles, CA: Sage.

Hognestad, H. K. (2012). What is a soccer Fan? In R. Krøvel & T. Roksvold (Eds.), *We love to hate each other. Mediated Soccer fan culture* (pp. 25–44). Göteborg, Sweden: Nordicom.

Hutchins, B., & D. Rowe (2012). *Sport beyond television: The Internet, digital media, and the rise of networked media sport*. New York, NY: Routledge.

Jansson, A., & Falkheimer, J. (2006). Towards a geography of communication. In J. Falkheimer & A. Jansson (Eds.), *Geographies of communication. The spatial turn in media studies* (pp. 9–25). Gothenburg, Sweden: Nordicom.

Kennedy, E., & Hills, L. (2009). *Sport, media and society*. Oxford, UK: Berg.

King, A. (2010). The new European stadium. In S. Frank & S. Steets (Eds.), *Stadium worlds. Soccer, space and built environment* (pp. 19–35). New York, NY: Routledge.

Kurtzman, J. (2005). Sports tourism categories. *Journal of Sport Tourism, 10*(1), 15–20.

Lury, K. (2005). *Interpreting television*. London, UK: Bloomsbury Publishing.

Månsson, M. (2011). Mediatized tourism. *Annals of Tourism Research, 38*(4), 1634–1652.

Milne, M. (2016). *The transformation of television sport. New methods, new rules*. London, UK: Palgrave.

Nishio, T., Larke, R., Heerde, H., & Melnyk, V. (2016) Analysing the motivations of Japanese international sports-fan tourists. *European Sport Management Quarterly, 16*(4), 487–501.

Proctor, D., Dunne, G., & Flanagan. S. (2018). In league? Destination marketing organisations and soccer clubs in the virtual space. *Journal of Sports & Tourism*. doi: 10.1080/14775085.2018.1523744.

Rodman, R. (2010). *Tuning in. American narrative television music*. New York, NY: Oxford University Press.

Rowe, D. (2014). New screen action and its memories: The 'live' performance of mediated sport fandom. *Television & New Media, 15*(8), 752–759.

Rudkin, S., & Scharma, A. (2017). The impact of *soccer* attendance on tourist expenditures for the United Kingdom. *Munich Personal RePEc Archive*. Retrieved November 1, 2018, from: https://mpra.ub.uni-muenchen.de/81427/2/MPRA_paper_81427.pdf.

Scannell, P. (1996). *Radio, television and modern Life*. Oxford, UK: Blackwell Publishers.

Scannell, P. (2014). *Television and the meaning of live*. Cambridge, UK: Polity Press.

Schäfer, M. S., & Roose, J. (2010) Emotions in sports stadia. In S. Frank & S. Steets (Eds.), *Stadium worlds. Football, space and built environment* (pp. 229–244). New York, NY: Routledge.

Smith, A.C.T., & Stewart, B. (2007). The travelling fan: Understanding the mechanisms of sport fan consumption in a sport tourism setting. *Journal of Sport & Tourism, 12*(3–4), 155–181.

Spirou, C. (2010). Cultural policy and the dynamics of stadium development. *Sport in Society, 13*(10), 1423–1437.

Tottenham Fan Chris Cowlin (2013). Youtube. Retrieved November 25, 2018, from: https://www.youtube.com/user/cowlinchristopher/about?disable_polymer=1

Turner, M. (2013). Modern 'live' soccer: Moving from the panoptican gaze to the performative, virtual and carnivalesque. *Sport in Society, 16*(1), 85–93.

Turner, V. (1988). *The anthropology of performance.* New York, NY: PAJ Publications.

Urry, J., & Larsen, J. (2011). *The tourist gaze 3.0.* London, UK: Sage.

Visit Britain (2015). *Soccer tourism scores for England.* Retrieved November 1, 2018, from: https://www.visitbritain.org/sites/default/files/vb-corporate/Documents-Library/documents/2015-9%20VisitBritain%20Report_Football%20tourism%20scores%20for%20Britain.pdf.

Weed, M. (2007). The pub as a virtual soccer fandom venue: An alternative to 'being there'? *Soccer & Society, 8* (2–3), 399–414.

Weed, M. (2010). Sport fans and travel – Is 'being there' always important? *Journal of Sport & Tourism, 15*(2), 103–109.

Whannel, G. (2014). The paradoxical character of live television sport in the twenty-first century. *Television & New Media, 15*(8), 769–776.

Wittek, H. (2013, 26 May). Microphone usage for sports broadcasting. *Live Production.* Retrieved November 1, 2018, from: http://www.live-production.tv/casestudies/production-facilities/microphone-usage-sports-broadcasting.html.

11

POP IDOLS, MEDIATIZED PLACES, AND IDENTITY-ORIENTED PERFORMANCES OF FANS AS DOMESTIC TOURISTS IN JAPAN

Yunuen Ysela Mandujano-Salazar

Introduction

Popular media contents, through their representation of space, create mediatized places that have the potential to influence the meanings people give to them and, by means of familiarization and association with specific stories or characters, impact how they perform as tourists in such contexts (Edensor, 2001; Månsson, 2015). Edensor (2001, p. 64), differencing between enclavic and heterogeneous tourist spaces, notices that in the former activities to be performed are mostly fixed, while in the latter, "transitional identities may be performed alongside the everyday enactions of residents, passers-by and workers." He also proposes that, in contrast with tourists who present direct performances—this is, they behave as expected—when performances are identity-oriented, actions, gestures, and props are used to symbolize and communicate such identity, distinguishing these tourists from others.

This chapter follows these postulates and it is positioned within the Cultural Studies. Its aim is to discuss the ways in which the media activities of the Japanese idol group Arashi and their explicit and implicit promotion of services and products anchored to specific spaces in Japan have re-signified certain places for audience. In particular, it intends to show that a special segment of domestic tourists is created among audience and these people, who are influenced by the group's media representations, perform in different ways than visitors who are not. To achieve these objectives, a qualitative methodological approach was designed. First, documental analysis was used to understand the insertion of Arashi in diverse policies implemented by the Japanese government and private corporations aiming at the promotion of national tourism. Then, through the interpretative textual analysis of a comprehensive sample of official and authorized media contents focused on the group, the dominant representations and narratives related to products and services anchored to specific places are retrieved to find the dominant messages that promote domestic tourism. Finally, drawing on fieldwork performed for a total of 17 months between 2012 and 2018 predominantly in Tokyo and Osaka, but including short visits to other regions of Japan, I discuss the impact of those media representations on how audience and fans make sense of and perform in the places mediatized by Arashi's contents.

Understanding the role of idols as social influencers in contemporary Japan

The relevance of media personalities in contemporary societies cannot be overlooked. They circulate and are consumed in such an avid way that they become one of the most powerful tools by which people make sense of the world these days (Marshall, 1997). In Cultural Studies, a celebrity is understood as a text located in the meta-reality—a midpoint between reality and fiction—which is read and interpreted by audience according to the cultural background and historical context in which is read. Although not necessarily all members of audience—even within the same cultural and historical background—interpret the text in the same way, there is a preferred or dominant meaning that is extracted by most people who are exposed at the same time in the same context (Hall, 1980, 1997). Thus, to extract the dominant meanings embodied by a media personality, it becomes essential to consider the intertextuality of the whole repertoire of imagery and narratives related to that text, as well as its intersection with other texts and with the culturally and historically situated reader. It is in this intertextuality where the symbolism of a celebrity is activated and acquires all its discursive potential. Because of this, the following paragraphs explain in a succinct manner the most relevant elements surrounding Arashi as a text.

In Japanese media, pop idols are omnipresent. They perform a wide range of roles which depend less on their actual talents than on the demand from audience to see more of them. The essential symbolism that distinguishes idols from other media personalities is their energetic, clean, healthy, honest, and socially proper image, as well as the approachability and the familiarity that audience develops towards them through the countless details about their private life which they candidly share with audience by means of all contents they appear in (Aoyagi, 2005; Darling-Wolf, 2004; Sakai, 2003). These symbolic elements are embedded in all Japanese idols and allow them to "attract people and perform as lifestyle role models (...) [informing] their viewers about appearances and personal qualities that are considered socially appropriate and trendy" (Aoyagi, 2005, p. 3).

Arashi is an all-male idol group that made its official debut in 1999, when the five members were still in their teenage years. Noteworthy symbolic characteristics related to the members are their attitudes and social interactions displayed in media, which are congruent with Japanese traditional vertical and group-oriented society and represent hegemonic masculinity values, placing great importance on their social roles, on loyalty and devotion towards their groups and company, and on the display of their continuous effort to comply with social expectations (Mandujano-Salazar, 2014).

Arashi's media activities expanded consistently since its debut to include musical releases, tours, presentations, and hosting of music and variety television and radio shows; acting in television dramas, movies, and stages; regular appearances in idol magazines; and the endorsement of a wide variety of products and services from the national industry. By the end of 2008, the group was at the top of the music sales rankings and domestic media began to refer to it as "national idol group" (Mandujano, 2013). In April 2010, the group's national representativeness became officially acknowledged. The Japan Tourism Agency (JTA)—a bureau part of the Ministry of Land Infrastructure Transport and Tourism (MLIT)—designated the group and its members as Tourism-oriented Nation Navigators, a title that implied their new and voluntary role as ambassadors for the promotion of domestic and inbound tourism (Kankōchō, 2010a). Following the words used by the JTA in official press releases, Arashi began to be called "face of Japan" by domestic media.

Soon, the group turned into a social phenomenon inside the country, increasing its influence among a wide range of population strata. Since then, their media appearances have occupied primetime spaces and have diversified to attract wide audiences, reaching most social sectors. Arashi became particularly influential among children and teenagers, women in general, and young families (Mandujano-Salazar, 2014). Since the group began to be linked to several commercial and non-commercial campaigns relying on the notions of Japan and the national in their narratives, Japanese government, organizations, and corporations have been capitalizing on the symbolic status of the group to appeal to the national sentiment of people. Although Arashi is popular also in other Asian countries and its appointment intended to attract foreign tourists to Japan, the popularity of the group and its influence among Japanese people made it particularly relevant for the promotion of domestic tourism through the mediatization of places.

Idols and the mediatization of places amid policies for tourism promotion and branding of the national

In order to illustrate how places are being mediatized and promoted through idols' media contents in Japan, a few examples will be presented stressing the cooperation among governmental, corporative, and media entities.

In June 2010, just after Arashi's designation by the JTA, the Japanese government, through the Ministry of Economy Trade and Industry (METI), established the Office of Cool Japan to implement soft-power strategies aimed at the revitalization of national economy through the promotion of cultural products and services—creative industries— inside and outside the country (Keizai Sangyōshō, 2010). Domestic business elites were relying on the appeal of national culture to increase the value of Japan as a brand and attract domestic and foreign clients and consumers. On this aim, five cultural industries were acknowledged as crucial for those policies: skilled manufacturing and regional specialties, media contents, fashion, food, and tourism (Ministry of Economy, Trade and Industry, 2012). Thus, the strategies of Cool Japan to propagate the branding of national cultural products and services required the coordination of the METI with other ministries, particularly the MLIT, not only because one of the creative industries was tourism itself, but also because the others were interrelated with it.

Accordingly, official strategies to promote domestic tourism by branding national culture were strengthened. For example, in September 2010, the JTA published a photobook that distributed to the libraries of public elementary, middle, and high schools in Japan with the stated objective of inspiring among young Japanese the love for their country and the desire to know more about it (Kankōchō, 2010b). The book's title was *Nippon no Arashi* (Arashi, 2011). It contained essays written by the members of the idol group, transcripts of conversations they had with regular people, artists, and craftsmen of different regions of the country, and, most importantly, pictures of them visiting and admiring distinctive places and experiencing diverse local cuisine, services, and handcrafts. Throughout the book, the group expressed opinions and expectations on Japan and Japanese culture in general, highlighting qualities of the national identity that are supposedly present or represented in every place, cultural product or service. Except for the covers, the multiple photographs in the book are candid-looking shots focused on the interaction of the members with the people they visited and their enjoyment of the diverse places. Looking through the book gives the sense of going through a personal photo album. Nevertheless, it gives very detailed information on the places visited by the members, making it easy, for everyone interested, to go there; thus, turning them into mediatized places that, for those who read the book, had a symbolism related to the narratives found in it and to Arashi.

At the same time, Japanese corporations, which have a long tradition of collaborating with the government to attain better results in their own industries, were also using the advancement of this branding to commodify the notion of the national. In the context of the implementation of Cool Japan policies and the ongoing promotion of domestic tourism, many companies began employing narratives focused on Japan and the exaltation of the national in their public activities. In their role as "face of Japan," Arashi was increasingly looked after to endorse places, products, and services and relate them to the discourse of national representativeness that the group embodied.

One relevant example is the case of Japan Airlines (JAL), the flag carrier of the country, which had to implement an aggressive restructuration plan to avoid bankruptcy in 2010. Among diverse measures, a renewed marketing strategy was implemented by JAL to rebuild its domestic clientele. For this, building on the image of the group as the Tourism-oriented Nation Navigators, the corporation signed Arashi to be the face of its national advertising campaign. The press release from JAL stated that the group had been chosen to be the image of the airline because it shared the purpose of the company of sending a positive and joyful message to Japan (Japan Airlines, 2010). Since then, the group and members' image began to be used not only in commercial contents, but also, for limited periods, on the exterior surface of some of the planes—so-called *JAL Arashi JET*—that serve domestic destinations; also, music of the group has been used in television commercial spots and as part of the basic repertoire inside those planes (Japan Airlines, 2010, 2011, 2012, 2015) (Figure 11.1).[1] This

Figure 11.1   Above: JAL Arashi JET at Haneda Airport in 2012. Below: People at the airport taking pictures of the plane

Source: Yunuen Ysela Mandujano-Salazar.

advertising strategy, extensively covered by regular news and wide-shows, has transformed the airports that are served by the *JAL Arashi JET* and the airplanes themselves into mediatized places particularly attractive for the group's fans.

After the Great Eastern Japan Earthquake, which hit the country on March 11th, 2011, media, cultural producers, and corporations increased their projects relying on a discourse about the solidarity and resilience of Japanese people, encouraging national audience to consume domestic products and services and to explore and rediscovery their own country, thus working as a campaign for the promotion of domestic tourism (Mandujano-Salazar, 2014).

In this context, Arashi's involvements with national endeavors thrived. In June 2011, *Nippon no Arashi* was published for its selling around the country to raise funds for the victims of the earthquake (Oricon Style, 2011). Becoming a best-seller, the book allowed the images of the places represented in it to reach wider audiences within the country. Also, from November 2011 to December 2013, Japanese public broadcaster NHK produced a series of documentaries—*Arashi no asu ni kakeru tabi*—where the members visited different places around the country—identified in detail—interviewed locals, and showed the attractions of the region, using a discourse that stressed the richness of national traditional and contemporary culture—including arts, cuisine, technology, and a special relation with nature expressed by architecture and the arrangement of spaces. In 2013, the group also hosted an annual charity event from private broadcasting corporation NTV—*24 Jikan terebi*—which had as theme "Japan…? The nature of this country"; in it, the members were shown travelling around Japan and introducing places and local culture (Mandujano-Salazar, 2018).

This kind of intensive promotion of Japan through Arashi has continued and become intertwined with the group's regular entertainment contents, functioning as a permanent campaign for national tourism promotion through the mediatization of places and the products and services they offer. Regular variety shows hosted by the group have incorporated segments focused on presenting the members visiting Japan's touristic locations, going to major shopping malls, and tasting a wide variety of food from all around the country. The narrative promoting domestic tourism is indirect but evident. Members rarely invite audience explicitly to visit or consume what they show. It is the fact that they have been in those places, experiencing certain services or consuming specific products what makes people—particularly fans—want to go and try them. One relevant example of this subtle but pervasive campaign and its impact on domestic tourism is found through the evolution of NTV's show *Arashi ni Shiyagare*.

This Saturday's primetime variety show began broadcasting nation-wide in April 2010 and continues to date. During its first year, the show was entirely filmed in a studio in Tokyo with a live audience and focused on Arashi interviewing and interacting with another male celebrities. However, by the second half of 2011, when the celebrities who were invited had been raised in other regions of the country besides Tokyo, a special segment showed the invitee taking Arashi in a virtual tour by the places memorable for him, or teaching them about products or cuisine from his hometown accompanied by images and general details of the places where they could be found. This was not a very specific or hard promotion, because neither Arashi nor the guests were shown there; however, they usually tasted the products in the studio.

Nevertheless, soon the TV show began promoting places and products from around the country in a more explicit and intense way. In 2012, the show debuted a segment called "Confrontation with the unknown." During this, invitees asked the members of the group to go to some specific place in Japan and do something that was new for them. Regularly, the segment was only a few minutes long, but during a one-hour special episode, Arashi's visit to Mount Takao—at the outskirts of Tokyo—was broadcasted. In this, the morning tour of the group, from their jumping into a cable car to their eating of snacks at the different stages of the climbing, was shown. Mount Takao was already a tourist place frequented mostly by Japanese older people who likes hiking—even the members were shown joking about it. Yet, as I was able to confirm during fieldwork, the mediatization of the diverse spots found along the road would serve to attract numerous young people and families alike even a few months after the episode was broadcasted.

In 2014, one of the members of Arashi began hosting a segment called "Camouflaged trip" in which, accompanied by other celebrities, visited some famous touristic place in Japan, but they had to disguise to not be identified by people around; if they were recognized, they had to stop the trip. In 2016, this segment became "Rediscovering Japan" and, in 2018, "Sho Sakurai goes to…"; the only change being that they no longer had to disguise and could interact with locals to get references about the attractions around the place. These segments were an obvious promotion of touristic places and attractions of very diverse genre. Their appearance in the show, however, boosted the general interest as they became informally endorsed by Arashi.

In 2013, the same TV show released an irregular segment called "Tokyo's good and annoying restaurant," in which Arashi visited a place in the city that was said to have very tasty food, but of which the owner was eccentric in some way, making the experience of going there more than just eating. These restaurants had the characteristic of being little-known places; and, many of them, inexpensive and family-owned. By 2015, the segment disappeared, but another called "Death match" was introduced and continues to date. In it, Arashi and the celebrities invited to the show are in the studio and a recorded video displays five different dishes prepared in popular restaurants or food-related stores around the country. Then, there is a quiz and those who answer correctly can eat the dishes, which are taken to the studio, and give their review. During the recorded video, name and location of the restaurants, and prices of the dishes are shown. Although the variety of food ranges from local specialties to international cuisine, all providers are small- to medium-sized places located somewhere in Japan and owned by Japanese people. Consequently, in both segments, the appearance in the show of specific eating places becomes a national-reaching type of advertising that those owners would be unable to reach by their own means; but, also, this mediatization adds a symbolism that is particularly meaningful for the show's audience and Arashi's fans.

Audience and fans' performances in Arashi-related mediatized places

During fieldwork, between 2012 and 2018, I visited numerous of the places shown in Arashi-related media contents. To those located in Tokyo, I went at some point during the month following the broadcasting; I visited touristic places, stores, and restaurants outside Tokyo sometime between one and five months after broadcast. In most cases, I found that the specific places where the members of the group had been shown and/or the food or products they had tried were informally signaled by vendors with colorful papers and in handwriting (Figure 11.2); and, there was always people—mostly women—taking pictures of themselves standing in those places or eating the products signaled as consumed by Arashi.

In each of these visits, I randomly interviewed between five and ten people who were there as customers, all of them Japanese. Following a semi-structured format, I asked them about their motivations for going there and, if they mentioned Arashi or one of their shows, I requested more information. I interviewed a total of 283 people—176 women,

Figure 11.2   Food vending spot at Mount Takao. The paper at the centre indicates that one of the
members of Arashi tasted that soup in the TV show *Arashi ni shiyagare*

Source: Yunuen Ysela Mandujano-Salazar.

107 men—from ages 14 to 74, with a mean age of 37. From the total of respondents, 212
expressed that, after watching the place, food, or product in some of the TV shows of the
group, they were curious and wanted to try it. From those, 107, all of them female, said
to be fans of the group and that it was specifically because they wanted to be at the places
that the members of Arashi had been, to consume what they had consumed, and—very
importantly—to take pictures to upload on social media. It was also relevant to find that
more than 90% of the people interviewed went out of their usual areas of transit and even
travelled from other regions of the country to visit there.

People who had watched the place and products in some Arashi-related content, but who
were not fans, bought what it had been shown in those contents, but performed just as any
other customer and did not take pictures of the group's references. They were influenced by
media representations of the quality of the products or the ambience offered by the place,
but they were not intending to distinguish themselves from others around and they acted as
expected from a customer or visitor of such place.

On the other hand, core fans of the group were easily identified among regular customers or
visitors because they went in small groups, were taking numerous pictures not only at, but of the
exact places where the members of the group had been and were displaying on their belongings
something related to Arashi—mostly official goods sold at concert venues. I noticed that carrying
those items, taking pictures of the places and products consumed by Arashi's members, and consuming the same exact products were part of their identity display as fans of the group.

At Mount Takao, I observed many young women walking with difficulty the winding dirt roads, wearing clothes, and shoes not designed for hiking, but taking pictures of all the places marked as those visited by Arashi. I interviewed a group of three high-schoolers who were wearing their uniform skirts and t-shirts of some Arashi concert. They told me they had travelled from Shizuoka—about two hours by train—because they wanted to be at the places they had seen their idols visit.

This Arashi-fan performance was also evident at Haneda Airport. I visited it one week after the 3rd version of *JAL Arashi JET* was released in 2012. At the aircraft viewing area, around the time the airplane was scheduled to arrive from Fukuoka, there were about 60 people with their cameras ready. Most of them were adult women who were talking in couples or small groups and were holding JAL advertising pamphlets featuring the idols. Interviewing 14 of them, I found out that eight were not just there to take pictures, but they were going to travel on the *Arashi JET* at some point the same day. It was a recurrent answer that the reason for travelling to Fukuoka or Osaka—places served by the JET—was just to experience being on that airplane. A couple of young adult women who identified themselves as fans of Arashi since the group's debut, told me that they had tried to get airplane tickets for the first two versions of the JET, but they were sold-out quickly. For this third version, they had been able to buy tickets on a morning weekday flight to Osaka, so they had taken a day off their jobs and they would go just to eat, do some shopping and go back the same night in another airplane. They were holding the group's latest concert bags and hand towels. When asked if they usually carried those items, they answered they only did when they went to "Arashi tours," by which they meant visiting places related to the idols' media activities.

It was evident that the performance of Arashi's fans in places that had been mediatized supported by the idols' image was clearly identity oriented. Fans visited those places with props and performing in particular ways that distinguished them from other visitors, and this was also part of their personal experience. They felt part of a bigger community of fans and there was a symbolic value to be at the places their idols had been or to consume what they had consumed.

Conclusions

The case of Arashi reveals how the cooperation among government, corporations, and media has been encouraging the branding of "Japaneseness" through the mediatization of places, which are being commodified and turned prolific for all actors involved. The endorsing of private products and services by Arashi—as the case of JAL—allows companies to relate them to the notion of the national that the group embodies, improving the impact of their campaigns and the response from the public. Also, the informal endorsement of locations, small businesses, and products belonging to the national tourist industry, made through regular media contents hosted by Arashi, has a positive impact on their visibility and attracts new clientele from among the national audience. Hence, it is clear how the attraction power of Arashi among audience, particularly fans, has been used to promote domestic tourism not only through specific campaigns but also by the continuous display of them going to diverse places around the country—mediatizing them—and consuming different products in their regular entertainment shows.

The results of the observations and interviews support the argument that mediatized places have a strong influence in the ways tourists perceive them and how they decide to act in such contexts. It is evident that the mediatization of places through Arashi's formal and informal endorsements have turned them into heterogeneous tourist spaces where direct and identity-oriented performances of Japanese audience as domestic tourists are present. Mediatized places related to Arashi have created a segment of tourists, particularly among fans of the group, who decode the meaning of those spaces in a different sense than other visitors. Linked to such specific decoding, their performances aim to display their identity as fans of the group and to experience a symbolic closeness with their idols.

The use of idols—and celebrities in general—in the mediatization of places and the promotion of domestic tourism in Japan is not new. However, its impact has expanded as social media and communication technologies allow tourists to post pictures and exchange comments and reviews with other people. In this sense, tourists who are influenced by idols' contents to make sense and perform in a specific way in mediatized places may become themselves influencers within their own friends and social media followers, widening the meanings attached to such places.

Acknowledgements

The present work is the result of a long-term and wider research about Japanese media idols, national identity and media imagery on Japanese and foreigners, sponsored at two different stages by The Japan Foundation through the Japanese Language Program for Specialists in Cultural and Academic Fields (2013–2014) and the Japanese Studies Fellowship (2016).

Note

1 In 2020, by the time this chapter is being reviewed, Arashi is still the face of JAL national campaigns.

References

Aoyagi, H. (2005). *Islands of eight million smiles. Idol performance and symbolic production in contemporary Japan*. Cambridge, MA: Harvard University Asia Center.

Arashi. (2011). *Nippon no Arashi Poketto-ban [The Arashi of Japan Pocket Edition]*. Tokyo, Japan: M.Co.

Darling-Wolf, F. (2004). SMAP, sex, and masculinity: Constructing the perfect female fantasy in Japanese Popular Music. *Popular Music and Society, 27*(3), 357–370.

Edensor, T. (2001). Performing tourism, staging tourism. (Re)producing tourist space and practice. *Tourist studies, 1*(1), 59–81.

Hall, S. (1980). *Encoding/decoding*. London, UK: Hutchinson.

Hall, S. (1997). The spectacle of the 'other'. In S. Hall (Ed.), *Representation: Cultural representations and signifying practices* (pp. 223–290). London, UK: SAGE Publications.

Japan Airlines. (2010, September 4). *Tokubetsu tosōki 'Jal Arashi JET' ga shukō [The special coated plane 'JAL Arashi JET' goes in commission!]*. Retrieved April 1, 2013, from Japan Airlines: http://press.jal.co.jp/ja/release/201009/001620.html.

Japan Airlines. (2011, October 30). *Tokubetsu tosōki Kaibutsukun JET 11gatsu tsuitachi shukō*. Retrieved from Japan Airlines: http://press.jal.co.jp/ja/bw_uploads/JGN11098.pdf.

Japan Airlines. (2012, October 23). *Tokubetsu dekaruki JAL Arashi JET dai 3 dan honjitsu shukō*. Retrieved from Japan Airlines: http://press.jal.co.jp/ja/bw_uploads/MjAxMjEwMjNfSkdOMTIxM-jhfk8GVyoNmg0qBW4OLi0CBdUpBTJeSSkVUgXaR5jOSZSCWe5P6j0GNcS5wZGY.pdf.

Japan Airlines. (2015, June 26). *JAL FLY to 2020 tokubetsu tosōki ga kokunai ni shukō*. Retrieved from Japan Airlines: http://press.jal.co.jp/ja/bw_uploads/MjAxNTA2MjZfSkdOMTUwNjJfSkFMIEZ seSB0byAyMDIwIJPBlcqTaJGVi0BfLnBkZg.pdf.

Kankōchō. (2010a, April 8). *Arashi x Kankōchō 'Kankō Rikoku Navigator' to shite Arashi wo Kiyō [Arashi x agency of tourism, the appointment of Arashi as 'Japanese tourism promotion representative']*. Retrieved April 25, 2011, from Kankōchō: http://www.mlit.go.jp/kankocho/en/news01_000038.html.

Kankōchō. (2010b, September 1). *Kankō Rikoku Kyōiku ni shisuru tosho no sōfu to gakkō ni okeru sono katsuyō ni tsuite [Concerning the sending of books to contribute to the education designed to make the country a tourism destination and its use in school]*. Retrieved April 25, 2011, from Kankōchō: http://www.mlit. go.jp/common/000124160.pdf.

Keizai Sangyōshō. (2010, June 8). *Cool Japan Shitsuno Secchi ni tsuite [About the establishment of the Office of Cool Japan]*. Retrieved April 3, 2012, from Keisai Sangyōshō: http://www.meti.go.jp/policy/ mono_info_service/mono/creative/index.htm.

Mandujano, Y. (2013). The politics of selling culture and branding the national in contemporary Japan: Economic goals, soft-power and reinforcement of the national pride. *The Scientific Journal of Humanistic Studies, 5*(9), 31–41.

Mandujano-Salazar, Y. Y. (2014). *Media idols and national 'representation': Strengthening the national identity in contemporary Japan [Doctoral dissertation]*. Ciudad Juárez: Universidad Autónoma de Ciudad Juárez. Retrieved from http://148.210.21.138/handle/20.500.11961/511?show=full.

Mandujano-Salazar, Y. Y. (2018). Media idols and the regime of truth about national identity in post-3.11 Japan. In F. Darling-Wolf (Ed.), *Routledge handbook of Japanese media* (pp. 154–166). New York, NY: Routledge.

Månsson, M. (2015). *Mediatized tourism. The convergence of media and tourism performances [Doctoral dissertation]*. Lund University, Department of Service Management and Service Studies. Helsingborg: Lund University. Retrieved from https://lup.lub.lu.se/search/publication/6ea3a2df-89fb-48af-92e0-ee7d8cf3ac8d.

Marshall, P. D. (1997). *Celebrity and power: Fame in contemporary culture*. Minnesota: University of Minnesota Press.

Ministry of Economy, Trade and Industry. (2012, January). *Cool Japan strategy*. Retrieved December 6, 2016, from Ministry of Economy, Trade and Industry: http://www.meti.go.jp/english/policy/ mono_info_service/creative_industries/pdf/120116_01a.pdf.

Oricon Style. (2011, July 7). *Arashi 'Nippon no Arashi pokettoban' ga hatsubai yokka de 20 man koe kotoshi saikou no shukanuriage wo kiroku [The book of Arashi 'The Arashi of Japan pocket edition' sells more than 200000 in four days getting this year record of higher weekly sales]*. Retrieved March 20, 2013, from Oricon Style: http://www.oricon.co.jp/news/ranking/88340/full/.

Sakai, M. (2003). *Aidoru Sangyō [Idol industry]*. Retrieved September 13, 2008, from Stanford Japan Center Web site: http://www.ppp.am/p-project/japanese/paper/sakai-paper.pdf.

12

DO YOU FEEL THE WARMTH? THE ONLINE DESTINATION IMAGE OF SOUTHEAST ASIA

Maria Criselda G. Badilla

Introduction

Within the context of tourism marketing and mass media studies, technological advances have given rise to the development of new media. Digital media channels have required innovation and new strategies made to applications that were not previously present. This development accelerates tourism destination development and marketing, as more people are able to gather travel information faster. Literature points to events (Mendez, Oom Do Valle, & Guerreiro, 2011); films (Yen & Croy, 2013); brochures, postcards, and travel guides (Milman, 2011); brand logo (Hem & Iversen, 2004); travel agents, travel guidebooks (Beerli & Martin, 2004); television and documentaries (Ab Karim & Chi, 2010); websites and online digital guides (Bohlin & Brandt, 2014) as major sources of travel information.

Tourism scholars have considered social media as a boost to tourism as word-of-mouth heavily affects tourism behaviour. With friends posting travel experiences as a form of social capital, technology has afforded changes in social relations. With internet access on mobile phones becoming more accessible, tourists can post their experiences almost instantaneously. The gazers become gazed upon when friends see their social media posts. This act contributes to the formation of collective and mediatized gazes (Urry, 2002). Social media has also improved channels of communication, both one to one and one to many (Hays, Page, & Buhalis, 2013).

As social media and the Internet become more prevalent in one's practical and social way of life, tourist behaviour is highly influenced by information available in media. There is a need for destination marketing organizations (DMOs) to get into this social space in order to influence travel behaviour towards their destinations.

The Association of Southeast Asian Nations (ASEAN), an association formed in 1967 to promote political and economic co-operation within the Southeast Asian region, is used as case study. Its member-countries are Indonesia, Malaysia, the Philippines, Singapore, Thailand, Brunei, Cambodia, Laos, Myanmar, and Vietnam. ASEAN celebrated its 50th founding anniversary in 2017 and has declared the same year as 'the Visit ASEAN Year'. Tourism in Southeast Asia is in an upswing with a consistent annual growth rate of 6% (UNWTO Highlights, 2018). Outlined in its ASEAN Tourism Strategic Plan (2016–2025), South East Asia plans to have a strategic marketing plan in promoting South East Asia as a single destination using the Internet as its primary platform. The ASEAN regional branding 'Southeast Asia, feel the warmth' has been conceptualized and promoted since the beginning of 2011.

As a marketing and communication specialist and tourism consultant for government projects on tourism, I take great interest on how DMOs maximize the use of online media in promoting their destinations. This chapter looks into how websites and social media sites are used by South East Asian Destination Marketing Organizations, in their attempt to create a single destination image for Southeast Asia (SEA). Themes within the different regional and national websites and official social media accounts (Facebook, Instagram, and Twitter) were identified to find prevailing common themes and images. The study investigates how the SEA DMOs portray and present tourism images online in line with its single destination branding.

As tourists have become more sophisticated and independent in planning their own trips and booking vacations without the help of professionals (Ozturk, Salehi-Esfahani, Bilgiha, & Okumus, 2017), DMOs and tourists have become colonized by the ubiquity of the Internet and social media on how tourism images are portrayed and presented online within the member-states of the ASEAN in line with its single destination branding.

A textual analysis of the official websites, Facebook pages, Instagram, and Twitter accounts of the ten ASEAN countries and ASEAN Travel for the period of January 1 to June 31, 2017 was conducted. The texts and images of each entry were coded and processed electronically using NVivo to look into its common themes. The DMOs manage their websites and social media sites with the goal of presenting a positive image for their specific destination. However, in this study, I iterate the fact that aside from the individual branding of the member-states, an integrated single destination image should surface. I argue further that image formation and place representation have moved beyond the DMO's control as tourists now participate in the co-production and co-creation of value. The DMOs are posed with a great challenge of not being able to control online conversations but can only try to influence such conversations (Gretzel & Yoo, 2014).

Impact of media on tourism

Media plays an important role in tourism particularly for destination marketing and image formation. Digital media will continue to evolve and provide new tools for tourism marketing and management. Its interactivity and ubiquity enable organizations to re-engineer the entire process of developing, managing, and marketing tourist products and destinations (Buhalis & O'Connor, 2005). Corner iterates mediatization as displayed by the increasing use of social media in tourism decision-making has slowly transformed tourism marketing. Mediatization refers to the various ways in which media use is studied to shape social and cultural transformations. As such, various studies have been conducted on the effects of media use without making use of the concept of mediatization. As media use has become embedded into one's daily life, empirical work firmly rooted on the concept of mediatization is relatively scarce (Corner, 2018). Further, social media has naturalized new forms and norms of social interaction but also altered the very materiality of everyday life and information disclosure (Jansson, 2013).

The rapid evolution of digital media has revolutionized tourism with regard to its communication structures (Scherle & Lessmeister, 2013). With websites and social media (SM) channels, information about tourism experiences has become easier for aspiring tourists, images are formed and perception develops faster than prior generations. The Internet has changed how people access information, plan and book trips, and share their travel experiences (Hays et al., 2013). With its decentralized nature and low entry costs, there is breadth, variety, and diversity of information available in the Internet (Lee, Kim, & Chan-Olmsted, 2011). The Internet has optimized access to information and lowered search costs for information, on the demand side (Scherle & Lessmeister, 2013). The importance of the Internet seems to rest largely on its capabilities to execute marketing and communication functions better than more traditional media and marketing systems (Chan-Olmsted, 2002). Computer-mediated media have not only introduced new forms of media but also new methods of communicating (such as interactivity) as well as synchronous and asynchronous communication (Goneos-Malka, 2013). The Internet has proven its worth when it comes to availability and immediacy of information and services (Scherle & Lessmeister, 2013).

People are directly engaged and have become the media themselves (Thevenot, 2007). Information control has shifted from tourism marketers and institutions to tourists wherein the ultimate control over one's image no longer comes from the source (Hays et al., 2013). Marketers no longer control the entire media environment (Muhern, 2009). Social media empowers tourists to perform an increasing number of tasks by themselves such as checking reviews, comparing prices, schedules and services, making reservations, and purchases as well as acting as endorsers (Sotiriadis & Van Zyl, 2013) and critics. The sharing and giving of recommendations are free and easily accessible (Sotiriadis & Van Zyl, 2013). People can share their opinions, insights, experiences, and perspectives through texts, images, and videos (Thevenot, 2007) instantaneously.

As such, online media has greatly contributed to the promotion of touristic places. Media have colonized tourism behaviour as tourists have become more dependent on social media for travel decisions. Mediatization of places occur every time destination images are posted online. Mediatized spaces are represented by the use of online exchanges that are encoded and decoded differently. Tourists have started to portray a new role as agents of image formation (Camprub, Guia & Comas, 2013). Web 2.0 tools have revolutionized the way DMOs project destination image and how tourists search and gather information about tourism destinations (Camprub et al., 2013). The marketing paradigm has greatly shifted from the traditional means of crafting marketing messages from the DMO to the consumer. The Internet and social media marketing paradigm has evolved into an interactive marketing process that revolves around engagement, action, and loyalty of its consumers (Evans, 2008) through establishment of the destination's reputation from repeated engagement and conversations with its customers (Gretzel & Yoo, 2014).

South East Asia (SEA) as a single destination

The official DMO websites analysed were (1) www.malaysia.travel, (2) www.tourismthailand.org, (3) www.yoursingapore.com, (4) www.indonesia.travel, (5) www.vietnamtourism.com, (6) www.itsmorefuninthephilippines.com, (7) www.tourismcambodia.org, (8) www.bruneitourism.travel, (9) www.tourismlaos.org, (10) www.tourismmyanmar.org and (11) www.aseantourism.travel as well as the official Facebook, Twitter, and Instagram of the ten member states and ASEAN Tourism.

Empirical data is discussed from this point on to elucidate the various themes evident in the DMO websites and social media accounts. With the ritualized behaviour of posting regular status updates, media have become significant to the production of social space (Jansson, 2013). However, since mediatization does not occur as a media effect, but rather as a long-term social process of mutual amalgamation and accommodation (Jansson, 2013), an investigation of a series of posts can validate the discourse of mediatization and how people can make sense of media and its practices. This textual analysis of webpages and social media posts shows concrete ways in which media are used to organize social space and how they are positioned in relation to the course of daily life (Jansson & Lindell, 2018).

The messaging in each of the country sites as projected online is discussed in this next section. After which, a consolidated theme is presented to validate how media technologies and their logics can shape the ways the DMOs think and position themselves in social space (Jansson & Lindell, 2018).

Brunei's branding slogan is as 'A Kingdom of Unexpected Treasures', dubbed as Asia's well-kept secret. The heart of its website is its #discoverbrunei tab itemizes the different destination packages which will allow tourists to enjoy its natural attractions, diving and cruising to enjoy the flora, fauna, and rich mangroves as well as its majestic temples, gentle people, and golf (www.bruneitourism.travel). Its social media accounts noticeably make use of #aseancityofculture. The country neatly positioned itself with a specific comparative advantage over the other SEA countries for its rich culture.

Cambodia's branding slogan 'Kingdom of Wonder' does not seem to be very evident in the webpages of www.tourismcambodia.com except for features of its mosques and palaces. The website lacks stories and visuals of the place that can pull people to be in wonder of its attractions. Social media entries in Cambodia's official sites were used to direct them back to the website which has richer information. The social media sites (SMS) presented Cambodia as a value for money country that is ready for the influx of tourists.

Indonesia's website www.indonesia.travel succinctly highlights its slogan 'Wonderful Indonesia' as having a myriad of places to see with its scenic natural landscapes which blends with the various unique culture of its people. The website highlights five major wonders, namely (1) nature, landscape, and wildlife; (2) culinary and wellness; (3) arts, culture, and heritage; (4) recreation and leisure; and (5) adventure. In its SMS, Indonesia is packaging itself as the most beautiful country in the world, rich in natural and cultural wealth. Using mirrored posts in its three SMS, images of the sun, sea, and sand attractions are evident and consistent in the images as well as the reference to the term Vitamin Sea.

With its tagline 'Simply Beautiful', *Laos* features its natural attractions that includes rivers, elephants, temples, and caves in its portrayal of the term beautiful. It has the most collaborative website, www.tourismlaos.org in terms of SEA as a single destination campaign effort is concerned having references to ASEAN and the ten countries in its homepage. ASEAN cross promotion is also evident in its SMS. There were a variety of activities showcased. These predominantly showed natural resources such as rivers, waterfalls, and mountains. Historical sites such as war memorials, photos of local people in their traditional attire and way of life, showcasing culture and the night market were highlighted in the posts. Laos portrays itself as a country rich in natural resources and culture with splendid post-colonial architecture, festivals, and warm people – ingredients for a worthwhile tourist experience.

Malaysia is portrayed in its website, www.malaysia.travel as a bubbling, bustling melting pot of races and religions where different ethnic groups live in peace and harmony. Its slogan 'Malaysia Truly Asia' captures and defines the essence of the country's unique diversity and sums up the distinctiveness and allure of Malaysia. In its social media accounts, it seeks to project a vibrant and dynamic destination where all the colours, flavours, sounds, and sights of Asia come together in Malaysia, a reason to be dubbed Truly Asia. Iconic images of the Petronas Twin Towers and areas of Kuala Lumpur, Penang, and Langkawi's 99 islands were frequently mentioned in its SMS.

Myanmar's official website www.tourismmyanmar.org featured mostly photos, articles, facts, and features about the different attractions in Myanmar. Its slogan 'Let the Journey

Begin' is predominantly located at the centre of the homepage but is seemingly not translated within the website. There were comprehensive and lengthy videos that highlighted the influence of Buddhism in its pagodas and shrines, colonial buildings, lakes and rivers, silk weaving and festivals. In its SMS, the slogan is non-existent as it was not used in any of the photos nor in hashtags. The social media posts present a picturesque destination but mostly through reposts, shares, and re-tweets.

The website www.web.tourism.gov.ph highlights *the Philippines* as a top destination, with professionally photographed world heritage sites and beach sites. Picturesque landscapes of famous attractions are presented. There were very few photos with people. The Philippines' slogan is 'It's More Fun in the Philippines' describing the core of the Philippine travel experience. The imagery of the posts was of its white sand beaches, luscious flora, relaxation, and unique experiences. The only reference to SEA was the Visit ASEAN year campaign in 2017.

Singapore launched a new destination brand in 2017, 'Passion Made Possible'. The DMO is presenting a brand that tells a broader story of Singapore beyond tourism. The brand articulates what it stands for as a country and supports the telling of many stories about the destination and its people. The six core target markets of Singapore are highlighted on the website's cover page: foodies, explorers, collectors, socializers, action seekers, and culture shapers. The city's representation shows that it has been well planned with man-made gardens and lagoons which complement its majestic skyline. Its skyline is captioned as 'a city like no other', 'the most photogenic skyline in the world' and the skyline anyone can recognize. Singapore has not made explicit references to ASEAN or any of its neighbouring countries.

Thailand's website www.tourismthailand.org identifies it as 'a kingdom of wonder', filled with spectacular natural, cultural, and attractions with history, adventure, and relaxation as its core products. The SMS made use of the hashtags #UniqueThailandExperience and #AmazingThailand. The tapping of specific market segments such as the weddings market, single female market, foodies, and adventure seekers was evident in the posts. Activities that are attractive to the said market segments were featured extensively in the SMS. It showed consistent online projection across the websites and social media sites.

Vietnam uses Timeless Charm as its branding slogan. In its website, www.vietnamtourism. com, Vietnam positions itself as having the power of attracting tourists with its products revolving around culture, sea, natural and ecological attractions, sports, and recreation. The overall image being projected in Vietnam's social media accounts seem to be its rustic and rural appeal. With most of the images showing photos of rural lifestyle such as a rustic boat, life of locals, rice paddies, and the morning market. Transportation options on moving around Vietnam were either by boat, bike, and foot adds to this rural and rustic appeal. Timeless charm is evident in its social media posts that were mirrored across channels.

The destination image presentation of the individual countries highlights their individual unique selling propositions. Geographically bound and culturally alike, each country has maintained its individuality. Sadly, for South East Asia, the 'Feel the Warmth' regional branding is not seemingly evident in the individual country websites and social media accounts.

Collectively, the DMOs portrayed tourism in South East Asia to revolve around the diversity of its natural and cultural attractions. This diversity is translated into various natural and adventure activities, cultural immersion spruced by varied colonial influences, exquisite flavours, and festivals. There is diversity of natural attractions as reflected by photos that featured the flora and fauna, wildlife, land and water formations. Repeatedly, the images showed SEA to be a value for money destination particularly for shopping, retirement, and wellness.

The images and texts presented in the websites are within McQuail's (2010) context of preferential coding. This is how the countries' DMO planned to present their respective countries. Since websites are static in nature, there is no room for negotiation, feedback, and contestation. Their self-presentation is one-way and linear. The images are also dictated upon by the DMOs. The potential tourists are given a framed image, a single gaze, which everyone sees because it is the only image/photo/text that the DMOs release. Because the DMOs have control over what they present, their presentation of selves is harmonious to their countries' imaging and perceived benefits. The DMOs allow the tourists to look up only for a specific context. This is similar to situation where a house owner only allows his visitor to view certain portions of the house by opening doors to the rooms that his visitor is allowed to enter. Such is preferential encoding – the DMO has control over what places visitors are allowed to see.

However, with the interactive nature of social media, DMOs are subjected to immediate comments and reactions from its audience – the potential tourists. The audience response serves as the differential decoding, meaning there may be different ways in which the posts are received by its audience.

To keep the conversation with its customers moving forward, the DMOs showed the richness of SEA's culture and heritage through the colours, flavours, and people of SEA in its online media posts. The colours of SEA were evident through the colours of the sky, sea, mountains, the transition of cosmopolitan cities, the colourful skin of people of different ethnicities, the elegance and combination of colours for the batik cloth, the flavourful cuisine, the richness of the flora and fauna, festivals and unique festivals and events that only happen in SEA. In summary, the colours could represent what is significantly unique in SEA.

The flavours of SEA are displayed in its showcase of food and native delicacies. The manner by which the food was creatively presented and plated shows the distinct artistry and skilfulness of ASEAN people. The use of endemic materials such as banana leaves, coconut, rice (grains and noodle versions), and curry/spices in its cuisine were evident in the posts. As tropical countries, fruits and vegetables abound all year round. The social media posts about food enhance the tourist's expectation of a culinary experience. Contrary to the webposts which did not show any aspect of cuisine, the social media posts had numerous entries on food as this kind of posts received the biggest number of likes and comments.

Lastly, posts about the local people and their interaction with tourists were captured in the social media posts. The people of the region were shown either in their colourful traditional costumes or with their smiling faces. These enhance the feel and vibe of SEA as an attractive destination. Also, numerous festivals were featured showing that SEA has the most number of festivals in the world. The different countries celebrate similar festivals simultaneously as proof of the SEA's cultural similarities.

Differential decoding ensues when DMOs and tourists interact on SMS. Stuart Hall (1980) in his seminal work on encoding/decoding proposed that producers of media encode meaning into media text which carries a preferred reading for its intended audience. The audience can adopt either of three stances in decoding a message. The receiver may accept the dominant message, adopt an oppositional message, or have a negotiated position depending on his biases, ideologies, and backgrounds (McQuail, 2010). As DMOs use online media as its platform for conversation and engagement, a careful strategy should be adopted to enhance engagement and loyalty to SEA.

DMOs have their own strategy in presenting destination image. However, once tourists comment on the DMO's social media posts, the DMOs may adjust their succeeding posts as what is usually done in an actual conversation. This is evidenced by the comments, shares, likes, and other emojis shared in the social media posts. This occurrence strengthens the argument that image formation is drawn from the collective experiences of tourists. It is no longer controlled by the DMOs. The DMOs alter their messages based on the way their audiences received the messages in an attempt to manage the conversation. Negotiation ensues in this back and forth encoding and decoding of messages. Again, this shows that image formation is no longer controlled by the DMO but is co-created with its audience. This also reinforces how the other social processes in a broad variety of domains and at different levels become inseparable from and dependent on technological processes and resources of mediation (Jansson, 2013). As the Internet and social media afford places and spaces to be mediatized, the engagement between DMOs and potential tourists have vastly improved. However, the potential of social media channels has not been fully maximized as reflected by the limited social media presence and use of some countries involved in the study.

In the process of mediatization of images, the encoding and decoding process goes through both preferential encoding and differential decoding. The DMOs have specific ways of presenting their destinations on their websites and SMS based on how they plan to present destination image. However, social media afford user-generated content making destination image formation co-created with tourists who have experienced the destination.

As the travel experience and online media consumption occur simultaneously, online representation of destinations is no longer solely controlled by the DMO. As online messages can either build up or destroy a destination's image, a destination may benefit from a positive post that turns viral, in the same manner that it may enter into a crisis for a negative post that turns viral. These technological advances shape what media can do and thus shape the path for mediatization to occur. For South East Asia, where the core products are its sun, sea, and sand, findings show that there is a need for these to be articulated in the branding of SEA as a single destination.

SEA embodies a destination that has diverse natural and cultural attractions and is a good value for money destination. By comparing the website posts and social media posts of the DMOs, some similarities among the destinations have become evident. Their vibrant colours are shown in the visual and written representation of the DMOs. South East Asia is full of colour, its colours representing its rich natural resources, diverse culture, exquisite cuisine, and gracious people. However, these have yet to come together as a collective branding that is deliberate, evident, and consistent across media channels and across the member countries. The challenge for South East Asia is to show its uniqueness and product differentiation as a region while retaining the individuality of each country.

Conclusions

Websites and social media sites have given tourism marketers a cost-effective way to promote products and provided unlimited information to potential tourists at minimal costs. Given the intangibility of the tourism product, digital media has enabled the concept of travel to be more concrete and tangible to more people. Social media has mediatized the power of word of mouth. Online media channels, particularly social media, have allowed marketers to express and share their travel information by providing the most credible means of convincing potential tourists to travel, electronic word of mouth.

There is a need for DMOs to understand that online media especially social media marketing is fundamentally about participation, sharing and collaborations rather than straightforward advertising and selling (Gretzel & Yoo, 2014). The way messages are crafted and shaped in the websites and social media sites should highlight authenticity, participation, and continued conversation with the primary goal of closing the sale (Gretzel & Yoo, 2014). For SEA to succeed in pushing forth its single destination agenda using online channels as its primary tool, a deliberate strategy on how to maximize its use should be done.

With ASEAN tourism encouraging countries to collaborate and co-operate while still competing for the tourist market share, the challenge is for SEA to bring their commonalities together to draw a single destination image that will help increase their global market share. Similarities in terms of natural attractions, culture and cuisine have emerged, which if presented properly can reap benefits for South East Asia as a single destination.

Mediatization has occurred as a consequence of tourist's dependence on digital media. It has heavily influenced travel behaviour patterns before, during and after their travels. SEA DMOs, as directed by the ASEAN Tourism Marketing Strategy 2017–2022, have relied heavily on online media to promote the single destination concept. The digital media has provided a social space for DMOs and potential tourists to engage and interact. This validates the fact that the media can no longer be separated from the tourist experience because travel can no longer be fully experienced without the presence of media.

Future research on how DMOs use digital and online media in shaping destination image should be pursued since there is very little academic literature available on online image formation in the context of social media. Future research can also look into maximizing the use of online marketing, barriers for its adoption, process of selection of channel used, and monitoring of its impact can be further explored. The impact of cooperative marketing strategies can also be looked into to increase the participation of SEA DMOs in pushing the single destination agenda forward.

References

Ab Karim, S., & Chi, C. G. Q. (2010). Culinary tourism as a destination attraction: An empirical examination of destinations' food image. *Journal of Hospitality Marketing & Management, 19*(6), 531–555.

Beerli, A., & Martin, J. D. (2004). Tourists' characteristics and the perceived image of tourist destination: A quantitative analysis – a case study of Lanzarote, Spain. *Tourism Management, 25*, 623–636.

Bohlin, M., & Brandt, D. (2014). Creating tourist experiences by interpreting places using digital guides. *Journal of Heritage Tourism, 9*(1), 1–17.

Buhalis, D., & O'Connor, P. (2005). Information communication technology revolutionizing tourism. *Tourism Recreation Research, 30*(3), 7–1.

Camprub, R., Guia, J., & Comas, J. (2013). The new role of tourists in destination image formation. *Current Issues in Tourism, 16*(2), 203–209.

Chan-Olmsted, S. (2002). Branding and internet marketing in the age of digital media. *Journal of Broadcasting & Electronic Media, 46*(4), 641–645.

Corner, J. (2018). 'Mediatization': Media theory's word of the decade. *Media Theory, 2*(2), 79–90.

Evans, D. (2008). *Social media marketing: An hour a day.* Indianapolis, IN: John Wiley.

Goneos-Malka, A. (2013). Suggesting new communication tactics using digital media to optimise postmodern traits in marketing. *South African Journal for Communication Theory and Research, 39*(1), 122–143.

Gretzel, U., & Yoo, K. (2014). Premises and promises of social media marketing in tourism. In S. McCabe (Ed.), *The Routledge handbook of tourism marketing* (pp. 491–504). London, UK & New York, NY: Routledge

Hall, S. (1980). Encoding and decoding. In S. During (Ed.), *The cultural studies reader* (pp. 90–103). London, UK and New York, NY: Routledge.

Hays, S., Page, S. J., & Buhalis, D. (2013). Social media as a destination marketing tool: Its use by national tourism organisations. *Current Issues in Tourism, 16*(3), 211–239.

Hem, L. E., & Iversen, N. M. (2004). How to develop a destination brand logo: A qualitative and quantitative approach. *Scandinavian Journal of Hospitality and Tourism, 4*(2), 83–106.

Jansson, A. (2013) Mediatization and social space: Reconstructing mediatization for the transmedia age. *Communication Theory, 23,* 279–296.

Jansson, A., & Lindell, J. (2018). Media studies for a mediatized world: Rethinking media and social space. *Media and Communication, 6*(2), 1–4.

Lee, C., Kim, J., & Chan-Olmsted, S. M. (2011). Branded product information search on the Web: The role of brand trust and credibility of online information sources. *Journal of Marketing Communications, 17*(5), 355–374.

McQuail, D. (2010). *McQuail's mass communication theory* (6th Ed.). Los Angeles, CA: Sage Publications.

Mendez, J., Oom Do Valle, P., & Guerreiro, M. (2011). Destination image and events: A structural model for the algarve case. *Journal of Hospitality Marketing & Management, 20*(3–4), 366–384.

Milman, A. (2011). The symbolic role of postcards in representing a destination image: The case of Alanya, Turkey. *International Journal of Hospitality & Tourism Administration, 12*(2), 144–173.

Mulhern, F. (2009). Integrated marketing communications: From media channels to digital connectivity. *Journal of Marketing Communications, 15*(2–3), 85–01.

Ozturk, A., Salehi-Esfahani, S., Bilgiha, A., & Okumus, F. (2017). Social media and destination marketing. In M. Sigala & U. Gretzel (Eds.), *Advances in social media for travel, tourism and hospitality: New perspectives, practice and cases* (pp. 89–101). London, UK: Routledge.

Scherle, N., & Lessmeister, R. (2013). Internet cultures and tourist expectations in the context of the public media discourse. In J. Lester & C. Scarles (Eds.), *Mediating tourism: From brochure to virtual encounters* (pp. 91–103). Farnham, UK: Ashgate Publishing.

Sotiriadis, M. D., & Van Zyl, C. (2013). Electronic word-of-mouth and online reviews in Tourism services: The use of twitter by tourists. *Electronic Commerce Research, 13,* 103–124.

Thevenot, G. (2007). Blogging as a social media. *Tourism and Hospitality Research, 7*(3–4), 282–289.

UN World Tourism Organization. (2018). *UNWTO Tourism Highlights 2018.* Retrieved from https://www.e-unwto.org/doi/pdf/10.18111/9789284419876.

Urry, J. (2002). *The tourist gaze* (2nd Ed.). London, UK: Sage.

Yen, C., & Croy, W. (2013). Film tourism: Celebrity involvement, celebrity worship and destination image. *Current Issues in Tourism, 19*(10), 1–18.

13

TOURISM AND POPCORNS

The role of feature films in branding and marketing destination New Zealand

Natàlia Ferrer-Roca

Introduction

New Zealand (NZ) is well known as having created and implemented one of the most success-ful destination marketing and branding strategies of all times with '100% Pure New Zealand' as its main tagline. It has been running for 20 years – since 1999. By harnessing of 'effective marketing, industry partnerships and non-traditional media', Tourism New Zealand (TNZ) has been able to create a 'powerful travel destination brand, positioned as an appealing niche player in the global tourism industry' (Morgan, Pritchard, & Piggott, 2002, p. 336). However, it has also received lots of criticism (Anderson, 2012; Cropp, 2017; Stuff, 2017), because – obviously – no country can 'promise a 100 per cent pure environment' (Bradley, 2018). This controversy has been creating tension within NZ for quite a while. A critical moment was when environmentalist Dr. Peter Nuttall lodged a complaint with the NZ Advertising Stan-dards Authority (ASA) in March 2013 for considering the campaign 'misleading' and 'un-substantiated' after 'research into NZ's environment showed a degradation of its beaches, waterways and biodiversity' (Henry, 2018). However, the Complaints Board considered that:

> '100% Pure New Zealand' was a positioning statement used to promote the unique experience New Zealand offered international tourists rather than a claim about New Zealand's environmental purity. The advertisements did not imply that the environ-ments featured were 100% pure, rather they implied that these scenes and places were a part of the unique visitor experience.
>
> *(Henry, 2018)*

Indeed, TNZ claims that 'pure' becomes synonymous with 'genuine' and 'authentic', and so it is a mere 'positioning statement rather than an absolute claim of environmental purity' (Henry, 2018). Nevertheless, one thing is what TNZ claims that the campaign intends to communicate, and another very different one is how the public perceives the overall mes-sage. Therefore, whether 'the general public regards the campaign as literally meaning that everything in NZ is 100% pure' is indeed debatable (ibid.). Perhaps that is why, after five years, TNZ has decided in 2018 to start 'changing tack with its 100% Pure campaign by emphasizing people as well as the place' (Bradley, 2018).

According to TNZ Chief Executive Stephen England-Hall, the '100% Pure New Zealand' slogan does not 'showcase New Zealand's unique people and culture, our way of being, our warm welcome' (Bradley, 2018). This view is also shared by consultant Brian Richards, one of NZ's most celebrated branding professionals. He argues that, although NZ has a 'significantly successful international image [...] the challenge is that it's a default image of envy from the frustrations of where one comes from' (TPBO, 2018). Thus, there seems to be a shared perception among NZ branding professionals that the tendency is a move towards telling the stories of the people and the place. Indeed, place branding has to be rooted in cultural origins (ibid.). As Richards points out, one of the country's most celebrated branding professionals: 'it's less about the backdrop and more about the foreground personality of the people in the story' (ibid.). We will see if – and how – TNZ will adapt the '100% Pure' campaign in the near future by telling the stories of its people and place.

Be it as it may, NZ has already experienced extensively in telling 'stories' to the wider general public. One of the most successful story-telling techniques that NZ has been using in recent years is filmmaking in order to position the country and strengthen its reputation internationally. Little has been researched about the relationships between feature filmmaking and destination marketing and branding, economic development and country reputation (D'Alessandro, Sommella, & Viganoni, 2015). This chapter thus provides an innovative cross-disciplinary point of view to fill this research gap, as pointed out by place brand practitioners (Govers, Kaefer, & Ferrer-Roca, 2017). Building on Ferrer-Roca (2015a), Kaefer (2014), and Morgan et al. (2002, 2003), this chapter applies an institutional political economy perspective and draws on findings derived from review of academic literature and secondary data, policy analysis, archival research, and expert interviews with key personnel in industry and state agencies.

Moreover, this chapter also supports the theoretical framework based on the notion of 'mediatized media' introduced by Månsson (2011), through which we can better understand and analyse how media processes interact with tourism. Precisely, the case study of NZ exemplifies that popular feature films can be used strategically for destination marketing and branding. Although this has already been examined from a management perspective (Morgan et al., 2002, 2003), this chapter is original, as it offers a media studies viewpoint.

The NZ feature film industry

Many scholars have recognized that screen productions have the power to create new place images with the potential to promote tourism (Beeton, 2010; Bolan & Williams, 2008; Buchmann, 2010; Connell, 2012; Croy, 2010; Frost, 2010; Jones & Smith, 2005; Kim & Richardson, 2003; Leotta, 2011; Månsson, 2011; Riley & Van Doren, 1992; Riley, Baker, & Van Doren, 1998; Tooke & Baker, 1996). Nevertheless, only big feature film productions that get into the international circuit are able to attract prolonged and international tourist flows by changing the collective perception of a destination and, therefore, are able to have a significant and long-term impact (D'Alessandro, et al., 2015, p. 185). The NZ case is paradigmatic and, precisely, the trilogies of *The Lord of the Rings* (LOTR) and *The Hobbit* are so far the most relevant for the case at hand.

In order to add clarity to existing distinctions between the different types of film production occurring in 21st-Century NZ, the 'three-tier structure' framework is used, which was coined by Jim Booth in 1984, the then *New Zealand Film Commission* (NZFC) director (Dunleavy & Joyce, 2011, pp. 84–85), and developed and applied for the first time in academia by Ferrer-Roca (2017). Bottom-tier features are small-budget films with a stronger proportion of NZ content as they are made with the domestic audience foremost in mind, for which the NZFC is the main investor and, in general, face significant economic challenges when confronted with a small domestic media market (Ferrer-Roca, 2015b, 2017, 2018, 2019). Middle-tier productions are medium budget films, including foreign co-productions, which are made with NZ financing, primarily from the NZFC, and significant offshore investment (Ferrer-Roca, 2017). Finally, top-tier productions describe large-budget films that are primarily, if not entirely, financed by foreign companies (ibid.). In this context, domestic feature films comprise bottom- and middle-tier productions, while international productions include middle- and top-tier features.

The NZ government, through its agencies, departments, and ministries, has played – and is still playing – a key role within the institutional ecology of the NZ film industry in supporting, funding, and sustaining all three-tiers of feature film productions (Ferrer-Roca, 2018). One of the main pillars is the role of the NZ government in regulating the NZFC, the main public funder, through the NZFC Act 1978. The NZFC works closely together with three NZ government ministries by administering their film funding schemes: the Ministry of Business, Innovation and Employment (MBIE), the Ministry for Culture and Heritage (MCH), and the Ministry of Foreign Affairs and Trade (MFA) (see also Ferrer-Roca, 2018, pp. 385–393; NZFC Act, 1978).

As institutional political economy theory argues, institutional and policy arrangements are often subject to competing preferences and policy pressures. For example, the MBIE clearly has economic priorities, while the MCH's first concern is cultural outcomes. Therefore, its developments are a direct consequence of 'contextual compromises among institutional actors' (Ferrer-Roca, 2017, p. 15). Taking this perspective into account, it can be suggested that the current institutional ecology of the NZ film industry, sustained by the three-tier structure, the several funding initiatives of these three Ministries and together with the central role of the NZFC, is 'both a strategic and creative response to the challenges of creating and sustaining a feature film industry in a small English-speaking country' (ibid.). This does not mean, however, that this institutional framework has been consciously arranged and implemented, but rather it is the result of constant tension and conflicting priorities among institutional actors. In terms of destination branding, this institutional ecology can be identified as part of the place-making strategy that has allowed NZ to take advantage of internationally acclaimed feature films in order to strengthen its positioning and country brand and, more precisely, its brand as tourism destination.

'100% Pure' NZ brand

From a tourism perspective, NZ is a 'geographically disadvantage destination', because it is situated 2.000 kilometres east of Australia and lies 'half way between the equator and the South Pole' (Morgan et al., 2002, p. 342). A little bigger than the UK (268.000 square kilometres) with one-fourteenth of the population, there are only 4.8 million New Zealanders. Tourism is nowadays 'New Zealand's largest export industry in terms of foreign exchange earnings', directly employing '7.5 per cent of the New Zealand workforce' (TNZ, 2016). According to TNZ, the total tourism expenditure in 2016 (last data available) was NZ$34.7 billion (or 20.5 billion Euros), an increase of 12,2% from the previous year generating 5,6% of GDP (ibid.).

NZ has 'the oldest tourism board in the world', which was established in 1901 as the Department of Tourist and Health Resorts (Morgan et al., 2002, p. 342). Known as TNZ as the trading name for the New Zealand Tourism Board (NZTB) since 1999, it launched the first-ever marketing campaign '100% Pure New Zealand' market by market between July 1999 and February 2000, with the main aim of intending to double the country's tourism expenditure of international inbound visitors by 2005 (Morgan, et al., 2003, p. 289). The vision of the campaign was 'to position NZ as the ultimate destination and for the world to know about it', while trying to recover part of the market lost to Australia, NZ's closest competitor (Morgan, et al., 2002, p. 343).

The brand essence and positioning were to present NZ as 'a relatively undiscovered, untouched land [...] shaped by its inhabitants over time' (Morgan, et al., 2002, p. 347). The founder partner of M&C Saatchi, Maurice Saatchi, who developed NZ's original campaign, described it as follows:

> As the world becomes increasingly 'manufactured', the world's nations have become more and more homogeneous. It's become almost impossible to find meaningful differentiation. But New Zealand is different. It's an authentic country. New Zealand doesn't come pre-packaged or prepared. New Zealand is real.
>
> *(as cited in Morgan, et al., 2003, p. 292)*

Following the framework of this strategic positioning, the tagline '100% Pure New Zealand' was developed in order to attempt to 'qualify a number of experiences and scenes as being "typically" or 100 per cent Pure New Zealand – what the New Zealand brochure describ[ed] as a place of "awesome sights, breathtaking vistas, indelible experiences"' (Morgan, et al., 2002, p. 348). Because a destination is not a single product but a combined structure consisting of a bundle of different components, including 'accommodation and catering, establishments; tourist attractions; arts, entertainment and cultural venues; and the natural environment' (Buhalis, 2000; cited in Morgan, et al., 2002, p. 337–338), the brand showcased under the '100% Pure' tagline the large variety of New Zealand's landscapes, people, culture, and tourism activities. In other words, TNZ did not pretend to claim 'environmental purity' or set any 'environmental standard', even though this is the main perception and, therefore, criticism that this successful campaign is receiving after 20 years of its release (RNZ, 2018).

Feature films strengthening the NZ destination brand

Just after the release of the '100% Pure' campaign, TNZ identified the strong potential of internationally acclaimed feature films, such as *The Lord of the Rings* trilogy, to strengthen the country's positioning and its brand internationally. Powered by the worldwide success of *The Fellowship of the Ring* (2001), *The Two Towers* (2002), and *The Return of the King* (2003), TNZ developed the website's promotion 'New Zealand, Home of Middle-Earth', as the trilogy was solely filmed in NZ and it was hoped that the scenery portrayed in the films would build associations around the world with adventure and 'Middle-Earth' (Carl, Kindon, & Smith, 2007; Croy, 2004; Lawn & Beatty, 2006). The text of the website explained:

> With stunning photographs and interviews with the cast and crew of the films, learn why New Zealand was the only landscape on earth that could have provided the locations for Tolkien's Middle-Earth. This behind the scenes journey is broken into five parts which focus on the interesting parallels between the fictional world and the real life country and people that helped realise it. Experience New Zealand, Home of Middle-Earth.
>
> *(as cited in Morgan, et al., 2003, p. 295)*

According to TNZ, during the peak of tourist interest between 2000 and 2004 (the period during which the *LOTR* films were released), NZ's visitor numbers increased an average of 7% (Easton, 2013). Of this increase, 1% – which translates to approximately NZ$33 million in terms of the total tourist spend – affirmed that the *LOTR* trilogy was their main or only reason for visiting NZ (TNZ, 2013a).

Although in cultural terms 'Tolkien's narrative is not accepted as a "New Zealand story"' (Jones & Smith, 2005, p. 934), the NZ-produced film adaptations of his books are often seen 'as exemplary of the emerging NZ national imaginary' (ibid., p. 928). This kind of reaction can be interpreted as an example of 'production fetishism', which happens when 'a nation proudly appropriates as "ours" an entity that is largely owned and controlled overseas' (Lawn & Beatty, 2006, p. 54). However, the illusion of local control and national productivity obscures the reality of foreign capital, along with the transnational distribution and earning-flows of this top-tier type of feature film production (Appadurai, 1990; Muñoz Larroa & Ferrer-Roca, 2017).

Be it as it may, in general terms, there is a shared perception that top-tier productions deliver positive economic spin-off to the overall NZ economy because, among other benefits, they add value to NZ's international destination brand (Jones & Smith, 2005; Morgan et al., 2003; RadioNZ, 2013, Mason, personal communication, July 30, 2013; Tzanelli, 2004). Linked to what is known as 'film tourism' by mediatizing places and spaces, the main goal of TNZ is to capitalize 'on opportunities generated by films made [in NZ], to promote NZ as an exceptional travel destination' (TNZ, 2018). In other words, the consumption of media content has an effect on tourist and on the tourism sector. The rationale behind this strategy is associated with the argument that 'when films are produced in NZ it provides an opportunity for TNZ to gain access to quality content, marketing opportunities and high impact media channels to promote New Zealand as a tourism destination' (TNZ, 2018).

After the *LOTR* promotion with 'New Zealand, Home of Middle-Earth', TNZ also decided to capitalize on *The Hobbit* film series by drawing upon the well-established '100% Pure New Zealand' campaign so as to create the offshoot of '100% Middle-earth' (Manhire, 2012). The NZ government anticipated that the tourism spin-offs from *The Hobbit* would exceed those of *LOTR* (Edwards, 2012) in terms of the marketing of NZ. However, as Minister Joyce declared, it is 'a very hard thing to finally substantiate in terms of the actual [tourism] numbers' (Hubbard, 2013).

Just after its release, TNZ (2013b) explained that the marketing strategy '100% Middle-earth, 100% Pure New Zealand' aimed 'to take advantage of that global profile by showing how easily the fantasy of *The Hobbit* movies can become reality in the form of a New Zealand holiday' (TNZ, 2013b). Nevertheless, in 2018, the focus of the marketing strategy has slightly shifted. It is not based anymore on the direct association of '*The Hobbit* films' and 'New Zealand holiday' but rather emphasizes NZ's landscapes, its people, and all the activities that any visitor can enjoy during their holidays in NZ (TNZ, 2018). Indeed, in the case of NZ, 'landscape is at the very core of the country's proposition' which, according to Morgan et al. (2002, p. 348), 'it will never be possible to change this, and fundamentally it is what brings the majority of visitors to the country', which emphasizes the importance of mediatized places and spaces.

Apart from TNZ, another active stakeholder in the promotion of NZ as home of Middle-earth has been the national carrier Air New Zealand (AirNZ). Christopher Luxon, CEO of AirNZ, is aware of the significant and essential role that AirNZ plays in not only supporting the country's marketing campaign but also making sure that 'New Zealand live up to the brand promise that it creates' (TPBO, 2015). They aim to showcase the best of NZ to the world through their customer experience:

> Our values – welcome as a friend, be yourself, can do and share your New Zealand – are based on NZ style service delivery at its best, so we actively celebrate the kiwi personality. We also showcase much of the best of New Zealand – we serve 100% NZ wine, the menus in our premium cabins are designed by kiwi chef Peter Gordon, and our stunning uniform is designed by Trelise Cooper. Plus we have long running partnerships with kiwi icons like the All Blacks.
>
> *(TPBO, 2015)*

As the national airlines AirNZ carries 'the destination's name in its branding and is key to the country's revenue in terms of both GDP and export earnings, as well as employment' (Morgan, et al., 2003, p. 291). AirNZ has a natural affiliation with the tourism board and, therefore, it partners many NZ campaigns across the world. As Luxon explains, AirNZ has

> a very significant partnership with TNZ, and lead much of the destination promotion in international markets – our campaign leveraging the Hobbit film trilogy, including two in-flight safety videos and two Hobbit themed livery's on 777 aircraft is a great example.
>
> *(TPBO, 2015)*

Named on YouTube as 'The Most Epic Safety Video ever Made #AirNZSafetyVideo', it has received almost 20 million views since it was uploaded on October 2014 (AirNZ, 2014). The association between *The Hobbit* Trilogy and Middle-earth is omnipresent throughout the safety video:

> As the official airline of Middle-earth, Air New Zealand has gone all out to celebrate the third and final film in The Hobbit Trilogy – The Hobbit: The Battle of the Five Armies. Starring Elijah Wood and Sir Peter Jackson; we're thrilled to unveil The Most Epic Safety Video Ever Made. #airnzhobbit.
>
> *(AirNZ, 2014)*

As CEO of AirNZ, Luxon recognizes that the 100% Pure New Zealand campaign has created 'significant demand for travel to New Zealand', and it is the AirNZ job 'to convert that to actual behaviour [...] to ensure New Zealand is well served from markets where demand exists' (TPBO, 2015). In relation to the controversy of the perception of the '100% Pure' campaign linked with a high expectation of environmental performance, Luxon seems to share the same impression:

> We absolutely believe we have a role to play in helping New Zealand live up to the 100% Pure brand promise, not just because of the reputational risk of poor environmental performance, but because it's what New Zealanders expect, and it's what we as a business expect. Of course that starts with ensuring our own house is in order, and we're working hard to make that the case, building a sustainability programme inherently linked to New Zealand's environmental performance.
>
> *(TPBO, 2015)*

Indeed, it seems that one of the most international NZ companies – AirNZ – puts a lot of effort in making sure that NZ live up to the brand promise of '100% Pure' environmental performance. Undeniably, the natural environment is the backdrop for NZ, because 'it gets people through the door from a tourism perspective' (TPBO, 2018). However, consultant Richards asks 'what is the real experience like on the ground?', and adds:

> At present we are chasing volume as opposed to value [...] In recent surveys, visitors are saying that the natural environment is wonderful, but they are critical of our built environment, accommodation standards and the urban experience. Our food doesn't live up to their expectations nor does our architecture or our approach to waste management.
>
> *(TPBO, 2018)*

Conclusions

The main contribution to this Companion is the implementation of a political economy of communication perspective within a media studies context by adding to the current discussion on representations of places and cultures through media. In this regard, the purpose of this chapter has been twofold. First, it has explored the use of media – feature films – in relation to tourism and tourist practices of destination branding. Second, it has included the notion of 'mediatized media' to understand the interactions between media processes and tourism within a three-tier of feature film production framework.

From a political economy of communication perspective, the marketing strategy '100% Middle-earth, 100% Pure New Zealand' produces two different kinds of tensions. One is the association that has been fostered between NZ as Middle-earth, an imagined place of rural, idyllic lifestyles and unspoiled landscapes. As a representation aimed at and with appeal to tourists, this exists in sharp contrast with the image of NZ that has been advantageous for its screen institutions to promote, as a highly developed technological 'hub' for high-end filmmaking, a representation that provides reassurance to film investors (TNZ, 2018).

The other tension is the expectation that some top-tier films that feature NZ landscapes enhance 'brand NZ' with a form of 'product placement' and through mediatizing the country's places and spaces. According to Columbia University's scholar Joe Karaganis, the country's indelible association with Middle-earth is a fragile and temporary one because 'lots of things that require dramatic landscapes this year will be done entirely on computers in five years' (Hubbard, 2013, para. 41). Indeed, in the *LOTR* films as well as *The Hobbit*, 'digital manipulation and modification often made the familiar unfamiliar' (Lealand, 2011, p. 263).

One wonders what this change might mean for the future of 'product placement' as an argument to justify film subsidies in NZ. This question is especially valid for the *LOTR* and *The Hobbit* trilogies as productions that involved a significant use of digital imaging and effects. However, the problem for 'product placement' through mediatizing the country's places and spaces arguments in relation to the depiction of NZ landscapes by these films is that the 'natural landscapes [they featured] were enhanced and manipulated by digital computer technology and special effects by Weta Digital and Weta Workshop' (Jones & Smith, 2005, p. 938; see also Lealand, 2011). This extent of digital modification reduces the usual associations between depictions of specific landscapes and the potentials for audience recognition of these, because it becomes more difficult for 'tourists to perceive a "realistic" sense of place' (Carl et al., 2007, p. 60).

In relation to these argumentations, there are several questions that can be further explored in future research. Regarding the mediatization of places and spaces, one aspect that merits further research is in which way mediatized tourists places evolve over time, both in terms of destination branding and destination development. Another question that could be addressed regarding the political economy of communication might consider how the power is exercised among the main stakeholders involved in implementing a destination marketing and branding strategy. Finally, one also wonders how digital technology will change the justification for top-tier film subsidies, considering that a destination landscape may hardly be distinguishable and, thus, to which extent will the destination brand be enhanced through mediatizing the destination's places and spaces. This is just one study and more research is therefore needed to add to the expanding body of knowledge linking media and tourism.

References

AirNZ. (2014). The most epic safety video ever made #AirNZSafetyVideo. *Air New Zealand*. Retrieved from https://www.youtube.com/watch?v=qOw44VFNk8Y

Anderson, C. (2012). New Zealand's green tourism push clashes with realities. *The New York Times*. Retrieved from https://www.nytimes.com/2012/11/17/business/global/new-zealands-green-tourism-push-clashes-with-realities.html

Appadurai, A. (1990). Disjuncture and difference in the global cultural economy. In M. Featherstone (Ed.), *Global culture, nationalism, globalization and modernity* (pp. 295–310). London, UK: Sage.

Beeton, S. (2010). The advance of film tourism. *Tourism and hospitality planning & development, 7*(1), 1–6.

Bolan, P., & Williams, L. (2008). The role of image in service promotion: Focusing on the influence of film on consumer choice within tourism. *International Journal of Consumer Studies, 32*(4), 382–390.

Bradley, G. (2018). Tourism New Zealand to change tack with its 100% Pure campaign. *New Zealand Herald*. Retrieved from https://www.nzherald.co.nz/business/news/article.cfm?c_id=3&objectid=12047130

Buchmann, A. (2010). Planning and development of film tourism: Insights into the experience of Lord of the Rings Film guide. *Tourism and Hospitality Planning & Development, 7*(1), 77–84.

Buhalis, D. (2000). Marketing the competitive destination of the future. *Tourism Management, 21*(1), 97–116.

Carl, D., Kindon, S., & Smith, K. (2007). Tourists' experiences of film locations: New Zealand as 'Middle-earth'. *Tourism Geographies: An International Journal of Tourism Space, Place and Environment, 9*(1), 49–63.

Connell, J. (2012). Film tourism: Evolution, progress and prospects. *Tourism Management, 33*, 1007–1029.

Cropp, A. (2017). Environment election 2017: Is 100% Pure New Zealand a big lie?. *Stuff.co.nz*. Retrieved from https://www.stuff.co.nz/business/96242308/environment-election-2017-is-100-pure-new-zealand-a-big-lie

Croy, W.G. (2004). *The Lord of the Rings, New Zealand, and tourism: Image building with film*. Melbourne, Australia: Monash University.

Croy, W. G. (2010). Planning for film tourism: Active destination image management. *Tourism and Hospitality Planning and Development, 7*(1), 21–30.

D'Alessandro, L., Sommella, R., & Viganoni, L. (2015). Film-induced tourism, city-branding and place-based image: The cityscape of Naples between authenticity and conflicts. *Journal of Tourism, Culture and Territorial Development, 6*. Alma Tourism Special Issue *4*, 180–194.

Dunleavy, T., & Joyce, H. (2011). *New Zealand film & television; institution, industry and cultural change*. Bristol, UK: Intellect.

Easton, P. (2013). Journey's end for Hobbit movies. *The Dominion Post*. Retrieved from http://www.stuff.co.nz/dominion-post/culture/8971392/Journeys-end-for-Hobbit-movies

Edwards, B. (2012). Political round-up: The politics of The Hobbit. *The New Zealand Herald*. Retrieved from http://www.nzherald.co.nz/opinion/news/article.cfm?c_id=466&objectid =10850288

Ferrer-Roca, N. (2015a). *Small country, big films: An analysis of the New Zealand feature film industry (2002–2012)*. Doctoral thesis, Media Studies, Victoria University of Wellington, New Zealand.

Ferrer-Roca, N. (2015b). Multi-platform funding strategies for bottom-tier films in small domestic media markets: Boy (2010) as a New Zealand case study. *Journal of Media Business Studies, 12*(4), 224–237.

Ferrer-Roca, N. (2017). Three-tier of feature film productions: The case of New Zealand. *Studies in Australasian Cinema, 11*(2), 102–120.

Ferrer-Roca, N. (2018). Feature film funding between national and international priorities. How does New Zealand bridge the gap? In P. C. Murschetz, R. Teichmann, & M. Karmasin (Eds.), *Handbook of state aid for film – Finance, industries and regulations* (pp. 383–401). Cham, Switzerland: Springer Verlag (Media Business and Innovation Series).

Ferrer-Roca, N. (2020). Art against the odds: The struggles, survival and success of New Zealand local cinema. In: A. Rajala, D. Lindblom, & M. Stocchetti (Eds.), *The political economy of local cinema: A critical approach* (pp. 175–203). Bern, Switzerland: Peter Lang.

Frost, W. (2010). Life changing experiences: Film and tourists in the Australian outback. *Annals of Tourism Research, 37*(3), 707–726.

Govers, R., Kaefer, F., & Ferrer-Roca, N. (2017). The state of academic place branding research according to practitioners (Editorial). *Place Branding and Public Diplomacy Journal, 13*(1), pp. 1–3.

Henry, H. (2018). 100% Pure New Zealand: Not all claims in advertising are meant to be taken literally. Retrieved from https://www.heskethhenry.co.nz/insights-opinion/100-pure-new-zealand-00047/

Hubbard, A. (2013). NZ a star paying to act in a supporting role. *The Dominion Post.* Retrieved from http://www.stuff.co.nz/dominion-post/news/8197506/NZ-a-star-paying-to-act-in-a-supporting-role

Jones, D., & Smith, K. (2005). Middle-Earth meets New Zealand: Authenticity and location in the making of the Lord of the Rings. *Journal of Management Studies, 42*(5), 923–945.

Kaefer, F. (2014). *Credibility at stake? News representations and discursive constructions of national environmental reputation and place brand image: The case of Clean, Green New Zealand.* Doctoral thesis, Management Communication, University of Waikato, Hamilton, New Zealand. Retrieved from http://researchcommons.waikato.ac.nz/handle/10289/8834

Kim, H., & Richardson, S. (2003). Motion picture impacts on destination images. *Annals of Tourism Research, 30*, 216–237.

Lawn, J., & Beatty, B. (2006). On the brink of a new threshold of opportunity: The Lord of the Rings and New Zealand cultural policy. In E. Mathijs (Ed.), *The lord of the rings: Popular culture in global context* (pp. 43–60). London, UK: Wallflower.

Lealand, G. (2011). The 'Jackson effect': The late 1990s to 2005. In D. Pivac, F, Stark, & L. McDonald (Eds.), *New Zealand film: An illustrated history in association with the film archive* (pp. 259–281). Wellington, New Zealand: Te Papa.

Leotta, A. (2011). *Touring the screen: Tourism and New Zealand film geographies.* Chicago, IL: Intellect, The University of Chicago Press.

Luxon, C. (2015). *Interview with Christopher Luxon, Air New Zealand.* The Place Brand Observer. Retrieved from http://placebrandobserver.com/interview-air-new-zealand-ceo-christopher-luxon-on-destination-branding/

Manhire, T. (2012). Hollywood reporter on The Hobbit: Peter Jackson's '$1bn gamble'. *New Zealand Listener.* Retrieved from http://www.listener.co.nz/commentary/the-internaut/hollywood-reporter-on-the-hobbit-peter-jacksons-1bn-gamble/

Månsson, M. (2011). Mediatized tourism. *Annals of Tourism Research, 38*(4), 1634–1652.

Morgan, N., Pritchard, A., & Piggott, R. (2002). New Zealand, 100% Pure. The creation of a powerful niche destination brand. *Journal of Brand Management, 9*(4/5), 335–354.

Morgan, N., Pritchard, A., & Piggott, R. (2003). Destination branding and the role of the stakeholders: The case of New Zealand. *Journal of Vacation Marketing, 9*(3), 285–299.

Muñoz Larroa, A., & Ferrer-Roca, N. (2017). Film distribution in New Zealand: Industrial organisation, power relations and market failure. *Media Industries, 4*(2).

NZFC. (1978). New Zealand Film Commission Act 1978. Retrieved from http:// www.legislation.govt.nz/act/public/1978/0061/latest/DLM23018.html?search=ts_act_New+Zealand+Film+Commission+Act+(1978)_resel

RadioNZ. (2013). Graeme Mason – Outgoing head of the Film Commission. *Nine To Noon.* Retrieved from http://www.radionz.co.nz/national/programmes/ninetonoon/audio/ 2566746/graeme-mason-outgoing-head-of-the-film-commission

Riley, R., Baker, D., & Van Doren, C.S. (1998). Movie induced tourism. *Annals of Tourism Research, 25*(4), 919–935.

Riley, R., & Van Doren, C. (1992). Movies as tourism promotion: A push factor in a pull location. *Tourism Management, 13*(3), 267–274.

RNZ. (2018). 100% Pure a 'marketing strategy… not an environmental standard. *Radio New Zealand*. Retrieved from https://www.radionz.co.nz/news/political/357017/100-percent-pure-a-marketing-strategy-not-an-environmental-standard

Stuff (2017). New marketing campaign 100% Pure New Zealand puffery. Retrieved from https://www.stuff.co.nz/the-press/opinion/94781950/editorial-100-pure-new-zealand-puffery

TNZ. (2013a). 100% Middle-earth, 100% Pure New Zealand. *Tourism New Zealand*. Retrieved from http://www.tourismnewzealand.com/sector-marketing/film-tourism/100percent-middle-earth

TNZ. (2013b). Film tourism fast facts. *Tourism New Zealand*. Retrieved from http://www.tourismnewzealand.com/sector-marketing/film-tourism/fast-facts

TNZ. (2016). About the industry. *Tourism New Zealand*. Retrieved from https://www.tourismnewzealand.com/about/about-the-industry/

TNZ. (2018). Film tourism. *Tourism New Zealand*. Retrieved from https://www.tourismnewzealand.com/markets-stats/sectors/film-tourism/

Tooke, N., & Baker, M. (1996). Seeing is believing: The effect of film on visitor numbers to screened locations. *Tourism Management, 17*(2), 87–94.

TPBO. (2015). Destination branding through national airline: Example air New Zealand. Interview with Christopher Luxon. *The Place Brand Observer*. Retrieved from https://placebrandobserver.com/destination-branding-national-airline-example-new-zealand/

TPBO. (2018). Interview with Brian Richards on place branding success strategies and New Zealand. *The Place Brand Observer*. Retrieved from https://placebrandobserver.com/brian-richards-interview/

Tzanelli, R. (2004). Constructing the 'cinematic tourist'. The 'sign industry' of The Lord of the Rings. *Tourist Studies, 4*(1), 21–42.

14

OFFICIAL DESTINATION WEBSITES

A place's showcase to the world

José Fernández-Cavia

Introduction

Tourism is an important economic sector for most countries, not only because it accounts for a significant share of GDP but also because it constitutes a key symbolic asset for nations. It can also be a catalyst for the development of places, including infrastructure, media attention, and residents' self-pride. Cities, regions, and countries compete for resources, and tourism can be conceived as a key resource and a driving force for the pursuit of wealth and prosperity. In this global scenario, media and communication tools play a central role since the public image of a place is built on the messages conveyed through television, newspapers, or the Internet. Now more than ever tourism is mediatized in a context of information abundance and battle for attention (Hjarvard, 2008). Tourism can be understood as a symbolic negotiation between a place and a person, and this negotiation is made partially through what Krotz (2017, p. 103) defines as a 'computer-controlled digital infrastructure'. Media provide people with the representation of tourist spaces (Månsson, 2011), and lately this role is increasingly played by online media –websites, travel blogs, video platforms, and social networks. In this context, the aim of this chapter is to understand how destinations use one of these communication tools – Official Destination Websites (henceforward ODWs) – in order to mediate an attractive and persuasive image of the place.

Tourism communication has deeply changed in the past few years, and media has imposed their logic and dynamics to the sector. Following Schulz's description of mediatization (2004), ODWs contribute to social change through:

a   *extension*: 'extending the natural limits of human communication capacities', as the potential visitor can navigate places throughout the world from a personal computer or a smartphone;

b   *substitution*: as official websites substitute a wide variety of personal interactions with friends, agents, or companies;

c   *amalgamation*: as online behaviour combines with real interactions; and

d   *accommodation*: as the institutions in charge of promoting destinations decide their actions and activities taking into account their potential online impact.

Thus, it can be asserted that media in general and ODWs in particular have become – using Hepp, Hjarvard and Lundby's (2015) term – 'co-constitutive' of the tourism field. ODWs are the first 'digital window' through which some tourists get to know and experience a tourist destination. In the following sections, from a communication and branding perspective (Fernández-Cavia, Kavaratzis, & Morgan, 2018), ODWs will be analysed in their symbolic, technical and functional facets to demonstrate why they constitute a crucial tool for the promotion of destinations.

Searching for a place to visit

Leisure travelling is responsible for almost 60% of international arrivals around the world (European Tourism Trends, 2018), and the way people are inspired to choose a destination, prepare the trip and book and buy related products and services has changed dramatically since the dawn of the Web. The tourism industry has been revolutionized due to the intense application of technology over the last few decades (Hjalager, 2015). Indeed, in this respect, information and communication technologies have been placed in the centre of the playing field (Buhalis & Law, 2008).

In this context, the search for information becomes critical to the tourist's destination choice, especially if the place has not been visited before. Here, magazines, brochures, travel agencies, and acquaintances have lost relevance compared to online communications. The Web has become the primary source of information for most travellers in the world (Pan & Fesenmaier, 2006), especially in Western countries and developed economies (Korneliussen & Greenacre, 2018). The Web not only provides information, it also offers virtual experiences on destinations and interaction with prior visitors. When it comes to travel decisions, anticipating a trip should be seen as part of the enjoyment of the adventure. At the same time, diverse web platforms (ODWs, social networks, recommendation sites) contribute decisively to the formation of a destination image (Llodrà-Riera, Martínez-Ruiz, Jiménez-Zarco, & Izquierdo-Yusta, 2015).

Destination branding

In order to promote a place as a tourism destination, cities, regions, and nations in the world use their names as if they were product brands, and set up advertising campaigns and public relations techniques to attract the attention of potential visitors. This task is usually performed by Destination Marketing Organizations (DMOs), public or public-private entities devoted to enhancing destination competitiveness (Pike, 2012). Among their duties, DMOs are responsible for designing and implementing marketing plans, and for communicating the main characteristics, identity, positioning, and promise of a destination, as part of a process that is usually known as destination branding.

Destination branding can be defined as the application of branding strategies and tools to the promotion of a tourism destination. The name of the place acts as if it were a commercial brand and tries to evoke a number of positive associations in the mind of the public in order to persuade them to visit it. Destination brands play two fundamental roles: identifying the main characteristics and attributes of a city or region and distinguishing it from its competitors (Qu, Kim, & Im, 2011).

One of the communication activities that DMOs usually carry out in order to generate a positive destination image is the creation and maintenance of an official website that works as a place's showcase to the world. This sort of websites acts as a public introduction to the place and as a meeting point for public managers, suppliers, and the media, as well as potential and actual tourists. Owing to their accessibility and cost-effectiveness, ODWs are crucial for building and disseminating a destination brand, in order to gain awareness and facilitate the traveller's choice (Fernández-Cavia & Castro, 2015).

The official destination website

An ODW is a powerful communication channel that allows DMOs to provide information, virtual experiences and interaction with actual and potential visitors (Lee & Gretzel, 2012; Li & Wang, 2010; Luna-Nevarez & Hyman, 2012). In this respect, according to the *Handbook on e-Marketing for Tourism Destinations* published by the World Tourism Organization (WTO, 2008), a quality website is crucial for a destination's effectiveness. In their recent study on online destination platforms, Molinillo, Liébana-Cabanillas, Anaya-Sánchez, and Buhalis (2018, p. 125) conclude that the ODW 'is a very interesting platform for creating the destination image, although it requires a high level of attention from users and only very moderately explains the intention to visit'.

From the standpoint of the DMO, the official website is a relatively affordable tool with a worldwide impact, dynamic and easy to update, accessible from everywhere –provided a network connection and a portable device are available – and it also guarantees control of information. From the standpoint of the traveller, ODWs are also an attractive source, as they are free, easy to access, and trustworthy, inasmuch as they only convey official information.

A text and a channel

ODWs can be considered a text and channel at the same time. A text inasmuch as an ODW composes a complete message in itself which can be read by Internet users. This vast, diverse, dynamic, and interactive message is written by DMO professionals in order to feed users with information, to inspire and persuade prospective tourists to visit the place, and to convey the most suitable planned image and brand personality for the destination (Vinyals-Mirabent & Mohammadi, 2018). Here, 'vast' means that the website is an extremely large message in itself, made up of thousands of web pages with a myriad of different texts, pictures, and videos available to the user. Huge amounts of information can be provided in an organized, easy-to-use manner in order to meet the needs of the tourist. 'Diverse' means that an ODW can address a wide variety of readers – from different countries, from different cultures, using a number of languages – providing specific content for each one. An effective destination website is able to adapt to the user's profile and offer customized information (see Figure 14.1). It should be prepared to meet a broad range of needs too.

'Dynamic' means that the website content must be reviewed, updated, and renewed on a regular basis in order to avoid outdated information. An ODW is a different text every day and must be carefully looked after. The agenda, events, and accommodation are sections that require special attention. Erroneous information, outdated news, or broken links can cause a negative impression and thus prevent potential tourists from choosing the destination. 'Interactive' means that the website is able to allow users to tailor the message they receive, to respond in different ways to the stimuli presented – selecting or avoiding it, rating it, subscribing to it – to incorporate their own content – stories, opinions, photographs, or videos – and to connect with the DMO managers or other users as well.

However, in parallel, an ODW may be seen as a communication hub that channels and allows other communication tools to be incorporated. The destination website can act as a

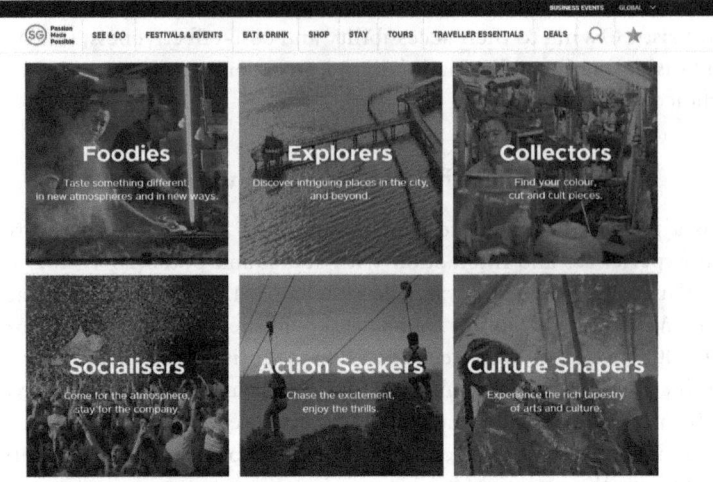

Figure 14.1 ODW and the selection of user profiles
Source: Visit Singapore, 2019.

meeting point for the official profiles on different networks; as an audiovisual platform for commercials, short documentaries, or testimonials; as a distributor of radio podcasts, news-letters, and blogs or as a provider for mobile applications.

ODWs compared to other communication tools

The organizations responsible for the management of a destination should use all kinds of communication tools at hand to promote the place and to attract visitors, depending on the resources available and the marketing objectives. DMOs from all over the world in-vest an enormous budget in media advertisements, public relations campaigns, publicity, mega-events, social networks, film tourism; all have their advantages and their limitations. The search for information becomes critical to the tourist's destination choice, especially if the place has not been visited before (Ekinci, Sirakaya-Turk, & Preciado, 2013; Llodrà-Riera et al., 2015). Here, traditional sources such as magazines, brochures, travel agencies, and ac-quaintances have lost relevance compared to the Internet (Kim, Xiang, & Fesenmaier, 2015).

Of all these communication tools, ODWs should be seen as the 'official' voice of the destination on the Web. One basic advertising strategy is to meet people where they are, and there are a lot of people browsing the Web every minute, every day. A great number of leisure trips are assumed to start with or be inspired by a simple word or combination of words in a search engine box (Pan & Fesenmaier, 2006).

In an online survey conducted in 2017 on a sample of 1,600 tourists from four different countries (Fernández-Cavia, Vinyals-Mirabent, Fernández-Planells, & Pedraza-Jiménez, 2020), half of the respondents stated that they used the Internet at some point to prepare their trip and that ODWs were the fifth most popular communication channel used in order to choose their destination. According to data provided by the ITB World Travel Trends Report 2015–2016 (IPK International, 2015), 75% of international travellers use online information as part of their trip planning, of whom 35% declare to use the ODW, 35% websites of accommodation companies, and 30% social media.

However, if we look online for Sidney, Spain, San Francisco, or Indonesia on the web, hundreds or thousands of millions of results are offered. This is why positioning and visibility are so important. Media news, recommendation platforms, Wikipedia, social networks, blogs, and unofficial destination websites all compete for attention against ODWs (Molinillo et al., 2018). Nevertheless, one clear advantage is that ODWs not only offer complete, useful information for travellers, they also provide the most up-to-date, reliable information available. From the traveller's perspective, an ODW is a convenient information source as it is free, available at anytime from anywhere – even being accessible from mobile devices, thanks to responsive web design, highly reliable since it is a message controlled and guaranteed by a public organization – and because it provides a virtual experience of the place and compiles a broad range of information in one single interconnected source (Li, Robinson & Oriade, 2017).

From the perspective of the DMO, an ODW is a highly convenient tool also because it is relatively affordable, assures an online presence worldwide, allows direct dialogue with actual and prospective tourists, and guarantees control over the information provided (Bonjisse & Morais, 2017).

The functions of the ODW

In his useful guide to tourism destination marketing, Morrison (2013) indicates nine main roles for ODWs, from the most basic one of providing information to building relationships with tourists and travel trade, and communicating the destination's positioning and branding.

Gathering, organizing, controlling, and getting information

Web 1.0 was basically composed of static pages, the foremost goal of which was to show content, usually in text format. Even though this primary stage has long been surpassed, information is still the key ingredient in any type of communication, as indeed it is for an ODW. DMOs need to compile a huge volume of data about the destination, its characteristics, the attractions that can be visited, sites and venues, restaurants, hotels, tours, its history, infrastructure, tourist products and services, including accommodation, transportation, food, weather, plans, and so on (Fernández-Cavia & Castro, 2015).

This vast array of information must then be verified, selected, organized carefully to make it easy to browse, and, lastly, published on the website. A substantial team effort is needed to undertake this hard task on a daily basis. However, at the same time, destination managers exert a control function through ODWs because they decide what is included in the official discourse about the place and what is not, and they decide the order, hierarchy, and prominence in which it is presented to the user.

Last but not least, the ODW amasses data about the people browsing the website, and frequently offers contests, newsletters, brochures, coupons, or other advantages in exchange for registering on a database. This can help DMOs to build potential visitor databanks and to implement certain market research techniques in order to better choose and tailor their communication activities (Morrison, 2013).

Contacting potential and actual travellers

Advertising scholars have coined the concept of 'touchpoint' to refer to any circumstance where the user or consumer encounters the brand. These contacts include contexts, moments, and mechanisms that are under the control of the brand and also include contexts, moments, and mechanisms that are not (Wind & Hays, 2016). In this light, ODWs can be seen as touchpoints controlled by the brand managers, connecting potential and actual travellers to the destination.

Every time a person browses the web and ends up on a destination website, it is a clear opportunity to introduce the place, explain its multiple charms, and take the browser one step further on the consumer journey. As Tassi (2018) explains: 'it is a major concern for the transmitter of a message not only to establish contact but also to attract the attention of and hold the interest of the recipient and make her want to go a step further' (p. 54). According to this author, contact with content should draw attention and attention, in turn, should generate interest and arouse emotion. ODWs must be capable of drawing attention, raising interest, and triggering emotion in order to create a bond between the tourist and the destination. But they also address other audiences that must be taken into account, such as tourism organizations, business travellers, service suppliers, the media, and locals.

Information, persuasion, and commercialization

Although information is its essential component, an ODW must also be conceived as a persuasive device, capable of generating expectations and convincing the potential visitor to choose the destination (Lee & Gretzel, 2012). For first-time visitors, the perceived risk of undertaking a journey to an unfamiliar location may be reduced or mitigated by the virtual experience, while, at the same time, the destination website provides a convincing array of reasons that can help when making a decision. Accordingly, text and images must be carefully designed and selected with the aim of persuading the user to visit the destination. The

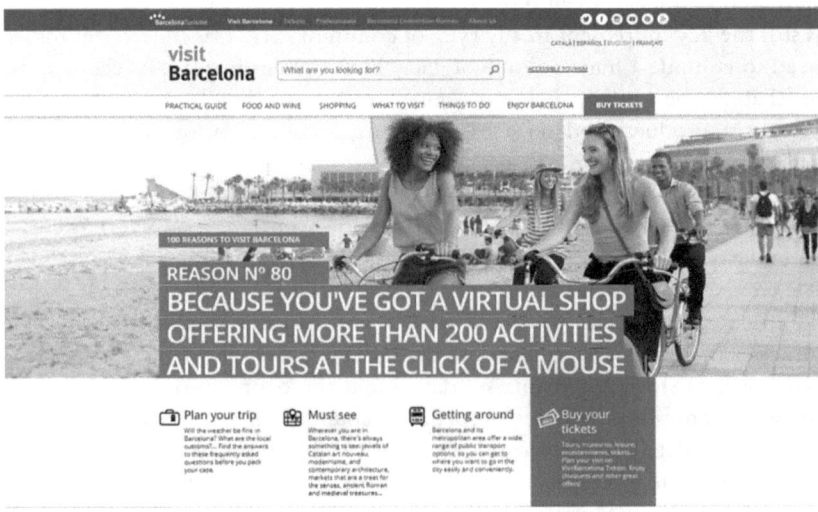

Figure 14.2 ODW offers reasons to visit as well as an online ticket service
Source: Visit Barcelona, 2019.

ODW is an effective brand proponent, a conveyor of the destination's main attractions and brand values and personality (Vinyals-Mirabent, Kavaratzis, & Fernández-Cavia, 2019), as well as a one-off opportunity to convert web surfers into clients, ambassadors, or fans. But a destination website can be considered, in parallel, an online marketplace where products and services can be examined, assessed, and acquired (see Figure 14.2).

With regard to this matter, DMOs adopt different strategies. Some of them refuse to market products or services on their websites and limit their activity to providing information and options but without directly leading to an actual purchase. Other DMOs provide booking or buying options but not within the official website itself; rather, by redirecting the user to third-party external websites, either belonging to individual providers or to general booking sites. Lastly, some DMOs opt for e-marketing of a number of selected products and services directly from their websites.

Indeed, in the first strategy, DMOs can be seen as a more neutral agent, without actively engaging in sales. But the second and third strategies allow DMOs to further supervise the business generated and to exert closer control on demand.

ODW features

All DMOs must create, maintain, and update an appealing official website in order to be competitive, but this is not an easy task (Fernández-Cavia, Marchiori, Haven-Tang, & Cantoni, 2017). An effective approach to this communication tool should not only focus on excellent technical development but should also consider other important factors:

- Providing and organizing all information a user would expect to find concerning a destination, arranging it in a comprehensive, understandable, up-to-date manner for the general public to use (Piñeiro & Igartua, 2013).
- Reflecting the image, character, and identity of the place (Lee & Gretzel, 2012) based on the understanding that a tourist destination is a destination brand.
- Promoting dynamic operation, interactivity, and the creation of user-generated content fostering communication and engagement between users and the organization managing the place.

Technical and informational quality as well as relational and persuasive features are all crucial for an effective destination website, as a number of recent scientific studies have demonstrated (Fernández-Cavia, Rovira, Díaz-Luque, & Cavaller, 2014). First of all, technical parameters assure correct operation of an ODW. Information architecture is critical in order to ensure a smooth user experience, and the website must be cleverly organized and structured to make it possible to find the information needed as soon and as directly as possible. Usability makes navigation and browsing more pleasant and it results in a better image both for the website and for the destination (Kim & Fesenmaier, 2008) while accessibility enables people with disabilities to use the ODW. Lastly, but no less importantly, Search Engine Optimization plays a crucial role. Only if the website is duly retrieved and positioned within the first organic results by search engines when carrying out a search can it work its magic; otherwise, the huge volume of work and resources devoted to the creation of the website will remain almost invisible for users.

Secondly, the amount and quality of information provided is fundamental for the ODW's performance (Park & Gretzel, 2007). The variety of content as well as the ability to meet tourists' needs is very important. Moreover, if we consider that a typical navigation route would begin on the website homepage, we can therefore agree that the design and attractiveness of this gateway may be deemed decisive. Visitors form a website impression within a mere few seconds (Luna-Nevarez & Hyman, 2012), so if a website wants to prevent them from leaving the site almost straight away, an appealing, impressive homepage must be built. Also, all this information should not be provided in just one language as a one-fits-all solution, in a uniform way, but in different languages and adapted as much as possible to the visitor's origin, culture, and profile.

Thirdly, a destination website must suitably convey the destination brand. The identity and the narrative of the place portrayed through the website must be aligned with those projected through the remaining communication channels. Functional and emotional, cognitive and symbolic attributes need to be coherent and consistent. If a destination wants to be perceived as creative and dynamic, creativity and dynamism should be apparent on the website. If a city adopts passion as one of its main identity traits, only a passionate website design –whatever that may be– would be convincing. Rhetorical and argumentative skills will be useful to achieve this (Lee & Gretzel, 2012).

Lastly, interactivity has been deemed a crucial dimension for brand websites. It helps users to process information, encourages longer periods of browsing, and improves attitudes toward the brand (Sicilia, Ruiz, & Munuera, 2005). This is why a destination website is no longer conceived as a billboard on a screen but rather as a conversation between the user and the place. Nonetheless, ODWs still have limited interactive features, and although user-generated content is claimed to be one of the most important factors in tourism decision-making (Hvass & Munar, 2012), it is scarce within this kind of website. User-message interactivity may be developed through tools such as trip planners, free downloads, mobile applications, or virtual tours. Interactivity between users is generally diverted outside the website to social media platforms. User-DMO interactivity is also very limited, as many of the sites do not allow comments, votes and do not offer a chat forum. As most DMO managers explain, this shortage of interactivity features is mainly due to fear or mistrust in giving the floor to users. ODWs are generally conceived as institutional informational platforms whose speech is designed and controlled by the DMO. Accordingly, giving power of speech to tourists is still seen more as a risk rather than an opportunity (Míguez-González & Fernández-Cavia, 2015).

The quality and metrics of an ODW

It has been empirically proven that the quality of an ODW indirectly affects potential tourists' intention to visit a destination (Chung, Lee, Lee, & Koo, 2015). But the concept of quality itself, applied to destination websites, is not an easy one. In this specific case, website quality can be seen:

1 as involving appropriate operation of the site;
2 as its capacity to be persuasive and generate positive attitudes among users – i.e., basically, intentions to visit;
3 as the potential to attract traffic and, ultimately, to generate actions or conversions.

ODWs must be constantly and carefully monitored in order to assure maximum efficacy. This is why a number of researchers have put forward different methods of assessment. They usually arise from adapting the concept of service quality to a context of technological interaction, but there's not a single universally accepted model.

Park and Gretzel (2007) summarized previous contributions to highlight what they considered key dimensions in website quality, namely, ease of use, responsiveness, fulfilment, personalization, visuals, quality of information, trust, and interactivity. This way of assessing websites – building up some general dimensions of analysis and their corresponding specific indicators – has been widely popular in academia, as the studies of Tran and Yan (2014), Fernández-Cavia et al. (2014), Loureiro (2015), or Zanna and Xuedong (2016) can confirm.

Nevertheless, this academic evaluation framework tends to neglect two alternative and powerful perspectives: that of the objective data, as number of visitors, average time spent, conversion rates, or subscribing visitors; and that of the users' subjective perception of the site. The former is often dismissed because of the difficulty of accessing real data; the latter poses some tough methodological challenges, as experimental techniques must be applied. There is still room for proposing new and more elaborate methods of ODW assessment, ideally combining website feature analysis, user subjective experience, and performance indicators. But it remains clear that the importance of this tourism communication tool justifies new research efforts in order to determine both how it functions and what are its effects in the process of tourism mediatization.

Conclusions

The tourism economy is highly impacted by mediatization. Many tourist experiences begin today by an inspirational spark in the media – a scene in a feature film, a documentary, a report in a magazine, a photograph in a social network – and end in the same way. This chapter has demonstrated how ODWs are a central part of mediatization, as they provide the 'official' portrait of the place, adapted to the visitors' needs, motivations, and expectations.

ODWs are powerful communication tools at the service of DMOs and destination brands. With the increasing use of the Web by tourists to seek inspiration for their travels, to choose a destination, and to plan a trip, ODWs have gained relevance and have been placed by managers at the centre of the destination's digital strategy.

The process of a destination's mediatization presents in this kind of websites the official version of the place, as a result of the branding professional activities carried out by the DMOs. A virtual image of the place mirrors its real identity with the aim of attracting new or repeated visitors. Nevertheless, ODWs are complex devices with multiple purposes, and this is why they have requirements, features, and specifications that must be coordinated and wisely planned so as to take advantage of their full potential. A destination website is the place's showcase to the world, accessible by everyone, at any time and everywhere; hence, they deserve all the attention, care, and creativity a DMO can devote.

References

Bonjisse, B. J., & Morais, E. P. (2017). Models for evaluating tourism websites. *Journal of Internet and e-Business Studies, 2017*, 1–15.

Buhalis, D., & Law, R. (2008). Progress in information technology and tourism management: 20 years on and 10 years after the Internet—The state of eTourism research. *Tourism Management, 29*(4), 609–623.

Chung, N., Lee, H., Lee, S. J., & Koo, C. (2015). The influence of tourism website on tourists' behavior to determine destination selection: A case study of creative economy in Korea. *Technological Forecasting & Social Change, 96*, 130–143.

Ekinci, Y., Sirakaya-Turk, E., & Preciado, S. (2013). Symbolic consumption of tourism destination brands. *Journal of Business Research, 66*, 711–718.

Fernández-Cavia, J., & Castro, D. (2015). Communication and branding on national tourism websites. *Cuadernos.info, 37*, 167–185.

Fernández-Cavia, J., Kavaratzis, M., & Morgan, N. (2018). Place branding: A communication perspective. *Communication & Society, 31*(4), 1–7.

Fernández-Cavia, J., Marchiori, E., Haven-Tang, C., & Cantoni, L. (2017). Online communication in Spanish destination marketing organizations: The view of practitioners. *Journal of Vacation Marketing, 23*(3), 264–273.

Fernández-Cavia, J., Rovira, C., Díaz-Luque, P., & Cavaller, V. (2014). Web quality index (WQI) for official tourist destination websites. Proposal for an assessment system. *Tourism Management Perspectives, 9*, 5–13.

Fernández-Cavia, J., Vinyals-Mirabent, S., Fernández-Planells, A., & Pedraza-Jiménez, R. (2020). Tourist information sources at different stages of the travel experience. *El profesional de la información, 29* (2), e290219.

Hepp, A., Hjarvard, S., & Lundby, K. (2015). Mediatization: Theorizing the interplay between media, culture and society. *Media, Culture and Society, 37*(2), 314–324.

Hjalager, A. M. (2015). 100 innovations that transformed tourism. *Journal of Travel Research, 54*(1), 3–21.

Hjarvard, S. (2008). The mediatization of society. A theory of the media as agents of social and cultural change. *Nordicom Review, 29*(2), 105–134.

Hvass, K., & Munar, A. M. (2012). The takeoff of social media in tourism. *Journal of Vacation Marketing, 18*(2), 93–103.

IPK International (2015). ITB World Travel Trends Report 2015/2016. Messe Berlin GmbH. Retrieved from: https://www.itb-berlin.de/media/itbk/itbk_dl_all/itbk_dl_all_itbkongress/ itbk_dl_all_itbkongress_itbkongress365/itbk_dl_all_itbkongress_itbkongress365_itblibrary/ itbk_dl_all_itbkongress_itbkongress365_itblibrary_studien/ITB_World_Travel_Trends_ Report_2015_2016.pdf

Kim, H., & Fesenmaier, D. R. (2008). Persuasive design of destination web sites: An analysis of first impression. *Journal of Travel Research, 47*, 3–13.

Kim, H., Xiang, Z., & Fesenmaier, D. R. (2015). Use of the internet for trip planning: A generational analysis. *Journal of Travel and Tourism Marketing, 32*(3), 276–289.

Korneliussen, T., & Greenacre, M. (2018). Information sources used by European tourists: A cross-national study. *Journal of Travel Research, 57*(2), 193–205.

Krotz, F. (2017). Explaining the mediatisation approach. *Javnost: The Public, 24*(2), 103–118.

Lee, W., & Gretzel, U. (2012). Designing persuasive destination websites: A mental imagery processing perspective. *Tourism Management, 33*, 1270–1280.

Li, S., Robinson, P., & Oriade, A. (2017). Destination marketing: The use of technology since the millennium. *Journal of Destination Marketing & Management, 6*, 95–102.

Li, X., & Wang, Y. (2010). Evaluating the effectiveness of destination marketing organisations' websites: Evidence from China. *International Journal of Tourism Research, 12*, 536–549.

Loureiro, S. M. C. (2015). The role of website quality on PAD, attitude and intentions to visit and recommend island destination. *International Journal of Tourism Research, 17*, 545–554.

Llodrà-Riera, I., Martínez-Ruiz, M. P., Jiménez-Zarco, A. I., & Izquierdo-Yusta, A. (2015). A multidimensional analysis of the information sources construct and its relevance for destination image formation. *Tourism Management, 48*, 319–328.

Luna-Nevarez, C., & Hyman, M. (2012). Common practices in destination website design. *Journal of Destination Marketing & Management, 1*, 94–106.

Månsson, M. (2011). Mediatized tourism. *Annals of Tourism Research, 38*(4), 1634–1652.

Míguez-González, M., & Fernández-Cavia, J. (2015). Tourism and online communication: Interactivity and social web in official destination websites. *Communication & Society, 28*(4), 17–31.

Molinillo, S., Liébana-Cabanillas, F., Anaya-Sánchez, R., & Buhalis, D. (2018). DMO online platforms: Image and intention to visit. *Tourism Management, 65*, 116–130.

Morrison, A. M. (2013). *Marketing and managing tourism destinations*. New York, NY: Routledge.

Pan, B., & Fesenmaier, D. (2006). Online information search. Vacation planning process. *Annals of Tourism Research, 33*(3), 809–832.

Park, Y., & Gretzel, U. (2007). Success factors for destination marketing web sites: A qualitative meta-analysis. *Journal of Travel Research, 46*, 46–63.

Pike, S. (2012). *Destination marketing. An integrated marketing communication approach.* New York, NY: Routledge.

Piñeiro, V., & Igartua, J. J. (2013). El análisis formal de sitios web y su papel en la promoción del e-turismo. *Revista Internacional de Comunicación Audiovisual, Publicidad y Literatura, 1*(11), 82–98.

Qu, H., Kim, L. H., & Im, H. H. (2011). A model of destination branding: Integrating the concepts of the branding and destination image. *Tourism Management, 32*(3), 465–476.

Schulz, W. (2004). Reconstructing mediatization as an analytical concept. *European Journal of Communication, 19*(1), 87–101.

Sicilia, M., Ruiz, S., & Munuera, J. (2005). Effects of interactivity in a website. *Journal of Advertising, 34*(3), pp. 31–45.

Tassi, P. (2018). Media: From the contact economy to the attention economy. *International Journal of Arts Management, 20*(3), 49–59.

Tran, D. T., & Yan, Z. (2014). Website evaluation for destinations: The application of an extended theoretical framework. In R. Egger & C. Maurer (Eds.), ISCONTOUR 2014. *Tourism research perspectives. Proceedings of the international student conference in tourism research* (pp. 137–147). Norderstedt, Germany: Herstellung und Verlag.

UNWTO (2018). *European tourism trends.* Retrieved from https://www.e- unwto.org/doi/book/10.18111/9789284419470

Vinyals-Mirabent, S., Kavaratzis, M., & Fernández-Cavia, J. (2019). The role of functional associations in building destination brand personality: When official websites do the talking. *Tourism Management, 75*, 148–155.

Vinyals-Mirabent, S., & Mohammadi, L. (2018). City brand projected personality: A new measure to assess the consistency of projected personality across messages. *Communication & Society, 31*(4), 91–108.

Wind, Y., & Hays, C. F. (2016). Research implications of the 'beyond advertising' paradigm: A model and roadmap for creating value through all media and non-media touchpoints. *Journal of Advertising Research, 56*(2), 142–158.

World Tourism Organisation and the European Travel Commission (2008). *Handbook on e-marketing for tourism destination.* Madrid, Spain: Europe-World Tourism Organization.

Zanna, M., & Xuedong, G. (2016). DMO tourism website's success evaluation model and framework. *European Journal of Economics and Management Sciences, 4*, 61–76.

15

DOING AS DIRECTED

Analysing representations of travel in contemporary Bollywood cinema

Apoorva Nanjangud

Introduction

Cinema and the socio-cultural fabric are known to constantly inspire and feed back into each other. In fact, Croy and Heitmann (2011, p. 190) argue that film 'creates, modifies and challenges the norms of the society'. Similarly, Bollywood cinema has also regularly inspired various trends in the Indian society, including also the way in which Indians travel, due to its many representations on-screen (Biswas & Croy, 2018; Mittal & Anjaneyaswamy, 2013). Especially in contemporary Bollywood cinema, the representation of travels is increasingly evolving, impacting footfall to places cinema popularizes due to being embedded in many emerging narratives that the audience may find interesting. As Dudrah's seminal work indicates, the representations of border crossing and travels have indeed undergone a paradigm shift ranging from the Indo-Pak border crossing, representations of diaspora, and the depiction of foreign landscapes (2012). Dudrah further elucidates the fact that cinema, and in particular Bollywood, is consumed by millions of people of the diaspora that makes it a 'globalized media phenomenon that traverses multiple locations and possibilities' (Dudrah, 2012, p. 2). His sociological reading of border crossing in Bollywood cinema (2012) is an important contribution and to some extent provides this chapter with a perspective to further study this trend.

In the current representation of place in Bollywood, a kind of formulaic portrayal of resident Indians travelling abroad for leisure is very evident. Previously, many films provided this escape through plots, stories, and narratives alone; However, it is observed that the escape has become more literal with elaborate vacationing narratives rising (for example, The film *Zindagi Na Milegi Dobara*). This has impacted the way Indians perceive tourism, and to some extent – I argue in this chapter – the way Indians perform travel. For example, films such as *Queen* (2013) or *Zindagi Na Milegi Dobara* (2015) brand young Indian travellers as an urban, mobile and globalized generation, travelling abroad, trying new cuisines, speaking foreign languages, etc. and undergoing transformation in the process. This chapter reflects on the narratives of a selection of three films that popularized and motivated certain travel behaviours such as road tripping, backpacking, and cruising. The chapter also examines the ideas that these films represent, that may percolate into the manner in which film-tourists perform travel onsite.

Apart from a selected few studies (see Angmo & Dolma, 2015; Bandyopadhyay, 2008; Biswas & Croy, 2018; Gyimóthy, 2018; Mittal & Anjaneyaswamy, 2013; Nanjangud, 2019), Bollywood tourism research is rather rare. Additionally, while studies analysing the narratives of travel on screen are minimal, fewer studies evaluate the socio-cultural complexities of class, Indianness, and rituals within these representations. However, more recently, Laing and Frost (2018) made an important contribution by analysing narratives of three predominantly western films set in India to understand the depiction of western tourist experience. However, as yet this kind of an analysis with respect to indigenous Bollywood narratives is still lacking.

Aitken and Zonn (1994, p. 5) note that 'the way spaces are used and places portrayed in film reflects prevailing cultural norms, ethical mores, societal structures and ideologies' and vice versa. It is such ideas that will be analysed in the narratives of these films. Therefore, the central aim of this chapter is to understand the narratives of travel represented in contemporary Bollywood productions, and in the process understand various socio-cultural values that they endorse from a critical cultural studies perspective. This chapter also contributes to the overarching concept of mediatized spaces and places by shedding light on the manner in which mediatization of places translates into, and may have bearing on the behaviour of tourists on-location.

Film and tourism: the journey from art to a destination promotion tool

Cinema facilitates transcultural exchanges, globalization, and exposure to different cultures and vice versa. Bollywood, also regarded as India's soft power (Thussu, 2016), has been effectively employed to boost India's cultural economy on a global scale, creating a large corpus of young upwardly mobile Indians with global attitudes. The narratives of films discussed here in, for example, catered towards the upwardly mobile Indian audiences are commonly visible in contemporary Bollywood films. As previously discussed, in a cinephilic country like India, cinema has a great degree of influence on people's cultural consumption practices, including tourism. Because of which, Indian tourists are increasingly drawing inspiration from locations popularized by cinema and conduct their tourist visitations to relive their onscreen experiences (Harjani, 2011; Khanna, 2017).

It is known that both the industries – cinema and tourism – are inherently intertwined to a great extent due to locations gaining prominence in filmmaking. This along with the fact that cinema is nothing but moving images means the two industries constantly feed back into each other (Beeton, 2015). Riley and Van Doren further argue that 'there's no finer publicity than that generated by a major motion picture' (1992, p. 267). Acknowledging that the touristification of cinema and filmification of tourism is inevitable, the collaboration of the two is increasingly becoming more pronounced (Croy & Heitmann, 2011). While this holds true globally, there are also remarkable examples within Bollywood film-tourism that indicates its influence in implicit place promotion. For example, Turkey received increased footfall after being extensively featured in the film *Dil Dhadakne Do* (2015). As this study finds, the film uses both dialogues and backdrops to highlight important cultural specificities, almost providing a visual catalogue for Turkish destinations. In fact, Turkey's changing cultural policy invites film crews including Bollywood filmmakers to shoot in Turkey by providing them with various concessions in exchange for place promotion (Ray, 2017).

Another film, *Zindagi Na Milegi Dobara* (2011), which was produced in co-operation with the Spanish tourism board, is worth discussing. Highlighting Spanish cultural elements via a mass medium like cinema made foreign culture accessible and led to a 32% increase in Indian tourists travelling to Spain, especially to the on-screen filming locations (Harjani, 2011). The solid association drawn by the audiences between Spain and the film is visible all over social media, for example, on platforms like Quora (2016, 2018). Developments such as popular tour companies providing packages to Spain, that highlight all the activities shown in the film including the Tomatina festival, are indicative of this association (Mittal & Anjaneyaswamy, 2013). This indicates that film indeed is used as an ideal carrier of destination images and aid in destination promotion, thereby boosting tourism to those film locations.

The role of film as art that organically motivated tourism to destinations is now not as simplistic. Cinema's role in destination promotion is to reiterate the role of film in tourists' experience, both pre - and onsite, create awareness, form an image, delve into the information about the location, develop expectations, and influence decisions (Croy & Heitmann, 2011, p. 194). Bollywood cinema, with its fascination for showcasing foreign locales, has introduced Indian mass audiences to places such as New York, London, Paris, Iceland, Switzerland, to name a few from popular Bollywood productions in the past. While not all films were an effort in destination promotion, it still led to an implicit awareness of the locations and a resultant association (Bandyopadhyay, 2008; Gyimóthy, 2018; Mittal & Anjaneyaswamy, 2013).

As both the film and tourism industries are becoming more sophisticated, the effects of such mediatization of places are seen in active collaborations and bilateral relations between the filmmakers and the locations; three examples are discussed in this chapter.

Bollywood tourism: whose travel is it anyway?

Bollywood tourism is defined as 'the act of conducting a touristic visitation under the influence of Bollywood cinema and/or its by-products' (Nanjangud, 2019). This has its basis in various conceptualizations of film-induced tourism (Beeton, 2005), media tourism (Reijnders, 2011), or contents tourism (Seaton, Yamamura, Sugawa-Shimada, & Jang, 2017). As established in the previous section, film acts as an ideal vehicle to disseminate images and sounds of a destination. 'Bollywood, the world's largest film industry, produces approximately 1,000 films a year and exports to 70 countries, forms the primary cultural factor, attracting consumers and audiences beyond their traditional constituency' (Dhanapalan, 2014, p. 617). However, it is worth questioning why although Bollywood cinema has a mix of profound and light-hearted storylines, it is often the cinema with pompous song and dance, and extravagant production values, that are retained around the image of Bollywood amongst the mass audiences, especially that of the diaspora. In addition, as this paper further indicates, this Bollywood lens is also often coloured with the notions of class, is celebratory in nature, and is often increasingly set in elite or foreign locations. This persuasive rhetoric of the Bollywood lens provides people with the impetus to explore these locations as shown in film and conduct tourism to those locations.

Leisure is one of the most lucrative industries in India, given the increasing purchasing power of the emerging middle class (Kerschner & Huq, 2011). However, who exactly are these people conducting film-induced tourism? As Clifford (1989, p. 35) raises a pertinent question of 'How do different populations, classes and genders travel?' Who dominates the travel narratives and can the intended audience afford reliving those on-screen experiences?

While there are several films made in Bollywood that talk about travel and journeys, this paper looks at the recent films which have had considerable impact on the real time travel trends in contemporary India as understood by the tourism figures, news, and social media. This study, therefore, attempts to go beyond mere analysis of travel representations and understand how its effect can be seen in the way film-tourists may perceive and perform travel.

Methods

This section includes a close reading of the films examined based on the storylines, representation of travels, character sketches, metaphorical references, and transformative experiences of those characters. Essentially, it also tries to examine what ideas do these films promote and relate those to how Indian tourists perform travel. I assess these representations by conducting narrative analysis of three blockbuster Bollywood movies sampled purposively between 2010 and 2015. A critical cultural perspective is adopted (Andsager & Drzewiecka, 2002) focusing on how 'meaning is constructed through pictures, narratives and other objects and language' (p. 404). Firstly, I will provide short descriptions of the three cases discussed in this chapter: *Zindagi Na Milegi Dobara*, ZNMD, from 2011; *Yeh Jawani Hai Deewani*, YJHD, from 2013; and *Dil Dhadakne Do*, DDD, from 2015. After this, a thematic discussion is conducted based on the analysis.

Case 1: Zindagi Na Milegi Dobara (ZNMD)

The film *Zindagi Na Milegi Dobara* from 2011 (English: You only live once), directed by Zoya Akhtar and produced in collaboration with the Spanish tourism promotion agency (The Local, 2016), is a story of three friends taking a bachelor's trip across Spain to live the final days of bachelorhood of Kabir, who is about to get married. With this background of the roadtrip, various nuances of relationships are explored.

> One of the reasons *Zindagi Na Milegi Dobara* has been so widely celebrated is for placing Spain as an attractive tourist destination in the imagination of India's emerging middle class and successfully boosting number of Indian tourists in the Iberian Peninsula.
>
> *(Tobias, n.d., p. 271)*

Many central characters speak Spanish, with Arjun also being fluent in Japanese depicting them as highly globalized Indians with high exposure and mobility. For example, the character of Laila is half-Indian, half-American living in London, which again indicates the globalized Indian diaspora.

Case 2: Yeh Jawani Hai Deewani (YJHD)

A film by Dharma productions, *Yeh Jawani Hai Deewani* from 2013 (English: This youth is crazy), was shot across Manali, Kashmir, Udaipur, and Paris. The story is about a group of youngsters exploring life issues, through being and travelling together. The major part of the film takes place as a flashback to Naina (Deepika Padukone), one of the central characters of the film who reminisces moments of friendship in the context of the travel experiences with her friends. A big part of this story is also the context of the destination weddings. This is further discussed in the analysis.

Case 3: Dil Dhadakne Do (DDD)

Dil Dhadakne Do from 2015 (English: Let the heart beat), by filmmaker Zoya Akhtar, is a story of modern-day dysfunctional Indian family that takes place primarily on a cruise ship. A central couple takes a bunch of their friends on a cruise trip to celebrate their wedding anniversary. A metaphor that can be understood is that the cruise is used as sort of a mobile island to trap all these dysfunctional characters and cause an interesting bunch of experiences. The cast is an ensemble with multiple actors. The film opened up gates for filmmakers to shoot actively in Turkey (Ray, 2017). In the context of the family story, there are many shots depicting the sailing of the cruise, thereby reintroducing the idea of cruising in contemporary popular Hindi cinema. Issues of class, Indianness and place promotion and so on will be elaborated in the analysis.

Analysing the discourse of travel in contemporary Bollywood cinema

During the analysis, five key themes were identified, that together indicate how cinema has been instrumental in inspiring, instructing, and moulding the Indian tourist's behaviour. These themes together address the following questions: What context to travel? Where to go? What to do? How to behave? Moreover, what are the underlying ideologies communicated in the process? These themes are discussed in detail below.

Film as cinematic destination catalogues

Firstly, I focus on the function of contemporary Bollywood cinema of providing a catalogue of destinations through cinematic narratives. It is understood that these films play an active role in destination promotion. For example, in the film ZNMD, various Spanish elements such as FC Barcelona t-shirts, the iconic La Sagrada familia, pristine beaches, the Flamenco dance with a Bollywood twist, the dish Paella are depicted. The Tomatina festival is shown in an interesting light, introducing foreign rituals to Indian masses. Barcelona, Costa Brava, Bunol, Valencia, and Seville are discussed at various points almost providing a brochure like description of Spain while mentioning the local specialities. This road trip is done in a vintage car further intensifying the European imagery. This, as (Tobias, n.d.) puts, is an attempt to appease Indian middle-class' longing for 'sanitized paradises' to escape the chaos of Indian cities. This film shot in association with the Ministry of Tourism, Spain, introduces the Indian audiences to various locations and various cultural elements, along with the possibility of adventure tourism in Spain.

Similarly, in DDD, the film depicts a detailed account of the major Turkish places of interest. From Izmir, where Turkish elements such as the hammam bath, the evil eye, Turkish tea, etc., are seen, to the skyline of Istanbul from the deck of the cruise makes for a compelling cinematic catalogue of Turkey. Through the film, the cruisers travel to various parts of Istanbul when the cruise ports. Places such as Taksim Square, symbols such as the flag of Turkey, the public transport of Istanbul such as the Istanbul Electric tramway are clearly depicted with their iconography. Along with Izmir and Istanbul, the beach town of Kusadasi, hot air balloons of Cappadocia are utilized to convey the diversity of landscapes Turkey offers. Other cultural elements include the Sufi Dervish, a traditional dance. The standee in one of the frames reads 'explore Turkey like never before', which indicates a possible collaboration with the ministry of culture and tourism, Republic of Turkey, as was the case. These films also act as feature-film length advertising for these destinations.

Travel as celebration

Another interesting finding is that these films portray resident Indians travelling in search of experiences to mark a new phase in life or a celebration. In all three films discussed above, travel has been seen as a way of marking a new beginning. More importantly and interestingly, weddings attain a prime focus in all the films. Whether it was a road trip across Spain to celebrate final days of bachelorhood as seen in ZNMD, or a trip to Udaipur, Rajasthan, for a destination wedding as seen in YJHD, or a luxury cruise trip taken to mark a 30th wedding anniversary in DDD, each of these films intertwined the ideas of travel with that of a celebration, and of ritual with location.

The way travel has unified with the wedding discourse in Bollywood cinema has also reshaped the Indian wedding market, where destination weddings have become commonplace (Economic Times, 2014; Wedmegood, 2016). A destination wedding is the idea of travelling to a particular, often scenic location to conduct the wedding rituals. This is also a huge impetus to the tourism industry; Multiple stakeholders such as the destinations, hotels, local workforce, and tour operators facilitating such large-scale weddings also benefit from the large influx of tourists who organize and attend such often upscale-weddings. Bollywoodized wedding discourses have entered the tourism industry and entertainment industry alike. This also explains that since the trend of Bollywood weddings is catching up, the wedding industry has expanded and is growing rapidly (Maran, 2016).

For example, in Udaipur, many local palaces have been converted into wedding venues due to a high demand for being showcased in YJHD (Economic Times, 2014). In the film, Udaipur's imagery is shown to be royal and exquisite, with rustic structures and sunsets. The wedding is an extravagant event organized in a palace, replete with dance and music, grand rituals, finery, and champagne. These kinds of weddings idealized by Bollywood are highly present in the emerging wedding market in India, where people travel to specific places to get married there (Maran, 2016). This finding is interesting to understand how Bollywood's idealized wedding discourses inspire tourism flows and is worth exploring and researching further.

Disruptive practices of the onscreen Indian tourists

The third theme concerns the preservation and display of Indianness in these films. In films such as ZNMD, or DDD, this is clearly observable. In places abroad, the depiction of the behaviour of Indian tourists did not change much in accordance to the country they were travelling to indicating less sensitization or travel etiquette. While occasionally it is shown that the characters are experiencing and exploring authentic local cultures, they were limited to its use for destination promotion, for example, the Turkish hammam or even the Tomatina festival. In ZNMD, a case of cultural appropriation is also seen in the attempt at Indianizing the Flamenco dance to make the culture more translatable to the Indian audiences, or even when the Tomatina festival is compared to the Indian Holi festival.

In DDD, on a cruise ship sailing in Turkey, the banter and robust dance sequence is in Hindi language, with foreigners merely assuming positions as background dancers. In ZNMD, the camaraderie between 'The three Musketeers' as they are called keeps the narrative young, relatable, and humorous. However, it is noted that at many times, the three friends behave in the same way as they did in school back in India, by tricking some local Spanish travellers for fun or by cracking inside jokes, or flirting with women. It also gets them jailed and puts them in an embarrassing situation, indicating that their behaviour does not change according to the rules and mannerisms of different countries. A point of critique here by (Tobias, n.d.) is also the lack of Spanish characters in the film. The very few Spanish characters that the protagonists encounter are just the 'victims of their constant pranks or backdrop characters of big events they take part in' (p. 282). This depiction of disruptive practices of these tourists on-screen – I argue – may reinforce poor codes of conduct.

Similarly, in DDD, when Kabir misbehaves with the cruise crew demanding them to stop the sailing cruise because he wishes to chase his love interest. As the crew disagrees because of their policy, he takes severe steps to get his way out, leading to an embarrassing situation and causing inconvenience to the entire ship. I argue, this portrayal of actions of Indians abroad seemingly may have an impact on the way the film-tourists perform travel. This finding is in line with the work of Frank (2016) and Gyimóthy (2018) who write about the disruptive and disorderly behavioural tendencies of Indian tourists when visiting abroad. Additionally, Frank (2016) pointed out that place-images disseminated through cinema, structure the performances of the Indian tourist within a local space. According to Gyimóthy (2018), 'Local operators are concerned about the disruptive performances of new travelers' (p. 385) especially Indians, and this indicates how certain characters depicted on screen may translate into how Indians perform travel globally.

The class factor

Across all three films, various narratives of class were found underpinning the travel narratives. It was found that most characters carried surnames like Talwar, Thapar, Mehra, Dewan, Saluja, etc. all interestingly upper-class (North-Indian) surnames. At various moments in the films, the affluence is depicted by the immediate decision-making power regarding travel. This makes it important to understand who really dictates, designs, and disseminates the travel discourse via contemporary Bollywood cinema and to what extent is this transferable to the Indian middle-class film-tourists.

These travel narratives are all set in primarily upper-class contexts. For example, in ZNMD, the opening shot establishes a mansion with Kabir down on his knees asking his girlfriend to marry him. The diamond ring, the ambience, flutes of champagne, all symbolize an upper-class affair. Similarly, with the friends renting vintage cars, independent villas, and personal diving lessons, a reflection of privilege is visible. There is a literal class distinction drawn between Kabir and Arjun who travel business class, and Imran who travels in the middle-seat of an economy class to their common destination, Barcelona. Interestingly, an expensive Hermès bag gains character of its own, becoming a constant source of humour. The bag becomes symbolic of upper class, and Imran constructs a satirical character out of the bag by calling it 'Bagwati' (madame bag).

Similarly, in DDD, among many elements that contextualizes the story in modern, affluent India, the most striking one is the very idea of taking a bunch of people on an all-expense paid cruise trip to Europe. While it is true that cinema inspires tourism, one must question to what extent is it plausible to do so.

However, it is interestingly noted that there is at least one character that is from a relatively lower class, and travel becomes a unifying factor amongst all the characters. This social cohesion is seen, for example, in Farhan's character in both ZNMD and DDD who is shown explicitly from a relatively lower class. Another example is that of Kabir from DDD, a rich businessperson's son who is hosting a cruise party, falls in love with Farah, a dancer on the cruiseship, who is not only from a working class but also from a different religion. This depiction of social-cohesion is not just rich-poor but is also religious and structural and is seen that as the journey continues, travel gradually equalizes these differences. This is interesting to identify, as it can have positive implications on a super-diverse country like India.

Transformative journeys

The final theme delves into the transformative nature of travelling and the transformative effects it has on Indian middle-class aspirations. In all the films discussed here, all central characters undergo transformations due to both the physical and emotional journeys they undertake during the course of the film. For example, in the film ZNMD – Arjun, from YJHD – Naina and Bunny, and from DDD – Kamal and Ayesha show stark transformations in their characters in terms of their ideologies, outlook, and also their relationships.

In due course of these adventure trips in ZNMD, each of their vulnerabilities is explored; fear of heights, fear of water, and fear of death; and so on. These adventure sports resonate with the audiences (Harjani, 2011) and create a new idea of travel. This is in line with the work of Laing and Frost (2018), where the central characters go through transformation in the process. This is an important finding and must be explored further to understand how these emerging travel narratives transform travel behaviours of the Indian-middle class.

Conclusions

Through this chapter, an attempt was made to analyse various ideas represented under the backdrop of emerging travel narratives in Bollywood cinema. All the three films analysed provide insight into how these contemporary Bollywood productions lead to the creation of mediatized narratives of places by providing a brochure-like view of the countries it represented and in turn inspire tourism, indicating a shift of Bollywood cinema from an art form to a legitimate destination marketing tool. As discussed previously along the enumerated themes, travel narratives have provided a potential backdrop for placing various values within these narratives that percolate into the socio-cultural fabric of contemporary India.

A general inclination towards European destinations is identified across all movies, which corroborates the findings of Josiam et al. (2015) suggesting that the perceived image of European destinations among a large section of the Indian population is strongly impacted by Bollywood films. It indicates that the desire for idyllic European landscapes is still what is fed to the audiences.

We know that tourism as an embodied performance is largely shaped by its mediatized narratives (Frost, 2009). This chapter suggests that the preservation of Indianness in the depiction of tourist behaviour globally, which – this paper argued – may have an impact on the way Indians perform travel onsite. I argue that films not only inspire potential tourism destinations but also promote certain tourism behaviour. For example, the trekking scene in YJHD received critique from (Arora, 2013) on the way Himalayan trekking was depicted in the film. While the hiking equipment and clothing was deemed inappropriate for the altitudes, it was also critiqued that self-decided missions or consuming alcohol on higher altitudes can lead to mis-adventures, which also raises questions about the responsibility of cinema.

Cinema and society constantly feed back into each other, and that is clearly made visible through the proliferation of the wedding market as idealized by Bollywood. Through these films, it is noted that by reiterating popular global destinations, an attempt is made to both tease and appease the aspirations of upwardly mobile urban Indians. It will be fruitful to individually explore these themes further in order to fully grasp the manner in which contemporary Bollywood cinema inspires, instructs, and moulds tourism behaviour of the middle-class Indian society.

Acknowledgement

This project has received funding from the European Research Council (ERC) under the European Union's Horizon 2020 research and innovation programme (grant agreement No 681663).

References

Aitken, S., & Zonn, L. (1994). *Place, power, situation and spectacle: A geography of film.* London: Rowman and Littlefield Publishers.

Andsager, J. L., & Drzewiecka, J. A. (2002). Desirability of differences in destinations. *Annals of Tourism Research, 29*(2), 401–421.

Angmo, T., & Dolma, K. (2015). Mass media and film induced tourism in Leh District, Jammu and Kashmir, India. *International Journal of Science and Research (IJSR), 4*(9), 301–305.

Arora, B. (2013, June 14). A reality check on trekking. Retrieved from: https://indiahikes.com/a-reality-check-on-trekking/ [Accessed 8th July 2019].

Bandyopadhyay, R. (2008). Nostalgia, identity and tourism: Bollywood in the Indian diaspora. *Journal of Tourism and Cultural Change, 6*(2), 79–100.

Beeton, S. (2005). *Film-induced tourism.* Clevedon, UK: Channel View Publications.

Beeton, S. (2015). *Travel, tourism and the moving image.* Clevedon, UK: Channel View Publications.

Biswas, J., & Croy, G. (2018). Film tourism in India: An emergent phenomenon. In S. Kim & S. Reijnders (Eds.), *Film tourism in Asia* (pp. 33–48). Singapore: Springer.

Clifford, J. (1989). Notes on travel and theory. In J. Clifford, & V. Dhareshwar (Eds.), *Traveling theories, traveling theorists inscriptions 5* (n.p.). University of California: Center for Cultural Studies.

Croy, W. G., & Heitmann, S. (2011). Tourism and film. In P. Robinson, S. Heitmann, & P. Dieke (Eds.), *Research themes for tourism* (pp. 188–204). Wallingford UK: CABI.

Dhanapalan, B. (2014). Communicating India's soft power: Buddha to Bollywood. *Asian Journal of Communication, 24*(6), 617–618.

Dudrah, R. (2012). *Bollywood travels: Culture, diaspora and border crossings in popular Hindi Cinema.* Contemporary South Asia Series, London, UK: Routledge.

Economic times, (2014). Udaipur: The place where Kalki Koechlin's wedding in 'Yeh Jawani Hai Deewani' was shot. [Online] Retrieved from: //economictimes.indiatimes.com/articleshow/45451973.cms?from=mdr&utm_source=contentofinterest&utm_medium=text&utm_campaign=cppst [Accessed 6th July, 2019].

Frank, S. (2016) Dwelling-in-motion: Indian Bollywood tourists and their hosts in the Swiss Alps. *Cultural Studies, 30*(3), 506–531.

Frost, W. (2009). From backcloth to runway production: Exploring location and authenticity in film induced tourism. *Tourism Review International, 13*(2), 85–92.

Gyimóthy, S. (2018). The Indianization of Switzerland. Destination transformations in the wake of Bollywood films. In C. Lundberg & V. Ziakas (Eds.), *The Routledge handbook of popular culture and tourism* (pp. 376–387). London, UK: Routledge.

Harjani, P. (2011). India's tourists flock to Spain. CNN, 19 September [online]. Retrieved from: travel.cnn.com: http://travel.cnn.com/mumbai/life/indian-movie-boosts-spanish-tourism-694426/ [Accessed 4th July 2019].

Josiam, B. M., Spears, D. L., Pookulangara, S., Dutta, K., Kinley, T. R, & Duncan, J. L. (2015). Using structural equation modeling to understand the impact of Bollywood movies on destination image, tourist activity, and purchasing behaviour of Indians. *Journal of Vacation Marketing, 21*(3), 251–261.

Kerschner, E. M., & Huq, N. (2011). *Asian affluence: The emerging 21st century middle class.* n.p.: Morgan Stanley Smith Barney.

Khanna, S. (2017). Adventure calling: Yeh Jawaani Hai Deewani Style! [online]. Retrieved from: https://www.makemytrip.com/blog/adventure-calling-yeh-jawaani-hai-deewani-style [Accessed 4th July 2019].

Laing, J., & Frost, W. (2018). Imagining tourism and mobilities in modern India through film. In S. Kim & S. Reijnders (Eds.), *Film tourism in Asia* (pp. 21–33). Singapore: Springer.

Maran, M. (2016). Bollywood weddings and conspicuous consumption in the emerging wedding sector in India. (Unpublished Master's thesis). London School of Economics and Political Science, London, United Kingdom.

Mittal, N., & Anjaneyaswamy, G. (2013). Film induced tourism: A study in Indian outbound tourism. *Journal of Tourism Studies, 8*(2), 37–54.

Nanjangud, A. (2019). Bollywood tourism in Japan, current challenges, potential directions. *International Journal of Contents Tourism, 4*, 1–11.

Quora (2016). How much will it cost to take a tour to Spain as in Zindagi Na Milegi Dobara? [online]. Retrieved from: https://www.quora.com/How-much-will-it-cost-to-take-a-tour-to-Spainas-in-Zindagi-Na-Milegi-Dobara [Accessed 7th August 2018].

Quora (2018). Do I have to save to plan a trip as shown in the Bollywood movie Zindagi Na Milegi Dobara? [online]. Retrieved from: https://www.quora.com/How-much-money-%E2%82%B9-do-I-have-to-save-to-plan-a-trip-as-shown-in-the-bollywood-movie-Zindagi-Na-Milegi-Dobara [Accessed 7th August 2018].

Ray, S. (2017). Turkey to woo Bollywood to shoot in its exotic locations. *Hindustan Times*, 10 May [Online]. Retrieved from: www.hindustantimes.com: https://www.hindustantimes.com/world-news/turkey-to-woo-bollywood-to-shoot-in-its-exotic-locations/story-HRm7TEz Wy3aLAmTwBzUC4L.html.

Reijnders, S. (2011). *Places of the imagination: Media, tourism, culture*. Farnham, UK: Ashgate Publishing.

Riley, R., & Van Doren, C. (1992). Movies as tourism promotion: A push factor in a pull location. *Tourism Management, 13*, 267–274.

Seaton, P., Yamamura, T., Sugawa-Shimada, A., & Jang, K. (2017). *Contents tourism in Japan: Pilgrimages to 'sacred sites' of popular culture*. Amherst, NY: Cambria Press.

The Local (2016, 20th June). Spain joins forces with Bollywood to boost tourism. Retrieved from: https://www.thelocal.es/20160620/spain-joins-forces-with-bollywood-to-boost-tourism [Accessed 6th August, 2019].

Thussu, D. K (2016). The soft power of popular cinema – Te case of India. *Journal of Political Power, 9*(3), 415–429.

Tobías, C. I. (n.d.) Zindagi Na Milegi Dobara: Exotic Spain, film tourism and the imagined geographies of the Indian middle class (Indo-Spanish Cultural Encounters: 1956–2016; 2017) Retrieved from: https://www.academia.edu/32620446/Zindagi_Na_Milegi_Dobara_Exotic_Spain_Film_Tourism_and_the_Imagined_Geographies_of_the_Indian_Middle_Class_Indo-Spanish_Cultural_Encounters_1956-2016_2017_

Wedmegood (2016). Wanna get married on the sets of Yeh Jawaani Hai Deewani? 10 Bollywood Locations that Make Great Wedding Destinations! Retrieved from: www.wedmegood.in: https://www.wedmegood.com/blog/wanna-get-married-on-the-sets-of-yeh-jaawani-hai-deewani-10-bollywood-locations-that-make-great-wedding-destinations/ [Accessed 6th July 2019].

Filmography

Zindagi Na Milegi Dobara (Zoya Akhtar, 2011)
Queen (Vikas Bahl, 2013)
Yeh Jawani Hai Deewani, (Ayan Mukherji, 2013)
Dil Dhadakne Do (Zoya Akhtar, 2015)

16

REPRESENTATION OF FOOD AND TOURISM IN LEGACY MEDIA

Rediscovering the roots

Francesc Fusté-Forné and Pere Masip

Introduction

The study of food, cuisine and gastronomy means the study of key cultural components. During recent years, food journalism has become a field of increasing interest (Hughes, 2010; Jones & Taylor, 2013; Naulin, 2012; Voss, 2012). The coverage of food issues is abundant in the current media landscape (Fusté Forné, 2017). Within this context, newspapers have been the first form of legacy media in reporting on food issues, aiming at large audiences. Here, different reasons are outlined with regard to the popularisation and democratisation of a food-based journalistic storytelling. Among them, the understanding of food as a daily leisure practice, and the motivation for food tourism experiences – as a channel for the discovery of local ways of living and further understanding of different cultures.

According to Månsson (2011), tourists' actions are influenced by media. Media representations provide tourists with information of places and destination's features (i.e. food) which have an impact on tourists' perceptions and consumption. While legacy media continue owning a huge influence in this context, nowadays tourists have moved from being passive media consumers into a new active role. This role is primarily observed in online social media networks, which exemplifies the convergence of mediatized tourism. This convergence shows that 'tourists are both consumers and producers of media products' (Månsson, 2011, p. 1635). In this sense, current media representations of (food) tourism count on travellers as new producers. In the new media hybrid system, food and traveller experts are able to generate content, which is broadly spread on amateur websites and social networks. Moreover, some legacy media also shelter them through columns, sections, and interviews.

Nonetheless, little previous research explored the role of media products in communicating the relations between food and tourism. From the comprehension of food and gastronomy as a crucial cultural attractions and drivers, the objective of this chapter is to analyse the representations of food and its geographical and territorial importance in contemporary Spanish media. Particularly, the current research investigates the narratives of food tourism in the four main Spanish newspapers, during the period 2005–2015. The methodology used is focused on both a quantitative and qualitative analysis that serves to identify the journalistic narratives focused on food tourism in Spain, which contributes to the fields of cultural communication, media studies, and tourism research.

The relationship between food, tourism, and media

Food, cuisine, and gastronomy are meaningful expressions of a culture. They are significant cultural identity markers. Sims (2009) affirms that food is a symbol of territorial identity and that eating and drinking 'enables the visitor to connect with the place and culture' (p. 333). Food conveys the natural and cultural landscapes of a territory and reflects the diverse lifestyles of a community. In this sense, the relationship between food and culture, and specifically the rediscovery of culture through food, is a valorised lifestyle and leisure practice (Jones & Taylor, 2013; Ravenscroft & Van Westering, 2001).

Within this context, tourism is a vehicle for the communication and enhancement of place identity (Hall, 1998; Wang & Xu, 2015). From a tourist perspective, there is an increasing food motivation for travelling (Hall, 2016) and thus there is a boom of food tourism practices which serve to explore cultures and lifestyles (Long, 2004). Hall and Sharples (2003) defined food tourism as the journey to gastronomic regions, with recreational and entertainment purposes, which includes visits to food producers, gastronomic festivals, food fairs, events, farmers markets, cooking shows and demonstrations, or other tourist activities related to food. This definition shows the diversity of culinary tourism practices, which may have the power to transmit a local sense of place to the customers. Here, while authors like Vaugeois and Predyk (2016) suggest the important role of products and local food traditions, other authors like Getz (2008) emphasises the significance of food festivals and events in terms of their contribution to food tourism growing. Products, tours, or events promote food tourism. Through them, the rediscovery of the roots is the rediscovery of the authentic products of a territory – its cultural and natural landscapes.

Food is a key component of sense of place (Smith, 2015). Local foods and culinary traditions, when presented as tourist experiences, play a critical role in terms of showing cultural distinctiveness. As reported by Germann Molz (2007), 'in the stories round-the-world travellers recount, it is clear that food is seen as symbolic of particular places and as a way of getting close to or consuming the essence of those places' (p. 88). This is because foodscapes communicate territorial identities (Brulotte & Di Giovine, 2014; Littaye, 2016), and its associated social and cultural contexts, which are particularly informed through food narratives. Also, Abarca and Colby (2016) point out that 'it is not food that defines our social and cultural subjectivities but the stories we tell about our food practices' (p. 7). Food, and its associated narratives, is a vital cultural expression (Lum & Ferrière le Vayer, 2016) that divulgates lifestyles and ways of life (Sutton, 2001).

As mentioned earlier, this study aims at analysing how media – particularly newspapers – represent food tourism. While previous research has studied the representations of food in magazines and press (Aguirregoitia & Fernández, 2015; Fusté Forné, 2017; Hughes, 2010; Jones and Taylor, 2001; Naulin, 2014; Voss, 2012) and showcased to what extent food topics are well arrayed in legacy media, this chapter contributes to the understanding of the close relationship between media consumption, food-based leisure, and tourism. Thus, media are a source of influence, inspiration, and of motivation for tourists (Butler, 1990; Kim & Richardson, 2003; Månsson, 2011; McLennan, Becken, & Moyle, 2017; Scheufele & Tewksbury, 2007). According to Kim (2012), media creates an appeal for particular geographical areas, which extends to its cultural and natural idiosyncrasies. Here, it is logical to affirm that media construct cultural, leisure, and social activities and events which people use and enjoy in their daily lives (Kim, 2012). In consequence, people use media to create their own gaze (Urry, 1990) with regard to identity aspects such as food. Later, while travellers visit destinations, this is not the only way to discover food and cultures. Beyond food tourism practices themselves, there are additional ways that do not include travelling but allow to symbolically incorporate all the aforementioned food-based resources.

Lifestyle and travel journalism (Hanusch, 2012; Pirolli, 2019), food journalism (Fusté-Forné & Masip, 2018; Hughes, 2010; Jones & Taylor, 2013; Naulin, 2012), and, recently, also the field of armchair tourism (Damkjær & Waade, 2014) are concepts that refer to the discovery of the connections between food and tourism through media. Consequently, within the frame of lifestyle studies, and particularly lifestyle media studies, authors such as Fürsich (2012), Hanusch (2012), or Hartley (2000) advocate that lifestyle journalism is a popular journalism that enables audiences to acquire knowledge about useful aspects related to their daily lives. Therefore, food is among one of these cultural and leisure issues (Cole, 2005; Craig, 2016; Fürsich, 2012; Hanusch, 2012). Similarly, Kristensen and From (2012) highlight that

> today the coverage of food includes good advice, recipes, reviews and expressions of taste and lifestyle, and the subject is therefore approached not only as guidance to cultural and/or gastronomic products or experiences (e.g. restaurant reviews) but also, like fashion, as a representation of ways of life and a symbolic marker of taste and lifestyle.
>
> *(p. 34)*

This can be applied to the case of travel journalism. Travel journalism as an asset of lifestyle media allows people to travel without physical mobility. Within this context, the term armchair tourism shows the fact that tourism and travel are crucial topics of media entertainment focused on lifestyles cultures (Damkjær & Waade, 2014). Reading a magazine and watching a documentary are also part of the domains of food tourism practices. According to Damkjær and Waade (2014), travel journalism is increasingly present in magazines, newspapers, television, and online media. For example, arenas such as film tourism-based research were revisited with regard to the study of the relationships between media entertainment and tourist consumption patterns (Buchmann, Moore, & Fisher, 2010; Connell, 2012). Similarly, studies on travel series (Damkjær & Waade, 2014) or soap operas (Kim & Long, 2012) illustrate domestic daily life narratives.

Among the topics developed by lifestyle and travel journalism, again food is one of the most important items. Here, food journalism as a specific type of journalism – cultural, lifestyle, and travel journalism – refers to the reporting of food from a variety of perspectives that may include products, dishes, restaurants, events, or the gastronomy as a social fact (Fusté-Forné & Masip, 2018). In particular, in terms of food writing, Voss (2012) affirms that writing about culinary topics refers to lives, traditions, and memories of people (p. 76). Thus, food journalism serves to describe how a place is and how a place tastes through its food features.

Method

The objective of this research is to study the narratives of food tourism in the four Spanish newspapers with highest circulation (OJD, 2016), namely, El Mundo, El País, El Periódico, and La Vanguardia, during the period 2005–2015. Drawing from a quantitative and qualitative analysis, the methodology used aimed at drawing the journalistic discourse around food tourism in Spain. Within the defined study period, odd years were selected and a constructed composite week sample was created (Neuendorf, 2002; Riffe, Aust, & Lacy, 1993).

A sample of 104 print issues of each newspaper, and year, was obtained. This means a total of 2,496 newspapers, whose hard copies were reviewed manually. As a result of the data collection, articles dealing with food and gastronomy topics were selected. For the whole period, a total of 860 articles from El Mundo, 708 articles from El País, 1,464 articles from El Periódico, and 1,312 articles from La Vanguardia were analysed. This means a total of 4,344 journalistic pieces. The articles were coded following a codebook tested and used in previous research (Fusté Forné & Masip, 2017, 2018).

The codebook included three main categories and up to eleven subcategories. Thus, three big thematic areas were defined, namely, production (products, dishes, and chefs); distribution (restaurants, selling points, and events); and gastronomy as a social fact (arts, society, nutrition, destinations, and media). From a quantitative point of view, data showed the significance of each of the themes: production (25.30%), distribution (46.62%), and gastronomy as a social fact (28.08%). As a result of the identification of the themes developed in each article, a discourse analysis was carried out, particularly focused on the drawing of the narratives dealing with the relationships between food and tourism in Spanish newspapers.

Food and tourism in Spanish newspapers

The relationships between food and tourism in Spanish newspapers are observed in a narrative that primarily develops products. These products are regarded as a hook from where a tourist motivation is created with regard to the discovery of a territory, its landscapes, and the cultural and social activities and events which, based on food produce, emerge as common leisure, lifestyle, and tourist practices.

Communication of national identity products

In communicating products, media talk about landscapes and heritage. Both cultural and natural environments are conveyed through local products. In this sense, all the newspapers offered a narrative were products are found to be a vital content. Products and their relation with territory – where they are cultivated, where they grow – are drawn from elements like tradition. Thus, gastronomy promotes cultural and territorial identity from a series of attributes. This relationship with the land is predominantly based on product quality and also on the ability to convey what is traditional in terms of food heritage.

First of all, with regard to 'quality', a discourse focused on products with quality labels is repeatedly observed. Here, Protected Designations of Origin (PDO) were found to be especially relevant. PDO is a label that allows to identify the significance of food produce in relation to a specific territory. It is thus a platform to enhance both products themselves and their environment. Newspapers' narratives highlight products such as Urgèlia cheese (El País, 6 August, 2005), Lleida pears (La Vanguardia, 31 August, 2013), or Asturian cider (El País, 28 December, 2005).

'Tradition' is another central element of the studied discourse. This means that there are a series of products that media report as a result of the combination between cultural identity and their geographical context. Thus, this relationship between culture, territory, and produce was observed in articles that developed for example the 'sobrasada' of Mallorca

(El Mundo, 27 November, 2009) or the spider-crabs of Galicia (El Mundo, 29 November, 2011; El País, July 9, 2005; El Periódico, 13 December, 2007).

This relationship between food and traditional culture is also extended to products associated with practices that are part of the annual cultural and religious calendar. This is the case of seasonal products such as mushrooms, chestnuts, or sweet buns, linked to autumn and the All Saints Day, widely represented in all the newspapers. The same pattern is observed in relation to the most typical products of Christmas period, where 'turrons' are one of the food landmarks of Spanish culinary and cultural roots. Later, through gastronomic fairs and festivals, a tourist value is given to all these products – events hence serve to popularise products, as it is explained below.

On the other hand, artisanal products in general terms are also part of a narrative that largely links gastronomy to identity and tradition. The search and tasting of local, traditional, and artisanal products is one of the key motivation factors for consumers – both locals and tourists – since these products build and communicate an authentic sense of place (Figure 16.1). This is observed in artisanal and handmade breads (El Mundo, 11 March, 2011; El Periódico, 25 October, 2007; La Vanguardia, 19 January, 2005), beers (El Mundo, 6 March, 2009; El Periódico, 6 September, 2007; La Vanguardia, 31 May, 2015), or cheeses (El País,

Figure 16.1 Local vegetables in a municipal market, Catalonia
Source: Francesc Fusté-Forné.

16 July, 2007; El Periódico, 18 January, 2013; La Vanguardia, 5 October, 2007), made by local and small producers. All of them have a media presence because a range of reasons such as their historical value, the unique characteristics of the making processes, or the application of innovations – as it is observed in the case of sustainable productions, an element that is increasingly valorised in newspapers' storytelling.

The wine-based narrative

Among the products highlighted throughout the study sample, wine is the most significant. In a similar direction than what was mentioned above, narratives around wine are based on two aspects: tradition and quality. Consequently, wine is narrated as a result of the relationship between both issues, which is manifested through the promotion of wines with quality labels.

Tradition and wine are intimately related because of the meanings attached to wine. These meanings rely on identity, landscape, culture, and territory. This is observed in journalistic narratives from abundant descriptions and references to wines and cavas belonging to different Protected Designations of Origin, both Catalan and Spanish. Among the examples of wines and areas of wine production in Spain, media discourses highlight major viticulture regions such as Catalonia, La Rioja, and Galicia. Also, as a result of the great presence of wine critics, there are many references to wines from other regions and countries that are also identified by their quality and international recognition, such as France, Italy, and California, United States.

In addition, articles approached winemaking from the point of view of ecological production or the use of new methods of production, such as biodynamic viticulture. All these perspectives show that wine is again a product closely linked to the landscape and the territory where it is produced. In particular, articles showed at the same time the importance of different geographical contexts for wine production, such as the case of Catalan regions of Penedès (El País, 25 August, 2007; El Periódico, 26 September, 2015; La Vanguardia, 3 August, 2011) and Priorat (El Mundo, 9 December, 2009; El País, 27 August, 2005; El Periódico, 3 July, 2009; La Vanguardia, 2 March, 2007), which are the most mentioned wine areas. Land provides wine with a unique aroma and texture, and this is what is transmitted when reporting on wines with a name and a surname, where the surname is always the place where wine is made – the landscape where wine grows and where wine is symbolically fed.

Experiencing food fairs and gastronomy festivals

Experiences around food and gastronomy are the main platform to mediatize the traditional products of a place. Thus, both fairs and festivals have large amounts of visitors, which include both local residents and tourists, who can discover how a territory is and tastes from an event perspective. In this sense, products' appeal is enhanced through fairs. These products are valorised because of their tradition, and their quality, as acknowledged above. In consequence, newspapers are communicating products' territorial roots, their social and cultural context, together with a set of associated lifestyles.

Knowledge around the foods a region produce – and the products that media in that geographical area inform – drives to events and food tourism practices, which again are examples of cultural and leisure practices that offer an understanding of local lifestyles based on food. Events serve to communicate gastronomic culture. Being a vital part of social life and tourism, events put products in a social and leisure context. Newspapers' narratives around events transmit and position culture and nature – the territory. Events are used to add value

Figure 16.2   Olive oil festival, Catalonia
Source: Francesc Fusté-Forné.

to traditional products (Figure 16.2). Thus, when products are the main hook of fairs and festivals, they are put into the core of the social life, which seems to be the main reason why a media attention is created around them.

Products' events that stand out in Spanish media refer mainly to identity products. Chestnuts (El Mundo, 22 October, 2007; El Periódico, 19 October, 2011; La Vanguardia, 26 October, 2009), mushrooms (El Mundo, 1 November, 2011; El Periódico, 24 October, 2005; La Vanguardia, 29 September, 2011) and 'turrones' (El País, 11 August, 2005; El Periódico, 22 November, 2009; La Vanguardia, 12 December, 2005) are basic examples observed in all the newspapers. In this sense, the case of mushrooms encapsulates both a large cultural tradition and a leisure practice that go far beyond the single product. It includes activities such as mushroom picking – which is linked to a wide knowledge of natural environment and mushroom cooking, a cultural practice where recipes are part of a collective inherited heritage.

Furthermore, agricultural products like 'calçots' – a Catalan type of scallion – (El Mundo, 11 March, 2009; El País, 31 January, 2005; El Periódico, 29 January, 2007; La Vanguardia, 26 January, 2013) are widely represented through local fairs and festivals. So, events are also used to inform about Spanish agricultural tradition, which have been sustaining the national economy for centuries. While its economic influence 'per se' is much lower nowadays, its relationship with tourism is a way to gain both impact and visibility. In consequence, gastronomy becomes an anchor for tourism. People are increasingly keen on eating out; they seek and enjoy tasting local foods and drinks; they aim at exploring the roots of a particular culture through its cuisine. Regions that show a close relationship with a product are usually organising local themed fairs around it and through fairs, these regions have engaged with tourism.

The identification of food tourism resources and destinations

Gastronomic tourism and its practices, beyond local fairs themselves, are the last character-istic of the 'food and tourism' narratives in Spanish newspapers. While fairs and festivals are undoubtedly the main identifying feature of the storytelling, other examples also appear as part of food tourism mediatization, namely, tours, museums, or markets.

In this sense, tours serve to contextualise products in a territory. Newspapers report on tours in several locations – all of them from the basis of the discovery of local landscapes and with regard to their contribution to regional development. This is the case of a cider tour in Asturias (El País, 9 July, 2011; El Periódico, 29 October, 2015), a green peas tour (La Vanguardia, 23 March, 2007), and a bread tour (La Vanguardia, 14 August, 2005) – which included the visit to medieval wood ovens in the mountain region of Pallars Sobirà, Catalan Pyrenees. This activity again shows the mediatization of local products as symbols of tradi-tion and history, in this case, in the context of Catalan Pyrenees.

These tours are also developed in urban environments, where 'pintxos' (El Mundo, 25 November, 2011; El País, 15 July, 2007) and 'tapas' tours (El Periódico, 9 November, 2011; La Vanguardia, 11 August, 2005) are cornerstones of journalistic narratives. Urban destinations are gaining attention because of their gastronomic appeal. This is the case of markets as co-shared spaces between locals and tourists, where La Boqueria market is high-lighted as one of the main food landmarks of Barcelona.

Food tourism therefore appeared as a growing phenomenon for the discovery and enjoyment of gastronomy. This is based on products and events – which were already discussed in previous sections – and examples deal with different destinations, mainly large cities or regions, both in Spain and abroad. Particularly, gastronomic destinations are promoted from both their restaurant offer (distribution) and their local produce (pro-duction) – especially wine, which has a great impact, with proposals such as Walk and Wine, or Wine Roads.

In addition, Spanish press informs about trends that promote leisure and tourism through gastronomy. The most significant examples are slow food and street food. For example, slow food brings people the opportunity to 'recover' products and gastronomic practices, such as the making of artisanal bread, earlier acknowledged, or the traditional consumption of vermouth (El Mundo, 21 August, 2009; El País, 7 May, 2015; El Periódico, 27 May, 2009; La Vanguardia, 24 September, 2005). These practices largely rely on tourism – for example, they reach a tourist status through street food activities. Thus, there are increasing numbers of festivals, not only food based, that include a street food offer. Here, food trucks have al-ready landed in Spanish cities, where they became a new way of consumption, a pathway to establish further social relationships, and a chance to discover cultural identities through foods and dishes.

Conclusions

Within the study of how newspapers draw place meaning, it is important to consider how their narratives may help in boosting tourism through food. This will certainly require fur-ther analysis since this research did not explore the audiences' behaviours, which is an im-portant field of research. Nonetheless, previous studies supported that consumption of media representations of a place drives people to the building of a contextualised knowledge of that place (Edensor, 2001; Kim, 2010; Urry, 1990). In this sense, media appropriates 'place'

(Connell, 2012) for food tourism purposes and creates an expectation on what people may experience (Kim, 2012) when they carry leisure activities or tourism practices based on food.

Narratives offered by Spanish dailies with regard to food tourism are based on products and experiences, which include food fairs and festivals, tours, markets, and, finally, culinary destinations themselves. All of these examples communicate a food identity. In this sense, food journalism as an example of lifestyle and travel journalism, and close to the concept of armchair tourism, allows audiences to gain knowledge about food issues. In particular, food is presented as a leisure practice which people is interested in because it forms a vital part of our daily lives. Within this framework, this ever-growing leisure culture offers a range of economical, social, and identity building impacts (Hanusch, 2012; Pirolli, 2019).

Findings of this research show that wine is a key topic when mediatising the relationship between food and tourism. Journalistic storytelling around food tourism also pays attention to the rediscovery of products' sense of place and the organisation of food fairs and local events that attract visitors and tourists to discover how a particular region tastes. Furthermore, some food tourism niches are regarded as emerging topics, which matches with the fact that they are also experiencing an increasing academic attention.

In particular, as a result of reviewing how Spanish newspapers reveal tourism practices through food, it appeared to be crucial the significance of products as regional identity markers. Particularly, there is a huge media attention paid to quality products (i.e. certified produce), sustainability issues (i.e. ecological production), and, definitely, a large variety of artisanal and traditional products – which later become niche food tourisms – which transmit the sense of place and allow people to rediscover the roots, that is, both local cultural and natural idiosyncrasies of a territory.

References

Abarca, M. E., & Colby, J. R. (2016). Food memories seasoning the narratives of our lives. *Food and Foodways, 24*(1–2), 1–8.

Aguirregoitia Martínez, A., & Fernández Poyatos, M. D. (2015). La gastronomía en la prensa española del siglo XIX. *Estudios sobre el Mensaje Periodístico, 21*(1), 17–33.

Brulotte, R. L., & Di Giovine, M. A. (2014). *Edible identities: Food as cultural heritage.* Aldershot, UK: Ashgate Publishing Ltd.

Buchmann, A., Moore, K., & Fisher, D. (2010). Experiencing film tourism. Authenticity and fellowship. *Annals of Tourism Research, 37*(1), 229–248.

Butler, R. (1990). The influence of the media in shaping international tourist patterns. *Tourism Recreation Research, 15*, 46–53.

Cole, P. (2005). The structure of the print industry. In R. Keeble (Ed.), *Print journalism: A critical introduction* (pp. 21–38). Abingdon, UK: Routledge.

Connell, J. (2012). Film tourism: Evolution, progress and prospects. *Tourism Management, 33*(5), 1007–1029.

Craig, G. (2016). Political participation and pleasure in green lifestyle journalism. *Environmental Communication, 10*(1), 122–141.

Damkjær, M. S., & Waade, A. M. (2014). Armchair tourism: The travel series as a hybrid genre. In F. Hanusch & E. Fürsich (Eds.), *Travel journalism: Exploring production, impact and culture* (pp. 39–59). Basingstoke, UK: Palgrave Macmillan.

Edensor, T. (2001). Performing tourism, staging tourism: (Re) producing tourist space and practice. *Tourist studies, 1*(1), 59–81.

Fürsich, E. (2012). Lifestyle journalism as popular journalism. *Journalism Practice, 6*(1), 12–25.

Fusté Forné, F. (2017). *Food journalism: Building the discourse on the popularization of gastronomy in the twenty-first century.* Barcelona, Spain: Universitat Ramon Llull.

Fusté-Forné, F., & Masip, P. (2017). Quin és el discurs mediàtic al voltant de la gastronomia? Narratives culinàries a la premsa catalana. *XXVII Conferència de la Societat Catalana de Comunicació "Congrés Internacional de Recerca en Comunicació".* June 30, 2017, Girona (Spain).

Fusté-Forné, F., & Masip, P. (2018). Descifrando la información periodística especializada: la gastronomía en la prensa diaria española. *Observatorio (OBS*), 12*(2), 108–121.

Germann Molz, J. (2007). The cosmopolitan mobilities of culinary tourism. *Space and Culture, 10*(1), 77–93.

Getz, D. (2008). Event tourism: Definition, evolution, and research. *Tourism Management, 29*(3), 403–428.

Hall, C. M. (1998). *Introduction to tourism: Development, dimensions and issues.* Melbourne, Australia: Longman.

Hall, C. M. (2016). Heirloom products in heritage places: Farmers markets, local food, and food diversity. In D. Timothy (Ed.), *Heritage cuisines: Traditions, identities and tourism* (pp. 88–103). Abingdon, UK: Routledge.

Hall, C. M., & Sharples, L. (2003). The consumption of experiences or the experience of consumption?: An introduction to the tourism of taste. In C. M. Hall, L. Sharples, R. Mitchell, N. Macionis, & B. Cambourne (Eds.), *Food tourism around the world: Development, management and markets* (pp. 1–24). Oxford, UK: Elsevier.

Hanusch, F. (2012). Broadening the focus. *Journalism Practice, 6*(1), 2–11.

Hartley, J. (2000). Communicative democracy in a redactional society: The future of journalism studies. *Journalism, 1*(1), 39–48.

Hughes, K. (2010, June 19). Food writing moves from kitchen to bookshelf. *Guardian Review,* pp. 2–4.

Jones, S., & Taylor, B. (2001). Food writing and food cultures: The case of Elizabeth David and Jane Grigson. *European Journal of Cultural Studies, 4*(2), 171–188.

Jones, S., & Taylor, B. (2013). Food journalism. In B. Turner & R. Orange (Eds.), *Specialist journalism* (pp. 96–106). New York, NY: Routledge.

Kim, C. (2010). Place promotion and symbolic characterization of new Songdo City, South Korea. *Cities, 27*(1), 13–19.

Kim, H., & Richardson, S. (2003). Motion picture impacts on destination images. *Annals of Tourism Research, 30,* 216–237.

Kim, S. (2012). The relationships of on-site film-tourism experiences, satisfaction, and behavioural intentions: From the film-tourism perspective. *Journal of Travel and Tourism Marketing, 29*(5), 472–484.

Kim, S., & Long, P. (2012). Touring TV soap operas: Genre in film tourism research. *Tourist Studies, 12*(2), 173–185.

Kristensen, N. N., & From, U. (2012). Lifestyle journalism: Blurring boundaries. *Journalism Practice, 6*(1), 26–41.

Littaye, A. Z. (2016). The multifunctionality of heritage food: The example of pinole, a Mexican sweet. *Geoforum, 76,* 11–19.

Long, L. (2004). *Culinary tourism: Exploring the other through food.* Lexington: University of Kentucky Press.

Lum, C. M. K., & Ferrière le Vayer, M. de (2016). *Urban foodways and communication: Ethnographic studies in intangible cultural food heritages around the world.* London, UK: Rowman and Littlefield.

Månsson, M. (2011). Mediatized tourism. *Annals of Tourism Research, 38*(4), 1634–1652.

McLennan, C. J., Becken, S., & Moyle, B. D. (2017). Framing in a contested space: Media reporting on tourism and mining in Australia. *Current Issues in Tourism, 20*(9), 960–980.

Naulin, S. (2012). *Le journalisme gastronomique. Sociologie d'un dispositif de médiation marchande.* Paris, France: Paris 4.

Naulin, S. (2014). Le plaisir affranchi de la nécessité? la représentation de l'alimentation dans le magazine Cuisine et Vins de France (1947–2010). *Sociologie et Sociétés, 46*(2), 109–131.

Neuendorf, K. A. (2002). *The content analysis guidebook.* Thousand Oaks, CA: Sage Publications.

OJD (2016). *Información y Control de Publicaciones.* Oficina de Justificación de la Difusión. Retrieved from: https://www.ojd.es.

Pirolli, B. (2019). *Travel journalism: Informing tourists in the digital age.* Abingdon, UK: Routledge.

Ravenscroft, N., & Van Westering, J. (2001). Wine tourism, culture and the everyday: A theoretical note. *Tourism and Hospitality Research, 3*(2), 149–162.

Riffe, D., Aust, C. F., & Lacy, S. R. (1993). The effectiveness of random, consecutive day and constructed week samplings in newspaper content analysis. *Journalism Quarterly, 70,* 133–139.

Scheufele, D. A., & Tewksbury, D. (2007). Framing, agenda setting, and priming: The evolution of three media effects models. *Journal of Communication, 57,* 9–20.

Sims, R. (2009). Food, place and authenticity: Local food and the sustainable tourism experience. *Journal of Sustainable Tourism, 17*(3), 321–336.

Smith, S. (2015). A sense of place: Place, culture and tourism. *Tourism Recreation Research, 40*(2), 220–233.

Sutton, D. E. (2001). *Remembrance of repasts: An anthropology of food and memory*. Oxford, UK: Berg.

Urry, J. (1990). *The tourist gaze: Leisure and travel in contemporary societies*. London, UK: Sage.

Vaugeois, N., & Predyk, J. (2016). *Consuming places through food tourism: Insights on the food artisan sector from BC*. Retrieved from: https://scholarworks.umass.edu/ttracanada_2016_conference/14.

Voss, K. W. (2012). Food journalism or culinary anthropology? Re-evaluating soft news and the influence of Jeanne Voltz's food section in the Los Angeles Times. *American Journalism, 29*(2), 66–91.

Wang, S., & Xu, H. (2015). Influence of place-based senses of distinctiveness, continuity, selfesteem and self-efficacy on residents' attitudes toward tourism. *Tourism Management, 47*, 241–250.

17

LIMINALITY AND THE STRANGER

Understanding tourists and their landscapes through *True Detective*

Hazel Andrews and Les Roberts

Introduction

In this chapter, we approach the mediatization of tourism through a particular lens: the symbolic significance of the stranger in narratives of travel and tourism and the anthropological and geographical import of liminality. We do this by offering reflections on the critically acclaimed first series of the HBO (Home Box Office) television drama *True Detective* (2014), demonstrating ways in which both the stranger and liminality are prominent themes running throughout the narrative, but also – and more crucially – exploring how these themes speak to wider issues in the spatial anthropology of travel and tourism. In this respect, the chapter is not about *True Detective* per se, but rather uses it as an illustrative tool to unpack salient themes and conceptual frameworks that remain underdeveloped in interdisciplinary studies that focus on the interplay between media practices and those particular to tourism. Situating ourselves with the symbolic landscapes of the liminal terrain in which the drama is both shot and set, we argue that the mediatization of place is an ongoing and multifaceted process of production and consumption that can rarely be isolated to a given instrumental text, demanding, as it does, closer analysis of the historical, cultural, and anthropological contexts that surround that text, and which do not start or end with it.

The stranger in a strange land

The opening of O'Rourke's (1987) well-known documentary film *Cannibal Tours* begins with the following statement 'There is nothing so strange, in a strange land, as the stranger who comes to visit it'. O'Rourke is referring to the group of white, wealthy Westerners on a holiday tour along the Sepik River in Papua New Guinea. The film follows the group as they 'bargain hunt' for indigenous art from the villagers they meet; it shows them gliding along the Sepik on a motorboat, aboard their luxury yacht accommodation and seeks their views about the people they encounter and the way they live. At the same time, the film shows the puzzlement (and at times discomfort) of local people as they are forced to sell their artwork and themselves, in the form of photographs, to earn money. They question why they are visited. One local man reflecting on the purpose of the tourists' visit ponders the role of the village's Spirit House, which is an attraction for the visitors that they pay to enter. He asks 'is this what they come for? I cannot understand'. The film highlights many issues that are relevant to understanding socio-cultural processes and underpinnings related to touristic practice, including commodification, authenticity, tourism as development, and disparities in the so-called 'host–guest' relations and encounters with otherness. However, it is that opening statement that guides us here – the *stranger in a strange land* and the strangeness attached to that.

The landscape of the Sepik River appears strange, unharnessed for rational production and as such contrasts to the spaces that doubtless many of the tourists in the film ordinarily inhabit. The immediate environs are characteristically swampy with lush, densely vegetated river banks, tall trees, and hanging vines. It is a marshy world with buildings on stilts and specially constructed wooden walkways across the boggy ground. Birds and insects can be heard in the background. The juxtaposition of this wild, untamed setting alongside the modernity of the tourists' transport and accommodation in conjunction with the obvious differences in ethnicity and wealth of the two opposing sets of protagonists (visitors and visited) could not be more striking. The film encourages the viewer to think of strangeness, of otherness, of being out of one's normal place, that is associated with the quotidian world of work and home life. The tourist has often been thought of as a searcher for difference or the unusual, to be found in unfamiliar places, without, perhaps thinking of how they themselves might appear as strange to the people in the settings that they, the tourists, find curious.

Tourism then, is, in part, premised on meetings of the strange, of otherness. In this respect, it is also inextricably linked to hospitality in terms of the basic tenet of welcoming an outsider and all that entails – at the very least providing shelter and sustenance. In the commercial world of tourists, and thus commercial hospitality, the commodification of the underlying principles of hospitality can arguably call the notion that it is hospitality into question (Andrews, 2000). Nevertheless, we can still understand and reflect on the rules that govern our way of thinking about what hospitality means, how it is practised and the significance of when it is blatantly not offered.

This chapter takes its cue from *Cannibal Tours*' opening statement to consider ideas of the stranger in a strange land in what is both a different, but not entirely dissimilar, setting. It considers the first series of HBO's highly successful *True Detective* (2014) created by Nic Pizzolatto and starring Woody Harrelson as Detective Martin 'Marty' Hart and Matthew McConaughey as Detective Rustin 'Rust' Cohle. Set in the strange landscape of the remote bayous of rural southern Louisiana alongside the lower Mississippi River in the US, the series is used as part of the marketing of the region by LouisanaTravel.com to attract visitors. Under the banner *Investigate These Three True Detective Filming Locations*, the website highlights key filming sites connected to the first series including Oak Alley Plantation where the detectives find the body of Dora Lange, whose murder they are investigating. Another of the show's settings – the mysterious Carcosa – is, in reality, Fort Macomb, home to a 200-year-old maze of catacombs. Fort Macomb is located on the Creole Nature Trail All-American Road, which consists of several intersecting roads spanning 100 miles along the Gulf Coast and is also referenced in the TV programme. LouisianaTravel.com lists several other locations connected to the filming of the programme.

The landscape of Louisiana depicted in the show is both similar to that of the Sepik River and different from it. Both O'Rourke's *Cannibal Tours* and Pizzolatto's *True Detective* have landscapes dominated by a huge river, the surrounding landscape is also marshy and green with dense vegetation in places, wooden walkways, and buildings constructed high above the water line in such a way as to avoid potential flooding. Papua New Guinea and Louisiana obviously differ in terms of economic development. In Louisiana, there is an industrialised landscape consisting of crude oil refineries and petrochemical plants with the associated pol-luting nature having taken its toll on the area's inhabitants. The 80–100 miles of industrial corridor is so heavily concentrated with polluting industries, it has earned the nickname 'Cancer Alley' (Byrnes, 2015). The industrial images seen in the TV series remind us that modernity is not far away, as in *Cannibal Tours*, the tourists' boats remind us that western modernity is within reach. We do not wish to overstretch the idea of the similarity between the two landscapes; however, both could be described as liminal. By liminal we are referring to the concept expounded by Van Gennep and later developed by the anthropologist Victor Turner which looked at ideas of structural in-betweenness in social and cultural life, most notably in relation to rites of passage and the ambiguities and uncertainties that underwrite particular forms of movement and mobility, whether socially or spatially. It is both the social and spatial significance of liminal phenomena that we are particular concerned with here (see Andrews & Roberts, 2012, 2015).

The stranger and strangeness in *True Detective*

The first season of *True Detective* follows Detectives Cohle and Hart as they investigate the ritualistic murder of a young woman, prostitute Dora Lange, in the 1990s. The action is set in the format of a time slip narrative between the present and flashbacks to their investiga-tion of the murder. In the present, both major protagonists are being interviewed regarding a more recent crime that shares similarities to the Lange murder. The story follows the en-deavours of Cohle and Hart over a 17-year period (1995–2012), as they investigate not only the murder of Lange but other young women in what appear to be occult-related murders.

Both Cohle and Hart work for the Louisiana State Homicide Unit into which Cohle had transferred from Texas the year before he begins work on the murders with Hart. Cohle is seen as a talented detective but is also the stranger in a strange land, being seen as odd by his co-workers. He has a troubled past, marred by tragedy that eventually led to his own drug addiction and breakdown. Following a period of recuperation, he turned down the offer of paid retirement and was subsequently transferred to where we find him as the story begins, over 550 miles away in the state of Louisiana. He lives alone, mainly friendless and has an uneasy relationship with his work colleagues, including Hart. Cohle has a different modus operandi underpinning his work compared to his fellow detectives, signalled in his use of a large notebook, or ledger, to record his investigative findings. The oddity associated with his meticulous note taking and the use of a large, rather than small, or, no notebook, earn him a derogatory nickname: 'The Taxman'. Eventually, following a violent altercation with Hart, Cohle resigns his post.

In the present-day context of the narrative (set in 2012), Cohle continues to jar with the setting, appearing as an unkempt, beer drinking and cigarette smoking oddball against the clean cut, respectable appearances of Hart and the interviewing officers. As the inter-views progress, it becomes apparent to both Cohle and Hart that the former is suspected of the murders he has been investigating. Eventually reconciled with Hart in their pursuit of the truth, they follow the evidence that Cohle has diligently collected, and finally identify the perpetrator.

True Detective has many twists and turns, it is disturbing and the characters complex, espe-cially Rust Cohle, not least because of his world view. His philosophical pessimism is at odds with Hart's approach to life and exacerbates his strangeness. The lack of insider knowledge that Hart attributes to Cohle as someone from outside of Louisiana manifests in a scene when the two detectives are observing a religious gathering. Hart asks Cohle 'What'd you know about these people?' Cohle replies, 'just observation and deduction' and proceeds to recount some characteristics he believes can be attributed to the group. In this respect, Cohle the stranger carries a certain degree of objectivity compared to Hart the insider.

In Simmel's influential essay, the titular stranger is a member of a group and participates in group life but is essentially distant or separate from the main group. Although Cohle shares some commonality with the people around him – job, gender, ethnicity, some (albeit limited) social activities – he remains distant, which for Simmel is marked in the stranger based on her/his origins. Cohle's strangerhood is not only apparent in his different behaviour and approach to detective work compared to his colleagues, but he is also not a local, both geographically – being from Texas and not Louisiana – and in his general approach to life. In addition, Cohle is seen as a type, a 'Texan', as a difficult person, an odd character, rather than as an individual. His strangeness is associated with danger, as one interview scene with a young woman suggests when she says, 'You're kinda strange.... like you might be dangerous', as well as in the aforementioned suspicion of him as the murderer.

In his writing, Simmel draws a distinction between the stranger and the outsider and the stranger and the wanderer. For Simmel, the wanderer 'comes today and goes tomorrow' (1950, p. 402) but the stranger is one who comes and stays. In this respect, Simmel's concept of wanderer is nearer to the leisure/pleasure tourist than the stranger because generally a tourist's presence is ephemeral. The character of Rust Cohle is more akin to Simmel's concept of the stranger because he is 'the person who comes today and stays tomorrow' (1950, p. 402). His position 'in this group is determined, essentially, by the fact that he has not belonged to it from the beginning, that he imports qualities into it, which do not and cannot stem from the group itself' (1950, p. 402). However, he has the potential to wander off because he does not really belong where he is and has no commitments to the people around him.

Building on Simmel's work, phenomenological sociologist Schutz was interested in how the stranger becomes incorporated into society whether that be a new club, new place to live or to work, noting that the stranger and the new group do not share a common history. For Schutz, this affords the stranger a unique position to question the assumptions of the group she/he approaches, as Cohle does at the religious gathering. To become part of their new world, the stranger needs to adapt to the cultural norms of the group. Failure to do so means they will remain at odds with their new situation. In the case of Cohle, his unwillingness to fit in and his strange philosophising mark him out as different from the Louisiana in which he finds himself. His apparent unwillingness to adapt to his new group means he is subjected to 'the reproach of doubtful loyalty' (Schutz, 1944, p. 507), culminating, as discussed, in his becoming a suspected murderer.

This brief outline of early sociological theories relating to the idea of the stranger focuses attention on the idea of the outsider and perhaps the difficulties that might be encountered in understandings between 'insiders' and 'outsiders'. The tourists' stay is time-limited affording little opportunity to engage with local people to achieve assimilation with the visited group – not that this appears often to be the purpose of their visit. The strangeness of Cohle in *True Detective* invites reflection on the relationships between tourists and locals that can occur in touristic practices, especially in situations such as that in *Cannibal Tours* where the cultural differences appear so marked between the incomers and the visited. As one of the locals explains, 'we don't know very much about the tourists, or from where they came.... We can't understand European people'.

Schutz argues 'Strangeness and familiarity are... general categories of our interpretation of the world' (1944, p. 507). In other words, we order the world based on what we think is unusual, different, familiar, known. Strangeness can be attractive, and it can also be frightening, puzzling, or make us feel uncomfortable. Social anthropologist Pitt-Rivers remarks that 'the unincorporated stranger cannot be abided' (1977, p. 105). The stranger cannot remain an outsider, they must become part of the pre-existing in-group. However, in *True Detective* Cohle remains unincorporated, and this makes him, as already noted, unlikeable, suspicious, and potentially dangerous. Tourists also usually remain unincorporated, never fully assimilated into the local culture. Hospitality as a basis to the development and maintenance of social life, because of its link to reciprocity, means the stranger is made welcome with the idea that at a future point the welcome will be returned. The laws of hospitality that may guide the treatment of the tourist stranger in a strange land have not been, it would appear, universally applied to Cohle in Louisiana. In the end, as an unincorporated stranger, he is able to bring a form of retribution to those who attempted to thwart his endeavours by solving the murders.

The liminality of landscape

As we have explored elsewhere (Andrews & Roberts, 2012, 2015; Roberts, 2018), when thinking about the way liminality both structures and infuses the encounter between tourist/traveller and other, it is instructive to keep in mind not only the psychosocial dynamics that underwrite such encounters but also the spatial configurations by which 'otherness' is rendered phenomenologically present as a real-and-imagined landscape through which the traveller charts his or her passage. Putting aside, for a moment, the sociocultural particularities attached to *True Detective* as a work of narrative fiction, or the threads of political and ecological critique (Byrnes, 2015; Kelly, 2016) that bind themselves to the story (but which by no means determine it), it is the materiality of landscape that we are left contemplating; a space that tugs on the embodied imagination (see Andrews, 2017), calling on it to breach a threshold that demands a degree of careful orientation lest it engulf us. The nature of such a landscape is to disorientate, ideal for the procedural drama in which protagonists are pitched against unknown forces that are there to be made knowable, or at least navigable. If the element of mystery is a property of the landscape itself, might the task of the detective be likened to that of a cartographer or surveyor, gauging the lay of the land in order to then gain mastery over it? And, if so, does such 'mastery' serve to drain the liminal qualities that make the landscape what it is in the first place: an ineffable, perhaps otherworldly domain that, once conquered, can be more readily dispensed with (or neutered) as a concretisation of abstraction? These are questions that, with some minor re-calibration, can just as well be applied to tourists themselves, who, as 'unsung armies of semioticians' (Culler, 1981, p. 127), become complicit in processes of transparency whereby all that is there to be 'read' is decoded to the point of exhaustion, like a well-worn paperback novel left for others to pick up whenever the fancy takes them. If the detective 'solves' the landscape mystery, or the tourist 'decodes' it, then what is left that remains unsaid or unmapped? To which, the answer is: *everything and anything* once the filtering lens of the rational subject (tourist, detective, cartographer) is allowed to be muddied once more.

Liminality permeates the unstructured and undisciplined terrain of the traveller once his or her command over that space is held in abeyance, either purposefully or by dint of a landscape that, in one way or another, has remained undisciplined, irrespective of the intentionality of the human agents who are thrown within it. The starting point in all of this are the liminal landscapes and 'taskscapes' (Ingold, 2000, p. 197) where we, as viewers of *True Detective*, or as travellers drawn towards the materiality of margins and fringes, brace ourselves for action. It is, then, as wayfinders, not cartographers, that those undertaking these journeys go about their business, but only insofar as the distinction is reflexively inscribed on the action that unfolds.

Drawing parallels between the wayfinding practices invested in the character of Cohle – in many ways, as we have seen, an archetypical stranger – and those that speak more fully to the liminal and experiential worlds of the traveller and tourist, it is the reflexive dispositions of both detective and tourist that are our key point of focus here. To the extent that the course of action of each is closely bound up with their status as stranger, it is as a 'stranger in a strange land' that their constitutive strangeness is more pointedly brought into play: they embody strangeness as much as the landscapes they move through do. In the case of *True Detective*, it is a landscape that carries the weight of a symbolic burden that far outstrips the capacity of the mise-en-scène – the locations and settings we are privy to in the series – to unload itself of all the meanings that we, as viewers, might take it upon ourselves to solicit and extract. In the first instance, the drama is set and shot in a landscape that is heavily mythologised in the cultural imaginary of the American South. As swamp and bayou, the wetlands of Louisiana conjure an array of symbolic associations, some of which dovetail, albeit obliquely, with *True Detective*, but which are more diffusely the cumulative product of folktales, myths, and imagined geographies that can be found across literature, popular culture, films, and television. However, despite this richly symbolic provenance, the landscape itself offers few concessions to the stranger passing through it. Byrnes comments on the dearth of roadside signage throughout the series. First and foremost, it is local knowledge that is privileged, forcing the detectives, Cohle and Hart, to repeatedly stop and ask for directions. In MacCannell's (1976) terms, the landscape is 'unmarked', there is little in the way of symbolic resources with which the detectives are able to reliably secure their bearings: '[their] investigative skills are predicated on the modernizing knowledges of roads, freeways and maps, while the criminal cabal's occult knowledges of the ever-shifting bayous elude the detectives' cartographies' (Byrnes, 2015, p. 101).

Entering the field of action equipped with a 'map' (whether literally or in the sense of a 'cartographic imagination': the rational prerequisite by which a given space of representation is rendered legible prior to one's embodied immersion within it) offers the travellers no obvious advantage. The paths they need to follow geographically are as knotted, changeable, and uncertain as those that define the emotional journey they are embarked upon. In both respects, for much of the series they remain lost; the drama takes shape around the tensions that build up as the detectives attempt to find their way through the 'labyrinths' (Franks, 2014, p. 6) that beset them at every turn. For Franks, exploring the Gothic underpinnings to the landscapes invoked in *True Detective*, 'The Louisiana bayous offer a contemporary re-imagining of the dark forest: the ideas of destruction and decay festering just below the surface of the waters' (ibid.).

The otherness of the landscape – its elemental strangeness, darkness, illegibility, even 'toxicity' (Byrnes, 2015; Kelly, 2016) – extends to that which might otherwise lay open the possibility of its touristic consumption. It is 'other' insofar as its constitutive liminalities preclude the neat packaging of place in ways that can profitably sustain an economy wedded to the influx of strangers with requisite disposable income. It lies beyond the scope of this chapter to explore the many points of connection between the wetland liminalities of *True Detective* and those that bear a close family resemblance in terms of their topographic characterisation (the siting of heavy industry such as power plants or oil refineries; the threat of imminent danger; a locus of the occult or supernatural; precariousness and instability; a social demography aligned with the economically marginalised, the educationally backward, or the lawless – see Andrews & Roberts, 2012; Roberts, 2018). But a quality that typically ties together liminal landscapes such as marshes, bogs, bayou, and swampland is a unique sense of atmosphere that pervades the flat, sprawling terrain, infused in no small part by an affective spaciousness in which light and sound can often be felt to move and resonate differently – a subtle shift in timbre, hue, and sonority as immersion in the landscape begins to work its sensorial magic. In this respect, the rural landscapes of *True Detective*, while not conventionally attractive in the aesthetic mould of the picturesque, pastoral, or sublime – any vestiges of wild(er)ness on offer remain nothing other than human in their primordial otherness – do nevertheless possess the power to enchant.

Towards a spatial anthropology of mediatized tourism

Building on the preceding discussion, the case can be made for attempting to 'read the runes' of these enchanted and immersively liminal landscapes, by which is meant to ethnographically flesh-out the imagined geographies that tourists (actual and virtual) and others bring with them into the field of practice. In the case of *True Detective*, such an approach would start by examining more closely the place-myths that have clustered in and around the key locations where the series was filmed; it would involve mapping the locations, or rather, identifying mappings of the locations posted online by fans of the show or by organisations, such as Louisiana Travel, keen to exploit its undoubted tourism potential. A spatial anthropology of *True Detective* would also set about charting the cinematic geographies of other texts that have drawn from the mythopoeic landscapes of the Louisiana bayous: for example, *The Reaping* (2007), *The Skeleton Key* (2005), *Angel Heart* (1987), or *Live and Let Die* (1973), all of which engage with ideas of the occult, supernatural, or demonic (or, in a different vein, Jim Jarmusch's *Down by Law* (1986), in which the bayous are the haunt of outsiders and low-life drifters). Having established some of the important contextual groundwork to a cultural mapping of the locations, the next steps would be to follow through with fieldwork: reading – and feeling/sensing – the material and symbolic landscapes firsthand; conducting ethnographic research with tourists who are drawn to these locations; following film and television trails, such as the many *True Detective* tour itineraries that have sprung up in response to the interest generated by the first series; gaining qualitative insights into local perceptions in order to understand how the mythic and symbolic fabric of these cultural imaginaries play out in the everyday lives of those caught in their sociocultural web.

Plundering the Southern Gothic (Byrnes, 2015; Demaria, 2014) sensibilities of *True Detective*, or following the folkloric threads of association which writer Nic Pizzolatto introduces with reference to the cult of 'The Yellow King' (Kelly, 2016, pp. 49–50; Laycock, 2015, p. 221), while instructive, can only yield so much when set alongside the wider webs of meaning that inform how these symbolic landscapes are consumed, imagined, and mythologised by tourists and others. The ritual significance of the Yellow King, a reference to an 1895 collection of gothic short stories by Robert Chambers that sits in the 'weird fiction' genre, alongside writers such as HP Lovecraft, is woven into the narrative of *True Detective* as an elusive demonic entity that stalks the landscape, whose summoning holds the key to the killings and the general toxicity infecting the social, political, and natural environment. As embodied in the form of the killer, Errol Childress, the illegitimate son of a prominent Christian minister and nephew of the state governor, 'the yellow king' represents, according to Kelly, 'the ugly face of systemic political and economic corruption' (2016, p. 51). While the textual anchoring of this mythic figure in the symbolic landscapes of *True Detective* is pivotal to the narrative and how audiences confer meaning on it, when venturing beyond the text its significance shifts and evolves. By seeking to unpack the performative geographies that interlace with and are propagated by the narrative fiction, it is the wider constellations of myths and meanings, and the way these create palpable atmospheres and affects of place and space, that are of more pressing import. By placing emphasis on the liminal character-istics of the wetland setting and the socio-culturally marginal spaces that are nested in and around these deeply mythopoeic locations, it is the capacity of the textual geographies of *True Detective* to merge with liminal spaces that are coextensive with, but crucially, not rig-idly bound by, those of the drama that command our attention. The ritual performances that are rooted and routed in these wider fields of practice – including, but not limited to, those of the tourist – are what spatial anthropology endeavours to speak to.

Conclusions

In this chapter, we have set out some provisional ideas for the anthropological study of me-diated landscapes in which the symbolic significance of the *stranger in a strange land* has been examined alongside considerations of both stranger and landscape as liminal phenomena. For the purposes of our analysis, the *in-betweenness* of the host–guest encounter has been performatively embodied in the character of Rustin Cohle, whose responsibility, as assigned to him by his office and his moral and philosophical values, is to invoke the otherness of a marginal social world and that of a radically liminal landscape that cannot, by definition, be straightforwardly 'known' or mapped. It is his inherent strangeness that is the key to unlock-ing a space that allows for insightful reflection and illumination (in his more precise case, for solving the murders). By ritually paying heed to the liminal qualities of the environment in which he finds himself, Cohle becomes receptive to the affects and cadences of landscapes that give of themselves more than can be semiotically nailed down (corresponding to what some geographers refer to as 'more-than-representational' knowledge). It may be stretching things to imagine Cohle as an anthropologist immersively engaged in fieldwork, but the analogies that may be drawn do nevertheless point to the detective work that comes with the task of rigorously engaging with an object of research by going beyond the boundaries set by what is deemed knowable or 'navigable'. Liminality is fuelled by uncertainty, imprecision, at times even danger. By recognising the fluidity and liminality that exudes beyond the frame of representation (beyond the mediatization of the landscape in which the tourist is invited to dwell), what can be said about a place or a landscape can be as open and as unbounded as that which can be said about the media practices by which that landscape is framed.

References

Andrews, H. (2000) Consuming hospitality on holiday. In C. Lashley & A. Morrison (Eds.), *In search of hospitality. Theoretical perspectives and debates* (pp. 235–254). Oxford, UK: Butterworth-Heinemann.

Andrews, H. (2017). Becoming through tourism: Imagination in practice. *Suomen Antropologi, 421*, 31–44. Retrieved from: https://journal.fi/suomenantropologi/article/view/59154/26229.

Andrews, H., & Roberts, L. (2012). Introduction: Re-mapping liminality. In H. Andrews & L. Roberts (Eds.), *Liminal landscapes: Travel, experience and spaces in-between* (pp. 1–17). London, UK: Routledge.

Andrews, H., & Roberts, L. (2015). Liminality. In N. J. Smelser & P. B. Baltes (Eds.), *International encyclopedia of the social and behavioral sciences* (pp. 131–137). Oxford, UK: Elsevier.

Byrnes, D. (2015). "I get a bad taste in my mouth out here": Oil's intimate ecologies in HBO's *True Detective. The Global South, 9*(1), 86–106.

Culler, J. (1981). The semiotics of tourism. *American Journal of Semiotics, 1*, 127–140.

Demaria, C. (2014). *True Detective* stories: Media textuality and the anthology format between remediation and transmedia narratives. *Between, 4*(8), 1–25.

Franks, R. (2014). Fear of the dark: Landscape as a gothic monster in HBO's True Detective. In D. Cavino (Ed.), *No escape: Excavating the multidimensional phenomenon of fear* (pp. 19–30). Oxford, UK: Inter-Disciplinary Press.

Ingold, T. (2000). *The perception of the environment: Essays in livelihood: Dwelling and skill.* London, UK: Routledge.

Kelly, C. R. (2016). The toxic screen: Visions of petrochemical America in HBO's True Detective. *Communication, Culture and Critique, 10*(1), 39–57.

Laycock, J. P. (2015). "Time is a flat circle": *True Detective* and the specter of moral panic in American pop culture'. *Journal of Religion and Popular Culture, 27*(3), 220–235.

MacCannell, D. (1976). *The tourist a new theory of the leisure of the class.* London, UK: The Macmillan Press.

O'Rourke, D. (1987). Director, *Cannibal Tours.* Canberra, Australia: O'Rourke and Associates.

Pitt-Rivers, J. (1977). *The fate of schechem, or the politics of sex. Essays in the anthropology of the mediterranean.* Cambridge, UK: Cambridge University Press.

Roberts, L. (2018). *Spatial anthropology: Excursions in liminal space.* London, UK: Rowman & Littlefield.

Schutz, A. (1944). The stranger: An essay in social psychology. *The American Journal of Sociology, 49*(6), 499–507. Retrieved from: http://www.jstor.org/stable/2771547.

Simmel, G. (1950). The stranger. In K. H. Wolff (Ed.), *The sociology of Georg Simmel* (pp. 402–408). Glencoe, IL: The Free Press.

PART III

Circle of representation

18

CO-CREATION CONSTRAINED

Exploring gazes of the destination on Instagram

Cecilia Cassinger and Åsa Thelander

Introduction

In relatively short time, visual social media have become key to Destination Marketing Organisations (DMOs) to convey favourable images of places to attract visitors. Tourism studies has a long tradition of studying the role of visual practices in the tourism experience. Everyday photography has been considered in relation to the consumption of tourist destinations and products (Gretzel, 2017; Schroeder, 2002). The interest for visuals in tourism studies is not least attributed to Urry's (1990) seminal work on the tourist gaze and the visual consumption of the tourist experience (see also Edensor, 2001, 2008; Urry, 2002). Recent research demonstrates the way that tourist photography and tourists' gazes are affected by digital technology and social media (Gretzel, 2017; Urry & Larsen, 2011). Dinhopl and Gretzel (2016) observe how the tourist gaze is intertwined with digital technologies for visually representing places and the self before social media publics. Urry and Larsen (2011) refer to the mediatized gaze to underscore the significance of different types of media for defining tourists' images and expectations of destinations. Marketers contribute to the mediatization of destinations through branding and promotional efforts. Recently, the local population has become increasingly involved to represent and co-create images of destinations, especially in digital media campaigns (cf. Pamment & Cassinger, 2018). Promotional strategies involving user-generated content on visual social media are employed by DMOs to promote and enhance images of destinations. Expectations among practitioners and academics are generally high in regard to what such strategies may accomplish in terms of representing destinations in new and exciting ways (Thelander & Cassinger, 2017). Nevertheless, there is scarce knowledge on what user-generated content actually do with and to images of destinations.

In this chapter, we explore a destination marketing initiative based on user-generated content on Instagram. The aim of the study is to examine the mechanisms that govern user-generated content in visual social media and, in a more general sense, the co-creation of destination image. In addition, the study queries whether user generated visual content enable novel images and new ways of 'seeing' the destination. As a particular case in point, we analyse a so-called Instagram takeover in which locals of a destination located in the southern part of Sweden curate the content. Our analysis focuses on locals' practices, representational practices, and the social conventions governing such practices. The Instagram take-over, studied here, was launched by the destination to counter negative media portrayal and images. The aim with the initiative was to reimagine the place by engaging locals. The study contributes to with new knowledge on visual aspects of the mediatization of destinations in social media (see also Wise in this volume).

The paper is organised in the following way. First, we present the extant research on media, destination, and image. Second, we develop a theoretical framework consisting of a practice-based approach to Instagram photography and the concept of the tourist gaze. Third, we account for the methodological assumptions of the study and explain how the empirical material was collected and analysed. Finally, we discuss the findings in relation to the theoretical framework suggesting that there is a social media gaze, which could further develop our understanding of social conventions in visual social media.

Destination image and social media

In recent years, visual-based social media have become central in destination marketing communications (Gretzel, 2017). Traditional assumptions that tourists' photographic practices can be controlled by DMOs and that tourists' images of a destination often converged with those of marketing are today questioned in view of the emergence of smartphone photography and social media. It is becoming increasingly obvious that social media images of destinations are difficult to control and that co-creation processes profoundly influence the image and reputation of destinations.

The digitalisation of personal photography in recent years has several consequences for the co-creation of destination image. First, a greater number of photographs are taken, which means that more images of destinations are in circulation. The expectation is greater on continuously producing and updating images. Social media users are overwhelmed by images. Second, planning is reduced in photography and images are produced more spontaneously, particularly when smartphones are used (Van House, 2011). The quality of photos has generally improved along with the possibility to adjust the photograph immediately after shooting. Third, everyday life situations are mainly photographed in social media images, which means that what is seen as photo worthy has changed from the spectacular to the mundane. Fourth, possibilities of editing and altering photographs are greater than ever. The large range of filters provided by Instagram facilitates modifying and improving photographs. Moreover, the Instagram platform makes it easy to share and store large numbers of photographs.

The role of social media has especially been researched within the areas of tourist and family photography (Chalfen 1987; Rose, 2014). In these areas, photography has traditionally been viewed as a 'natural activity'. The critic Susan Sontag (1977, p. 9) that 'it seems positively unnatural to travel for pleasure without taking a camera along. Photographs will offer indisputable evidence that the trip was made, that the program was carried out, that the fun was had'. In photo-sharing activities in social media, however, personal photography changes from a memory object to a source of identity formation. We 'use digital cameras for live communication instead of storing pictures of life' (van Dijck, 2008, p. 58). Hence, visual technologies change the meaning of photography. Dinhopl and Gretzel (2016) argues that more focus is directed to social relations which implies a changed relation to the destination. Less research concerns domains where photography is less 'natural' or where photography is part of someone else's aims and interests, for instance, when photography is integrated in destination marketing activities. Previous studies have recognised the significance of photography in community intervention projects with the aim of strengthening, changing, or improving the identity of a neighbourhood or area (see Packard, 2008). In these studies, analogue photography and the photographs per se have mainly been studied. Pink (2011), however, calls for an expanded definition and view of photography and photography practice, taking into consideration the experience of being involved in projects and activities where photography and visual social media is used. In focusing on participants practices and resulting gazes, this chapter is in part a response to this call.

Disciplining co-creation – gazing and gaze

The present study is informed by theories on practice and gaze in order to grasp how images of a destination are co-created by locals participating in a destination marketing initiative on the photo-sharing platform Instagram. In this section, we give a brief overview of these theories.

A practice-based approach

Even though practice theory is scattered across different disciplinary fields and empirical study objects, there are a few shared basic assumptions in regard to practice (see e.g. Postill, 2010; Schatzki, 1996). For example, that social life is performative; that practices are collectively organised and coordinated by shared understandings, procedures, and engagements; and that practices are relational and change over time. Practice theorists differ in regard to the relation between structure and agency. Applying practice theory to consumption, Randles and Warde (2006, p. 229) emphasise the role of structure in claiming that 'nor are practices understandable without regard to the broader political, infrastructural and technological environments in which they are sustained'. Hence, practices are viewed as combinations of mental frames, artefact, technology, discourse, values, and symbols (Orlikowski, 2007; Schatzki, 1996). A particular combination of these different building blocks constitutes practice, which, for example, can be 'routinized ways in which bodies are moved, objects are handled, subjects are treated, things are described and the world is understood' (Reckwitz, 2002, p. 250). Here, we regard personal photography as an everyday practice that is carried out within the confinements of destination marketing campaign. With Schatzki's (1996) terminology, we are studying an 'integrative practice', that is to say, a practice that is embedded in a particular domain of social life in contrast to dispersed practices, which can take place within and across domains. The different activities are only implicitly related to the implication of structure for practice. Therefore, we consider gaze as an analytical concept to identify and discuss social conventions of photography practice in social media.

Gaze and gazing

The concept of gaze originates from Foucauld and is closely connected with discursive practices (e.g. Foucault, 1977). In his studies of society's institutions, such as the medical clinic and the prison, Foucault outlines gaze as a form or power or discipline. Thus, gaze is a way of disciplining a particular subject. The medical gaze, for instance, refers to ways on which objects of medical knowledge and practice are viewed and understood (Heat, 1999). The gaze is not fixed but develops over time, and several gazes may co-exist. In late modern society, gaze is interiorised within the individual. Individuals are implicated in the disciplining and monitoring of themselves to a greater degree than before, thus gaze operates in a network of social actors and relations.

In tourism studies, the concept of gaze is foremost associated with Urry's (1990) work on the tourist gaze (see also Urry & Larsen, 2011). Urry (1990) introduced the tourist gaze to capture the visual nature of tourism. The tourist gaze holds that the way we see things is a collective and cultural ability. We learn how to see and look at the world. We never look at the object in itself, but object in relation to ourselves, and the world. The tourist gaze includes the distinct, striking, unusual, and photo worthy. Urry and Larsen (2011) write

> Just like language, one's eyes are socio-culturally framed and there are various 'ways of seeing'. 'We never just look at one thing; we are always looking at the relation between things and ourselves' (Berger 1972, p. 9). People gaze upon the world through a particular filter of ideas, skills, desires and expectations, framed by social class, gender, nationality, age and education. Gazing is a performance that orders, shapes and classifies, rather than reflects the world
>
> *(Urry & Larsen, 2011, p. 2)*

These values are used and reproduced in advertisements, destination marketing, and by tourists. However, it should be underscored that there is no such thing as a tourist gaze, as there are many different ways of seeing. Gaze is perhaps best understood in terms of the practice of gazing (Larsen & Urry, 2011). Gazing means to shape and order the world, rather than passively reflecting it. The practice is embedded in and framed by our personal experiences, memories, and learnt abilities. Ways of seeing and representing places are also mediated and therefore Edensor (2001, p. 71) writes 'When stepping into particular stages, pre-existing discursive, practical, embodied norms and concrete guides and signs usually choreograph tourists'. This definition of tourist gaze makes the concept useful when practices of photography are studied in order to discuss social conventions and norms in parallel with other visual norms and practices.

Translated into the present context of study, we may, then, understand gazing as a form of disciplining of images that are institutionalised with particular discourses (e.g. tourism, marketing, visual social media). Approaching Instagram photography as a discursive practice enables us to understand the ongoing making of images situated in intersecting practices and social relations. The approach focuses on how photography is performed and acknowledges the role of technique as well as social conventions and personal competence. Studying discursive practices involves paying attention not only to the meanings that participants ascribe photography but also how photography is disciplined by certain ways of gazing.

Capturing destination image in social media

The question of how user-generated content shape the image of destination is investigated through a qualitative study of an Instagram takeover launched to improve the public image of the city of Landskrona situated at the Swedish southwest coast. Landskrona currently struggles to reinvigorate its image in the eyes of the public. The population has declined, and unemployment is currently an urgent problem for the city. Social problems are well documented, and violent incidents (e.g. shootings, assaults, and honour crimes) have gained negative attention in the national media. Housing in Landskrona is highly segregated with the city centre being dominated by economically disadvantaged populations, whereas the middle-class live in the coastal area in the outskirts of the city. A study carried out by the municipality in 2012 revealed that common associations of Landskrona, among those living and working there, were unemployment, criminality, and multiculturalism (Zimmerdahl, 2013).

Landskrona's destination marketing campaign thus aimed to create and encourage alternative imaginaries of the city to counter the stereotypical view of the place. The project team anticipated that the representations on the Instagram account would provide an authentic image of the city, which in turn could serve as a positive contrast to the media's portrayal of the city as decaying (Zimmerdahl, 2013). Locals were invited to participate in the construction of a new place identity by way of curating the city's official Instagram account for one week. During the week, the participants could freely upload posts about the city. Participants were recruited on a voluntary basis. The official requirement for partaking in the initiative was that participants should have a personal Instagram account, which could be linked to the official one. For the most part, the participants were local politicians, celebrities, entrepreneurs, municipal employees, and friends of those working at the city's communication department. Hence, the participants belonged to a relatively homogenous group of people identified by the project team to be opinion leaders in Landskrona. Thus, not everyone was able to participate in the project, which had consequences for the representation of the destination.

In-depth interviews over photographs

In-depth interviews were used to prompt participants in the project to discuss and reflect on their participation. Interviews were carried out with 16 participants. The interviews lasted around 45 minutes and were verbatim transcribed. The project had been running for eight months when the interviews were carried out. Almost 30 citizens had participated. Half of this group were selected for an interview. In order to be able to understand practice and different practices, we wanted a heterogenous group of interviewees i.e. a strategic selection. The criterion used for the selection was age as it can be an indicator of different experience of photography and social media. Moreover, time for participation and character of photographs were used for the selection. We selected interviewees who participated in the very beginning of the project and those who participated the weeks before the interview took place as their interpretation of the task may differ. We also looked at the photographs and the interaction during different weeks. Interviewees who apparently took different types of photographs were pinpointed as well as those which photographs had generated many as well as few likes.

The photographs that the participants had taken and published on the official Instagram account served as input to the interviews. The procedure can be defined as an interview over photographs implying that the photographs were used as props in the conversation about the photographic practice. The focus of attention in the interviews was the experience of partaking in the campaign and how the participants took photos. To acknowledge the role of structure, we posed questions about technology use (e.g. smartphone cameras, Instagram, and other social and visual media) and ideals of photography. The interviewees themselves were highly aware of conventions. They talked about 'normal' behaviour and 'dos and don'ts' on social media platforms. Participants' reflections on their practices captured their understanding of the marketing campaign and the types of photographs that were posted. Moreover, in their reflections, we could trace and identify social conventions that guide their practice.

The analysis of the interviews and photographs was directed towards the participants' experiences of participating in the project and representations of the destination. Focus was on identifying typical ways of seeing and representing the destination, i.e. gazes. The analysis was informed by theoretical constructs concerning gaze and practice. Three typical gazes structuring participants' practices were identified based on participants' accounts and photography. These gazes were labelled the tourist gaze, the Instagram gaze, and the art gaze. It should be noted that the gazes are not static, but change in time and place. Moreover, as we demonstrate in the next section, they are relationally defined, and thus in part overlapping.

Gazes on the destination

The gazes directed attention towards certain modes of representation and performance. The gazes, we argue, discipline the practice and outcome in the Instagram takeover and in a wider sense the co-creation of destination image. The gazes, performances, and representations are summarised in Table 18.1.

Tourist gaze

> I did not put too much effort in terms of planning, but thought that I should show this, this and this. It should not be a boring week. /... / In some way I felt that I should show my favourite places. (Ingrid)

The *tourist gaze* structure the practices of locals who regarded themselves to be representatives of the destination and who presented favourable images before an imagined audience of visitors. This particular group wanted to show the positive side of Landskrona in order to counterbalance the destination's negative reputation. In this context, the tourist gaze refers to seeing the city where one lives through the eyes of the tourist. The locals depictured scenic spots, historical buildings, landmarks, and nature that they viewed as worthy to show potential visitors (Urry, 1990). Hence, they performed as tourists or tourist guides. They compared their photographs to tourist photography and used it as their point of reference and ideal. Gazing is here conditioned on past experiences of being a tourist and knowledge about what is appropriate to show the tourist (Urry & Larsen, 2011). A preference for the picturesque, landscapes, and buildings in a past temporal orientation are some of the characteristics which resembles the romantic and elite tourist gaze (Figure 18.1).

Even though the participants were aware of that similar scenic spots and landmarks were promoted by the local DMO, they reasoned that they personalised these familiar motifs by means of filters, tags, and comments. This type of personalisation strengthens the authenticity of the representation of the destination, as they are embodied in local practices.

Table 18.1 Gazes

Gazes	Tourist	Instagram	Art
Guiding principle	Beautiful photographs of landmarks, attractions, scenery.	Personal moments. Glimpses of backstage settings.	Critical eye.
Performance	Act like a tourist or tourist guide. Take photos of attractions.	Act as an individual. Integrated with personal social media practice.	Explore and question established images. Ignorant of number of followers and 'likes'.
Representation	Postcard aesthetics.	Imperfect, spontaneous, instant.	Personal, unique, complex.

Figure 18.1　The tourist gaze
Source: Photo by the participant.

Instagram gaze

> The private and the public blur in the project. I can easily upload the photographs on my work account during the weekend. I enjoy using smartphone, since you get such a direct response on what you post.
>
> *(Sofia)*

The second type of gaze identified is labelled the *Instagram gaze*. This type of gaze structures the practice of participants who draw on knowledge of visual social, particularly Instagram, to represent the destination. The Instagram gaze represents a way of seeing and experiencing the destination as ordinary and everyday (de Certeau, 1984). Locals who express to this type of gaze are highly experienced social media and Instagram users. They use Facebook and have several Instagram accounts, for private as well as for professional use. Therefore, imaging the destination was easily integrated to their everyday use of social media. The Instagram gaze is conditioned on the logic and rules of the Instagram platform. It resembles what Urry and Larsen (2011) term the post-tourist gaze. This gaze intersects with media culture and everyday life in a greater extent than the tourist gaze. The post-tourist is aware of the tourist but wants to create a distance towards – what is perceived as – an elite way of seeing.

The conventions on the platform that are considered especially important refer to instant and spontaneous moments. Spontaneous, natural, relaxed, and the real are words used for describing what they want to achieve. To be spontaneous and oriented towards the mundane are also ways to accomplish authenticity and trustworthiness. Their preference for the spontaneous and mundane as well as their acceptance of imperfect photographs is a way to obtain a glimpse of unexpected views of the city. The capturing of instant moments on Instagram is demonstrated by that photographs often represent the here and now. Photographs typically show people as they pass by on the street, sit in cafés, or talk outside shops. Moreover, local politicians, work colleagues, celebrities, shop owners figure on the photos. The ambition is

Figure 18.2 The Instagram gaze
Source: Photo by the participant.

not to image the destination through its built or natural environment but through its people. The focus on people resembles the logic of social media, which is based on making social relationships visible (Figure 18.2).

Another aspect of seeing involved in the Instagram gaze is the strong expectation on photographs to generate audience responses. The participants expressed that picturesque representations of the destination and well-known people generated positive responses from followers. Controversial photographs were considered inappropriate and avoided. Consequently, representations of the destination are viewed and evaluated in instrumental terms of the number of likes, comments, and followers. The instrumental aspect of seeing in the Instagram gaze thus frames the understanding of the authentic and mundane according to a social media logic.

The art gaze

> My purpose was to be a little subversive in presenting my view on Landskrona. I am averse to doing the same thing as everyone else. But it was not a popular strategy. People thought that I was annoying and that I did not meet the expectations. My photographs did not generate a lot of likes.
>
> *(Johan)*

The third gaze is labelled the *art gaze*. It structures practices enacted by participants with professional knowledge of or unconventional understanding of photography. By contrast to the tourist and Instagram gazes, the art gaze is not concerned with expectations of an audience. Seeing here becomes an issue of self-development and improvement. The art gaze involves breaking norms, social conventions, and expectations. Representations of the destinations

I #LANDSKRONA FINNER MAN SINA EGNA VÄGAR.

Figure 18.3 The art gaze
Source: Photo by the participant.

are therefore individualist and eccentric displaying a certain aesthetic. Representations explore critically what makes the destination attractive by means of crossing genres in disrespectful manners (Figure 18.3).

The art gaze is about seeing the city with new critical eyes and question taken for granted rules and ways of acting on Instagram. The destination is imaged in political and unexpected ways. The locals use photographs to express opinions and direct critic towards among other the destination, which is unthinkable in the other types of gazes. These photographs did not generate as many comments, likes, and followers, rather locals informed by this type of gaze stated that they did not care about their posts being liked.

Conclusions

In this chapter, we propose three types of gazes, that is to say, ways of looking at, sensing, or comprehending the destination, that structure the practice of taking photographs for the Instagram account and by extension the co-creation of the image of the destination. The gazes refer to the romantic tourist gaze, the Instagram gaze, and the art gaze. They represent the destination according to a postcard way of seeing the destination, a personal way of seeing the destination, and a critical way of seeing the destination. Those gazes imply that the participants 'steps into pre-existing discursive, practical, embodied norms' (Edensor, 2001) which guides their actions and results in different representations. In the Instagram takeover, we could identify that established patterns of representing a destination are reproduced. The gazes are relationally defined, and there are several overlaps between the romantic tourist gaze and the Instagram gaze in terms of privileging positive representations of the destination, blurring boundaries between the public and personal, and refraining from dealing with political and critical issues. Both gazes involve a social media logic – a concern for the followers and number of comments and likes. Consequently, the established and well-known aspects of the city are selected and published on the Instagram account. Instagram technology, like filters, is used to present it in the most favourable way. Nevertheless, the Instagram gaze was also identified as a form of post-tourist way of seeing resulting from increasingly mediatized tourist experiences (Urry & Larsen, 2011). The post-tourist gaze is characterised by a search for the elusively different elements of the tourist experience within a postmodern, globally interconnected world. This particular gaze does not reproduce images of the destination connected to heritage, as much as images revealing the staged local life and tourism as a spectacle for the masses. The third gaze was labelled the art gaze and was identified as a different way of seeing the destination through more critical eyes. Representations were often abstract and obscure and received little attention in the project by the DMO as well as other participants. Nevertheless, the representations guided by the art gaze stand for an alternative way of imaging the destination that may counteract the representations used for marketing purposes by the DMO. Hence, we may draw the conclusion that some gazes on the destinations are more dominant and influential than others. In this relatively limited project, we identified three gazes that guided practice and the representations; however, it may be expected that larger marketing initiatives could incorporate a greater repertoire of gazes. As ways of seeing become more institutionalised in Instagram photography, however, we may expect less variety in gazes on the platform.

Involving visual social media for co-creating a new image of the destination is not without problems. Our study demonstrates that it is the ideal, scenic, and attractive image of Landskrona that is predominately reproduced on Instagram. Such images converge with the dominant ways of representing destinations on the social media platform and in marketing communications. The participants have clear boundaries between private and public life, and they avoid being too private in their photographs. Hence, it is the positive, well-known, and impersonal image they want to present. The most popular photographs on Instagram are also the ones contributing to a favourable destination image. The most liked photographs are similar to the idealised photographs published in tourist brochures.

We began this paper with pointing to the strong belief in what can be accomplished through participation in the co-creation of image of a destination in visual social media. Can user-generated content by locals on social media reinvigorate images of destinations? A practice-based approach was used to study how citizens engaged in a rotation curation project initiated by a Swedish municipality. The practice-based approach to photography enabled us to understand how different gazes guided how locals participated in co-creating the image of the destination.

We conclude that the gazes result in the city being visualised in certain ways but not re-imagined. Hence, involving citizens in place promotion strategies does not necessarily mean that novel images of place are generated, but that the co-creation of images is disciplined or constrained according to different gazes.

References

Chalfen, R. (1987). *Snapshot versions of life*. Bowling Green, OH: Popular Press.

De Certeau, M. (1984). *The practice of everyday life*. Berkeley: University of California Press.

Dinhopl, A., & Gretzel, U. (2016). Selfie-taking as touristic looking. *Annals of Tourism Research, 57*, 126–139.

Edensor, T. (2001). Performing tourism, staging tourism: (Re)producing tourist space and practice. *Tourist Studies, 1*(1), 59–81.

Edensor, T. (2008). *Tourists at the Taj: Performance and meaning at a symbolic site*. New York, NY: Routledge.

Foucault, M. (1977). *Discipline and punish*. London, UK: Allen Lane.

Gretzel, U. (2017). The visual turn in social media marketing. *Turismos, 12*(3), 1–18.

Heat, J. (1999) The gaze and visibility of the carer: Foucauldian analysis of the discourse of informal care. *Sociology of Health & Illness, 21*(6), 759–777.

Larsen, J., & Urry, J. (2011). Gazing and performing. *Environment and Planning D: Society and Space, 29*(6), 1110–1112.

Orlikowski, W. J. (2007). Sociomaterial practices: Exploring technology at work. *Organization studies, 28*(9), 1435–1448.

Packard, J. (2008). 'I'm gonna show you what it's really like out here': The power and limitation of participatory visual methods. *Visual Studies, 23*(1), 63–77.

Pamment, J., & Cassinger, C. (2018). Nation branding and the social imaginary of participation: An exploratory study of the Swedish Number campaign. *European Journal of Cultural Studies, 21*(5), 561–574.

Pink, S. (2011). Amateur photographic practice, collective representation and the constitution of place. *Visual Studies, 26*(2), 92–101.

Postill, J. (2010). Introduction. In B. Bräuchler & J. Postill (Eds.), *Theorising media and practice* (pp. 1–26). Oxford, UK and New York, NY: Berghahn.

Randles, S., & Warde, A. (2006). Consumption: the view from theories of practice. In K. Green & S. Randles (Eds.), *Industrial ecology and spaces of innovation* (Chapter 10) (pp. 220–237). Cheltenham, UK: Edward Elgar.

Reckwitz, A. (2002). Toward a theory of social practices: A development in culturalist theorizing. *European Journal of Social Theory, 5*(2), 243–263.

Rose, G. (2014). On the relation between 'visual research methods' and contemporary visual culture. *The Sociological Review, 62*(1), 24–46.

Schatzki, T. (1996). *Social practices: A Wittgensteinian approach to human activity and the social*. Cambridge, UK: Cambridge University Press.

Schroeder, J. (2002). *Visual consumption*. New York, NY: Routledge.

Sontag, S. (1977). *On photography*. New York, NY: Delta Books.

Thelander, Å., & Cassinger, C. (2017). Brand new images? Implications of Instagram photography for place branding. *Media and Communication, 5*(4), 6–14.

van Dijck, J. (2008). Digital photography: Communication, identity, memory. *Visual Communication, 7*(1), 57–76.

Van House, N. A. (2011). Personal photography, digital technologies and the uses of the Visual. *Visual Studies, 26*(2), 125–134.

Urry, J. (1990). *The tourist gaze: Leisure and travel in contemporaries societies*. London, UK: Sage.

Urry, J. (2002). *The tourist gaze* (2nd Ed). London, UK: Sage.

Urry, J., & Larsen, J. (2011). *The tourist gaze 3.0*. London, UK: Sage.

Zimmerdahl, H. (2013). *Merparten beskriver Landskrona med två ord: naturskönhet och arbetslöshet*. Sydsvenskan, January 7.

19

REPRESENTATIONS OF A GREEN IRELAND

A case study of global franchises *Star Wars* and *Game of Thrones*

Pat Brereton

Introduction

This chapter focuses on the green branding of transnational franchises which actively represent touristic images of the island of Ireland. Drawing on environmental studies of eco-media and the appeal of travel to a distant past or an austere future which presents new visions of homeland, this chapter explores the textual features that attract audiences in large numbers through such powerful representations of the country. These audio-visual signifiers of Ireland help to speak to audiences and tourists, who seek out new modes of engagement with landscape, which in turn open up new possibilities for touristic representation.

Within Irish cultural studies, green-tourism is often equated with an overly romantic set of images that has been actively used to sell the country abroad. Co-opting notions like the tourist or environmental gaze, coupled with an endorsement of sustainable eco-tourism; media representations can present a potentially more progressive vision of nature and landscape. Facilitating such a transformation, the topography of rugged scenery can be re-purposed, while remaining forever etched on the tourists' imagination.

Buell (2004) most notably tried to define this re-purposed form of literary (alongside cinematic) imagining, as reminiscent of nature writing in the Thoreauvian tradition and suggested that the reorientation of human attention and values to a stronger ethic of care for one's habitat could help make the world a better place for all life on the planet. An audio-visual exploration of landscape appears to have the power to 'express what is otherwise inexpressible' (Lefebvre, 2006, p. 12): in particular involving the creation of moods, emotions, and other engaging forms of nostalgia. In earlier analysis, I questioned the effectiveness of using landscape for environmental and educational purposes, where the perceived 'real Ireland' was located firmly in the rural west, creating an appealing narrative that had global appeal through its filmic imaginaries (Brereton, 2006). One could argue that contemporary touristic imaginations through powerful media representations have become more multi-faceted and reflexive, either going back centuries to ancient times, recalling more generic mythological legends as visualised in *Game of Thrones*, or alternatively projecting futuristic *Star Wars'* myths and allegories.

All the while, the Irish Tourist Board continues unashamedly to tap into and provoke the romantic and mythic pleasures of the Irish landscape and also affirming the uniqueness of its people wherever it can. Recalling for instance a 1966 tourist documentary *Ireland Invites You*, which begins with the slogan

> this is Ireland, a green island set in the seas like a gem of rare beauty, a haven of undisturbed peace in a restless world, a land of infinite variety of scenes, an ageless, timeless place where old beliefs and customs live on besides the spreading tide of human progress.

This thick residue of nostalgic language and imagery continues up to the present day with sumptuous invocations of the potency of Irish land and seascape being magnified through its representation across these internationally successful film and television franchises.

Drawing on the tensions inherent in eco-tourism, while sampling classic evocations of representations of Irish landscapes, this chapter will strive to both environmentally situate and frame these internationally renowned franchises. For instance, recalling classic rural narratives with their striking wide-angle shots of romantic and nostalgic landscapes, audiences are left with the privileging of a touristic vision (Brereton, 2006). But how to marry this broadly utopian, yet one-dimensional vision with a more sustainable and contemporary environmental model of landscape management and representational engagement remains a challenge. The cursory filmic analysis presented in this chapter seeks to outline how some textual strategies can be effectively deployed to illustrate the power of filmic representations to help tourists as well as audiences generally, appreciate such places alongside their precious environmental habitats.

By teasing out various audio-visual strategies used to promote a form of touristic branding, together with supporting the environmental greening of the country, new challenges for representation are raised. Remembering how conventional rural images of Ireland, from *The Quiet Man* (1952) to *The Field* (1990), for example, have become re-purposed, we can trace more contemporary re-branding of Ireland as a tourist destination. This progression is encapsulated by the recent successful Irish tourist initiative 'The Wild Atlantic Way', which in many ways corresponds with the active promotion of the *Star Wars* franchise.

Greening Ireland: touristic allegories of home

Greening media practice ostensibly involves transitioning to a complementary and co-dependent relationship with all living systems. Certainly, the translucent greenness of living matter directly alludes to active environmental concerns, while recalling the focus of much 'first wave' environmental literature (Branch & Slovic 2003). This movement privileges the pure romantic/sublime and the celebration of wild pristine nature, while at the same time promoting a deep form of biophilia or love of nature. An environmental strategy which all but mirrors the Irish tourist board's historical and well-formulated image of the island as an idyllic site of escape. This approach is visualised and dramatised using multiple close-ups and wide-shots of the rugged beauty of the landscape, encouraging tourist to physically come and see with their own eyes.

Remembering back to *The Quiet Man*, which involved an epic retelling of the diasporic journey of returning home from America, this story is reminiscent of touristic promos like 'The Irish in Me' made for Bord Fáilte (Irish Tourist Board) that portrayed a character Shéila, who embodies the experience of Ireland as one of spiritual home-coming. The touristic scholar MacCannell (1976) speaks most powerfully of recapturing a lost, if 'staged authenticity', which the Western world has believed to have existed in its own 'golden age' of Edenic innocence. This appears to continually inspire many tourists in their choice of destination. The Irish Tourist Board unashamedly sets out to provoke this form of mythic, albeit vicarious pleasures of the Irish landscape and its people. All of which is re-purposed and re-modelled with an environmental veneer and encapsulated by a powerful global evocation of place and homeland that *Star Wars* suggests, with its evocative use of Skellig Michael as site of retirement for its main protagonist.

All these feed off the current mythologisation of the tiny island, alongside the popular historical fantasy of the multi-cited *Game of Thrones* franchise. Yet one could argue the appealing colours and hues presented and foregrounded in these contemporary tales tend to move away from over-privileging a one-dimensional, green-screen image of Ireland – as evident across earlier filmic and touristic representations of the island. Often this extended palate of colours and representational approaches to the landscape highlight the gothic underbelly of conventional horror and dystopic historically troubled landscapes that are created for *Game of Thrones* in particular, coupled with the multi-faceted pre-Christian mythology of the monastic/bird-sanctuary and The United Nations Educational, Scientific and Cultural Organization (UNESCO) site, encapsulated by the uniquely cinematic visualisation of a small island off the south west coast of the country and adopted for *Star Wars*.

Reflecting on this multi-coloured palate of colours and hues, coupled with the synthetic construction of benign birds, or the creation of malevolent mythical dragons, all of these stylistic and narrative elements help to call attention to the artificially constructed filmic ecosystems and how they might be re-imagined through an environmental lens. One might even hope they can potentially signal the roots of a new representational template that would enable touristic audiences to better understand their need to tread lightly on the planet. This alternative environmental and sustainable transformation, which ostensibly appears at odds with the touristic agenda, can be foregrounded through new forms of digital media and an evolving film-making process. As Lukinbeal and Zimmermann (2006) suggest, film geography connects the 'spatiality of cinema with the social and cultural geographies of everyday life' and legitimates the notion that the representation of filmic material is packed with cultural additions that have the power to transform a real place into a fictitious environment.

Witness for instance the enormous touristic boon for New Zealand, as a result of being chosen to represent the varying sublime topographies and magical spaces encapsulated by the *Lord of the Rings* phenomenon. On a much smaller scale, here in Ireland for instance, Wicklow's wild but easily accessible landscape was used to capture the mythological *Excalibur* (1981), together with big budget Hollywood stories like *King Arthur* (2004). Furthermore, well-targeted Government tax incentives facilitated the plains of Meath being re-purposed for the battle sequences in *Braveheart* (1995), while an Irish beach in Wexford doubled as the memorable opening sequences of the WWII American landings in France for *Saving Private Ryan* (1998). At the same time, more conventional, yet nonetheless authentic Irish landscapes were (re)presented through such diasporic nostalgic heritage films like *The Quiet Man* (1952), *Ryan's Daughter* (1970), and even the risible Irish-American epic *Far and Away* (1992). Ostensibly, this touristic approach has led to a filmic positioning of Ireland as essentially signifying home, with all the connotations of the familiar and being hospitable, which in turn strongly appeals to the touristic imagination.

Characteristics of such on-screen tourism as stimulus have identified several strategies for leveraging active engagement, with for instance one in five oversees UK visitors to Ireland claiming that film or television shows wholly or partially motivated their travels (Steele, 2008). There is much to explain, however, in teasing out any direct correlations between fictional screen images and various signifiers of Ireland, coupled with actual tourists who visit the country, sparked by such vicarious mediated experiences. More fundamentally, as is also highlighted by several scholars (McKercher, 2007), questions constantly need to be raised regarding the real demand and direct appeal of film as a tourism motivator. Undoubtedly, film tourism also reveals a range of often well-hidden assumptions, conflicts, and contradictions especially around tensions between the pre-imagined vision of a touristic place provided by such fictional imaginary and the subsequent 'real' experience of visiting the place (Connell, 2012, p. 1008). Most especially, we must take into account how and where specific notions of place are created, especially through the viewer's personal and emotional attachment with a combination of themes, story, character, and place; all of which end up creating what Tuan (1974) terms a form of recognisable topophilia. While Tooke and Baker (1996) usefully suggest that the attraction of filmed locations in Ireland and elsewhere is greater when plots and locations are interrelated and emotional connections are formed between the viewer and the spectacle. This conjunction between place and plot remains especially resonant across these two global franchises. But first lets continue to tease out the appeal of audience expectations for such touristic and representational experience.

Reception of environmental and branded touristic messages: horizons of expectations

Jauss (1982) invented the term 'horizons of expectation' which has everything to do with a readers' position and nothing to do with projected sunset and sunrise times constituting a landscape for instance, as coded within much early reception studies of media and touristic pleasures.

A defining question posed by Jauss' theory regarding environmental issues in particular focuses on how landscapes and touristic images might speak to the user's sense of place and by extension how this might encourage increased environmental engagement, while bringing with them expectations of what they wish to experience within a visited landscape. These can incorporate audiences' expectations of how their observation of some representations or actual visits might in turn influence what they saw or heard. Furthermore, one can examine the readers' expectations about how to interpret what they are reading or viewing, while at the same time taking into account an understanding of the media's genre or style, the time they live in, their cultural and political values and identities, and so on. To put it another way, 'horizons of expectations' simply means uncovering what the reader 'expects' of the represented images, much less what they might expect of the resultant touristic experiences, and discovering how this will add to the overall impression of Ireland in this case. It is believed such mediated tourism can create a strong emotional tie between a tourist and their destination (Kim & Richardson, 2003).

How such theorising of audience engagement might apply to new cycles of tourist-driven films and televisual adaptations based in Ireland remains a preoccupation for this chapter. Certainly, in their differing ways, both *Star Wars* and *Game of Thrones* franchises seek to construct quality and coherent brand identity, by co-opting the unique touristic habitus of Ireland's potent iconic symbols of beautiful places. In branding a place, Keller (1993) usefully identifies various stages of development: firstly, by establishing brand identify that involves creating customer's top-of-mind awareness about the brand, which is closely related to tourists' negotiation and use of information sources. This in turn is embedded by recognisable

Figure 19.1 Skellig Michael
Source: Ms Ester Toribio-Roura.

signature music, including memorable opening credits and storylines; all of which are effectively triggered within these two franchises. Second, it is important to establish brand meaning by supporting customer brand association and focus on their ability to identify and evaluate brand attributes and benefits, followed by evaluating customer response and establishing positive brand attitudes. Finally, the process coalesces around establishing customer relationships aimed at creating brand attachment and effective brand activity in securing long-term loyalty. All of these aspects are constantly reinforced by repeated use of powerful landscape iconography, using a rich coloured palate and other textual elements as part of a tourist branding strategy, embedded within the media franchises under discussion (Figure 19.1).

Co-opting of Skellig Michael to promote eco-tourism in *Star Wars*

'A growing body of literature provides evidence that cinematic film and television drama productions can influence people's travel decisions and entice them to visit particular destinations they have seen on screen' (Bolan & Kearney, 2017, p. 2149). Commercial film productions travel around the world with relative ease and are often courted by local governments, who welcome the prospect of inward investment. Certainly, transnational media productions help global citizens understand how the national relates to the global in several different ways. For instance, Mazdon contends transnational cinema 'should not be reduced to international co-productions or an accumulation of national cinemas. Understanding cinema as transnational, means being aware of its porosity, its intersections with others (including the national), its indeterminacy and its contingency' (cited in Fisher & Smith, 2016).

According to a piece by John Daly in the *Irish Examiner* (9th Dec. 2017) on the film being the best advert for Ireland ever made, he quotes JJ Abrams the recent director in charge of *Star Wars* as looking for somewhere 'otherworldly and sacred', while fleshing out a new iteration of this transnational franchise. This desire finally led to Naoise Barry – who worked for the Irish Film Board back in 2014 – taking Disney/Lucas scouts to a remote island off the south west coast of Ireland and helped negotiate a deal.

The decision by Lucas Film to turn the rocky island of Skellig Michael into the futuristic planet Ahch-To, as recreated so far in both *The Force Awakens* (2015) and *The Last Jedi* (2017), has already brought a huge surge in Irish tourism. Many have asked what are the unique attributes of this small uninhabited island that appeals to such an international franchise? Patrick Nugent in a piece in the *Irish Times* (19th Dec 2017) notes how the literary writer George Bernard Shaw described Skellig Michael as an 'incredible, impossible mad place' that is part of 'our dream world'. So perhaps it makes sense that such sites are now becoming part of the modern global fantasy folklore of *Star Wars* with their trans-generational appeal. Monks historically started building on the island in the year 600 AD, at a period in history when it would have involved a treacherous journey on small boats called curraghs. The religious inhabitants desired extreme isolation to bring them closer to God, but their visit unfortunately was often short lived. They mostly died very young, according to many reports, including Nugent's review, as 'many got rheumatism, the damp was insidious, as lighting a fire on the island was impossible – there was nothing there to burn. And the diet consisted entirely of birds' eggs, fish and seaweed'.

This religious mode of living constituted the ultimate manifestation of a frugal, anti-materialist philosophy, reflecting at one extreme a deep ecological form of living. Escape from modernity/civilisation and its easy comforts and luxuries continues to reflect a niche aspect of eco-tourism. While the more defined appeal of a spiritual pilgrimage, seeking out areas that are untouched by any form of modern civilisation, has also had a long history on this sacred site. By any measure, Skellig Michael fits the bill and has become a fulcrum of global mythic engagement, sparked by being chosen for the *Star Wars* franchise, all the while evoking a pure form of idyllic spiritual identification.

The brief but transformative sequence show-casing Ireland at the conclusion of *The Force Awakens* calls attention to the rare beauty of the island in such a visceral manner. This is aided by the stunning cinematography of Steve Yedlin with several evocative aerial and low angle shots of the tourist island. Furthermore, John Williams' sumptuous musical score helps recreate for audiences the audio-visual splendour of this rugged landscape, which effectively situates the homeplace for now retired and monk-like Luke Skywalker (Mark Hamill), who has finally retreated to the island for solace. However, Skywalker is literally shaken out of his torpor, by the unexpected visit of his protege Rey (Daisy Ridley) who needs the magic of the force that he has developed over the years. Daly's (2017) review effectively captures the magic of the island and the film's representational power to promote this unspoilt touristic space.

> As the chopper arcs and wheels 1,000 feet above the azure blue of Dingle Bay, we've left the normal world far behind on a flight path edging us closer to the un-earthly *Star Wars* domain of Ahch-To, sanctuary of The Last Jedi.

For global audiences and fans, the unspoilt uniqueness of the landscape and its bleak topography is effectively dramatised through this climatic closing sequences of returning home for this universally cherished filmic franchise.

Synthetic birds and the prospect of crass eco-tourism

Unlike the co-opting of the island landscape for touristic, if not spiritual purposes, the Porg toy was designed simply to mimic Puffins and serve as a piece of explicit franchising for the film. Such iconic birds are found all around the island but unfortunately became too difficult to blot out of the cinematic *mise-en-scene*. Not surprisingly, there was much environmental criticism of the film's crude commodification of the simulated birds as a technical solution, which in turn could be more easily visualised to fit into the diegesis of the fantasy world-view. In an interview with creature concept designer Jake Lunt Davies, it is revealed that the Porgs 'were only invented to cover up the *actual* adorable creatures that turned out to be a huge pain in the ass' for writer/director Rian Johnson. 'By law, *Star Wars: The Last Jedi* was not permitted to mess with the puffins' (www.starwars.com). Hence, even if the might of the global franchise could do what it liked by re-imagining this iconic island with its austere habitat and topography; protecting rare species while recalling globally ascribed notions of conservation was another matter. One could argue 'the force' appears on the side of the birds, if not with their precious habitat, whose opaque aura could more easily be re-constructed and re-imagined as a site for transnational audiences and tourists to identify and empathise with.

Meanwhile, other scholars like Fletcher (2015) more caustically explore the difficult ethical role of ecotourism and alternatively perceive it embedded within the neoliberalisation of environmental education that is defined by *The International Ecotourism Society* as 'responsible travel to natural areas that conserves the environment and improves the well-being of local people'. Notwithstanding, ecotourism in all its various guises has become among the fastest growing segment of a global tourism industry. Yet, relatively little discussion from a media and eco-tourism perspective has been carried out concerning the development of pro-active environmental education through such potent representation (Sander, 2012), despite the fact that this is commonly considered one of the core components of sound ecotourism practice (Honey, 2008). Honey goes on to assert that 'ecotourism means education, for both tourists and residents of nearby communities' (p. 30). One might hope, at least for future iterations, some more defined sense of environmental learning could take place for *Star Wars* fans around the flora and fauna of this unique habitat?

Certainly, Sayan and Blumstein's (2011, p. 103) recent call to fix the 'failure of environmental education' calls on ecotourism as a potentially fruitful delivery system for this form of learning. They observe that operators are increasingly making sure their programs teach ecotourists how to promote protection of various habitats. One might hope that in spite of the risible commodification of its bird population, the filmic representations of the eco-tourist experience of *Star Wars* and its evocation of Skellig Michael could in some small way help to emulate this touristic learning process. Meanwhile, *Game of Thrones* speaks through a very different register of new generational tourist engagement, yet at the same time, one could argue also evoking new forms of environmental and place representations through the series multiple storylines and its global appeal (Figure 19.2).

Figure 19.2   Pathway to Ireland
Source: Ms Ester Toribio-Roura.

The historical fantasy of *Game of Thrones*: 'Winter is coming'

The hugely successful televisual long-play franchise series began development in Northern Ireland in 2007 concentrated in the Titanic studios in Belfast and the Linen Mill studios in County Down. Some more 'exotic locations' were added, using Malta, Morocco, Croatia, Iceland, and most recently Spain. Key landscapes around Westeros (the fictional setting of the show) are also filmed in Northern Ireland, featuring a range of countryside and coastal locations. Many business and tourist scholars such as Beeton (2005) and Hudson & Ritchie (2006) allude to the importance of developing movie maps and film trails, 'both to entice film tourists and adding value to their experience when visiting such destinations'. Going much further in representational terms than *Star Wars* in the south with its visceral final sequences, the franchise actively branded the whole of Northern Ireland as 'Game of Thrones Territory' (Bolan & Kearney, 2017, p. 2152).

This reading will concentrate on the opening credits sequence which remains a constant signifier in this ongoing franchise. A memorable clockwork vignette set to a pounding drumbeat shows a three-dimensional game-like map of the lands that the various would-be Kings and Queens are fighting over. This animated sequence is reused across every episode to whet audience expectations and also helps to brand the series core representational identity. Navigating the usual haunts where the drama will unfold throughout the series, starting with the capital of King's Landing, before zooming up to Winterfell, the ex-Stark stronghold of the North and then up to the iconic Wall that most significantly keeps the mystical and gothic White Walkers out of Westeros. High production values and quality game play are exuded by this unique (analogue) board-game credit sequence, contrasting with more conventional audio-visual (digital) filmic modalities evident in the opening credits of *Star Wars*.

Furthermore, this sense of quality yet vicarious experience is forcefully represented by an intertextual promotional video released for the first episode, which talks of the great effort and expense taken to recreate a snow-covered forest region. The crew needed ten weeks to re-create this wintery environment in Northern Ireland, featuring the mythical White Walkers, who remain a pervasive spectre seen throughout all of the subsequent series. First of all, the pristine forest was covered with a membrane and then dressed in snow-like particles to represent the illusion of a very inhospitable climate. This permanently harsh winter environment is adapted and frequently used across many of the storylines, which constantly recalls a pervasive fear announced by numerous protagonists that 'winter is coming'. A tagline which could be re-interpreted as a foreboding of contemporary climate change.

More creatively, the crew in the DVD bonus-feature documentary talk of how all of the rest of the country can often get 'four seasons in a day'; hence, it's particularly difficult to secure consistent artificial lighting coming through the forest. Nonetheless, in spite of this, they felt the overall look and aesthetics which was developed turned out to be really beautiful and all were very pleased with the overall result. This further recalls a unique environmental appeal of Irish tourism and its ever-changing climatic conditions, which visitors often notice. Such technical commentary further reflects the active manipulation of real authentic habitats and the problems as well as opportunities that result from a re-imagining of habitats within the more controlled audio-visual media production process – similar in ways to the representation of birds that had to be tamed in the reconstruction of Skellig Michael for *Star Wars*. Here, however, the trees and forest and later the demons and dragons are simply co-opted as both short-hand signifiers and narrative devices towards actively placing those mythic storylines inside a 'believable' environment, foreshadowing the evil deeds that take place when nature is in peril. The globally successful storylines convincingly dramatise these so-called mythical creatures with their piercing eyes, while constantly mixing horror with fantasy.

This mode of fantasy representation appears a long way from the one-directional nostalgic and romantic evocation of landscape and Irish culture presented across earlier Irish film. Such contemporary globally addressed touristic narratives convincingly tap into dark myths and other topographical tropes buried deep in the Irish zeitgeist – as evident through centuries of historical conflict seeping into more contemporary political eruptions and 'Troubles'.

Conclusions

While these snippets harvested from two extensive franchises might appear a long way away from the authentic and more sustainable representations of Irish landscape and culture valorised by a deep ecological mindset, nonetheless such fantasies can either be co-opted as a fruitful eco-representation of the country and island, or simply dismissed as crudely exploitative, promoting a crass model of global tourist promotion.

For instance, many scholars cite the *Star Wars* slogan 'let the force be with you' as symptomatic of a post-religious form of popular spirituality (Davidsen, 2016). A contrasting but dissipated force could also be applied to *Game of Thrones* with its pre-Christian secular mode of superstition and spirituality, serving to justify conflict and recalling more local and tribal forms of patriotism. Many protagonists in the series for instance concurrently talk of a place they might go after they die – having little apparent fear of violent death – as being paradoxically 'the ultimate vacation destination'. Conjoining these half-cooked philosophies of the after-life, alongside more secular heroic/utopian parables, remains I suspect a key pleasure for contemporary audience engagement.

While at a more prosaic and tangible level, a primary feature embedded in both mediated franchises calls attention to a growing environmental preoccupation with promoting effective stewardship of the land and especially re-discovering values and beliefs worth fighting for. The main protagonists in both these fictions have to constantly fight evil forces and uncover the right course of action. But recalling discussion around the foregrounding of synthetic birds, or for that matter the use of mythical dragons to enact violence and retribution, one wonders if real or authentic pro-social environmental triggers are being activated across any tangible level of audience engagement?

In a survey paper entitled 'Nature Tourism and Irish Film' (2006), I explored the notion of an environmental gaze remaining undetected within much analysis of fictional nature media; one wonders now with the development of such contemporary globally successful franchises, has the greening of Irish media become more extensively anchored and in turn become more globally recognisable? Alternatively, as also alluded to in this reading, might this appeal remain more superficially addressed through a crudely monetised tourist sensibility and representational register. Earlier film and Irish media generally valorised nature and landscape as part of an ongoing romantic nationalist project. By all accounts, this touristic agenda has appeared to be transformed with these mega-global franchises. Nevertheless, such trans-national images of Irish spatial identity no doubt continue to serve as a direct touristic stimulus, helping to keep our landscape culturally vibrant (Brereton, 2006, p. 416). Scholars can still approach these two global audio-visual and touristic franchises from various angles to explain how contemporary audiences think and imagine our ongoing relationship with the complex ecological structures of the world, where audiences and filmic representations become intertwined (Frymer et al., 2010, p. 216).

Of course, none of these dilemmas and questions can be fully addressed without primary empirical research to test and evaluate such a hypothesis linking touristic branding and active environmental engagement. But at least, by posing these questions in a new way, such (Irish) franchises have done some good work in bringing tourism and environmental concerns into sharp focus, through such provocative modes of representation. Eco-film scholars like Weik von Mossner (2017) and others rightly believe that fictional film and franchise like these examples can speak to large international audiences and even trigger awareness of environmental issues, while also as many scholars suggest by hailing audiences to pro-actively visit the sites of such productions. Such film franchise and powerful place-based representations can speak directly to global audiences with their evocation of landscape as touristic sites for travel, reinforced by engaging plots and immersive narratives. Certainly, filmic representations and eco-tourism continue to have a synergistic relationship, which is evolving all the time.

References

Beeton, S. (2005). *Film induced tourism*. Clevedon, UK: Channel View Publications.

Bolan, P., & Kearney, M. (2017) Exploring film tourism potential in Ireland: From *Game of Thrones* to *Star Wars*. *Revista Turismo and Desenvolvimento, 27/28*, 2149–2156.

Branch, M., & Slovic, S. (2003). *The ISLE Reader: Ecocriticism 1993–2003*. Athens, Greece: University of Georgia Press.

Brereton, P. (2006). Nature tourism and Irish film. *Irish Studies Review, 14*(4), 407–420.

Buell, L. (2004). *From apocalypse to way of life: Environmental crisis in the American century*. New York, NY: Routledge.

Connell, J. (2012). Film tourism: Evolution, progress and prospects. *Tourism Management, 33*, 1007–1029.

Daly, J. (2017). Skellig Michael's star turn in Star Wars: The Last Jedi is 'the best advert for Ireland ever made'. Retrieved from: https://www.irishexaminer.com/lifestyle/features/skelligmichaels-star-turn-in-star-wars-the-last-jedi-is-the-best-advert-for-ireland-ever-made-464160.html

Davidsen, M. A. (2016). The religious affordance of fiction: A semiotic approach. *Religion, 46*(4), 521–549.

Fisher, A., & Smith, I. R. (2016). Transitional cinema: A critical roundtable. *Frames Cinema Journal*. University of York. Retrieved from: http://eprints.whiterose.ac.uk/100112/1/transnational_cinemas_a_critical_roundtable.pdf.

Fletcher, R. (2015). Nature is a nice place to save but I wouldn't want to live there: Environmental education and the ecotourist gaze. *Environmental Education Research, 21*(3), 338–350.

Frymer, B., Kashani, T., Nocella, A. J., & Van Heertum, R. . (Eds.). (2010). *Hollywood exploited: Public. Pedagogy, corporate movies and cultural crisis*. New York, NY: Palgrave Macmillan.

Honey, M. (2008). *Ecotourism and sustainable development, who owns paradise?* New York, NY: Island Press.

Hudson, S., & Ritchie, B. (2006) Promoting destinations via film tourism: An empirical identification of supporting marketing initiatives. *Journal of Travel Research, 44*(4), 387–396.

Jauss, H. R. (1982). *Towards an aesthetic of reception*. Minneapolis: University of Minnesota Press.

Keller, K. (1993). Conceptualisation, measuring and managing customer-based brand equity *Journal of Marketing, 57*(1), 1–22.

Kim, H., & Richardson, S. (2003). Motion picture impacts on destination images'. *Annals of Tourism Research, 30*, 216–237.

Lefebvre, M. (Ed.) (2006). *Landscape and film*. New York, NY: Routledge.

Lukinbeal, C., & Zimmermann, S. (2006). Film geography: A new subfield. *Erdkunde, 60*(4), 315–325.

MacCannell, D. (1976). *The tourist: A new theory of the leisure class*. London, UK: McMillan.

McKercher, B. (2007). Phantom demand: How some research 'proves' demand where none really exists. In proceedings of the 5th *DeHaan Tourism Management Conference* 'Culture, Tourism and the Media' (pp. 5–26). Nottingham University Business School, 12th December. England.

Sander, B. (2012). The importance of education in ecotourism ventures: Lessons for Rara Avis Ecolodge, Costa Rica. *International Journal of Sustainable Society, 4*(4), 389–404.

Sayan, C., & Blumstein, D. T. (2011). *Failure of environmental education: And how to fix it*. Berkely and Los Angeles: University of California Press.

Steele, J. (2008). *Cracking the code: How visit Scotland's PR activity capitalised on the phenomenon of The Da Vinci Code*. Edinburgh, UK: Visit Scotland.

Tooke, N., & Baker, M. (1996). Seeing is believing: The effects of film on visitor numbers to screened locations. *Tourism Management, 17*, 87–94.

Tuan, Y. F. (1974). *Topophilia: A study of environmental perception, attitudes and values*. Englewood Cliffs, NJ: Prentice Hall.

Weik von Mossner, A. (2017). *Affective ecologies: Empathy, emotion and environmental narrative*. Columbus, OH: The Ohio State University Press.

20

REPRESENTATION OF THE UAE AS A TOURISTIC DESTINATION IN NAT GEO ABU DHABI

An analytical study

Alyaa Anter

Introduction

The UAE has considered the effectiveness of communication to manage its image as a tourist destination and has been transformed from oil producers to tourism hotspots. It boasts projects aimed at globalisation that enhance its reputation as a unique tourist destination (Meethan, 2011; Stephenson & Ali-Knight, 2010). The UAE's mass media usage brings to mind the 'soft power' concept, the ability of the state to achieve its objectives based on its wealth of culture, values, strategies, credibility, and reputation (Nye, 2008). Therefore, the UAE is keen on using Nat Geo Abu Dhabi to convey a nation's positive image. It is a free documentary channel in Arabic launched in 2009 based on the partnership between Abu Dhabi Media and National Geographic Channels International (NGCI) (Abu Dhabi Media, n.d.). Abu Dhabi Media's strategic decision of partnership derives from many rational objectives. For one, The UAE sought to capitalise on the experience of NGCI to promote tourism through documentary cultural production, which is an indirect approach, unlike direct commercial advertising. Besides, TV documentaries may be more credible and trustworthy because viewers see them as portraying reality. Documentary films as a kind of reality programmes have improved the UAE's branding as an ideal tourist destination. According to Sinha Roy (2007, p. 569), NGCI constitutes 'nation-branding' and teaches viewers how to navigate the ideological spaces of home and the world and draw a visual map of new and developing areas across the globe. Consequently, this study argues that mass communication and tourism (mediatized tourism) have mutual effects, especially mass media reflects not just one facet of the tourist destination but all aspects—economic, political, cultural, and security—which all help shape its image.

This study analyses the first stage of mediatization, drawing on the meeting point between the concepts of mediatized tourism and framing. While some researchers examine the relationship between mediatization and framing in the field of political news only (e.g., Strömbäck et al., 2011), I investigate the possibility of integrating both the mediatization concept and framing in the field of tourism communication as a theoretical background to the study. I aim to examine the UAE's representation as a tourist country, determine the role of documentaries in the portrayal of its tourist destinations. For the theoretical background, I explore the practices of media productions in tourism promotion. Moreover, framing analysis provides a means of examining the narratives used to frame the UAE as a tourist destination that affects how people perceive it as such.

The study belongs to the first stage in both framing analysis and mediatization—they meet in the representation of the UAE as a tourist destination—and it attempts to answer how the institutions produce content and interact with media surroundings to affect society. Regarding the factors affecting mediatized tourism in the stage of constructing the messages that affect investors and tourists' motives, it argues that the country's touristic image cannot be isolated from other contexts of services, politics, economy, and security.

This analytical study explores the Nat Geo Abu Dhabi channel's production and investigates how documentaries reflect the state policy that aims to promote a sustainable economy. The content analysis sample includes (N=125) episodes on the UAE. Both quantitative and qualitative methods are used to analyse the framing of The UAE as a tourist destination via Nat Geo Abu Dhabi's YouTube channel.

Representation of tourist destinations: is it the crossroads of mediatization and framing?

Researchers apply the mediatization concept in various contexts; for example, in politics, Strömbäck (2008) concluded that politics and society are not independent of the media; on the contrary, they are becoming more mediatized. Mediatization concepts have the merit of applicability in interdisciplinary communication studies and can be used to examine the interlaced relations between media and social systems. Although the idea of mediatization is not sufficiently clear, the concept will continue in the future; due to its substantial impact on media and communication studies, researchers who adopted the concept have attempted to convert it into a theory for media research (Ampuja, Koivisto, & Väliverronen, 2014, p. 111).

In the field of tourism, Månsson (2009) applied the concept of mediatized tourism to denotes all mutual relations between media products and tourism industries that ultimately affect tourists' behaviour. The circulation of signs, a distinguishing factor in mediatized tourism, is similar to what is called, in political economy theory, cultural signs production (Murdoch, & Golding, 2016) and, in framing analysis, frames focusing, which includes some ideas and excludes others (Entman, 1991). These intended signs (frames) affect consumers' images of destinations and, consequently, their decision-making.

Mediatized tourism is a process represented in media convergence such as social networking sites that make even a consumer an active and interactive agent in circulating and marketing tourism production (Månsson, 2011). Mediatization constitutes a process of 'self-regulation' of organisations as amendments based on media changes. It starts from the inside of organisations that are trying to adapt to media technology, considered a vital resource to develop the organisations' structure (Eskjar, 2018, p. 104). In addition, from the framing perspective, organisations or owners try to control media content and manage their representations therein.

Regarding framing analysis, most studies concern such an analysis of the countries' news and how it affects people's perception of the countries as touristic destinations (Ma, 2016). Meanwhile, the researchers pay little attention to framing analysis as a method of analysing documentaries. Thus, I argue here that framing analysis is suitable to identify latent frames of countries as touristic destinations in documentaries. Strömbäck et al. (2011) attempted to tie mediatization to framing concepts, finding similarities between both concepts, and making use of framing to examine how the media represents European Parliamentary Election campaigns. Accordingly, here, this study benefitted from the intersection of both concepts, using the mechanisms of framing by concentrating on specific traits, words, and verbs to represent the UAE as a whole.

Film-induced tourism: fiction or non-fiction?

Film tourism, film-induced tourism, and popular culture are synonyms for using film to develop tourism. The research field of film-induced tourism is an emergent and expanding one as films become an indirect marketing strategy. Although the concept is broad, including documentary films and TV reality series, most previous studies have been focused on fictional drama such as soap operas and movies (Beeton, 2016; Martin-Jones, 2014; O'Connor & Bolan, 2008; O'Connor, Flanagan, & Gilbert, 2008) rather than non-fiction documentaries and reality TV programs. Films within the tourism management strategy are mainly aimed at increasing awareness of the destinations' positive traits (Croy, 2010). Because movies play a significant role in shaping viewers' images and perceptions of destinations (O'Connor et al., 2008), film-induced tourism has excellent potential for branding destinations and opening new markets, even when the film is fictional (O'Connor & Bolan, 2008). The placement of tourist destinations in movies and TV shows is considered an attractive indirect marketing tool that grows tourist numbers and economic developments (Hudson, & Ritchie, 2006, pp. 394-395).

Less attention is paid to reality programs, although documentaries may have a substantial impact on people, especially viewers who see them as truthful. Furthermore, TV documentaries that support tourism include not only the genres of heritage, food, and travel but also business, industrial, geological, and scientific documentaries. Thus, travel documentaries increase audiences' engagement with tourist destinations and eagerness to visit them. They give integrity to the tourist events, making people trust what they see (Lopriore, 2015, pp. 216-217). Therefore, all countries seek to possess the media that portrays the truth, such as news and documentary channels, based on its enormous potential to inspire belief among the viewers and, consequently, to manage the countries' image.

Methodology

The study focuses on both quantitative and qualitative methods to analyse documentaries on Nat Geo Abu Dhabi's YouTube channel. It relies on YouTube for various reasons, including the difficulty of retrieving recordings of programmes previously aired on television, and because it aimed to analyse documentaries exclusively featuring the UAE and produced during the past three years. Moreover, YouTube videos allow for observation of viewers' reactions. The programme represents the unit of analysis, which may be in a series or a standalone movie-type show. Hence, the content analysis sample includes (N=125) episodes on the UAE.

The reliability test was conducted through the consistency between two coders, meaning the arrival at the same results via the employment of the same categories and units of analysis to the same content, by analysing 10% of the analytical study sample. The Holsti formula for reliability was applied and yielded 0.91.

The analytical study aimed to identify the mainframes of the UAE offered to viewers, the coverage's tone, viewers' interactivity, genres of UAE tourism, and the tourist attractions as depicted on the channel to attract tourists.

Framing means to feature certain aspects, information, and data while ignoring others. This study does not depend on ready-made categories of frames as used in previous deductive research but rather on inductive methods through investigation of the narratives, themes, verbs, adjectives, sentences, testimonials, and images upon which TV documentaries have been focused on framing the UAE. Determining categories of frames is done by analysing 10% of the sample with two coders, then using the concluded frames to assess the whole sample. Moreover, each programme or video may have more than one frame.

Concerning the duration of the episodes, running times varied as follows: 45–50 minutes (70%), 30–44 minutes (28%), and 2.5–5 minutes (2%). In terms of language, Nat Geo Abu Dhabi is keen on using both English and Arabic in the production of TV documentaries about the UAE. Many documentaries are broadcast via both the International Nat Geo channel and Nat Geo Abu Dhabi in tandem with Arabic dubbing (and in some cases vice versa), because usually an English version is aired on International Nat Geo Abu Dhabi (English), directed at an international audience, and an Arabic version is broadcast on the Arabic channel, directed to an Arabic audience. In seeking to portray a positive representation of the UAE as a tourist destination and showcase the developments that qualify it as one of the best destinations in the Middle East, thus Arabic dubbing accounted for 58.7% of the episodes, whereas Arabic dubbing accounted for 39.3% of the episodes.

Viewers' likes and comments on TV documentaries

Nat Geo Abu Dhabi applies media convergence to reach the most massive possible audience and to heighten interactivity via electronic platforms, including the channel's website, which is interlinked with its network of social media accounts (YouTube, Facebook, and Instagram). Interactivity in terms of viewers' likes of content watched and comments indicating to exchange of expertise on tourist destinations in the UAE was examined by observing the number of Nat Geo Abu Dhabi's viewers that clicked like on videos featuring the UAE. Regarding likes, 75.5% of the videos had 0–500 likes, with lowest being 6 likes; 21.8% of videos had 500–1000 likes; and 2.7% of the videos had 1,000–2,000 likes. The results indicate that liking is moderate and points to some engagement via YouTube. In terms of comments, Nat Geo Abu Dhabi YouTube viewers commented heavily on the videos; 83.2% of the episodes had comments, indicating to the channel's efforts to increase interactivity among people, to promote the image of the UAE on the levels of international, regional, and Arab viewers.

UAE tourist destinations garnered a variety of comments, with 80.8% of programmes eliciting a positive reaction. This result indicates the efficacy of video productions in transmitting a positive image of all kinds of tourism in the country. Examples of viewers' comments on the Khalifa tower video include the following: '1,000 years from now people will (look) at Khalifa tower as they look today upon the pyramids and wonder about how it was built'; 'One of the biggest hallmarks of Dubai; a gigantic building and an engineering marvel'. These comments prove that viewers perceive the tourist destinations from the innovation perspective upon which TV documentaries concentrate on as the main frame.

Comments on the Global Village included 'Every time I visit Dubai I go to visit the Global Village. It is a fun, amazing place'; 'Superb location and a fascinating family destination'. This shows that viewers are sufficiently satisfied to recommend a specific destination to others. Meanwhile, examples of viewers' comments on the Louvre Abu Dhabi episode included: 'The Louvre is one of the most beautiful places you will ever visit; it opens the door to knowledge about ancient civilisations and modes of living'; 'A beautiful, sublime museum containing myriad sections and paintings galore from around the world'. Here, comments highlight that viewers perceived the destination according to the culture and heritage frame upon which videos concentrated.

That said, 17.3% of programmes received neutral comments while negative comments were directed at the remaining 1.9% (although these were off the tourism topic). Nat Geo Abu Dhabi also employed a strategy centred on diversifying the nationalities and jobs of characters that hosted and appeared in documentary shows to affirm the cosmopolitan country image and to promote global investment in UAE, so as to send messages to the world on the global leaps accomplished by the country in the area of tourism investment. Concerning guests, specialised experts accounted for most of the personalities who highlighted this area, appearing in 52.8% of episodes, because they are much more capable of providing detailed information regarding the advantages of tourism enterprises. Employees came in second place, appearing in 42.4%, with the channel relying on staff testimonials from various nationalities as information sources in a documentary series of Dubai International Airport. Tourists came next, appearing in 32% of episodes; the channel interviewed them in documentary films on the Global Village, Dubai Mall, and the Sheikh Zayed Mosque. It tries to convey through real target audience a message to the world clarifying the UAE's deep-seated values of tolerance, peace, safety, and security. Next came leaders and state officials (15.2%), with the channel calling on senior diplomats for information and testimonies cited in archival footage. Then came ordinary citizens (8.8%) and finally officers and soldiers (5.6%), as in the series entitled *Emirati Sons on UAE National Service*. As military exercises, participation in sports activities, and patriotism and allegiance were among the themes featured in the show. That represents the state's attention of the army and security affairs, which give the impression of the quality of safety and security in society. It also reflects the patriotic spirit of Emareiti youth.

Types of tourism in the UAE

This analytical study shows that touristic packages in the UAE are varied and enhancing its attractiveness as a tourist destination. TV documentaries represent a diverse array of tourism products to meet all tourists' requirements. Consequently, business and investment tourism in the UAE has come to the forefront and takes several colours in multiple subsectors that draw tourists' interest. Business and investment tourism comprised the majority of tourism packages, featuring in 50.4% of the episodes. TV documentaries demonstrate the UAE's advancements, manifestations of which include the emergence of modern hotels and facilities that have massively elevated the country's hospitality. The facilities turned UAE into a much-sought destination for business people and entrepreneurs to close their deals. As well as the TV documentaries outline that UAE offers an excellent lifestyle suitable for all nationalities.

Next comes cultural tourism, featured in 18.6% of programmes, with a focus on aspects such as civilisation and heritage as documentaries devoted special attention to historical tourism to contextualise the civilisations of the Arabian Peninsula and the formation of the Union. Subjects included 'the Etihad Museum', which traces the UAE's history and remarkable achievements, and '*the Louvre Abu Dhabi*'. Nat Geo Abu Dhabi also covered cultural and artistic festivals in the UAE.

Beaches and parks tourism accounted for 14% of the programmes. The TV episodes highlighted the UAE's advantageous location overlooking the Arabian Gulf and its sandy beaches backed by premium resorts. The channel devoted an episode of its show Megastructures to Palm Islands in Dubai. Shopping tourism made up 9.6% of the episodes, with the one on

'*Dubai Mall*' reflecting the UAE's keenness on establishing world-class shopping centres that draw both tourists and locals.

Next, culinary tourism accounted for 5% of episodes via the Global Village series, which had eight episodes entitled the '*Taste World with Bader Najeeb*', with a run-time of no more than 2.5 minutes per episode. Each video clip featured the presenter visiting a new restaurant representing the food of a different country in the Global Village in Dubai; the show is much like a series of ads. It is noteworthy that each year, the UAE hosts the Dubai Food Festival and the Abu Dhabi Food Festival, attracting participation from restaurants the world over. These events play a crucial role in cementing the UAE's image as a significant culinary destination for global dishes. Finally, religious tourism, as shown in the episodes on '*the Sheikh Zayed Grand Mosque*', accounted for 1.6% of the episodes, while the Adventure Tourism appeared once in '*Ferrari World*' documentary program also accounted for 0.8% of the episodes of documentary programs. Adventure tourism percentage was low because it is still a new emergent genre of tourism.

One positive tone

The tone of videos is divided into positive, neutral, and negative. It is positive when discussing the attractiveness and merits of the UAE as a tourist destination; neutral when the treatment of the events, information, and data do not highlight either positive or negative points but concentrates on facts; and negative when concentrating on disadvantages, but this did not appear in the results of the content analysis, which indicated significantly that TV documentaries met the agenda of the owner by targeting promotion and creating a positive image. The UAE was portrayed positively in 85.6% of documentaries broadcast by Nat Geo Abu Dhabi. This demonstrates the government's determination to mobilise all its resources and make utmost efforts to develop and modernise all available touristic and urbanisation approaches. The programs showcased Dubai as one of the world's foremost tourist destinations. The channel bolstered UAE's reputation as a superb destination for the business, exhibition, and shopping tourism subsectors. It also concentrated on a superior government services package capable of drawing and satisfying tourists and making it both unique and appealing to the business community and tourists.

UAE frames pinpointed by TV documentaries: one frame is not enough

Nat Geo Abu Dhabi applied a variety of frames to tackle the issue of tourism investment and educate the global, regional, and Arab communities on the accomplishments achieved by the UAE on all fronts. The 'innovation in tourism' frame took first place with 50.4% of total episodes; it focused on producing new and unique products, services, designs, and technology in the tourist system, the inception of novel forms of tourism, such as adventure tourism, and the development of tourism practices and strategies that render the UAE one of the foremost destinations worldwide. TV documentary features using artificial intelligence in the enhancement of ventures and services. Innovation in tourism has three kinds of outputs—changes in product, process and organisation, and use of information and communication technology is pivotal to the implementation of operational processes and institutional innovations (Divisekera & Nguyen, 2018). Nat Geo Abu Dhabi has produced documentary films highlighting the innovations in tourism investment in the UAE, such as building modern-design skyscrapers. A host of statements evidenced the innovative design, including 'Khalifa Tower's design inspired by the Tulip; mimics minaret design' and 'Al Bahr Towers in Abu Dhabi are on the cutting edge of innovation because they can reduce heat by almost 50%'. The previous statements are testament to a new kind of sustainable tourism, which is defined as kind of sustainable responsible development aims to improve the quality of environment, maintain landscape, and use renewable resources (Niedziółka, 2012). Nat Geo Abu Dhabi produced documentaries highlighting the theme of innovation and novelty in the UAE, including a creatively crafted film about the Sheikh Zayed Mosque. Moreover, the channel made a series about the ongoing modernisation at Dubai International Airport and hailed Burj Khalifa.

The frame of 'megaprojects and achievements' followed in second place at 49.6% of the total used frames, concentrating on rapid and intelligent completion of giant projects, Nat Geo Abu Dhabi produced documentaries showcasing the UAE's accomplishments in terms of accuracy, speed, precision, and resolving of all obstacles, such as Strategic Tunnel Project. This category was accompanied by comments such as 'the backbone of underground sewage lines' and 'nonstop digging in a tunnel; completes work with pin-point precision'.

In the Palm Islands episode, commentary included: 'Mankind has never undertaken such an audacious project ever before in history' and 'engineering and workers are racing against the clock'. The channel was keen on highlighting features reflecting a positive media image that underpins the determination and resoluteness of the UAE, such as the construction of skyscrapers and infrastructure that draws the international business community and tourists. In the documentary series icons of the UAE, one episode was dedicated to Capital Gate Tower, Ferrari World, and Aldar Circular Tower. The show's preview ran the teaser 'En-gineering icons that brought the world's attention to the UAE'. The series Mission Everest: The UAE Military Team focused on the UAE national team's successful climb of Mount Everest and planting of the UAE flag atop it, as well as the honours bestowed upon the team by the country's leadership.

The frame of 'economic boom and sustainability' took third place, appearing in 44% of total episodes, comprising projects that testify to the diversification of production and sources of revenue and including industries follow criteria for green and clean industry. Nat Geo Abu Dhabi produced the series (*Made in the UAE*), which shed light on Dubai Industrial Park, credited with erecting some of the most technologically advanced and eco-friendly factories in the world that manufacture world-class products and draw new international in-dustries. The channel also made the film Jinan Rain Field that follows cooperation between the UAE and Sudan. The film's theme is the transformation of the desert into green spaces and the application of artificial rain. The frame was upheld by comments such as: 'Sustain-ability is of paramount importance for the UAE', 'eco-friendly green factory', and 'vast areas of dunes in north Sudan transformed into a green paradise'.

The 'culture and heritage' frame followed, accounting for 8% of the frames. It intends to the preservation of the location's historic character. The channel produced films showcasing expeditions such as that to Marouh Island in which the island was hailed as 'a historical, ar-chaeological achievement, that paints a truthful picture about the history of Abu Dhabi and the way of life of its early inhabitants'. The UAE also dedicates special attention to cultural and artistic festivals, as highlighted by the channel in a four-episode series on Al Dhafra Festival, a heritage festival with the theme of reviving old heritage and passing it down for posterity. The program contained statements such as 'In a country overflowing with cultural heritage, the Festival strives to strengthen ties between the UAE people's history and their present'.

The 'security and safety' frame followed at 3.2%, concentrating on security procedures and managing state security by introducing new technology to guarantee travellers' and tourists' safety, as Nat Geo Abu Dhabi produced a four-episode series on Dubai International Airport (DXB) in which it lauded security rules and laws at DXB. The purpose of the safety and security measures applied at DXB is to reaffirm the image of the UAE as a safe, stable, and secure country that possesses a vigilant security awareness within the bounds of the law. Watchwords on this theme included 'Meticulous searching of incoming passengers to Dubai', 'questioning and arresting suspects', and 'enforcing UAE laws upon them'. The analytical study demonstrates that in spite of the multiplicity of cultures of UAE residents, who are of more than 200 nationalities, UAE has achieved the highest rate of safety and security in the region, by applying smart security systems according to international standards.

Finally, the 'union and political stability' frame made up 2.4% of the frames, focusing on affirmations of the internal political understanding, cooperation, and stability.

The channel produces a special show each year celebrating the UAE National Day, as in the film (*The Emergence of a Union*). The film pays tribute to the Sheikh Zayed's role in the rise of the UAE and hails the Union as one that binds all the Emirates. In the words of the film, 'Seven Emirates came together and bonded to form a Union' and 'Sheikh Zayed devoted his life to the unification of isolated Emirates' and the consecration of the unity framework that binds and upholds stability, safety, and security throughout the UAE. The documentary's concentration on this frame is due to the relationship between political stability and economic stability, which draws tourism investments and reaffirms the UAE's role overseas.

Conclusions

The reality of the political economy has prompted the UAE to invest in the field of reality media as a tool to manage its touristic image globally. Mediatization process represents the UAE's window to the world, portrays the country's global lifestyle, manages media content, and distributes cultural signs. Nat Geo Abu Dhabi has broadcasted documentaries on world-class skyscrapers, touristic villages, colossal shopping malls, old heritage locations, and infrastructure developments. The channel has succeeded in presenting the UAE as a global destination for tourists from all world markets. Thanks to the host of advantages that documentaries have featured in terms of safety, security, stability, and diversity in tourism products. The channel provides tourists with a broad array of options, not to mention Nat Geo Abu Dhabi's great emphasis on governmental efforts to upgrade the tourism sector and provide all available resources for the achievement of sustainable development. This affirms that the UAE has benefitted from the experience of National Geographic Society on new kind of sustainable tourism called geotourism, which preserves locations' environment, heritage, beauty, culture, and the well-being of their inhabitants, and helps local small businesses maintain sustainable tourism based on the local geography (Dowling, 2015).

The UAE's ownership of an open documentary satellite channel is a signifier of the potential to control the production of mass media content and, consequently, to manage its touristic image. The case study here proves that the mass media impact on tourism cannot be isolated from the security, political, infrastructure, and economic contexts and how to reflect them via mass media. Therefore, it is not possible to study the image of tourist destinations without framing the security and economic situation that attracts not only tourists but also investors in the field of tourism and other fields.

The study affirms that promotion is more effective in several indirect ways known as promotional culture, such as TV documentaries. Accordingly, these films symbolically benefit the tourism industry (Davis, 2013, pp. 1-4). The results highlight similarities in the process and effects of the circulation of signs in the mediatization process and the framing of tourist destinations in mass media. The documentaries' content analysis has reframed the UAE away from the traditional frame of a petrol-exporting state to a broader scope of luxurious tourism and international investments' country. Nat Geo Abu Dhabi has featured tourism in its all aspects as a new investment.

The UAE's collaborations with international filmmakers are a pivotal matter for the growth of Emirati filmmaking and employing it as a soft power in branding tourist destinations (Saberi, Paris, & Marochi, 2018). This study illustrated connections between the tourism and film industries for the sake of branding the UAE and management of its positive global reputation. This case study reaffirms the concept of mediatized tourism by UAE and contemplates relations between investment in mass media (Nat Geo Abu Dhabi Channel) and tourism investment. The UAE responds to new ways in branding the country as a tourist destination and supporting its reputation as a cosmopolitan touristic country. It has employed TV Documentaries as a tool to sustain its tourist image and to represent many frames to promote sustainable tourism in the UAE.

References

Abu Dhabi Media. (2019). National Geographic Abu Dhabi. Retrieved from: https://www.admedia.ae/en/Channel/7/National-Geographic-Abu-Dhabi.

Ampuja, M., Koivisto, J., & Väliverronen, E. (2014). Strong and weak forms of mediatization theory: A critical review. *Nordicom Review, 35*(special issue), 111-123.

Beeton, S. (2016). *Film-induced tourism* (2nd Ed.). Bristol, UK: Channel View.

Croy, W. (2010). Planning for film tourism: Active destination image management. Tourism and *Hospitality Planning & Development, 7*(1), 21-30.

Davis, A. (2013). *Promotional cultures: The rise and spread of advertising, public relations, marketing and branding.* Cambridge, UK: Polity.

Divisekera, S., & Nguyen, V. K. (2018). Drivers of innovation in tourism: An econometric study. *Tourism Economics, 24*(8), 998–1014.

Dowling, R. (2015). Geotourism. In J. Jafari & H. Xiao (Eds.), *Encyclopaedia of tourism* (pp. 1-3), Cham, Switzerland: Springer International Publishing.

Entman, R. M. (1991). Framing U.S. coverage of international news: Contrasts in narratives of the KAL and Iran air incidents. *Journal of Communication, 41*(4), 6–27.

Eskjar, M. (2018). Mediatization as structural couplings: Adapting to media logic(s). In C. Thimm, M. Anastasiadis, & J. Einspänner-Pflock (Eds.), *Media logic(s) revisited: Modelling the interplay between media institutions, media technology and societal change* (pp. 85-109). Cham, Switzerland: Palgrave Macmillan.

Hudson, S., & Ritchie, J. (2006). Promoting destinations via film tourism: An empirical identification of supporting marketing initiatives. *Journal of Travel Research, 44*(4), 387-396.

Lopriore, L. (2015). Being there: Travel documentaries. *Saggi/Essays*, Fall(6), 216-228.

Ma, T. (2016). Framing China as a tourism destination: A study on media discourse. In 2016 TTRA International Conference (pp. 1-6). *Massachusetts: Travel and Tourism Research Association: Advancing Tourism Research Globally.* 9. Retrieved from: https://scholarworks.umass.edu/ttra/2016/Grad_Student_Workshop/9.

Månsson, M. (2009). The role of media products on consumer behavior in tourism. In M. Kozak & A. Decrop (Eds.), *Handbook of tourist behavior: Theory and practice* (pp. 226-236). New York, NY: Routledge.

Månsson, M. (2011). Mediatized tourism. *Annals of Tourism Research, 38*(4), 1634-1652.

Martin-Jones, D. (2014). Film tourism as heritage tourism: Scotland, diaspora and the Da Vinci Code (2006). *New Review of Film and Television Studies, 12*(2), 156-177.

Meethan, K. (2011). Dubai: 'An exotic destination with a cosmopolitan lifestyle'. In J. Mosedale (Ed.), *Political economy of tourism: A critical perspective* (pp. 175-188). New York, NY: Routledge.

Murdoch, G., & Golding, P. (2016). Political economy and media production: A reply to Dwyer. *Media, Culture and Society, 38*(5), 763-769.

Niedziółka, I. (2012).Sustainable tourism development. *Regional Formation and Development Studies, 3*(8), 157–166.

Nye, J. (2008). Public diplomacy and soft power. *The Annals of the American Academy of Political and Social Science, 616*(March), 94-109.

O'Connor, N., & Bolan, P. (2008). Creating a sustainable brand for Northern Ireland through film-induced tourism. *Tourism Culture & Communication, 8*(3), 147-158.

O'Connor, N., Flanagan, S., & Gilbert, D. (2008). The integration of film-induced tourism and destination branding in Yorkshire, UK. *International Journal of Tourism Research, 10*(5), 423-437.

Saberi, D., Paris, C., & Marochi, B. (2018). Soft power and place branding in the United Arab Emirates: Examples of the tourism and film industries. *International Journal of Diplomacy and Economy, 4*(1), 44-58.

Sinha Roy, I. (2007). Worlds apart: Nation-branding on the National Geographic Channel. *Media, Culture & Society, 29*(4), 569–592.

Stephenson, M., & Ali-Knight, J. (2010). Dubai's tourism industry and its societal impact: Social implications and sustainable challenges. *Journal of Tourism and Cultural Change, 8*(4), 278-292.

Strömbäck, J. (2008). Four phases of mediatization: An analysis of the mediatization of politics. *The International Journal of Press/Politics, 13*(3), 228-246.

Strömbäck, J., Negrine, R., Hopmann, D., Maier, M., Jalali, C., Berganza, R., & Róka, J. (2011). The mediatization and framing of European Parliamentary Election campaigns. In M. Maier, J. Strömbäck, & L. L. Kaid. (Eds.), *Political communication in European parliamentary elections* (pp. 161-174). Farnham, UK: Ashgate.

21

REWRITING HISTORY, REVITALIZING HERITAGE

Heritage-based contents tourism in the Asia-Pacific region

Philip Seaton and Sue Beeton

Introduction

While much of the research into film-induced tourism, literary tourism, and other forms of media-induced tourism examined in this volume and elsewhere discusses new patterns of tourism created by the release of original creative works, tourism may also be triggered by the re-presentation or recycling of old stories within a mediatized popular culture. Certainly, many classic stories receive multiple remakes or adaptations over time, and a popular contemporary work may quickly expand out into a franchise involving sequels, spin-offs, and fan productions. Whenever multiple works in various formats depict a narrative world that catches the imagination of fans, it becomes increasingly necessary to concentrate on 'the contents' – namely the narratives, characters, locations, and other creative elements – of a narrative world that attracts fans and induces them to travel.

This underpins the concepts of *kontentsu* (contents), which emerged in Japan's creative industries during the 1990s (particularly anime and manga), and *contents tourism*, which has been an important component of Japanese tourism policy since 2005, when the national government urged local authorities to cultivate the 'narrative quality' of their regions by treating local stories and creative culture as tourism resources to be exploited (Beeton, Yamamura, & Seaton, 2013, p. 146). Contents tourism has been defined as 'travel behaviour motivated fully or partially by narratives, characters, locations, and other creative elements of popular culture forms including film, television dramas, manga, anime, novels, and computer games' (Seaton, Yamamura, Sugawa-Shimada, & Jang, 2017, p. 3). However, the key characteristic of contents tourism, as opposed to other terms such as film-induced tourism or literary tourism, is that the 'contents' inducing the tourism have been disseminated across multiple media platforms or in multiple works, thereby making it impossible to identify a single work inducing the tourism. Instead, it is the 'narrative world' or 'contents' created by multiple works that are the centre of analysis.

The concept of 'contents' is closely linked to the idea of convergence in media and cultural studies as developed by Jenkins as follows:

> By convergence, I mean the flow of *content* across multiple media platforms, the cooperation between multiple media industries, and the migratory behavior of media audiences

who would go almost anywhere in search of the kinds of entertainment experiences they wanted. Convergence is a word that manages to describe technological, industrial, cultural, and social changes, depending on who's speaking and what they think they are talking about.

(Jenkins, 2006, pp. 2–3, italics added by authors)

Note Jenkins' use of the word 'content' in his key definition, supporting the premise that convergence is the process by which contents are created. When there is a conscious, active effort to expand the multimedia and content base of a particular narrative world, it may also be referred to as 'contentsization' (Yamamura, 2019, p. 18). Furthermore, as Jenkins notes, 'Convergence is both a top-down corporate-driven process and a bottom-up consumer-driven process' (Jenkins, 2004, p. 37), whereby media companies controlling the copyrights for particular sets of contents produce multiple works for sale in many formats, and fans produce amateur derivative works as part of their fan behaviours.

While convergence emerged as a concept within media and cultural studies, it has clear applications within tourism studies. In discussing the role of media on travel choices, Månsson (2011, p. 1635) writes:

> Media products accordingly converge and float around in people's awareness without demarcation in an ongoing circle of references. Hence, it is limiting to study a single media product—such as film—and its influence on tourists in isolation because *content* [italics added by authors] in film, novels and other media products continuously intertwine. This convergence process is referred to as 'mediatized tourism' [...]

There are interlinked processes of multiuse by media corporations, the production of derivative works produced by fans, and the development of related tourist sites. These processes combine and contribute to the expansion of 'the contents' or a 'narrative world' that people may visit either virtually, via media consumption, or bodily as contents tourists. The contents do not remain static but are continually evolving as new works, characters, storylines, locations, and other creative elements are added to the contents. And as the contents evolve, so too do the meanings associated with the contents, and thereby the influences on future representations of the narrative world in new works derived from the narrative world.

The fundamental premise of this chapter is that convergence and its related mediatized tourism phenomena, what we call contents tourism, does not need to be limited to fictional stories. History, myths, and legends are fertile seeds for the creation of new sets of contents. In particular, those stories, characters, and events that have entered national heritage or the collective memory tend to receive repeated adaptations within popular culture, ranging from historically accurate via semi-fictionalized to fantasy. The contemporary significance of historical events can be radically altered by their representation in works of popular culture, and there are numerous examples of such works also generating tourism, including *Braveheart* (Edensor, 2005 and Beeton, 2016), *The Killing Fields* (Beeton, 2015), and Japanese historical dramas (Seaton, 2015).

If there is a set of associated heritage sites (typically museums, monuments, battlefield sites, and so on), fans have a pre-existing heritage trail to follow, which may also be supplemented by shooting locations or other sites specifically related to the work/franchise in question. New works based on history (or set in specific historical times) may revitalize tourism at such heritage sites. Sometimes it is only a temporary 'shot in the arm' to tourism levels, but oftentimes new popular culture renditions of history have long-term implications for the sustainability and profitability of heritage sites. Previous research on heritage tourism and film tourism has identified that many fans of historical film tend to visit heritage sites rather than filming locations which are often in different places (Frost, 2006, p. 251; Seaton, 2019), while others may, in fact, visit both the filming location and historical place (Beeton, 2016). In the two case studies that follow, we identify that changing the metanarrative associated with the heritage site can be a key way in which works of popular culture alter or revive the heritage tourism experience. In one of our cases, this was achieved by reorienting the appeal of a destination from male travellers to female travellers, and in the other by shifting the place association from aboriginal stories to settler stories.

The use of the case study in tourism research is often criticised as merely descriptive, yet if one wishes to understand the tourism phenomenon in more depth, it is often only through taking a qualitative, descriptive approach via case studies where this can occur. This is supported by many researchers, including the 'father' of the case study method, Robert Yin who has amply demonstrated its efficacy over many years (see Yin, 1994; 2011). While not debated sufficiently in the field of tourism research, it has been discussed by Beeton (2005) and further supported by Xiao and Smith (2006) with more recent discussion from Huber, Milne, & Hyde (2017).

Our cases are drawn from two countries in the Asia-Pacific region: Japan and Australia. *Massan* was Japan's morning drama in 2014–2015 and was a semi-fictionalized retelling of the life of Japanese whisky-maker Taketsuru Masataka and his Scottish wife, Rita. *Picnic at Hanging Rock* is a story about a girl who goes missing in the Australian bush that has had multiple adaptations since the novel was released in 1967. Both case studies present the pre-existing heritage tourism, how the work re-presented or contentsized history/legend, the changes in tourism, and longer term implications. Taken together, these case studies demonstrate how the re-presentation and contentsization of history assigns new value and new values to related heritage sites, with major implications for their tourism potential in the long run.

Case study one: Nikka Whisky and *Massan*

Taketsuru Masataka went to Scotland in 1918 to study whisky-making before returning to Japan in 1920 with his Scottish bride Rita and establishing the Nikka Whisky Distillery in Yoichi, Hokkaido (Nikka Whisky, n.d.). Rita is remembered with great affection locally – for example, she has a kindergarten and a riverside walk named after her – because of the depth of her love and support for her husband. Travelling to Japan as a young bride in 1920 was a momentous act of dedication, and she stayed with her husband throughout World War II even as Japan and the United Kingdom were mortal enemies. This is acknowledged in the official corporate version of their romance:

> The young Scotswoman who, in 1920, embarked with her Japanese husband on a long voyage to Japan, adopted the ways of the distant land. She steadfastly supported her husband throughout their marriage, as he built Nikka and made it flourish, until her passing in 1961.
>
> *(Nikka Whisky, n.d.)*

Taketsuru chose Yoichi as the location for his whisky factory because it was the place in Japan where the climate most resembled that of Scotland. It became a great success and in recent decades the Nikka Whisky Distillery has become Yoichi's main tourist attraction.

However, the town could hardly be called a tourism destination. Most visitors do a day trip from nearby Sapporo or Otaru, or stop off while driving through the town en route to the picturesque Shakotan Peninsula or the outdoors sports mecca of Niseko. Hotel capacity in this small rural town famous for its orchards, fishing, and in more recent times its vineyards is limited, with no long-term need to expand due to the rural nature of the environment.

In 2014–2015, Yoichi experienced a major tourism boom. *Massan* was the morning drama on NHK (Japan's public broadcaster) from 29 September 2014 to 28 March 2015. Over 150 episodes of 15 minutes each, the drama told a semi-fictionalized, albeit largely historically accurate, version of the story of Taketsuru Masataka and Rita. However, for the purposes of the drama, the names of the central characters and companies were changed: for example, Masataka became Masaharu, and Rita became Ellie. This had two primary purposes: first, any historical inaccuracies or interpretations causing offence to the family and company could be dismissed as being 'just fiction'; and second, the local story was converted for copyright purposes into 'NHK's contents'.

The tourism booms that accompany the NHK morning drama and also the taiga drama (an epic historical biopic shown throughout the year on Sunday evenings) are well-known phenomena and constitute an established part of the Japanese tourism calendar (see Scherer & Thelen, 2017; Seaton, 2015). As indicated in Figure 21.1, *Massan* precipitated record tourism levels in Yoichi. The peak of 1.59 million total visitors in 2015 was double the level of the previous decade, and the number of people visiting the Nikka Whisky Factory tripled. Visitor numbers returned to near pre-boom levels after the drama finished and public attention moved to the locations associated with the next morning drama.

Despite record visitation levels, Yoichi town was unable to gain the full benefits of this tourism boom. The first reason was that, as noted above, hotel capacity is limited, so visitors at hotels were virtually unchanged compared to previous years, indicating that there was little spare capacity to absorb the influx of *Massan* tourists. Instead, the boom consisted

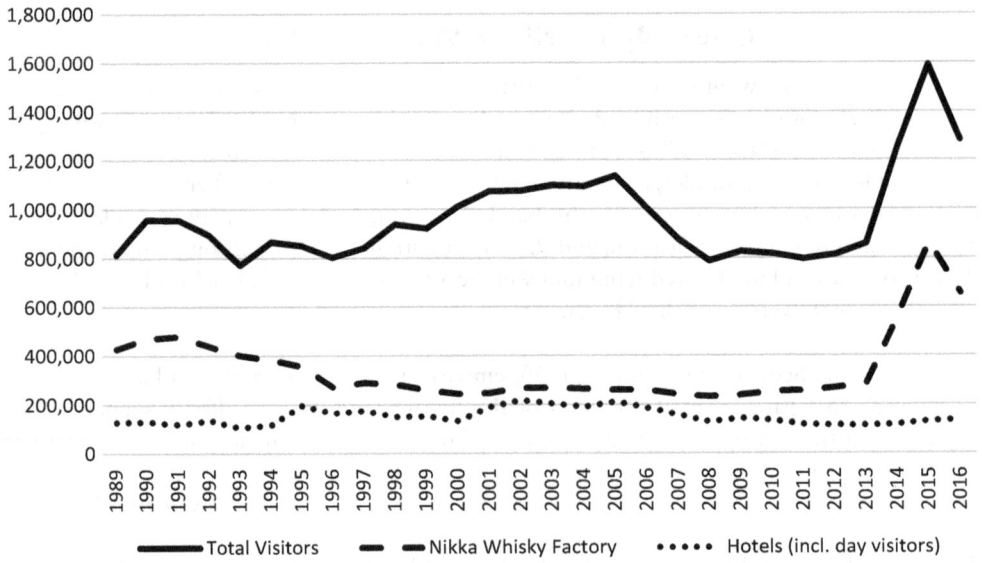

Figure 21.1   Visitor data in Yoichi, 1989–2016
Source: Compiled by the authors based on Yoichi Town (2018, p. 19).

primarily of day trippers who based themselves in nearby major cities, as has been witnessed in other parts of the world, such as Goathland in the UK (Beeton, 2016). This was where the greater economic benefits accrued, both at hotels and at the companies running day tours to Yoichi. Second, many of the profits from *Massan* merchandise also left Yoichi. Profits from increased whisky sales largely ended up at the Nikka Whisky head office in Tokyo; the extra demand for souvenirs was largely covered by confectionary producers outside the town; and any merchandise even just mentioning the drama's name had to gain a license from NHK Enterprises (NHK's business subsidiary that manages copyrights) and pay a 3% commission on sales.

Licensing issues also meant that the tourism board in Yoichi faced severe restrictions on using any images from the drama in its publicity materials. Ultimately, the main occasion on which the profits from the boom stayed in the local economy was when tourists ate at local restaurants, including the one in the Nikka Whisky Factory, which had its premises at the Factory but was an independent local business and not part of the Nikka Whisky group (Ito & Kobayashi, 2014). Other winners included local bakeries, which were also able to profit while circumventing licensing restrictions by producing popular lines of pastries and puddings based on the real Rita's recipes.

The long-running drama drew out the story over six months, providing an opportunity for local school children to learn about their local heritage with renewed pride (Ito & Kobayashi, 2014). The contentsization of the Masataka/Rita story via its re-presentation as the Ellie/Masaharu story also significantly shifted the orientation of the contents. Rather than Rita having a background supporting role as the wife of the 'Father of Japanese Whisky', in the drama Ellie took center stage. In particular, she was often depicted as picking up Masaharu when he was down, and the real strength behind him and Nikka Whisky.

This revision of history dramatically altered the potential tourist market for the Nikka Whisky Factory, and more broadly Yoichi. Prior to the drama, Yoichi was a very 'male' attraction: whisky is a drink commonly associated with male businessmen, and the corporate history was of a dedicated male entrepreneur. But, in the drama, Masaharu took second billing to Ellie, and *Massan* switched audience attention away from Masaharu's attempts to build his whisky factory and onto Ellie's life as a Western woman in 1920s, 1930s, and 1940s, Japan. This reorientation occurred because the target audience for the morning drama is married women, and the drama's main protagonist is always a woman. So, whereas the pre-drama heritage site presented a male-centric narrative and a tourism experience more likely to appeal to males, the drama re-presented the narrative as female-centric and gave a narrative quality to the town far more likely to appeal to female tourists, too.

In summary, a semi-fictionalized drama had fundamentally re-oriented the representation of actual historical events in the public imagination. By doing so, it also shifted the destination brand of Yoichi from a place more likely to appeal to men to being a place with broader appeal to both women and men. It is too early to tell the long-term effects of the drama on tourism in Yoichi, but the *Massan* boom was simultaneously an archetypal contents tourism boom triggered by historical contents, and an example of the power of popular culture to rewrite history, or at least reorient the image of history, in the public mind.

Case study two: *Picnic at Hanging Rock*

Hanging Rock, an enormous rock formation on the plains below Mount Macedon in the Australian state of Victoria, has been a significant site for thousands of years and remains sacred to the indigenous Dja Dja Wurrung, Taungurung, and Wurundjeri nations. Prior to

European colonisation, it was a ceremonial site used by both men and women as well as a gathering site for trade (Heritage Victoria, 2016). Post-colonisation, the site, being relatively close to the city of Melbourne, became known as a favourite recreational place for picnics as represented in William Ford's 1875 painting 'At the Hanging Rock' (Village Well, 2016).

In 1967, author Joan Lindsay released her book, *Picnic at Hanging Rock*, that was to change Australians' relationship with the site forever. This entirely fictional tale was built on early British settlers' fear of 'the bush', via its story of a group of schoolgirls in 1900 who went missing and were never found, and resonated with many Australians, quickly becoming a classic. The story was subsequently presented cinematographically in an evocative movie directed by Peter Weir in 1975 and a powerful gothic play adapted by Tom Wright which was first performed in 2016 and a number of times since, along with a recent Foxtel six-part TV series in May 2018. In fact, the series debuted as Foxtel's top non-sport program with an average of 122,000 viewers (which does not include subsequent streaming), placing it with some of the company's biggest international drama launches (News Corp, 2018). The popularity of the story and filmic imagery effectively erased the indigenous culture and heritage of the site from Australia's collective memory and imagination.

The story is a mystery around the unsolved disappearance of three girls from a very English girls school at a picnic on Valentine's Day in 1900, and while the mystery drove the narrative, the true power of the story lies in the landscape itself. Lindsay reflects dramatically on the colonial settlers view of Australia as wild and dangerous while they attempt (unsuccessfully) to impose their British way of life upon it. As Steele explains, '[f]ive centuries of aristocratic tradition and English history are reduced to insignificance when faced with the reality of the landscape' (Steele, 2010, p. 47).

The site of the story, Hanging Rock, is certainly real, as is the dramatic landscape in which it sits. The novel explores the reverberations that the girls' disappearance has upon the entire community, as 'the pattern of the picnic continued to darken and spread' (Tom Wright, in O'Neill, 2016, p. 8). The Australian Bush is the protagonist in this story, one that broods with silent yet evil intent, never giving up its truth. The Bush wins. As Matthew Lutton, Artistic Director of the play explains:

> The central character of Picnic at Hanging Rock however is nature. It releases and disturbs all the characters… [S]ometimes we see nature thinking in the sign over the stage, or glimpse a physical manifestation hanging in the shadows, or sense its infiltrating presence in the darkness.
>
> *(Matthew Lutton in O'Neill, 2016, p. 5)*

While the novel itself was popular in Australia, it was the movie that catapulted the story and place onto the world stage, with its ethereal imagery and soundscape. Peter Weir's film heralded a new and golden era of Australian cinema. The movie added to the novel's emotional landscape and subversion of colonial notions of rationalism through its careful use of sound, from the music to the sparse natural sounds of the Australian bush (Iner, 2017).

It is remarkable that such a story was so successful in blurring fact and fiction, that many still believe that the novel was historical non-fiction, as Wright intones:

> It is really interesting the amount of people who think these events actually happened, that people will still ask, 'did they ever find those girls?' For some reason this is one story that people have latched onto, perhaps because they want to believe it. And children did go missing in the Australian bush in the 19th century and early 20th century. And

that was a very frightening prospect. But Lindsay made this up. It is a constructed myth. When a story has such power that it is no longer viewed as a fabrication, and it assumes a significance to living, breathing people, then it's no longer just a story and it becomes a myth.

(Tom Wright in O'Neill, 2016, p. 8)

Much has been written about the book and movie (and more to come with the powerful re-interpretation presented through the play), from its gothic styled landscape and colonial narrative through to seeing a (somewhat tenuous and gratuitous) relationship in the movie with the Aboriginal 'Dreamtime' [sic] by writers such as Gauper (2001) and Catania (2012). As noted earlier, Hanging Rock is a sacred place for Australian indigenous people, but authors such as those noted above tend to ascribe somewhat fantastical and naïve 'native' elements to this connection which, while incorrect, have served to add to the place's mystery and appeal for many Europeans.

This convergence across multi-media platforms, from novels to cinema, stage, and even artwork has resulted in many visitors to the rock believing the story to be real. Even those who know differently acknowledge that it could be entirely possible. Indeed, there are those who remain convinced of its veracity, including author Janelle McCulloch, who claims that she has solved the real mystery of the disappearance (or at the very least presents her own theory) in her 2017 publication, *Beyond the Rock: The Life of Joan Lindsay and the Mystery of Picnic at Hanging Rock* (Koob, 2017). Such is the power of this tale.

As noted earlier, prior to Lindsay's novel, the site had been used by European settlers for recreational purposes, which continue today. Being within an hour's drive from Melbourne, the region experiences many day-trippers for picnics, or to attend events such as the New Year's Day horse races and even rock concerts, which were first held in 2010. There is also a cricket oval, tennis club, and petanque facilities that are regularly used by the local community as well as visitors.

However, the movie in particular put the site on to the world map, increasing the number of international visitors and tour groups. Fan fixation on Miranda (the main character in the story) was inspired by Weir's character who cried out 'Miranda!' when searching for the lost girls, with such fans doing the same when they visit the site (Koob, 2017). In fact, much like the cry of 'Stella!' in *A Streetcar Named Desire*, set in New Orleans, it is hard to resist. Place and story have become one. The gothic nature of the story and subsequent multiple media representations have had further impact on those visiting the site at Hanging Rock itself, with many moving off the main trails, either looking for Miranda or wanting to experience the ethereal nature of the rock themselves. Certainly, the crowds at the site at peak times do not allow for a lonely, ethereal experience encouraging people to move off the paths, impacting on the environment, and risking their own safety.

The colonialist theme of the story and its subsequent tourism has resulted in others calling for a re-imaging of Hanging Rock to incorporate an authentic view of the indigenous significance of the site. A campaign 'Miranda Must Go' was started in 2017 by Melbourne artist Amy Spiers who bemoans the fact that visitors to the site seem to all have some knowledge of the movie, but not of the Aboriginal people who lived there (http://www.mirandamustgo.info). While to some it may sound somewhat incongruous that they are proposing the removal of a fictional character from a real site, studies as to the power of such emotional connections via pop culture and contents tourism support the influence of such stories, especially when portrayed via film (Beeton, 2015; Beeton et al., 2013; Reijnders, 2011).

Conclusions

Most tourist sites these days are assuming ever more meanings as mediatized images and works of popular culture expand and personalise the associations with a particular place, which benefit from studying them from both media and cultural studies perspectives. The power of popular culture and tourism to re-orient historical memory is also clear. This chapter has adopted these disciplinary perspectives in order to analyse two examples of places that had particular meanings as heritage sites before mediatized works of popular culture reoriented and converged on the place narrative, opening up new tourism flows to the heritage sites. As such, the chapter exemplifies the circle of representation by which mediatized representations feed into and expand destination image. In one case the flows might be interpreted as broadly positive (increasing the interest among female travellers in a site previously of primarily male interest) and in another case the flows can be construed as broadly negative (the works of popular culture introduce narratives that overlay and obliterate the aboriginal stories associated with the site). To see both of these tourism effects simply in terms of increased levels of commercial activity, therefore, would miss a significant issue: the heritage or place narrative itself is being transformed, and that transformation may be both positive and negative.

The case studies also expose two closely linked questions: When the source material is historical, whose contents are they? In both cases, the original meanings of the site generated locally were largely displaced by powerful external media actors who whether by accident or design significantly reoriented the contents beyond the control of local people. The locality lost much agency and control of their place narrative. The second question is, who benefits financially? In the case of Yoichi, the city had some financial gains from the *Massan* boom, but it had to settle primarily for intangible benefits, such as increased local pride and identity, while the main financial gains accrued elsewhere. At Hanging Rock, the financial benefits do not accrue to aboriginal communities, and certainly not at a level that might compensate for the erasure of their narratives at the site caused by subsequent works of popular culture.

History is constantly being rewritten and re-presented, so it is unreasonable to expect heritage sites to remain immune to change over time. However, this chapter illustrates that a process of contentsization, particularly when its starting point is actual history or heritage, is never neutral. Popular culture may be all-powerful in reshaping the tourism experience at heritage sites. So while mediatized works of popular culture can act as drivers of tourism to heritage sites – phenomena known by various names including film-induced tourism, literary tourism, mediatized tourism, and contents tourism – this chapter has shone light on a different role: popular culture's power to change the very meanings that exist at those heritage sites.

References

Beeton, S. (2005). The case study in tourism research: A multi-method case study approach. In B. Ritchie, P. Burns, & C. Palmer (Eds.), *Tourism research methods. Integrating theory and practice* (pp. 37–48). Oxfordshire, UK: CABI International:

Beeton, S. (2015). *Travel, tourism and the moving image.* Bristol, UK: Channel View.

Beeton, S. (2016). *Film-induced tourism* (2nd Ed.). Bristol, UK: Channel View.

Beeton, S., Yamamura, T., & Seaton, P. (2013). The mediatization of culture: Japanese contents tourism and pop culture. In J. A. Lester & C. Scarles (Eds.), *Mediating the tourist experience: From brochures to virtual encounters* (pp. 139–154). Farnham, Surrey: Ashgate.

Catania, S. (2012). The hanging rock piper: Weir, Lindsay, and the spectral fluidity of nothing. *Literature/Film Quarterly, 40*(2), 84–95.

Edensor, T. (2005). Mediating William Wallace: Audio-visual technologies in tourism. In D. Crouch, R. Jackson, & F. Thompson (Eds.), *The media & the tourist imagination: Converging cultures* (pp. 105–118). Abingdon, UK: Routledge.

Frost, W. (2006). Braveheart-ed Ned Kelly: Historic films, heritage tourism and destination image. *Tourism Management, 27*, 247–254.

Gauper, S. (2001). Aborigine spirituality as the grounding theme in the films of Peter Weir. *The Midwest Quarterly, 42*(2), 212.

Heritage Victoria (2016). *Database report, Hanging Rock Reserve*, Report generated 26 July 2016.

Huber, D., Milne, S., & Hyde, K. F. (2017). Biographical research methods and their use in the study of senior tourism. *International Journal of Tourism Research, 19*(1), 27–37.

Iner, J. (2017). Sounding 'unstable terrain' in Peter Weir's 'Picnic at Hanging Rock' *Metro Magazine: Media & Education Magazine, 193*, 84–91.

Ito, J., & Kobayashi, E. (2014). Interview conducted by P. Seaton at Yoichi Tourism Office, 27 November.

Jenkins, H. (2004). The cultural logic of media convergence. *International Journal of Cultural Studies, 7*(1), 33–43.

Jenkins, H. (2006). *Convergence culture: Where old and new media collide*. New York, NY: New York University Press.

Koob, S. F. (2017). 'Miranda must go': Push to end Hanging Rock's enduring myth. *The Australian*, September 23, 2017.

Månsson, M. (2011). Mediatized tourism. *Annals of Tourism Research, 38*(4), 1634–1652.

News Corp Australia (2018). Picnic at hanging rock makes strong start in ratings, with streaming figures to come. *News Corp Australia Network*, May 7th 0218.

Nikka Whisky (n.d.). The founder. Father of Japanese whisky. Retrieved from: https://www.nikka.com/eng/founder/.

O'Neill, V. (2016). *Prompt pack for picnic at Hanging Rock*, Malthouse Prompt, Melbourne, Australia.

Reijnders, S. (2011). *Places of the imagination: Media, tourism, culture*. Farnham, UK: Ashgate Publishing.

Scherer, E., & Thelen, T. (2017). On countryside roads to national identity: Japanese morning drama series (*asadora*) and contents tourism. *Japan Forum*. Published online 20 December 2017.

Seaton, P. (2015). Taiga dramas and tourism: Historical contents as sustainable tourist resources. *Japan Forum, 27*(1), 82–103.

Seaton, P. (2019). On the trail of *The Last Samurai* (I): Taranaki. *International Journal of Contents Tourism, 4*, 12–24.

Seaton, P., Yamamura, T., Sugawa-Shimada, A., & Jang, K. (2017). *Contents tourism in Japan: Pilgrimages to "sacred sites" of popular culture*. New York, NY: Cambria Press.

Steele, K. (2010). Fear and loathing in the Australian Bush: Gothic landscapes in bush studies and Picnic at Hanging Rock. *Colloquy, 20*, 33–56.

Village Well (2016). *Hanging Rock vision paper*, Melbourne, Village Well.

Xiao, H., & Smith, S. L. (2006). Case studies in tourism research: A state-of-the-art analysis. *Tourism Management, 27*(5), 738–749.

Yamamura, T. (2019). Cooperation between anime producers and the Japan Self-Defense Force: Creating fantasy and/or propaganda? *Journal of War & Culture Studies, 12*(1), 8–23.

Yin, R. K. (1994). *Case Study Research. Design and Methods* (2nd Ed.). Thousand Oaks, CA: Sage Publications.

Yin, R. K. (2011). *Applications of case study research*. Thousand Oaks, CA: Sage.

Yoichi Town (2018). *Yoichi-chō kankō shinkō keikaku: Heisei 30-nendo ~ Heisei 34-nendo* Retrieved from: https://www.town.yoichi.hokkaido.jp/chousei/files/kankosinkokeikaku.pdf.

22

CHALLENGES OF FILM-INDUCED TOURISM IN CROATIA

From *Winnetou* to *Game of Thrones*

Božo Skoko and Katarina Miličević

Introduction

In this chapter, we analyse film as a public relations tool for creating global recognizability and an important element of the destination branding process. Furthermore, we focus on the analysis of the impacts of institutional incentives on the film tourism and film industry itself in these processes. This interdisciplinary field is connecting public relations and film as a part of the destination branding process, and destination image as an outcome of the destination branding process. The key question in this chapter is to what extent does the fact that a destination has become a film location contribute to its popularity and the development of tourism, and to what extent is this 'exploitation' dependent on the accompanying public relations and tourism destination management, i.e. the mediatization of tourism. The theoretical foundations and existing body of research on this topic will be presented in the practical example of Croatia as a popular European and Mediterranean destination. Croatia recently began using film strategically[1] as an economic tool for increasing revenues and export services, public relations, and destination promotion, i.e. the mediatization of tourism. Given that there were expectations that the shooting of popular series and films in Croatia would automatically contribute to its popularity, and thereby, to additional earnings from tourism, in this chapter, we analyse the efforts Croatia has undertaken at the institutional and legal level in order to drive its film-induced tourism, as well as the effects that resulted from them. The strategy, the advantages and disadvantages, and the effects of this approach will be analysed.

This chapter presents the steps taken by Croatian public institutions to attract film productions and the ways they are utilized to promote the tourism offer. It is also described how, a full 40 years after shooting the popular series (Winnetou), the shooting location started being utilized as a tourism product. On the other hand, *Game of Thrones* shot in the historic city of Dubrovnik demonstrates the power of this popular TV series in attracting new tourists to this part of Croatia. Although the number of tourist arrivals in Dubrovnik has grown, at the same time, the series has overshadowed the rich history and cultural heritage of this cultural monument inscribed on UNESCO World Heritage List. Furthermore, the Croatian example supports the hypothesis that destinations that do not play themselves on screen, but rather are the backdrop for other locations, have difficulties benefiting from films in tourism and promotion. In this context, much greater public relations efforts are necessary for people to associate films with the actual shooting locations. After serving as a shooting location for numerous film productions for decades, which is a fact still going unnoticed even amongst the local inhabitants, Croatia has finally decided to take advantage of these benefits. This resulted in linking the 'Filming in Croatia' project with the country's tourism promotion, i.e. the Croatian National Tourist Board, as the umbrella state branding institution.[2]

The power of film-induced tourism

Film is often called a 'seventh art'. However, besides its artistic values, film may also evoke other values by promoting certain principles, ideas, people, or countries. Due to the popularity of this genre of entertainment, it has often been used as a tool of public diplomacy in improving the image of a country or nation. It is therefore not surprising that tourism destinations – countries, regions, and cities – started offering different incentives to film production companies for shooting at their locations or choosing scenario themes that promote their country, lifestyle, history, natural and cultural resources, and other particularities (Skoko, Brčić, & Vidačković, 2013, p. 55).

Screen tourism attracts consumers of audiovisual media who respond to the opportunity to visit a place they have seen on screen and who wish to find more information about it. Their desire to visit such destinations converts them into tourists, with the content either providing the primary driver for visiting a place or providing something that gives added value to a region or place. The latter may result in the longer visitor stay, but it can also change the image as well as the awareness of the place for the tourist and make destinations more competitive. For some places, it might even result in a prolonged destination life cycle as new visitor segment. Another effect could be the expansion of the visitor season since screen tourism is not dependent on a specific season (Niziol, 2015, p. 149). By means of acting as push and pull factors films are ideal destination marketing tools. They provide a massive amount of exposure to destinations at little cost to the Destination Marketing Organizations (DMOs) (Hoffmann, 2015, p. 85).

Film tourism research has undergone rapid development over the last decades (Lee & Bai, 2016). Film-induced tourism and its related phenomena have not received much attention until the end of the 20th century. In her review article, Heitmann (2010) concluded that early discussion on film-induced tourism can be dated back to Boorstin (1962), but a more extensive research can be attributed to Tooke and Baker (1996) and subsequently Beeton (2001, 2004, 2005) as well as Busby and Klug (2001). Busby and Klug (2001) provide an extensive list of different forms and characteristics of movie-induced tourism: (i) a film location can be an attraction in its own right, either before or as a result of being depicted on film; (ii) movie tourism can be part of the main holiday; (iii) movie tourism can occur as the main purpose out of special interest; (iv) movie tourism packages can be created by the private sector, only elements of the film (icon, actors, natural scenery, historical background, storyline, symbolic content, human relationships) are the focal point of tourist interest; and (v) travel programs. Films' main advantage is that they tell stories in an entertaining, easily comprehensible fashion (Thompson, 2003). The promotional strength of the film has recently been the subject of a number of scientific and professional papers, analysing the causal relationship between the success of a film and the popularity of the destination measured by the increase in the number of visitors (Beeton, 2001, 2004; Busby & Klug, 2001; Hudson & Ritchie, 2006; Riley, Baker, & Van Doren, 1998; Tkalec, Zilic, & Recher, 2017; Tooke & Baker, 1996). The best cases of synergy between film and promotion of destination are *The*

Lord of the Rings trilogy (2001–2003) and New Zealand; television show *Sex and the City* (1998–2004) and New York (Niziol, 2015, p. 153); and the *James Bond* filmography (1962 until today) promoting Great Britain; and other, often exotic, destinations (Forbes, 2018).

During recent years, the role of mass media and its crucial position in the tourism information and decision-making process has increased. The decision to choose one-holiday destination over a myriad of available choices is just one of many steps in the tourist's decision-making process (Lexow & Edelheim, 2004, p. 51). The media have a crucial role to play in putting emerging destinations on the world tourism map. The relationship between tourism and the media is vital and complex. Tourism is highly dependent on media reporting because the vast majority of travel decisions are made by people who have never seen the destination first hand (Praveen Kumar, 2014, p. 189). Based on their research, the author claims that media contribute to 80% of tourism revenue and tourism contributes to 25% of the media's revenue (Praveen Kumar, 2014). Furthermore, it is important to point out that different media have different intensity of influence on the creation of tourism destination image (Šantić, Bevanda, & Bijakšić, 2016, p. 180). Govers, Go, and Kumar (2007) found that autonomous and 'covert-induced' agents, such as television, magazines, Internet, books, and movies, are more popular information sources than 'overt-induced' agents, such as advertising. In fact, a number of scholars assert that the best instrument to reach wider audiences with fewer investments and promotion is film, because even if films are not produced with the intent to attract tourists to a destination, they tend to influence viewers indirectly as a background to the movie's message without the perceptual bias of promotional material (Giraldi & Cesareo, 2016, p. 108). The exposure to a film or a television program gives a destination an advertisement viewed potentially by millions of people, an audience not reachable through specifically targeted tourism promotional activities (Giraldi & Cesareo, 2016, p. 109). A study by thinktourism Ltd., which focused on tourist-generated content on Croatia as a film tourism destination, further supported these claims in the context of Croatia. Among four movies and a total of 3312 posts analysed, *Game of Thrones* was by far the most frequently mentioned film, communicated in 3024 different posts. Mamma Mia was mentioned in 230 posts, Star Wars in 77, and Winnetou in 20 posts.[3] The majority of posts related to *Game of Thrones* (54%) were neutral in sentiment, while 42% had a positive sentiment and 4% had a negative sentiment. The words 'beautiful, love, amazing, fans, happy, top, blue, popular, famous and beauty' contributed most to the positive sentiment. While the content was much less filled with the negative attitude towards filming location, some notable negative mentions are 'mass tourism' and 'crowded'.

Film as a strategic government tool for tourism development

Croatia has a long tradition of hosting foreign film productions, dating from 1923. Some extremely popular films and series, including Spaghetti-Westerns, WWI and WWII films, historical and biblical dramas and sci-fi films, were filmed in Croatia, but for decades, the international audience, Croatian residents, and tourists were largely unaware of this. *Jadran film* was the largest and most famous film studio in Central Europe and, during that period, 145 international co-productions and 124 domestic films were shot, hosting some of the most important films, directors, and acting stars. However, shooting locations in Croatia often served to portrait locations in other countries. For instance, Zagreb, the capital of Croatia, was used for re-creating Austrian, Hungarian, French, Swiss, Czech, and German cities (Škrabalo, 2006; Skoko et al., 2013).

In order to bring back the rich tradition of shooting foreign productions, as well as further develop its own audio-visual production and ensure additional revenues to its economy and tourism, the Croatian Government has established a separate public agency responsible for audiovisual production and has set up a special tax incentive policy for foreign productions shooting in Croatia. In 2007, the Croatian Parliament passed a new Film Act in order to attract a larger number of foreign products to Croatia which represent a significant source of revenue for the creative industries in general but also to use films for additional global promotion of a country already recognized for its natural and cultural heritage (The Croatian Audiovisual Centre – HAVC, 2018). In doing so, it emphasized its commitment to boost the creative industries body – The Croatian Audiovisual Centre (HAVC). In 2010, the *2010–2014 National Program for Promotion of Audio-visual Creative Work* was adopted, which had special emphasis on encouraging investments in domestic films, the film industry, and the export of film services. Providing an on-going and full support to the development of film industry and film culture in Croatia and to attracting international film projects to the country, in 2012, the Government introduced the *Film Production Incentive Program*. It is available to international and local filmmakers in the form of a rebate of 20% on qualifying Croatian expenditure.[4] The Incentive Program is administered by the *Filming in Croatia* department of the Croatian Audiovisual Centre. Since its introduction, the Program, Croatia was chosen as the location for the production of numerous projects. More than 40 productions benefited from the scheme, originating from countries as varied as the US, the UK, France, Belgium, Germany, Switzerland, the Netherlands, Finland, Denmark, Sweden, and India. In 2014, Croatia attracted a series of high-profile international productions (TV series: *Borgia, Lost Treasure of Aquila, Game of Thrones, Dig, Jonathan Strange & Mr. Norrel*; films: *Full Contact, Winnetous Weiber* – Winnetou's Wives, *The Witness*).[5]

Croatia has recently become an attractive backdrop for the film industry. The authenticity of numerous locations in Croatia with their natural beauties and cultural heritage have become the backdrop for sceneries in many famous films and series. Those locations are also well-known tourism destinations. In this manner, new tourism products are generated, including shooting location tours, a variety of authentic souvenirs and gifts, and many interesting storytelling-based activities.[6] Being aware that the popularity of films shot in Croatia could be used for the promotion of the country, the Croatian Audiovisual Centre presented the film tourism project in October 2013, which received the support of the Croatian parliament. Key goals of the *Screen Tourism Strategy* are expanding the tourism season, attracting tourists from new markets, attracting culture tourists as a more lucrative market segment, unlocking the potential of unknown Croatian locations never featured on screen and little known to tourists.[7] The management team of the Croatian Audiovisual Centre believed that Croatia's Screen Tourism Strategy can lay the foundation for sustainable and ecologically friendly tourism. However, further development of film-induced tourism occurred sporadically for the most part and there was no central coordination of the project.[8] Therefore, the key role was played by small travel agencies, which began organizing tours to shooting locations of the most popular films and series. At the same time, synergy was lacking, i.e. a coordinated promotion of popular productions and destinations, which is why film-related themes often overshadowed the identity of the destination itself (Skoko, 2014, p. 188).

However, the growth in the number of tourists and revenues from production had an optimistic effect on state institutions. Namely, from 2012 to 2016, 34 international film and TV projects were realized, achieving a total consumption of 390 million Croatian Kuna (HRK) (Croatia paying back 73 million HRK in incentives). Foreign film and TV employees realized 118 thousand overnight stays in Croatia (Croatian Audiovisual Centre, 2017, pp. 20–21).

Encouraged by all those successes and with a desire to attract even more foreign film productions, in 2017, the Croatian Government increased the incentive from 20% to 25% to be even more competitive in comparison to surrounding countries. Unfortunately, the proposal to increase the incentive to 30% for those countries that use elements related to Croatia in their scripts (where Croatia is not simply the backdrop for another country but plays itself) was not adopted.

In 2017, the new *2017–2021 National Program for Promotion of Audio-visual Creative Work* was adopted, with the aim to increase the number of film projects. Also anticipated by the Program is building a film studio that would make Croatia even more attractive location for foreign productions.

Tourism activity influenced by popular film locations in Croatia

Winnetou and Paklenica National Park

Between 1962 and 1968, 11 films about Winnetou, based on the novels by the German author Karl May, were shot in Croatia as a co-production between Berlin's Rialto Film and Zagreb's Jadran Film. Croatian locations, such as the Dalmatian hinterland, Plitvice Lakes, Paklenica and Krka national parks 'starred' American Wild West, Velebit Mountain was one of the main locations for the Winnetou and Old Shatterhand films.[9]

After the film screening, the popularity of Croatia grew significantly in Germany, and even tourist tours to the film shooting destinations were organized by German travel agencies. However, the power of film tourism and the popularity of that series were largely underestimated in Croatia's tourism offer. Film tourism was completely neglected. Only in 2009 (40 years after the first screening of the film series), Winnetou Museum was opened in Starigrad Paklenica

> as a sign of gratitude to the film employees who recognized the beauty of the area, the film crews who shot there, fans of Karl May films, tourists who eagerly visit this area, as well as tourism industry employees from Starigrad who foster this tradition.
>
> *(Winnetou Museum, 2018)*

Also initiated was an excursion program called *Following the Trails of Winnetou* (Skoko et al., 2013, p. 61). It is an eight-day excursion, where visitors could visit the locations where the films were shot, try horseback riding, shooting with a bow and arrow, have meals in nature by a campfire, and much more. Winnetou Days in Starigrad Paklenica are held at the end of June when visitors have the opportunity to hang out with special guests. In 2011, the special guest was Marie Versini, who played the role of Winnetou's sister (Winnetou Museum, 2018). Furthermore, the first 'Indian reserve in Europe' was opened in Croatia, and all of this is due to the popularity of the films about chief Winnetou. The tourism project entitled Winnetou Land (Dnevnik.hr, 2009) is located in the village of Rakovica, and it allows visitors to acquire Indian skills and become familiar with customs that, until then, they could only watch on screen. Therefore, films on Winnetou have certainly left the largest and deepest mark among the films shot in Croatia (Skoko et al., 2013, p. 61).

Due to the immense popularity of the films on the fictional Indian chief, which has obviously not faded over the years, three new sequels were filmed in Croatia in 2015.[10] All of this has encouraged the Ministry of Tourism to include the 'Winnetouland' theme park, dedicated to the characters and topics from Karl May's novels in the *Strategy on the Development of Croatian Tourism from 2014 to 2020* (p. 88) among greenfield projects that should enrich Croatia's tourism offer.[11] This is an example of the later use of film for tourism needs, as well as an example of the strong relationship between the popularity of a film with the locations where it was shot (without additional investment in promotion and public relations for the destination).

Game of Thrones *and Dubrovnik*

Game of Thrones is an American fantasy drama television series created by David Benioff and D. B. Weiss. It is an adaptation of A Song of Ice and Fire, George R. R. Martin's series of fantasy novels, the first of which is 'A Game of Thrones'. It is filmed in Belfast and elsewhere in Northern Ireland, Canada, Croatia, Iceland, Malta, Morocco, Scotland, Spain, and the United States. The series premiered on HBO in the United States on April 17, 2011. The series concluded with its eighth season in April 2019. The four last seasons of the popular series Game of Thrones was partly filmed in the historic Croatian city of Dubrovnik, which is inscribed on UNESCO World Heritage List. Dubrovnik was used as one of the most important locations in the series – the capital of the seven kingdoms, King's Landing. Numerous global media reported on shooting the series in Croatia, such as Daily Mail, USA Today, CBS News, Sky, Telegraph, Huffington Post, etc. Thanks to the Game of Thrones, the website BuzzFeed, visited by over 80 million people monthly, in 2013 placed Dubrovnik shoulder to shoulder with 19 famous film locations, which were also popular tourism destinations. In addition to crowding, tour guides further point out that tourists in Dubrovnik are six to seven times more interested in Game of Thrones series than the history of the city itself, with some guests even convinced that the walls of Dubrovnik were built for the series shooting.

Conclusions

Films are 'pull' factors for destinations that served as sets for a specific film or TV series (Hoffmann, 2015; Macionis, 2004; Niziol, 2015; Riley & Van Doren, 1992). The majority of research on film-induced tourism still focuses on understanding the phenomenon, the motivation, and experience of film tourists or films as tools for destination marketing, by adopting a case-study approach and analysing the impacts of films on the success of tourism destinations. On the other hand, this chapter is focused on the analysis of the impacts of institutional incentives on film tourism and the film industry itself, and also provides a snapshot of the film-related content in the perception of Croatia as a tourism destination.

The strategic dedication of the state and state incentives for strengthening film productions can contribute to increasing the number of foreign productions and revenues but also indirectly to strengthening film tourism and popularizing a destination. In order to succeed in this, it is also necessary to work on destination promotion and public relations, which will link with the film or series. Furthermore, generalizations should not be made, since impacts change from one production to the next, i.e. depend on the concrete film and its connection with the destination, as well as on efforts to take advantage of the fact that a popular film was shot there. As we have demonstrated, the German film series Winnetou contributed to the popularization of Croatian destinations without additional investment in tourism promotion, due to the popularity of the film itself in the German-speaking area and the attractiveness of the locations that 'doubled' as the American Wild West. Regarding the Games of Thrones series, its popularity has, to a great extent, overshadowed the identity of Dubrovnik as a city of great history and transformed it into the background of a film. The reason is, obviously, insufficient public relations and proactive communication of this historic city. As for the popular musical Mamma Mia, which was filmed on the Croatian island of Vis, doubling as a Greek island, it is questionable who profited – Croatian or Greek tourism, because viewers might not be familiar with the fact that the beautiful scenery in the film are actually Croatian, and not Greek.

It is obvious that Croatia will have to invest additional efforts in order to take advantage of foreign productions for its own promotion because it often represents the backdrop for other real or fictional locations (Skoko et al., 2013, p.60). It is not enough to simply participate in the production, but rather it is essential to manage destination public relations, through which destinations would inform film-lovers about the film location itself, as well as its values and tourism offer that go beyond it. Namely, there is a possibility that viewers will not even find out where a certain film was shot and/or that the location appearing in the plot of a film becomes popular, not the actual shooting location. The locations used for shooting by numerous foreign productions fell into oblivion over the years and were not used at all for tourism purposes. This can be avoided by strengthening public relations and promotion campaigns in order to better connect its locations with popular films and series, as well as create new tourism products and services in destinations that are inspired by films and series. For instance, Zagreb, although frequently used as a backdrop in films, received its first guide *Zagreb – the Film Stage*, noting the popular films and locations where they were filmed, only in 2013. Zagreb had already become a very popular tourism destination, so the guide and accompanying tourist tours attempted to enrich the tourist offer. To take advantage of the promotional potential of the film industry, Croatian authorities can invest in attracting the interest of world-renown producers for Croatian stories, heroes, history, or locations, in this manner making Croatia an integral part of their scripts. Thus, the mediatization of tourism is indispensable for the development of tourism and the promotion of destinations. In this context, due to its popularity and influence, film has significant potential. However, Croatia's experience demonstrates that attracting foreign productions and shooting popular films and series at certain locations is insufficient. In order to exploit the opportunities offered by film-induced tourism, it is necessary to develop high-quality public relations that will link the destination to production and continually manage the destination's offer.

Notes

1 An institution was established to attract foreign film productions, while the Croatian Government has provided production companies with special tax incentives.
2 On November 15, 2018, the premiere of the Croatian film 'Osmi povjerenik' (The Eighth Commissioner) was held in Los Angeles. This film, whose plot takes place on a picturesque island in the Adriatic Sea, was also Croatia's official Oscar candidate. For that occasion, an event entitled 'Night of Croatian Film and Flavors' was organized as the first joint presentation by the Croatian Audiovisual Centre (HAVC) and the Croatian National Tourist Board (CNTB). It gathered representatives of American media and notables from the film industry, who had the opportunity to become acquainted with representatives of the Croatian film industry and the presentation of the destination where the film was shot while enjoying the accompanying culinary and oenological offer. The result was significant publicity for the film, as well as for Croatia as a tourism destination.
3 The results obtained in this research were in the English language. Having in mind the popularity of Winnetou on the German outbound market, the results in the German language might be completely different.

4 A qualified spend consists of the costs of goods and services purchased in Croatia and wages paid to Croatian tax residents for services carried out in Croatia.

5 Read more at http://filmingincroatia.hr/

6 Examples: https://www.total-croatia.com/game-of-thrones-croatia/; https://www.kingslanding dubrovnik.com/; https://www.wallsofdubrovnik.com/walking-tours/dubrovnik-game-of-thrones-tour/; https://gameofthronestravel.com/dubrovnik-game-of-thrones-locations/; https://www. theguardian.com/travel/2018/jul/15/vis-island-croatia-travel-mamma-mia-here-we-go-again.

7 Read more at https://www.havc.hr/infocentar/novosti/havc-predstavio-gospodarske-i-fiskalne-rezultate-audiovizualne-industrije-i-projekt-filmskog-turizma-u-hrvatskom-saboru.

8 Read more at https://www.total-croatia.com/game-of-thrones-croatia/.

9 Read more at https://www.dw.com/hr/50-godina-winnetoua-na-filmu/a-16444347

10 Philipp Stölzl is the director of all three Winnetou sequels, the chief scriptwriter is Jan Berger, and producer Christian Becker. In the leading roles are Wotan Wilke Möhring as Old Shatterhand, and Nik Xhelilaj as Winnetou.

11 Read more at: https://mint.gov.hr/UserDocsImages/arhiva/130426-Strategija-turizam-2020.pdf

References

Beeton, S. (2001). Smiling for the camera: The influence of film audiences on a budget tourism destination. *Tourism Culture & Communication, 3*(1), 15–25.

Beeton, S. (2004). *The more things change… A legacy of film-induced tourism.* Melbourne, Australia: Monash University.

Beeton, S. (2005). *Film induced tourism: Aspects of tourism.* Manchester, UK: Wordswork Ltd.

Boorstin, D. J. (1962). *The image: A guide to pseudo-events in America.* New York City, NY: Atheneum Books.

Busby, G., & Klug, J. (2001). Movie-induced tourism: The challenge of measurement and other issues. *Journal of Vacation Marketing, 7*(4), 285–301.

Croatian Audiovisual Centre (2017). *Nacionalni program promicanja audiovizualnog stvaralaštva 2017. – 2021.* Zagreb, Croatia.

Dnevnik.hr (2009). Retrieved from: https://dnevnik.hr/vijesti/hrvatska/drugaciji-turizam-winnetou-na-plitvickim-jezerima.html.

Giraldi, A., & Cesareo, L. (2016). Film marketing opportunities for the well-known tourist destination. *Place Branding and Public Diplomacy, 13*(2), 107–118.

Govers, R., Go, F. M., & Kumar, K. (2007). Promoting tourism destination image. *Journal of Travel Research, 46*(1), 15–23.

Heitmann, S. (2010). Film tourism planning and development—Questioning the role of stakeholders and sustainability. *Tourism and Hospitality Planning & Development, 7*(1), 31–46.

Hoffmann, Nicole Beate (2015.): On-location film-induced tourism: success and sustainability. Retrieved from: https://repository.up.ac.za/bitstream/handle/2263/50623/Hoffmann_Location_2015.pdf?sequence=1&isAllowed=y.

Hudson, S., & Ritchie, J. R. B. (2006). Promoting destinations via film tourism: An empirical identification of supporting marketing initiatives. *Journal of Travel Research, 44*(4), 387–396.

Lee, S., & Bai, B. (2016). Influence of popular culture on special interest tourists' destination image. *Tourism Management, 52*, 161–169.

Lexow, M., & Edelheim, J. R. (2004). Effects of negative media events on tourist's decisions. In W. Frost, G. Croy, & S. Beeton (Eds.), *International tourism and media conference proceedings.* 24th–26th November 2004 (pp. 51–60). Melbourne, Australia: Tourism Research Unit, Monash University.

Macionis, N. (2004). Understanding the film-induced tourist. In W. Frost, G. Croy, & S. Beeton (Eds.), *International tourism and media conference proceedings, 24–26 November 2004* (pp. 86–97). Melbourne, Australia: Tourism Research Unit, Monash University.

Niziol, A. (2015). Film tourism as a new way to market a destination. *Scientific Review of Physical Culture, 5*(4), 149–156.

Praveen, K. (2014.). Role of media in the promotion of tourism industry in India. *Global Review of Research in Tourism, Hospitality and Leisure Management, 1*(3), 187–192.

Riley, R., Baker, D., & Van Doren, C. (1998). Movie induced tourism. *Annals of Tourism Research, 25*(4), 919–935.

Riley, R., & Van Doren, C. (1992). Movies as tourism promotion: A push factor in a pull location. *Tourism Management, 13*(3), 267–274.

Šantić, M., Bevanda, A., & Bijakšić, S. (2016). Influence of media on creation of a tourist destination image. *Informatol, 49*(3–4), 180–189.

Skoko, B. (2014). Mogućnosti i načini jačanja brenda Dubrovnika kroz filmsku industriju i organiziranje događaja. *Zbornik Sveučilišta u Dubrovniku, 1*(1), 175–190.

Skoko, B., Brčić, T., & Vidačković, Z. (2013.): *Uloga igranog filma u promociji Hrvatske – dosezi i mogućnosti / The role of movie in promotion of Croatia – reach and opportunities. Media Studies, 4*(7), 54–74.

Škrabalo, I. (2006). *Hrvatska filmska povijest ukratko (1896.–2006.).* Zagreb, Croatia: VBZ.

Thompson, K. (2003). *Storytelling in film and television.* Cambridge, MA: Harvard University Press.

Tkalec, M., Zilic, I., & Recher, V. (2017). The effect of film industry on tourism: *Game of Thrones* and Dubrovnik. *International Journal of Tourism Research, 19*(6), 705–714.

Tooke, N., & Baker, M. (1996). Seeing is believing: The effect of film on visitor numbers to screened locations. *Tourism Management, 17*(2), 87–94.

Winnetou Museum. (2018). Retrieved from: https://www.hotel-alan.hr/en/karl-may-museum.aspx.

23

BEATEN TRACKS

Belatedness and anti-tourism in guidebooks

Tim Hannigan

Introduction

The travel guidebook is perhaps the oldest type of media directly associated with tourism. Its antecedent forms – the *periploi* and *navigationes* of ancient Greece and Rome; the pilgrimage guides of the Middle Ages – date back many centuries, while the concept of printed guides aimed specifically at leisure travellers emerged around the European 'Grand Tour' in the 18th century, culminating during the 1830s with the launch of the Murray's and Baedeker series (Parsons, 2007). These publishers established a format – listings of individual attractions, practical travel information, potted historical and cultural background, and regularly updated editions – that remains essentially unchanged in the contemporary output of Lonely Planet, Rough Guides, DK Eyewitness, Insight, and myriad other guidebook companies.

The rise of new media has somewhat diminished the commercial prominence of the traditional, printed guidebook in the 21st century. The total value of the guidebook market in the UK has fallen by almost half in recent decades, from a high point of £100 million a year (Jones, 2018). The downward trend does not appear to be unremitting, however, with modest increases in sales reported since 2014 (Dickinson, 2018). The major publishers continue regularly to release new and updated titles, and the basic imperative of the guidebook – to provide practical, actionable travel information – seems certain to endure, even if dispersed across multiple media.

There have been a number of scholarly case studies of guidebooks, including Laderman's examination of coverage of Vietnam (2002), and Bhattacharyya's (1997) work on Lonely Planet's India guide, as well as at least one recent broad overview (Peel and Sørensen, 2016). Generally, however, given their ubiquity, as Scott Laderman notes, 'it is startling how little scholarship exists on the [guidebook] genre' (2016, p. 258). This is despite a large body of academic work falling under the designation of 'travel writing studies', which has emerged during the last four decades, following the publication of Edward Said's seminal work, *Orientalism* (1978).

A possible reason for the lack of attention paid to guidebooks is revealed in the way scholars specialising in travel writing tend to define the object of their study. The following, from Tim Youngs, is typical:

> [T]ravel writing consists of predominantly factual, first-person prose accounts of travels that have been undertaken by the author-narrator
>
> *(2013, p. 3)*

Similar definitions are provided by Jan Borm (2004, p. 17) and Carl Thompson (2011, p. 26). Travel writing studies tend, then, to focus on texts in which both a temporal narrative and an authorial 'I' are explicit. It typically examines ways in which travel writers encounter and construct difference; negotiate, subvert or reiterate colonial discourses; and handle issues of power in their interactions with what Mary Louise Pratt (2008) calls 'the travellee' – that is, the people resident in the places that the traveller passes through. In guidebooks, which typically have only an 'implicit narrator' (Bhattacharyya, 1997, p. 374), such matters may seem less readily accessible for examination.

It is also possible that the guidebook may be seen as a less prestigious – and therefore less appealing – area of study. It has been suggested that first person-narrated travel writing itself suffers a lack of critical attention because of its 'uncertain status as literature' (Mee, 2009, p. 305). The guidebook's standing is surely still less certain for literary scholars, while conversely its superficial resemblance to conventional 'literary' texts (in format and mode of production) may make it an ambiguous object for those approaching from the direction of media or tourism studies. The guidebook exists, then, in a liminal zone between scholarly disciplines in which the theoretical stock of travel writing studies certainly remains a somewhat underutilised resource.

The aim of this chapter is to consider guidebooks with close attention to certain areas – particularly 'belatedness' and consequent 'anti-tourism' – that have often been discussed in relation to first-person travel narratives. The chapter also gains an unusual dual perspective from my previous background as a professional guidebook writer, working mainly on destinations in Asia for DK Eyewitness, Insight Guides, and Tuttle, amongst others. The following analysis of guidebooks covering Nepal takes in a text of which I was the (partial) author. In summary then, this chapter approaches its object from a point of departure in literary studies broadly, and travel writing studies specifically, while also drawing from direct experience of the practice of guidebook production.

Belated travellers

The idea of 'belatedness' as a feature of what I will here call 'narrative travel writing' (to differentiate it from the travel writing contained in guidebooks) has frequently been discussed by scholars. For some, such as Debbie Lisle (2006), it is a characteristic directly linked to various problematic elements of the genre. Briefly, the theory of belatedness has it that the very practice of narrative travel writing – going out to far-flung destinations and returning with an account of the journey – is fundamentally belated. The dissolution of European empire, increasing globalisation, and – perhaps most significantly in this discussion – the rise of mass tourism serves to render the figure of the narrative travel writer fundamentally anachronistic. While in previous centuries, narrative travel writing could be justified as presenting ostensibly 'new' information about a place little-known to its primary audience, it now faces obsolescence, and narrative travel writers must develop means by which to overcome their own belatedness.

Various such strategies have been identified. Narrative travel writers may 'maintain their relevance in a globalised world by mimicking their colonial forebears' (Lisle, 2006, p. 3). That is, they retain and reiterate – at times subtly and even unconsciously – colonial-era notions of 'us' and 'them' and operate with an inflated sense of narrative authority typically leading to a persistent inability to grant the travellee proper respect and agency. Alternatively, narrative travel writers may adopt a mode of cosmopolitanism, drawing attention to and celebrating facets of globalisation, and thus making 'deliberate efforts to distance themselves from the genre's implication in Empire by embracing the emancipatory possibilities created by an interconnected "global village"' (Lisle, 2006, p. 4).

Apparently connected to the first approach identified by Lisle is a widespread practice of attempting to locate a destination in a sort of rhetorical past. This process, as Elizabeth A. Bohls has pointed out, was already at play in the 19th century: 'By visiting or reading about places like the Scottish Highlands – or North America, Tahiti, or Africa – English tourists believed they could travel through time, as it were, and experience earlier stages of civilization' (2016, p. 249). Such an approach is always likely to exacerbate problematic power imbalances between traveller and travellee, and to result in reductive, stereotypical depictions.

For a genuinely adventurous traveller in the early 19th century, it was relatively easy rhetorically to 'past-ify' a distant and difficult-to-reach destination, be it Scotland or Tahiti. But for many travellers since, a particular challenge has been posed by others of their own ilk; for, as James Buzard writes, 'the tourist is an unwelcome reminder, to self-styled "travellers", of the modern realities that dog their fleeing footsteps' (1993, p. 8). Attempts to overcome this particular facet of belatedness led, in the 19th century, to the emergence of what Buzard terms 'anti-tourism'.

The word 'tourist' seems to have first appeared in the late 18th century as a neutral synonym for 'leisure traveller' (Kinsley, 2016). However, the concept rapidly developed pejorative connotations. With the end of international Napoleonic conflicts, leisure travel within Europe, and then beyond, became increasingly accessible to ever larger numbers of people, beyond the ultra-privileged elite which had previously undertaken the 'Grand Tour':

> Travel's educative, acculturating function took on a newly competitive aspect, as travellers sought to distinguish themselves from the 'mere tourists' they saw or imagined all around them. Correspondingly, the authentic 'culture' of *places* – the *genius loci* – was represented as lurking in secret precincts 'off the beaten track' where it could be discovered only by the sensitive 'traveller', not the vulgar tourist.
>
> *(Buzard, 1993, p. 6)*

(The terms 'authentic' and 'off the beaten track' should be noted here, as they will be returned to later in this chapter.)

Such anti-touristic sentiments have undoubtedly endured to the present day amongst self-styled 'travellers' – including travel writers. Some scholars of contemporary narrative travel writing argue that anti-tourism is not simply a means of overcoming belatedness, but also a way of shirking ethical responsibility for the possible negative impacts of travel on the places visited. As Holland and Huggan put it, '"travellers" seemingly plaintive need to dissociate themselves from "mere" tourists' functions as a strategy of self-exemption, whereby they displace their guilt for interfering with, and adversely changing, the cultures through which they travel onto tourists' (1998, p. 3).

Unlike a narrative travel book, however, a guidebook 'constructs the reading subject ("you") as a *potential traveler* and presupposes the realization of its addressee's desire for the [place described]' (Behdad, 1994, p. 41). And unlike a narrative travel book, a guidebook is explicitly designed to aid the practice of tourism. The idea of a guidebook as an anti-touristic text, then, seems contradictory. And the idea of a guidebook overcoming belatedness by rhetorically 'past-ifying' the places it describes seems to have at least the potential to undermine its need to be informational, accurate, and 'up to date'. So do guidebooks, like narrative travel books, deploy anti-tourism and other belatedness-driven strategies, and if so, how do they deal with these apparent contradictions? For the purpose of considering these questions, the remainder of this chapter will focus mainly on two recent guidebooks from two different publishers, both covering the same destination: Lonely Planet's 2018 edition of its Nepal guide (hereafter *LP Nepal*) and Insight Guides' 2014 *Nepal* (hereafter *IG Nepal*). I was the main author working on the latter title.

Narrativity, authorial presence, and the production of guidebooks

Simon Cooke argues that all forms of travel writing, including the guidebook, involve 'a dimension of life writing' (2016, p. 15). This is perhaps true in as much as that any single line of text ultimately has an author, whose stylistic and tonal decisions will be partly determined by a uniquely personal response to the experience or information that informs it. But it ignores the particularly complicated question of authorship in guidebooks.

LP Nepal is attributed on its title page to three authors (Bradley Mayhew, Lindsay Brown, and Paul Stiles). But it is also identified as the eleventh edition of the book. Clearly, the same three authors have not worked on all ten previous incarnations of *LP Nepal*, but equally they will not have produced entirely new text for this most recent version. A guidebook such as this will inevitably contain multiple layers of authorship, built up over many years, in which any 'dimensions of life writing' will be extremely fragmentary and deeply encrypted. This is in marked contrast to the record of singular private experience which largely defines narrative travel writing. As Lisle points out, 'guidebooks construct a collective authorial voice that is primarily identified by the publisher' (2008, p. 161).

Discussing the origins of narrative travel writing, Mary Baine Campbell notes that '[n]either power nor talent gives a travel writer his or her authority, which comes only and crucially from experience' (1988, pp. 2–3). The need to establish authority is arguably much more pressing for guidebooks than for narrative travel books. And yet the most obvious signal of experience – the first-person mode – is generally not available to guidebook writers. So how do guidebooks go about tackling this challenge? In the case of *LP Nepal*, one answer is to retain fragments of first-person narrativity. The listing for the Mahabouddha Temple in Patan contains the following line: '[the tower] was cloaked in heavy-duty scaffolding when we last visited' (*LP Nepal*, 2018, p. 145). The listing for another temple in Bhaktapur states that it 'was under reconstruction when we visited' (p. 155). And another temple, in Kirtipur, was awaiting repairs 'when we last visited' (p. 173). In these instances, a textual trace of an actual journey briefly emerges. The authorial 'I' is partly subsumed in a collective 'we'; but these are, in fact, momentary intrusions of a past-tense, first-person narrative.

Guidebooks, unlike narrative travel writing, typically use the present tense, which, according to Ali Behdad, 'is indicative of the way the guidebook claims the realisation of its reader's fantasy as an immediate and possible reality – in contrast with the retrospective discourse of the travelogue' (1994, p. 42). A guidebook, then, aims to create an impression of describing the situation as it is *now*. But with these references to personal visits, the past tense intrudes ('when we *visited*'). This alone may undermine the impression of being up to date. And if a reader arrives in Patan and finds that the tower of the Mahabouddha Temple is actually no longer surrounded by scaffolding, the book's authority is further undermined. The illusion that the guidebook describes a sort of perpetual actuality fails, revealing a hidden kinship with narrative travel writing: that the text is, to some degree, a personal record (or an amalgam and a layering of multiple personal records) of transient moments. Here, a key but little remarked tension in guidebooks is revealed – between a residual claim to the powerful authority of personal experience, and a contrary claim to the authority of immediacy and verifiability.

Unlike the Lonely Planet, *IG Nepal* does not identify individual authors on its title page; nor does the text contain any overt fragments of personal narrative of the kind identified above. However, in the book's small-print end-matter, I am identified as 'author', and it is stated that I 'travelled through the country researching this book' which is 'thoroughly updated' and 'comprehensively revised' (2014, p. 348).

I was commissioned in 2013 to work on what was to be the sixth edition of *IG Nepal*. I had not previously written about Nepal (though I had visited the country – as a tourist), but I had published travel journalism about neighbouring countries, and had worked on other Insight Guides projects in Southeast Asia.

Historically, guidebooks from major publishers have typically been revised every one to four years, though as the industry has contracted, revision cycles – particularly for less popular destinations – have lengthened, and few guidebooks now appear in annual editions. This particular commission was for a major overhaul, rather than a more cursory 'fact check'; the expectation being that I would substantially rewrite sections of the existing text, provide some entirely new content, and, as far as possible, visit key areas of the country in person. I was provided with an extensive brief, containing detailed instructions about structure and style, updating procedures, and highly prescriptive parameters for any new text (such as specific character counts for new side panels). This, in my personal experience, is normal procedure for guidebook updates. I then worked directly into Word documents containing the text of the previous edition, with all changes tracked for the reference of the editors. Again, this appears to be standard practice (in my experience, more cursory 'fact check' guidebook commissions may involve an author marking up suggested changes in a PDF of the previous edition). I spent one month on the ground in Nepal in the spring of 2013, roughly one year ahead of the book's eventual publication. Once again, based on personal experience and the anecdotal evidence of my peers (including those writing for Lonely Planet), this is fairly typical of guidebook production.

Authenticity and getting 'off the beaten track'

The terms 'authentic' and 'off the beaten track' feature in James Buzard's characterisation of anti-tourism, quoted above, suggesting that they may be significant signals of the phenomenon. Given this, I have investigated their occurrence in both *LP Nepal* and *IG Nepal*.

'Off the beaten track' (or variants thereof) occurs 13 times in *LP Nepal*, most notably in a section in the planning chapter where it is used as a heading over a list of individual destinations deemed to possess that apparently desirable quality. The introduction to this section is as follows:

Off the Beaten Track

It's not that hard to find a relatively little-visited corner of Nepal. You'll likely have the following towns [Budhanilkantha, Panauti, Kirtipur, Pharping, Ilam] all to yourself, especially if you visit outside October/November.

(LP Nepal, 2018, p. 21)

A synonym for 'off the beaten track' in the guidebook lexicon is revealed here: 'little-visited' (this term, or variants thereof such as 'rarely-visited' or 'seldom-visited', occurs 12 times in *LP Nepal*). But the passage is also striking in terms of its approach the Nepalese travellees. Readers are told that it is easy to find a 'little-visited corner of Nepal' without any qualifier. The places to which this designation is applied are all towns, sites of concentrated human habitation. Travellees do not seem to count, then, when it comes to defining a place as 'little-visited' and 'off-the-beaten-track'. Indeed, readers are told that they will likely have these destination 'all to themselves' – suggesting a clear relegation of the residents to some kind of subordinate, perhaps even nonhuman, status.

One of travel writing's most persistent and egregious ethical problems is revealed here: its tendency to a 'denial of coevalness' (Fabian, 2014, p. 35) for the travellee. Examining an earlier Lonely Planet guidebook to a neighbouring country (the 1993 India guide), Deborah Bhattacharyya found a tendency for the 'vast majority of Indians' (1997, p. 383) who were neither 'middlemen' (those directly involved in the tourist industry) nor 'tourees' (those whose cultural characteristics made them a direct object of tourist attention) to vanish in the text. But in this more recent instance from Nepal, *all* travellees, without distinction, are rendered invisible. It is worth noting at this point that none of the authors of either *LP Nepal* or *IG Nepal* is *from* Nepal, and while shifts in Lonely Planet's editorial policy following the company's acquisition by BBC Worldwide in 2007 'allowed for the inclusion of local voices in interview' (Butler & Aatkar, 2017, p. 212), this practice appears to have been discontinued, and no such box text interviews feature in *LP Nepal*. Local informants – hotel staff, tourism officials, and many others – will likely have provided information to the authors, but these contributions are not overtly acknowledged in the text.

It is interesting that variations of 'little-visited' appear markedly less frequently in *IG Nepal* (6 instances) than in the Lonely Planet book (12 instances). Also, three of the instances in *IG Nepal* do include a qualifier – 'Seldom visited by tourists' (2014, p. 202); 'seldom visited by foreigners' (p. 271); 'visited by only a handful of Westerners' (p. 303). These qualified uses still suggest a degree of implied subordinate status for the traveller, but it is less pronounced than in *LP Nepal* (where no such qualifier is included in any of the 12 instances). 'Off the beaten track' also appears far less frequently in *IG Nepal* (6 times, as opposed to 13).

A possible explanation for this may lie in Lonely Planet's origins as an 'alternative' publisher, mainly targeted at young, low-budget backpackers. Debbie Lisle suggests that Lonely Planet originally 'cultivated [as readers] a community of adventurers who define[d] themselves predominantly against mainstream tourism' (2008, p. 156). A tendency to a particular strain of anti-tourism was thus likely to be pronounced in its output. Lonely Planet's shift towards a broader demographic in recent decades is well attested, 'with the content and writing style being adapted to accommodate more mainstream tastes' (Iaquinto, 2011, p. 708). But a residue of the original anti-tourism may endure, arguably in contrast with Insight Guides, which has always targeted a more mature and upscale audience.

When it comes to the use of another possible anti-tourism marker, however, the pattern is reversed. 'Authentic' and its variants occur 19 times in *LP Nepal*. However, all but three of these instances appear in restaurant listings (where, interestingly, 'authentic' is not applied solely to Nepalese cuisine: Thai, Italian, and Korean restaurants in Kathmandu and Pokhara are all praised for their apparent 'authenticity'). In *IG Nepal*, variants of 'authentic' appear 22 times – sometimes in restaurant listings, but more frequently in descriptions of places or travel experiences. This pattern in itself suggests that 'authentic' may be seen editorially as a more desirable quality for the perceived Insight Guides audience (described as '30 plus' in the author's brief), while 'off the beaten track' is judged a key appeal for the traditional Lonely Planet demographic.

One use of an 'authentic' variant in *IG Nepal* appears in the section describing Thamel, the main tourist accommodation area of Kathmandu (p. 168). I directly contributed to this passage, but an examination of the original working Word documents reveals layers of multiple authorship. The following is the text of the previous, 2008 edition :

> Thamel is a world in itself, a budget paradise or hell, depending on your perspective. Its international flavour and bargains are intriguing, but it bears little resemblance to anything 'Nepalese'. (Insight Guides, 2008)

The updated 2014 text, meanwhile, reads as follows (with the deletions shown, and my additions indicated by italics):

> Thamel is a world in itself, a budget paradise or hell, depending on your perspective. Its international flavour and bargains are intriguing, ~~but it bears little resemblance to anything~~ *and though there is little authentically* 'Nepalese' *about the place, in its own way it is as much a part of Kathmandu as the ancient temples or modern middle-class districts.*
>
> *(IG Nepal, 2014, p. 168)*

I was evidently concerned by the tone of the previous version, hence my modifications. Notably, I equate Thamel to both 'ancient temples' and 'modern middle-class districts' (while also inserting a key anti-tourism signal, 'authentic', in the process). This may be symptomatic of an author to some degree aware of scholarly criticisms of travel writing. But it could also be viewed as an example of the second of the strategies for overcoming belatedness identified by Debbie Lisle: the ostensible embracing of contemporary cosmopolitanism. The precedent of a hidden intertext is also subtly hinted here, namely, Pico Iyer's *Video Night in Kathmandu* (1988), often regarded as a key example of 'cosmopolitan' contemporary travel writing (and which I had certainly read and probably had vaguely in mind when I worked on *IG Nepal*).

It is crucial to note that Lisle does not see the 'cosmopolitan' approach of writers such as Iyer (1988) as a successful means of overcoming travel writing's ethical problems. Indeed, she argues that it is 'the travel writers who enact a cosmopolitan vision who are most alarming, for they smuggle in equally judgemental accounts of otherness under the guise of equality, tolerance and respect for difference' (2006, p. 10). This view is echoed by Graham Huggan with specific reference to Pico Iyer, characterising *Video Night in Kathmandu* as a 'slightly queasy combination of too-cool-for-school postmodernism and bleeding-heart sentimentality' (2010, p.16). Whether a 'judgemental account of otherness' is involved in my own reworking of the Thamel listing is highly debatable; but it is certainly arguable that I am here deploying cosmopolitanism as a means of shirking ethical responsibility – both my own and that of the guidebook's putative readers – for the negative impacts of tourism.

Trapping a destination in past and present simultaneously

The following passage appears in the 'Kathmandu to Pokhara' section of *IG Nepal*, describing the 'Pokhara Trail', an alternative route to the main highway between these two towns:

> Few foreigners tackle this route, however, and it offers a fine insight to Nepalese life away from the tourist traps, as the trail passes through villages where old men sit cross-legged

sucking tobacco smoke through hookahs (water pipes), and women weave on looms stretched taut across their backs.

(2014, p. 211)

Again, an examination of the Word files from my work on this guidebook reveals multiple layers of authorship here. During my own 2013 research, I did not have time to travel this route myself, so I was unable to rewrite this section based on first-hand experience. I did, however, substantially rework the passage, introducing some clear signals of anti-tourism in the process ('Few foreigners tackle this route'; 'away from the tourist traps'). However, I retained two elements (italicised in the excerpt below) of an extended descriptive passage which appeared in the previous edition:

> Newars, Brahmans, Chhetris, Tamangs, Gurungs and Magars populate this corridor [...] *Old men sit cross-legged sucking tobacco smoke through a hookah (water pipe)*; schoolrooms full of boys and girls pronounce their lessons aloud; *women weave on looms stretched taut across their backs* – memorable scenes abound [Emphases added].

I suspect that I was somewhat uncomfortable with the 'human zoo' style listing of ethnic groups, hence its excision. But I evidently wished to retain some of the colourful depictions of 'memorable scenes'. Crucially, these images give the impression of being based on direct observation. This, then, is a signal demonstrating the text's basis in personal experience, though it is more subtle and deeply encrypted than the intruding past-tense, first-person instances in *LP Nepal*.

The 'traditional' activities (water pipes and weaving) clearly indicate a tendency to 'pastify' the destination here. Had these images been presented in a piece of narrative travel writing, implicit in the form would have been the idea that these were observations representing one transient moment and one unique experience, already past. But what is significant here in the guidebook is that they are described in the present tense. The text implies that these old men and women are a permanent, unchanging feature of the destination, as immutable as the surrounding mountains.

Strikingly, the passage as it appeared in the 2008 edition, elements of which I carried over to the 2014 edition, appears in exactly the same form (other than a slight difference in the final line: 'countless memorable scenes out of a slice-of-life scenario') in the very first edition of *IG Nepal*, published in 1991 (p. 196). The hookah-smoking old men and weaving women of Nepal's Middle Hills – who may actually have been observed in one passing moment by a researcher, probably in the late 1980s – have remained fixed, both in the rhetorical past and the perpetual present of a guidebook's discourse, for three decades.

Conclusions

Guidebooks, despite being intended to facilitate leisure travel, clearly contain traces of anti-tourism. This apparently contradictory phenomenon is likely sustained by reliance on readers' own self-construction, as summed up by Evelyn Waugh's oft-cited line: 'The tourist is the other fellow' (1930, p. 44). This may be particularly the case with Lonely Planet, given the legacy of its 'alternative' origins, even long after its move into a more corporate, 'mainstream' market. Readerly self-perception allows a tourism text to contain an anti-tourist discourse, apparently without irony.

Another means of overcoming belatedness (and indeed constructing exotic difference in general) – namely, positioning destinations in a rhetorical past – seems at first glance to be a more problematic strategy for a guidebook to deploy. Behdad characterises guidebooks as follows:

> [T]he tourist guide, like the journalistic statement, borrows its authority from a claim to verifiability, not from the experience of a speaking subject. [...] The tourist guide encourages its readers to check, to confirm or deny the validity of the information it provides, because it is precisely this acknowledgement of the possibility of error that makes the tourist believe what the guide claims, while perpetuating, of course, a circular system of exchanging information
>
> *(1994, pp. 43–44)*

So, if a reader of *IG Nepal* 'checks' the depiction of the villages in the hills between Kathmandu and Nepal, and finds modern houses, motorbike traffic, and digital connectivity more prominent than old men and women, smoking hookahs and weaving, the validity of the text will apparently be called into serious question. But guidebooks do, persistently, privilege 'timeless' or even 'medieval' qualities in their descriptive passages – qualities which may not always be particularly obvious in the actuality.

The most likely explanation for how this is sustained lies in authority of the guidebook itself. It has often been argued that tourists 'possess a predisposed understanding of a place due to a saturation of media images of a destination or attraction before seeing it for themselves' (McWha et al., 2016, p. 85). Edward Said, discussing the construction of Western discourse about the 'Orient', refers to travellers 'falling back on a text', because 'people, places, and experiences can always be described by a book, so much so that the book (or text) acquires a greater authority, and use, even than the actuality it describes' (1978, p. 93).

Bhattacharyya's (1997) analysis of Lonely Planet's coverage of India found a tendency to edit out the existence of local people who did not fit within the expectations of tourism as either middlemen or 'tourees'. The examination of *LP Nepal* here suggests a tendency to render the *entire* travellee population invisible when it comes to designating certain places as 'little-visited' or 'off the beaten track. It seems quite possible that guidebook users too may deploy a similar editorial process. They may filter their own lived experience according to the parameters fixed by the texts that mediate their journeys, which in turn allows those texts considerable scope for moulding the presentation of a destination in ways designed to overcome belatedness and express anti-tourist sentiments, even while retaining a supposedly strictly informational imperative.

References

Behdad, A. (1994). *Belated travelers: Orientalism in the age of colonial dissolution*. Durham, NC: Duke University Press

Bhattacharyya, D. P. (1997). Mediating India: An analysis of a guidebook. *Annals of Tourism Research, 24*(2), 371–389.

Bohls, E. A. (2016). Picturesque travel: The aesthetics and politics of landscape. In C. Thompson (Ed.), *The Routledge Companion to travel writing* (pp. 246–257). Abingdon, UK: Routledge.

Borm, J. (2004). Defining travel: On the travel book, travel writing and terminology. In G. Hooper & T. Youngs (Eds.), *Perspectives on travel writing* (pp. 13–26). Aldershot, UK: Ashgate.

Butler, R., & Aatkar, S. (2017). From "colour and flair" to "a corporate view": Evolutions in guide-book writing: An interview with Jenny Walker. *Studies in Travel Writing, 21*(2), 208–220.

Buzard, J. (1993). *The Beaten Track: European tourism, literature, and the ways to culture, 1800–1918*. Oxford, UK: Oxford University Press.

Campbell, M. B. (1988). *The witness and the other world: Exotic European travel writing, 400–1600*. Ithaca, NY: Cornell University Press.

Cooke, S. (2016). Inner journeys: Travel writing as life writing. In C. Thompson (Ed.), *The Routledge Companion to travel writing* (pp. 15–24). Abingdon, UK: Routledge.

Dickinson, G. (2018, June 29). Ignore the digital doomsdayers – The printed travel guide is here to stay. *The Telegraph*. Retrieved from: https://www.telegraph.co.uk/travel/comment/travel-guidebook-here-to-stay/.

Fabian, J. (2014). *Time and the other: How anthropology makes its object* (3rd Ed.). New York, NY: Columbia University Press.

Holland, P., & Huggan, G. (1998). *Tourists with typewriters: Critical reflections on contemporary travel writing*. Ann Arbor: University of Michigan Press.

Huggan, G. (2010). *Extreme pursuits: Travel/writing in an age of globalization*. Ann Arbor: University of Michigan Press.

Iaquinto, B. L. (2011). Fear of a lonely planet: Author anxieties and the mainstreaming of a guidebook. *Current Issues in Tourism, 14*(8), 705–723.

Insight Guides (1991). *Nepal* (1st Ed.). Singapore: APA Publications.

Insight Guides (2008). *Nepal* (5th Ed.). London, UK: APA Publications.

Insight Guides (2014). *Nepal* (6th Ed.). London, UK: APA Publications.

Iyer, P. (1988). *Video night in Kathmandu*. London, UK: Bloomsbury.

Jones, P. (2018, February 16). Journey's end. *The Bookseller*. Retrieved from: https://www.thebook-seller.com/blogs/journey-s-end-733816#.

Kinsley, Z. (2016). Travellers and tourists. In C. Thompson (Ed.), *The Routledge Companion to travel writing* (pp. 237–245). Abingdon, UK: Routledge.

Laderman, S. (2002). Shaping memory of the past: Discourse in travel guidebooks for Vietnam. *Mass Communication and Society, 5*(1), 87–110.

Laderman, S. (2016). Guidebooks. In C. Thompson (Ed.), *The Routledge Companion to travel writing*. (pp. 258–268). Abingdon, UK: Routledge.

Lisle, D. (2006). *The global politic of contemporary travel writing*. Cambridge, UK: Cambridge University Press.

Lisle, D. (2008). Humanitarian travels: Ethical communication in Lonely planet guidebooks. *Review of International Studies, 34*, 155–172.

Mayhew, B., Brown, L., & Stiles, P. (2018). *Nepal* (11th Ed.). Melbourne, Australia: Lonely Planet.

McWha, M. R., Frost, W., Laing, J., & Best, G. (2016). Writing for the anti-tourist? Imagining the contemporary travel magazine reader as an authentic experience seeker. *Current Issues in Tourism, 19*(1), 85–99.

Mee, C. (2009). Journalism and travel writing: From grands reporters to global tourism. *Studies in Travel Writing, 13*(4), 305–315.

Parsons, N. T. (2007). *Worth the detour: A history of the guidebook*. Stroud, UK: Sutton.

Peel, V., & Sørensen, A. (2016). *Exploring the use and impact of travel guidebooks*. Bristol, UK: Channel View Publications.

Pratt, M. L. (2008). *Imperial eyes: Travel writing and transculturation* (2nd Ed.). Routledge, UK: Abingdon.

Said, E. (1978). *Orientalism*. New York, NY: Pantheon.

Thompson, C. (2011). *Travel writing*. Abingdon, UK: Routledge.

Waugh, E. (1930). *Labels: A mediterranean journey*. London, UK: Duckworth.

Youngs, T. (2013). *The Cambridge introduction to travel writing*. Cambridge, UK: Cambridge University Press.

24

FILM TOURISM AND A CHANGING CULTURAL LANDSCAPE FOR NEW ZEALAND

The influence of Pavlova Westerns

Warwick Frost and Jennifer Frost

Introduction

A steam train heads over the Great Plains to the Rocky Mountains. Inside, a young girl reads a book, occasionally stealing glances at the rugged scenery. An older woman chaperoning her, comments, 'this will not be the same place you read of in books'. She urges the girl to return to England, but the girl is adamant that she wants to join her uncle on a ranch in the American West. However, this is indeed not the American West. It is not the same place that the audience has seen in films or read books about. The mountains are not the Rocky Mountains – they are the Southern Alps of New Zealand. This is the opening scene from the Pavlova Western *Good for Nothing* (2011).

Pavlova Westerns are set in the 19th-century USA and follow the conventions and style of Hollywood Westerns, even though they are completely filmed in New Zealand. Taking their inspiration from the *Spaghetti Westerns* of the 1960s, produced by Italians and filmed in Spain; films such as *Good for Nothing* and *Slow West* (2015) have been proclaimed as *Pavlova Westerns* after the iconic New Zealand dessert. This new style of film comes at an interesting time in the *circle of representation* of New Zealand. Jenkins (2003, p. 307) uses this term to refer to the idea that 'particular visual images circulate within a culture and become imbued with particular meanings, associations and values' and studies how the images used to promote Australia to backpackers might influence the kinds of photographs they take as tourists. The same idea can be applied to destination depicted within films, and the potential effect they might have on the places people visit. These images can also feed into destination marketing. For example, the success of *The Lord of the Rings* trilogy (2001–2003) resulted in increased tourism into New Zealand and helped shape a new imagining of the national identity (see Buchmann, Moore & Fisher, 2010 and other studies discussed later in this chapter). That effect, however, was at its strongest in the early years of the 21st century. While other block-buster movies were filmed in New Zealand around the same time, such as *The Last Samurai* (2003) and *The Lion, the Witch and the Wardrobe* (2005), none of these captured the popular imagination in the same way. Coming over a decade later, Pavlova Westerns may lead to a different way of looking at New Zealand's cultural landscapes. In this multi-disciplinary study, our aim is to examine how Pavlova Westerns contribute to the imagining and reimagining of New Zealand's landscapes and cultural heritage through films. Pavlova Westerns might also be viewed as an example of *mediatization*, where popular culture and tourism are essentially interlinked, and interest in one form of media, in this case film, affects and influences travel behaviour (Månsson, 2011).

Locational dissonance

As Frost and Laing (2015, p. 246) note: 'Like all cultural heritage tourism, that of the West is heavily mediated'. Western movies have been highly influential in shaping the image of the American West and encouraging tourists to visit film-related destinations and attractions. However, despite this strong geographic identification, there is a long history of transporting the imagery of the Wild West to other places, mostly for economic reasons. *The Magnificent Seven* (1960) was filmed in Mexico so that its star Yul Brynner could avoid paying taxes in the USA (Hannan, 2015). With the success of that film, the concept of *the runaway production* – in which to reduce costs, films were made in cheaper countries – was born (Frost, 2009). Later in the 1960s, this trend intensified with the Spaghetti Westerns (Frost & Laing, 2015). This was followed by shooting Westerns in Canada, starting with *McCabe and Mrs Miller* (1971): a pattern which has become stronger in the last few decades.

In the 1990s, researchers began to examine the phenomenon of film-tourism, with initial studies focusing on how individual films led to increased tourism. Given that some films were made in locations that were tricked-up to look like other places, this led to speculation as to how that might affect tourism. Tooke and Baker (1998), for example, highlight the example of the Sharpe television series, set in Napoleonic Spain, but filmed in the Ukraine. They speculated whether fans would travel to Spain or the Ukraine as a result of viewing the series. Jewell and McKinnon (2008) similarly provided a range of examples, including *The Last Samurai* (2003), set in Japan, but shot in New Zealand. Frost (2009), in his study of location, tourism, and runaway productions, utilised the example of *Braveheart* (1995), set in Scotland, yet filmed in Ireland. He noted that even though the Irish had developed Braveheart themed tourism trails, the main flows of tourists were still to Scotland. How does this affect the circle of representation of tourist images in cases where there is confusion or a lack of knowledge about where a film was made?

Frost's (2009) study concluded that while tourists were usually drawn to where a film was set, there were five instances where there might be tourism to the filming location. First, where the setting is general or fictional (as in Middle Earth). Second, where a critical mass of films have been made at the location, so that people have been made aware of the disconnection between the fictional locality and the place where the film-making occurred. Third, where there is a special appeal, such as in a cult film or a strong story about the stars or filming. Fourth, if there is a striking memorable shot, landscape, or still existing set; and fifth, where there is a persuasive marketing campaign or tourist operation such as guided tours or an attraction. Frost also highlighted the importance of multiple sources of information for attracting tourists to where a film was shot. These included word of mouth, guide books, general destination marketing, movie maps or trails, websites, and DVD special features.

Film commissions have a vital role in attracting film productions to a region and have been major players in encouraging the growth of runaway productions. In promoting a region as a film production destination, their activities are very similar to those that destination marketing organisations use to attract tourists. Both are attempting to gain economic benefit by persuading potential customers that their destination is superior to others. Indeed, in many cases, the two agencies are situated within the same umbrella structures of governments. To attract productions, film commissions often broker tax concessions and other financial incentives (Frost, 2009; Hudson & Tung, 2010). Of particular relevance to this study is the success of film commissions in New Zealand, Canada, Australia, and some states of the USA.

Two examples illustrate the connections between film commissions, runaway productions, and film tourism. The first regards the province of Alberta in Canada. As with adjoining British Columbia, the film commission of the provincial government has been very active in attracting productions that look like they have been filmed in the USA. This includes Western films such as *Unforgiven* (1992) and *Open Range* (2003) and television series *Fargo* (2014–) and *Hell on Wheels* (2011–). However, unlike British Columbia, Alberta has also used these productions in their destination marketing to tourists. With the release of the Brad Pitt Western *The Assassination of Jesse James by the Coward Robert Ford* (2007), a campaign entitled 'Explore Hollywood in Alberta' was released in 2008. The centrepiece of this promotion was the development of a touring route called the 'Cowboy Trail', 'a fabled route where the Western movie and the cowboy who rides tall in the saddle loom large' following Highway 22 for 700 kilometres down the inland side of the Canadian Rockies (Fisher, 2008). More recently, Alberta has been the site for filming of *Inception* (2010), *Interstellar* (2014), and *The Revenant* (2015). In announcing the filming of *Interstellar*, government interest in the connections between subsidies, industry development, identity, and potential tourism was made clear:

> Alberta Culture minister Heather Klimchuk said it's no surprise filmmakers keep choosing wild rose country for their projects and is pleased [director Christopher] Nolan is returning to the province for his latest one. 'It indicates the talent we have here in Alberta, Number 1, and part of it for me it just demonstrates all the hard work being done in this industry', she said. 'We know the diverse scenery we have in Alberta, you can look like you are filming back in the 1900s or you can look like you are filming in the future.' As well, Alberta has no sales tax, [and has] the Alberta Multimedia Development Fund.
>
> *(Schneider, 2013)*

Our second example demonstrates how tax concessions and credits for film financing can influence where films are made and how this may flow through to affecting the imagery they project to potential tourists. *The Light Between Oceans* was a 2012 novel set in Western Australia just after World War One. It was filmed in 2016 and was originally intended to be shot at the Western Australian coastal locations featured in the novel. However, as the director Derek Cianfrance explained 'at the last minute, the *Pirates of the Caribbean* crowd flew into town and took all the tax credit. So I could no longer afford to shoot the movie in Australia'. This meant that he could not emphasise the aftermath of the war and 'what that meant to that nation and the loss and grief and the loss of husbands and fathers and sons'. Given that there had been such high rates of volunteering and casualties from rural Western Australia, he 'thought it was important to keep it in the landscape of a country that was grieving'. However, having lost the tax credits, Cianfrance shifted to the South Island of New Zealand and changed the role of the landscape in the film. This new direction came as 'moving to New Zealand made me think about how to turn this into a primal landscape'. Actress Alicia Vikander added, 'I'd never seen such landscapes; it was overwhelming, almost claustrophobic, as you felt this big scale of nature closing in on you' (quoted in Bunbury, 2016).

The Lord of the Rings and New Zealand film tourism

In New Zealand, the success of *The Lord of the Rings* (LOTR) stimulated research on film tourism and authenticity. Jewell and McKinnon (2008) saw these films as effective in developing a new sense of place identity for New Zealand. Carl, Kindon, and Smith (2007) undertook an empirical study of participants in guided tours to LOTR film sites. They concluded that those surveyed experienced high levels of satisfaction, often exceeding their expectations, even though they knew that the film was set in a fictional world. This case seemed to exemplify that many tourists were happy to accept the landscape of New Zealand as an imagined representation of fictional Middle Earth.

Buchmann et al. (2010) undertook a similar study of participants on LOTR tours, but reached different conclusions. They argued that these tourists valued both existential and objective authenticity. The existential authenticity provided a more authentic version of the self through linking the books, films, and visitation of the filming locations, whereas there was still objective authenticity in that they were visiting those places where the processes of filming had actually taken place. In the same vein, Reijnders (2011) found that tourists were searching for 'places of the imagination' where the real and fictional worlds converged. In both these studies, the researchers argued that tourists sought to immerse themselves in cultural landscapes associated with their favourite fiction. Similar findings also came from a study of guided tours of the Hobbiton movie set from *The Lord of the Rings*. Most notable was that the tourists interviewed were fascinated by leftover production items, such as wooden marker stakes and ribbons. It was observed that 'many of them recorded photographic and videotaped images of these objects, which were described expressly [by the tourists] as "authentic," and one participant described the set as "a part of film history"' (Peaslee, 2011, p. 43). There is thus a circular representation of images from film to photographs and videos, which in turn inspire the promotion of New Zealand to potential tourists as a place from their cinematic dreams.

The concept of *associative landscapes* is important to this study. These are cultural landscapes in which there are cultural and artistic linkages, even though there may not be any material signs of those connections. Accordingly, even though the original books of *The Lord of the Rings* are English cultural productions, the films are now firmly associated with New Zealand and its landscapes (Carl et al., 2007; Jewell & McKinnon, 2008). In the cases of the Pavlova Westerns, the classic American Western narratives are matched by the cultural landscapes of Central Otago and Canterbury, which incorporate layers of Maori settlement, gold-mining, pastoralism, and modern-day film-making. In this chapter, our intention is to examine how these films create a new layer for that cultural landscape.

Western tropes

In *Good for Nothing*, Isabella (Inge Rademeyer) is travelling to her uncle's ranch when she inadvertently gets involved in a gunfight and is kidnapped by an outlaw known as 'The Man' (Cohen Holloway). As they journey through the West, they are followed by a posse who are seeking The Man for the murder of a sheriff. In *Slow West*, Scottish teenager Jay (Kodi-Smit McPhee) has come West to find his girlfriend Rose (Caren Pistorius). He is almost killed by soldiers but is rescued by outlaw Silas (Michael Fassbender). Silas agrees to be Jay's guide. Following close behind is a gang of outlaws led by Payne (Ben Mendelsohn). Both Silas and Payne know there is a reward for Rose, which Jay is unaware of. The two films are similar in following the conventions and tropes of the cinematic Western. The hero/heroine is a newcomer to the West and engages in a long journey through an unsettled and violent frontier with which they are ill-equipped to deal. There are encounters with Native Americans, outlaws, and eccentric characters, and the landscape is wild, beautiful, and dangerous. Both films feature an enigmatic and taciturn loner, who is a violent gunman with a mysterious past and ambivalent attitude towards the hero/heroine.

The journeys in the films are quests, with strong possibilities of personal transformation, but also with a strong risk of death (Frost & Laing, 2015; Hannan, 2015). In this respect, there are connections with *The Lord of the Rings*. In *Good for Nothing*, Isabella is seeking to escape the Man, while in *Slow West*, Jay's quest is to be reunited with Rose. Their journeys through the West can also be seen in terms of the *katabasis* from Greek mythology. This involves the hero venturing into nightmarish wasteland, often a descent as into hell, in order to effect a rescue. It is a common theme in Hollywood Westerns, as in the classic *The Searchers* (1956) (Clauss, 1999), and in using these mythic themes, the 'media – particularly, but not limited to, film – has created the myth of the West' (Frost & Laing, 2015, p. 241). In *Good for Nothing*, the climatic gunfight in which both the Man and Isabella are nearly killed by the posse takes place in a deep and gloomy canyon. In *Slow West*, the landscape is on fire when both Silas and Jay meet at the beginning and when they separate at the end. Furthermore, a number of characters refer to Silas as the Devil.

The narrative of *Slow West* is dominated by the common Western tropes of reversal, deceit and hidden identities (Frost & Laing, 2015). Jay does not know that there is a bounty on Rose's head and that Silas is only helping him to find the girl. The soldiers are murderous renegades and the vicar is a bounty hunter in disguise, while a young Swedish couple are actually robbers. Werner, seemingly a kind and learned German, steals Jay's horse and possessions. Rose does not love Jay – it was only a childhood infatuation – and when she shoots Jay, she does not even recognise him. Similarly, in *Good for Nothing*, the 'trusted men' sent by Isabella's uncle get her into a saloon fight, the sheriff is a rapist and the posse are trying to kill rather than rescue Isabella.

These complexities frame the transformations that occur for the lead characters in journeying through the West. In *Good for Nothing*, Isabella is kidnapped by the frightening and murderous Man and spends much of the film trying to escape. When he is badly wounded in the gunfight at the canyon, their positions are reversed. He is helpless and relies on her. She applies the knowledge of treating wounds that he taught her to save his life. As in the classic Westerns *Shane* (1953) and *The Searchers* (1956), the violent killer cannot re-enter society. The Man is partly redeemed and takes Isabella to her uncle's ranch but leaves her at the gate and rides off. In *Slow West*, the teenage Jay is a naïve dreamer, gazing at stars, reciting poetry, and clutching his upbeat guidebook *Ho! For the West!!!* (a real book, published in 1858). At first, Silas hardly engages with him, feeling Jay's foolishness will quickly get the boy killed. Silas is, after all, only using Jay to get to Rose and claim the reward. Jay, for his part, feels that Silas is a 'brute'. It is only after he is robbed by Werner that he realises he needs Silas to survive. From that point on, the two gradually become friends. Finally, Silas decides that he does not want the reward but instead will warn Rose that the bounty-hunters now know where she is. Intriguingly, Silas is completely ineffective in the gunfight. Shot early, he takes no place in the action and it is the mortally wounded Jay who kills Payne. Jay's quest has ended in his death, but Silas is redeemed and ends up with Rose. Indeed, once Silas joins Jay at the beginning of the film, the outlaw does not fire his gun again, proving that he can reform.

Such plots can be seen as operatic and completely in tune with those of the Classic Hollywood Westerns and Spaghetti Westerns. Indeed, the charm of these two Pavlova Westerns is in how they reference and pay homage to the conventions and tropes of past films.

What is also striking is how different they are to the ambivalent and realist modern 'existentialist' Westerns made by Hollywood at almost exactly the same time, most notably *Meek's Cutoff* (2010) and *The Homesman* (2014). They do, however, look like Westerns made in the USA. An emphasis on authenticity in clothing, horses and weapons is a common hallmark (Rosenstone, 1995) and this is well accomplished in these films. The various actors look Western in their costumes and demeanour and there are no anachronisms. Neither film could work as a convincing Western without this attention to detail. As discussed in the next section, the New Zealand landscapes have a major role in creating these illusions.

Imagining the Western Landscape

For the makers of these films, the challenge was to use the landscapes of New Zealand in such a way that it convincingly looked like the American West. In undertaking such a strategy, they were following the path of the Spaghetti Westerns, which successfully used the arid Almeira region of Spain. The New Zealand film-makers, however, had two major challenges. New Zealand is not particularly arid, and audiences may have seen some of these landscapes before, particularly in *The Lord of the Rings* trilogy and *X-Men Origins: Wolverine* (2009). Countering these potential negativities was that New Zealand and the American West share a common history, as locations of the great gold-rushes of the 19th century, settler societies and as sites of conflict between European newcomers and Indigenous peoples (Belich, 2009; Buchmann & Frost, 2011; Frost, 2016). There were, accordingly, resonances and connections between the two cultural landscapes.

Mike Wallis, the Writer-Director of *Good for Nothing*, commented about the choice of locations in the film, saying 'we are really proud of the landscapes. It is showing New Zealand in a way it has not been seen before' (quoted in Constantine, 2012). He utilised five New Zealand heritage sites/attractions to recreate the Wild West. Three are privately owned and operated tourist attractions. The first is Old West Town at Mellonsfolly Ranch, near Tongariro National Park on the North Island of New Zealand (Mellonsfolly Ranch, 2018). This recreation of an old western town is heavily based on cinematic imagery and caters for groups who like to dress-up and play at being in the Wild West. In the film, it is used for all the town sets. The second is Goldfields Mining Centre in the Kawarau Gorge between Cromwell and Queenstown in Central Otago (Goldfields Mining Centre, 2018). This tourism venture caters for day visitors and is located in a highly popular tourism area (the Kawarau Gorge is the home of bungee jumping). Its recreated Chinese Village is used for a Chinese mining operation in *Good for Nothing*. The third is the Kingston Flyer vintage steam train, which is used for the railway in the opening scene. This attraction, however, ceased operations in 2012, just after the film was made. Its future was uncertain (Mead, 2017), but it appears that it may be restored and relaunched (Chandler, 2018).

The use of historical theme parks and recreated pioneer towns is common for Westerns made in the USA. Attractions such as Old Tuscon (Arizona) and Buckskin Joe (Colorado) operate to business plans that include both film productions and tourism (Frost, 2009). It is less common outside of the USA. In Australia, for instance, the recreation of the 1850s Gold Rushes at the Sovereign Hill Outdoor Museum has only been used for one film, *The True Story of Eskimo Nell* (1975). The utilisation of three private tourism ventures as locations for *Good for Nothing* suggests there is future potential for such hybridisation of tourism/cinema locations in New Zealand.

Figure 24.1   Shot from the right angle, the Bannockburn Sluicings Heritage Area can suggest
America's iconic Monument Valley

Source: Warwick Frost.

Two key sites in the film are managed as free-entry attractions by New Zealand's Department of Conservation. Both of them are historical landscapes from the Central Otago gold
rushes which started in the 1860s. The first is the Bannockburn Sluicings Heritage Area,
where large terraces of gravel and clay were mined using hydraulic sluices. What remains
today is a treeless wasteland of cliffs and gullies. In *Good for Nothing*, the area is filmed to
suggest something like Monument Valley, an iconic location for many Westerns in the USA
(Figure 24.1). The second is the Bendigo Historic Reserve. This was an isolated moun-
tainous goldfield that is today a ghost town. Adding to the locational dissonance, this New
Zealand goldfield was named after the Bendigo goldfield in Australia in the 19th century. In
the film, one of the remaining stone buildings has been temporarily restored to serve as the
home of a Native American Medicine Man (Figure 24.2).

With a much bigger budget, *Slow West* did not utilise heritage attractions as sets. It was
shot primarily around Twizel in alpine Canterbury. With the action set in Colorado, the
film took advantage of the sweeping plains of the area, with the snow-capped Southern
Alps standing in for the Rockies. This area was also partly used in *Good for Nothing* and was
extensively featured in *The Lord of the Rings* trilogy. To the south of Twizel, the privately
owned Omarama Clay Cliffs were used to suggest the spectacular canyons and rock formations of the American West. This geological formation is due to the natural erosion of a
glacial terrace, as opposed to the man-made goldfields landscape of Bannockburn used in
Good for Nothing.

Figure 24.2 Stone building dating from the nineteenth century and utilised for *Good for Nothing*, Bendigo Historic Reserve

Source: Warwick Frost.

A New Zealand accent

The casting of these films complements the notion that we are in a different version of the West. Neither film has any American actors. In *Slow West*, the leads are Michael Fassbender (born in Germany, raised in Ireland), Kodi Smits-McPhee (Australian), Ben Mendelsohn (Australian), and Rory McCann (Scottish). For *Good for Nothing*, the cast are entirely New Zealanders. It is particularly noticeable in *Good for Nothing* that Inge Rademeyer has a strong New Zealand accent, even though she is playing an English woman. In both films, the roles of Native Americans are played by Maori actors. In the films, encounters with Indigenous peoples and Chinese miners present the issue of their marginalisation, reminding the viewer that a similar history occurred in New Zealand.

Both films come at a time when New Zealand voices are becoming more prominent in Hollywood productions. Examples of New Zealand actors using New Zealand accents in recent films include Julian Dennison (*Deadpool 2*, 2018), Taika Waititi (*Thor: Ragnarok*, 2017), Rhys Darby (*Jumanji*, 2017; *The Boat that Rocked*, 2009), and Zoë Bell (*The Hateful Eight*, 2015). It may be that viewers are now comfortable watching New Zealand actors in the Wild West context, particularly against the background of New Zealand landscapes. Indeed, in *The Hateful Eight* (2015), Quentin Tarantino posits that there were New Zealanders present in the American West of the 1870s.

Conclusions

In the case of *Good for Nothing* and *Slow West*, both films – we argue – engage in an interesting dialogue with the audience. Such a conversation occurs in all films which are made in different places to where they are set, raising issues of what the film-maker intends to project and what the audiences see, experience, and accept, which may in turn influence their desire to travel to these places. In these films, the conventions of the Western are followed faithfully, but it is accepted by the film-makers that the audience knows that this is New Zealand. Whereas *The Lord of the Rings* trilogy used New Zealand mining sites to represent the shattered ruins of an ancient civilisation, the Pavlova Westerns highlight them, using New Zealand's 19th-century heritage and reminding us of the shared history across the Pacific. The use of the Southern Alps as the American Rockies is an acceptable illusion, for they are suitably high, rugged, and snow-capped. Issues of European settlement marginalising Indigenous peoples and Chinese migrants – familiar tropes in the cinematic Westerns – are represented in these films with the understanding that these are also major components of New Zealand's frontier history.

There is a playfulness to both films. The cultural landscape that is projected is also one that is internationally known from its use in *The Lord of the Rings* trilogy. Accordingly, the film-makers know that they are reimagining these known landscapes in a new context. This is particularly important for the New Zealand film industry, as *The Lord of the Rings* trilogy is now nearly two decades in the past. For the industry to continue to develop, the familiar cultural landscapes have to be used again and again in different genres of cinema. In terms of New Zealand's national identity, these films demonstrate their skills as film-makers and add new layers to the cultural landscape. In the case of these two films, the new circle of representation is not of a fictional invention, as was the case in the depiction of Middle Earth in *The Lord of the Rings* trilogy. Instead, these Pavlova Westerns tap into ideas about similarities between the American West and New Zealand. These include the rugged wilderness, wide open spaces, and the shared frontier heritage. Through these associations, these films fashion a new vision of New Zealand's cultural landscapes, pointing to potentialities beyond *The Lord of the Rings* for promoting the country to visitors.

References

Belich, J. (2009). *Replenishing the earth: The settler revolution and the rise of the Anglo-world, 1783–1939*. Oxford, UK and New York, NY: Oxford University Press.

Buchmann, A., & Frost, W. (2011). Wizards everywhere? Film tourism and the imagining of national identity. In E. Frew & L. White (Eds.), *Tourism and national identities: An international perspective* (pp. 52–64). London, UK and New York, NY: Routledge.

Buchmann, A., Moore, K., & Fisher, D. (2010). Experiencing film tourism: Authenticity and fellowship. *Annals of Tourism Research, 37*(1), 229–248.

Bunbury, S. (2016). A dream cast adrift. *The Melbourne Age*, Spectrum section, 5 November, pp. 12–13.

Carl, D., Kindon, S., & Smith, K. (2007). Tourists' experiences of film locations: New Zealand as 'Middle Earth'. *Tourism Geographies, 9*(1), 49–63.

Chandler, P. (2018). Kingston Flyer getting back on track. *Otago Daily Times*, 13 December. Retrieved from: https://www.odt.co.nz/regions/queenstown/kingston-flyer-getting-back-track (Accessed 20 April, 2019).

Clauss, J. (1999). Descent into hell: Mythic paradigms in *The Searchers*. *Journal of Popular Film and Television, 27*(3), 2–17.

Constantine, E. (2012). 'Pavlova Western' Screens. *Otago Daily Times*, 28 April. Retrieved from: https://www.odt.co.nz/news/dunedin/pavlova-western-screens (Accessed 17 September 2018).

Fisher, M. (2008). Explore Hollywood in Alberta: Tour your favorite movies. *Travel Alberta*. Retrieved from: www.skicanadianrockies.com/alberta-winter-stories/hollywood-comes-to-alberta (Accessed 24 September 2008).

Frost, W. (2009). From backlot to runaway production: Exploring location and authenticity in film-induced tourism. *Tourism Review International, 13*(2), 85–92.

Frost, W. (2016). Challenges in story-based interpretation: Gold rush heritage and visitors. In L. Carpenter & L. Fraser (Eds.). *Rushing for gold: Life and commerce on the Goldfields of New Zealand and Australia*, (pp. 271–283). Dunedin, New Zealand: Otago University Press.

Frost, W., & Laing, J. (2015). *Imagining the American West through film and tourism*. London, UK and New York, NY: Routledge.

Goldfields Mining Centre (2018). Goldfields Mining Centre, Central Otago. Retrieved from: https://www.goldfieldsmining.co.nz/ (Accessed 30 September 2018).

Hannan, B. (2015). *The making of the Magnificent Seven: Behind the scenes of the pivotal Western*. Jefferson, NC: McFarland.

Hudson, S., & Tung, V. W. (2010). "Lights, camera, action …!" Marketing film locations to Hollywood. *Marketing Intelligence & Planning, 28*(2), 188–205.

Jenkins, O. (2003). Photography and travel brochures: The circle of representation. *Tourism Geographies, 5*(3), 305–328.

Jewell, B., & McKinnon, S. (2008). Movie tourism – A new form of cultural landscape? *Journal of Travel and Tourism Marketing, 24*(2/3), 153–162.

Månsson, M. (2011). Mediatized tourism. *Annals of Tourism Research, 38*(4), 1634–1652.

Mead, T. (2017). Uncertain future for Kingston Flyer vintage steam train, *Newshub*, 10 November 2017. Retrieved from: https://www.newshub.co.nz/home/new-zealand/2017/11/uncertain-future-for-kingston-flyer-vintage-steam-train.html (Accessed 30 September 2018).

Mellonsfolly Ranch (2018) Old West Town, Mellonsfolly Ranch. Retrieved from: http://www.oldwesttown.co.nz/ (Accessed 30 September 2018).

Peaslee, R. (2011). One ring, many circles: The Hobbiton tour experience and a spatial approach to media power. *Tourist Studies, 11*(1), 37–53.

Reijnders, S. (2011). Stalking the count: Dracula, fandom and tourism. *Annals of Tourism Research, 38*(1), 231–248.

Rosenstone, R. (1995). *Visions of the past: The challenge of film to our idea of history*. Cambridge, MA: Harvard University Press.

Schneider, K. (2013). Hollywood's hottest Alberta-bound this summer for filmmaker Christopher Nolan's new sci-fi film Interstellar. *Calgary Sun*, 2 June. Retrieved from: http://www.calgarysun.com/2013/06/02/hollywoods-hottest-alberta-bound-this-summer-for-filmmaker-christopher-nolans-new-sci-fi-film-interstellar (Accessed 7 March 2017).

Tooke, N., & Baker, M. (1998). Seeing is believing: The effect of film on visitor numbers to screened locations. *Tourism Management, 17*(2), 87–94.

25

TOURISTS' PLACE-MAKING PERFORMANCES THROUGH MUSIC

Jörgen Eksell and Maria Månsson

Introduction

Music is an important driver of tourism to destinations all over the world (Bennet, 2002; Xie, Osumare, & Ibrahim, 2007). Since the 1950s, particularly popular music has increased in significance and affected travel patterns attracting tourists to concerts and festival in global destinations (Sound Diplomacy and ProColombia, 2018). It is however only recently that an academic body on tourism and music has developed (for overviews, see Cohen 2007; Gibson & Connell, 2005; Lashua, Spracklen, & Long, 2014). One explanation that the field of music tourism has been established in recent times is that music regularly has been treated under the broader theme of culture (cf. Bolderman, 2018). Music as a popular culture phenomenon is highly media saturated and closely related to mediatization (Fornäs, 2014). In this chapter, we follow Krotz who argue that mediatization is 'a metaprocess grounded in modification of communication as the basic practice of how people construct their social and cultural world' (Krotz, 2009 cited in Couldry, 2014). Therefore, this definition of mediatization opens up for an understanding of music tourism in which tourists are seen as active in the creation of the experience. This standpoint is crucial as tourists in tourism studies traditionally have been seen as rather passive media consumers who only search for and imitate for example images seen before (cf. Jenkins, 2003; Urry, 1990). Thus, we address the performative aspects involved in the mediatized gaze (Urry & Larsen, 2011) by recognising that meditization can shape 'communicative figurations' i.e. communicative construction of culture and society (Hepp & Hasebrink, 2014). This standpoint brings the constitution of the communicative practices performed by tourists in their construction of music experiences into the limelight.

In addition, mediatization in relation to music and tourism has largely been treated as a media-driven practice that provides a priori understanding of the destination before a tourist journey (e.g. Bennet, 2002). In this chapter, we depart from a broad perspective on music including lyrics, music, and artists-related representation and involvement. There are limited studies of popular culture focusing on music and tourism with attention to the tourists and their experiences (cf. Bolderman, 2017). To further explore tourists' relationship to music, meditization, and place as they visit destinations, we approach tourists' actions as place-making practices. Hence, the purpose of this chapter is to problematise the performance of place-making practices by tourists as they engage with music. Furthermore, the performance perspective will be complemented with a ritual perspective that allows for an analysis of the mundane and repetitive place-making practices of tourists (cf. Collins, 2004; Goffman, 1959). Of particular importance to this chapter is that a ritual can transform a person, a relationship, a social position, and a place (van Gennep, 1960), but also create symbols of group membership and values (Collins, 2004).

Liverpool, UK, is used as a case in this chapter although the city is well studied in terms of how popular culture and especially music is important for a destination and for the visitors (e.g. Boland, 2010; Cohen, 2012). Even if Liverpool has been the focus of several studies in the last ten years, the mediatization of the place-making process from a tourist perspective is still underdeveloped. In this chapter, we depart from a strategic communication perspective while discussing this phenomenon.

Method

This chapter builds on research done in city centre locations of Liverpool during 2017 and 2018. The fieldwork and data collection were influenced by mobile data-collection techniques (Kusenbach, 2018). The techniques, also called go-alongs, refer to methods of participating in patterns of movements while conducting fieldwork (Büscher, Urry, & Witchger, 2011). In this specific study, the majority of data was primarily collected by *walking* in Liverpool. The urban geographers Pierce and Lawhon (2015, p. 656) argue that walking is not only 'the act of moving through the city on foot but also include related processes of standing, casual interaction, and observation'. The majority of the data was collected in a fieldwork consisting of eight days of work that was done on two occasions in March 2017 and June 2018. In addition, one of the authors of this chapter was a resident in Liverpool during this time and performed shorter and complimentary observations on ten occasions. The data collection with the means of walking was performed in an ethnographic approach (Clifford & Marcus, 1986; Tedlock, 2000) which formed the base for a flexible research design that evolved during the research process. Before the fieldwork started, we had identified a number of relevant locations in Liverpool and a few overarching research questions related to tourism, music, and place. During the fieldwork, we observed tourists' movements, interactions, practices, and media-related activities. The observations were documented by field notes and photographs. In this chapter, we analyse two situations in Liverpool that highlight a number of communicative practices performed by tourists.

In addition, data from digital platforms where tourists post text and pictures of their visits such as Instagram, Twitter, and TripAdvisor were collected by following the most important hashtags from March 2017 and June 2019.

Music tourism and the performance of place

Music tourism is a growing form of media tourism based on the relationship between music and a genre of music, or a song, artist or festival – and places. The relationship between music and place is established in research (for instance Carney, 1997; Fry, 2017; Long, 2014) and can be exemplified by the following examples. There is a connection between a birthplace of a music genre and a place, such as Memphis and Rock & Roll or New Orleans and Jazz. In other cases, the birthplace of an artist is indicated in order to distinguish a place, such as Salzburg and Mozart, or Liverpool and The Beatles. Furthermore, specific songs can be connected to places for instance the songs and places Penny Lane or Strawberry Field (cf. Keeling, 2011). These places are regularly subject of fan tourism, and tourists follow the artist's residence and movements. In addition, Bolderman (2018) argues that

> music-related locations are appropriated by multiple actors such as music listeners, the tourism, media and music industries, as well as locals, involving the interaction of multiple media, such as musical sounds, images, and texts of and about music and place.
>
> *(2018, p. 10)*

In order to unveil tourists' role and interactions as they visit places, we depart from perspectives that visualise the tourists' agency in relation to music-related activities.

The performance perspective (see for instance Edensor, 1998, 2000; Franklin & Crang, 2001) frames tourists as active place-making agents who act and perform when they move their bodies to new places (Hultman & Hall, 2011; Månsson, 2015; Sheller & Urry, 2004). In consequence, a place is something which is made and performed by tourists (Edensor, 2001) and the meaning of a place, for instance a city, is fluid and not fixed over time (Ek, 2006). Therefore, tourists' bodies are assumed to transform the visited places by the manner they engage with them (Crouch, Aronsson, & Wahlström, 2001). Thus, the place-making by tourists is therefore considered as an organic, incremental process driven by individual agency (cf. Lew, 2017). In this tradition, performances are regarded as a socially constructed process involving tourists' mobilities, practices, relationships, and interaction with a multitude of human and non-human entities (Murdoch, 2006; Sheller & Urry, 2004).

Previous studies from a performance perspective, in the context of music tourism, show that tourists are active performers of the visited place. The tourists' walk in 'the footsteps of musical icons, stand inside venues where music was performed, walk through studios where the songs that define our identity were first played and recorded and interact with geographical landscapes' (Fry, 2017, p. 4). According to Fry (2017, p. 58), the appeal of music tourism is in the heightened emotional experience realised through the performative act of physically being a part of the music and the creative process. Thus, it is about doing and not just listening and viewing.

In this chapter, tourists' place-making through music is also analysed as a ritual practice. Ritual theory originated in 20th-century functionalist anthropology (Durkheim, 1965[1912]; Mauss, 2000[1950]; van Gennep, 1960[1909]) and later in sociology (Collins, 2004; Goffman, 1959, 1982). A ritual can transform a person, a relationship, a social position, and a place (van Gennep, 1960) but also create symbols of group membership and values. The later has been advanced by Collins (2004) in his micro-sociological theory interaction ritual chain. Thus, the ritual shed lights not only to the agency of tourists and the symbolic significance of mundane, repetitive, and sometimes incidental place-making practices of tourists but also values relating to music and place.

Previous studies on the ritual in the context of place-making have covered a few of these aspects. For instance, Edensor (2001) discusses several modes of the ritual in regard to tourists' performances of place. One mode, he defines, is disciplined rituals. Thus, tourists almost follow a script and do the things that are expected of them such as see certain sights or take photos at specific places. In this mode, there is no room for improvisation. Another mode is improvised performances, in this sense, there is a range of performances possible at a place. Still, tourist performances can in some respect be influenced by scripts provided by, for example, music, on how to act at certain stages. Another study is Couldry's (2005) analysis of tourists to the Granada studio tour in Manchester, where the outdoor set of *Coronation Street*, Britain's longest-running soap opera, was filmed. Interestingly, Couldry notes that there is a power in place as some places request the performance of certain ritual practices. A final study is Fry (2017) who uses van Genneps concept of rite-of-passage in relation to tourists visiting Nashville. He argues that it is

> a transformative experience allowing music fans to detach from their former selves and momentarily perform, experience, and participate in a liminal community within a specific touristscape that signifies both one's identity and cultural otherness through the performance of the past'.
>
> *(2017, p. 70)*

Moreover, this transformative process is not just connected to the visited place since there are several preparatory rites such as planning before the performances can take place at the destination (Fry, 2017).

Tourist place-making in Liverpool

The number of possibilities for tourist to engage with music-related activities are numerous in Liverpool and they cannot all be covered in this chapter. Therefore, we will problematise how tourists perform place-making through music in this section by analysing two situations in the city centre of Liverpool. First, we will deal with the serendipitous meeting between tourists and street musicians. Second, we will deal with tourists' visit to a music quarter, in Liverpool.

Place-making performances: pacing, listening, clapping, and pay respect

Street musicians and street music are a notable part of urban street life in Liverpool. As has been noted by Bennet and Rogers (2014), street musicians and their performances have not been studied to a great degree. They recognise that street music historically and today encompass 'a number of different genres and the only unifying factor is the performance of the music by street musicians' (p. 454). In this section, tourists' place-making practices in relation to streets music are explored.

At Church Street in central Liverpool, music performances and other performances by street artists are succeeding one another (illustrated in Figure 25.1). As we approach a young man playing his guitar, we are forced to follow the stream of people and move left on the street. Reluctantly we stop as pedestrians in front of us halt to listen. People smile, a few people stamp a foot or clap hands. A young woman approach the young man and film the performance with her phone. Children are placed in the front of the crowd near the artist. Before the performance ends, a few people move on and break up the gathering. A woman with a tram leaves and people look irritated at her, as she makes her way through the crowd. Other pedestrians squeeze by our backs and move quickly down the street. As the performance ends and the guitar-playing man asks for a contribution, only a few people approach. They kneel at the case and leave a few coins or a bill. Others send their children with a contribution. The majority of the crowd move down the street to the next venue.

In this chapter, we interpret the performance of the street musician and the mundane practices of the crowd as a gift-giving ritual. Mauss (2000) argues that there are no free gifts and gift giving is always connected to a system of reciprocity – a gift demands action from a recipient. As the gift of the street musician – the music – to some extent is involuntary delivered to the passing people the system of reciprocity between the involved parties

Figure 25.1 Street musician in Liverpool
Source: Maria Månsson.

is intriguing. As is illustrated above, the music affects the people in the crowd to perform practices such as adapt to the crowd, stop, listen, stamp, and clap. Hence, the practices of the people create a focus on the street performer and as they conform each other roles they constitute a distinct group with external boundary. Collins (2004) argues that a boundary that separates insiders from outsiders is a decisive quality in a ritual, and in this situation, we see a boundary that to some extent is fluid. A few people pass through the crowd or leave during the performance. In addition, the practices establish the roles of the street performers as an artist and the crowd is transformed into an audience as they accept the music as a gift. As the gift is accepted, a bond of reciprocity is established between the different actors. A bond that in this situation is solved primarily by giving a monetary offering to the artist. However, not all people provide a monetary offering to the performer, but in social media channels, we see practices performed to resolve the bond of reciprocity. The practice of posting pictures and films of the performance in social media channels is prevalent on for instance on YouTube. The resolution to the bond of reciprocity is ordinarily performed in the presence of the recipient and a participating audience. However, in this situation, the mediatized practices performed by people in the crowd are not recognised by the street performer and participating audience until later, if ever. The mediatization of practices such as posting pictures and films of the musical performance on social media channels thus change the temporal sequence of the rituals – prolonging its execution.

The place-making practices by tourists in the context of street music are admittedly mostly incidental. Edensor (2001) calls this kind of ritual as *improvisational* performances. The place-making of tourists on pedestrian streets of Liverpool is improvisational, as they are difficult to forecast. Tourists that visit Liverpool do not know where street musicians perform, who performs, or what music genres and music are played on a particular day. In addition, the conditions for the place-making practices vary from day-to-day. In the depicted situation, we see that practices can be performed involuntary and not all pedestrians choose to participate. In consequence, the interaction between the performer and the people on the pedestrian street may never evolve into a ritual at all. As we will see in the next section, tourists' place-making practices through music in places connected to establish artists can follow a different pattern.

Place-making: holding arms, kissing, leaning and posting

Several recording artists make a notable expression in the city of Liverpool. There are numerous museums dedicated to artists and their music, such as The British Music Experience and The Beatles Museum, but there are a number of places of significance too in the city that are frequently visited by fans. Tourism mobilities to places known from literature, film, or TV-series are well documented in the field of tourism (see Reijnders, 2011 for an overview), and in this chapter, we recognise that fans of music and musicians perform *secular pilgrimage* (Collins-Kreiner, 2010; Fry, 2017; Reader, 1993) to Liverpool to follow their traces and visit places of significance. In this section, we analyse the tourists' place-making through music in the Cavern quarter.

The Cavern Club on 10 Mathew Street in Liverpool is allegedly the most famous club in the world. On their visit to Liverpool, Beatles' fans from all over the world visit the club that is associated with the bands break through. During its Golden Age in the 1960s, the Beatles played 292 times at the club. In 1973, the club was demolished to allow the construction of an underground railway ventilation duct. In 1984, a replica of the club was built on Mathew Street, it was built with 15,000 bricks retrieved from the original club site (cavernclub.com).

Moving down Mathew Street on a Saturday night, music are pouring from windows and doors of clubs and pubs of various sizes: beats from classical songs from the Beatles are mixed with contemporary hits from different artists fill the street. Close to the entrance of the Cavern Club are two life-size bronze statues: one of John Lennon, a member of the Beatles, and one of Cilla Black that worked as cloak room girl at the Cavern before she had a successful career (see Figure 25.2). People gather in groups at the statues of the artists. We notice a group of young women who are distinguished by notable hen-does: tiaras and veils placed on their heads. They take up their phones and film or take pictures of the statue and one or several tourists. Voices are high, cheeks are flushed, and foreheads are sweaty. When the picture session is completed, they move on and the next group proceed to the statue. People are talking loudly, the music is numbing, we continue down the street. As we return to Mathew Street in the early morning, groups of tourists are out and taking photos of themselves with the statues.

According to Collins (2004), there are two instances that need to be fulfilled in order to create an interaction ritual. First, there has to be a co-presence of the involved actors. Second, there has to be a boundary maintenance that separates insiders from outsiders (Collins, 2004). From observations at Mathew Street, it is evident that groups of tourists regularly created a group with the statue and a boundary that separate the group from themselves and other people in the street. In addition, the tourists always took a pose that is related to the statue. The tourists performed practices such as giving a kiss on the cheek, putting a relaxed arm on the shoulder, curling an arm round a waist, giving a curious gaze, or leaning a back or shoulder in a relaxed manner to the statue. The tourist creates a unity with the statue and performs a number of practices; they gaze, hold on to, lean to, and take pictures. In his research, Collins (2004) points to the recurrence in rituals as a necessity to create rituals that not only create energy for the involved parties but also symbols of unity and different values. In this case, we see that the posing and photographing of tourists and a statue are to

Figure 25.2   Tourists early in the morning interacting and documenting with a Cilla Black statue
Source: Jörgen Eksell.

agreat extent identical to practices we perform when we meet family and friends for a special occasion, such as birthday, an anniversary, a joint travel, or a baptism. It could be argued that these practices are part of recurring rituals in the tourist's life that aims to create a value of unity and familiarity of the participants as well as intimacy; they are performed as a means to strengthen the relationships of the involved parties including the statue.

In addition, the moment at the statue was always documented with pictures taken by persons in the poser's company (see Figure 25.2). The practices of photographing are performed in a similar manner by groups of tourists. Later, pictures were posted on social media platforms with hashtags placing their visit in the city, quarter, street, and a specific establishment such as #cavernclub, #cavernquarter, #cillablackstatue, #johnlennonstatue, #mathewstreet and #liverpool. From a ritual perspective, the practices are performed in an identical way, and it is clear that the performance is scripted. This is in line with Edensor (2001) who discusses that there are scripted performances that are inspired by external sources. However, tourists interpret this process differently that opens up for a range of place-making processes although it is scripted. Moreover, the ritual chain, the series of events, that has to be performed, has obviously been alleviated by the tourists' earlier use of social media platforms and possibly marketing material.

In his studies of tourist visiting the film studio of *Coronation Street*, Couldry (2004) argues that the ritual connects the 'media world' of the soap opera to the 'ordinary world' of the tourists. In the present study, we recognise that the place-making of tourists in connections to music, musicians and statues on Mathew Streets are performed both on the street and on social media platforms. Interestingly, the performance of these practices is usually done in small circles of tourists but as the pictures are mediatized and the media connect to a broader circle of sociability. It is now possible to interact and share with not only family, relatives, and friends but also earlier visitors to Mathew Street and Liverpool or organisations and companies following the hashtags. The tourists' performance of place through music therefore consists of practices that involve a surprisingly large number of actors that are not physically present on the street.

Conclusions

This chapter has shown that the performance of place-making practices by tourists as they engage with music is organic, multifaceted, spontaneous, and dynamic. The tourist performances problematised in this chapter focused on two different situations in the streets of Liverpool that highlighted a number of different practices. The first situation focused on the place-making on a pedestrian street in which tourists' serendipitous encounters with street musicians and their music performed practices that changed the crowd relationship with each other, the performance of mutual obligations, and different roles and status of the place. Even if the performance of the tourists can be seen as improvisational or even incidental, the practices have vital consequences for all involved actors in the street as they engage with the music in a more or less involuntary way. The second situation focused on the place-making practices performed on a street connected to established and known musicians and music. In this situation, the tourists performed practices that at least to some extent are a result of earlier knowledge of the city, the music, and the artist. In this situation, it is clear that the tourists' image of the destination is continually evolving and created by official destination marketing organisations communication as well as representation in media, popular culture such as guidebooks, televisions shows, and music listening as well as digital channels such as YouTube and Instagram.

It has been argued that tourists follow a circle of representation where only images seen before are encountered at the destination. However, as we have shown in this chapter, tourist performances of place are foundational to the visit, and place-making practices are active, formative, and integrative of the experiences. The a-priori image of the music city clearly creates a frame for these performances but the place is decisive to the process. As we have seen in Liverpool, a music interest in The Beatles might take the tourists to locations such as Mathew Street or to any of the statues that is possible to encounter in the city. Thus, many tourists visit the same attractions and take images while interacting with, for examples, different statues. This is similar to the circle of representation, however, even if it is the same type of rituals that is performed, it does not mean that these are the same places that are created. The two different situations problematised in the chapter show the dynamic between the planned and unplanned, and scripted and spontaneous performances.

Even in a situation where you have a scripted performance and tourists are interested in the same attractions, the tourist can still perform, create, and interact in order to make the place their own. Moreover, as they reach out to a wider social circle on different social media platforms, the place-making continues as people share, discuss, and confirm each other's experience. Does that mean that an unplanned and spontaneous performance gives the tourist more freedom? Not necessarily, the practices performed by the tourist on the pedestrians' street as they encounter a street musician and listen to the music are clearly unplanned and spontaneous, but the practices cannot be labelled as voluntary. Clearly, the situations require certain practices that the tourists perform. Couldry (2005) would have said that there is a 'power of the place' that requires acts. In the chapter, we have shown that the power of place can be constituted very different depending on the situations at hand, but the practices are united by their social and relational character informed by music.

Another aspect addressed in this chapter is the overall power of music for the place-making practices. Music tourism tends to focus on a special music interest, for example, The Beatles and Liverpool, ABBA and Stockholm or country music and Nashville, but we have shown that music in general have an impact on tourists' place-making process as they visit destinations. Music that is more mundane and unplanned such as, for example, street music have an impact on the experience of the place when people perform according to other peoples' patterns when walking, stopping, and for that matter blocking passing by. Thus, even music that is not consciously searched for have an impact on the place-making process.

Furthermore, there are practices performed by tourists in the presence of others that create a unity and closeness among present company. Likewise, there are performances that aim to reach out to a broader social circle of relationships through social media platforms. The practices are to establish relationships among a larger community that has performed these practices before or plan to perform them in the future. We therefore do not see the point in separating the digital from the real world as on-line practices have real place-making consequences. The place-making process is highly mediatized since tourists' interactions with music are intertwined with tourists' media performances such as creating images, sharing and comment them in an ongoing process blurring digital and physical performances. In consequence, the communicative figuration (Hepp & Hasebrink, 2014) of the two situations problematised in this chapter provide insights in the vital role mediatized practices have for tourists place-making performances when visiting a destination. To conclude, in this chapter, we have shown that performance of place through music consists of an intricate system of interaction between tourists, artefacts such as, for example, statues, music in different situations, and technology such as social media platforms.

References

Bennet, A. (2002). Music, media, and urban mythscape: A study of the' canterbury sound'. *Media, Culture and Society, 24,* 87–100.

Bennet, A., & Rogers, I. (2014). Street music, technology and the urban soundscape. *Journal of Media & Cultural Studies, 28*(4), 454–464.

Boland, P. (2010). 'Capital of culture—You must be having a laugh!' Challenging the official rhetoric of Liverpool as the 2008 European Cultural Capital. *Social & Cultural Geography, 11*(7), 627–645.

Bolderman, L. (2017). Have you found what you're looking for? Analysing tourist experiences of Wagner's Bayreuth, ABBA's Stockholm and U2's Dublin. *Tourist Studies, 17*(2), 164–181.

Bolderman, L. (2018). *Musical Topophilia. A critical analysis of contemporary music tourism.* Diss. Rotterdam: Erasmus Research Center for Media, Communication and Culture (ERMECC), Erasmus University, Rotterdam, The Netherlands.

Büscher, M., Urry, J., & Witchger, K. (2011). *Mobile methods.* Abingdon, UK: Routledge.

Carney, G. (1997). *The sounds of people and places: Readings in geography of American folk and popular music.* Lanham, MD: University Press of America.

Clifford, J., & Marcus, G. E. (Eds.). (1986). *Writing culture. The poetics and politics of ethnography.* Berkeley: University of California Press.

Cohen, S. (2007). *Decline, renewal and the city in popular music culture: Beyond the beatles.* Aldershot, UK: Ashgate

Cohen, S. (2012). Musical memory, heritage and local identity: Remembering the popular music past in a European Capital of Culture. *International Journal of Cultural Policy, 19*(5), 576–594.

Collins, R. (2004). *Interaction ritual chains.* Princeton, NJ: Princeton University Press.

Collins-Kreiner, N. (2010). Researching pilgrimage: Continuity and transformation. *Annals of Tourism Research, 37*(2), 440–456.

Couldry, N. (2005). On the actual street. In D. Crouch, R. Jackson, & F. Thompson (Eds.), *The media and tourist imagination: Converging cultures* (pp. 60–75). Abingdon, UK: Routledge.

Couldry, N. (2014). Mediatization and the future of field theory. In K. Lundby (Eds.), *Mediatization of communication* (pp. 227–248). Berlin, Germany and Boston, MA: De Gruyter Mouton.

Crouch, D., Aronsson, L., & Wahlström, L. (2001). Tourist encounters. *Tourist Studies, 1*(3), 253–270.

Durkheim, É. (1965) [1912]. *The elementary forms of religious life.* London, UK: Allen & Unwin.

Edensor, T. (1998). Tourism and performance. In T. Jamal & M. Robinson (Eds.), *The Sage handbook of tourism studies* (pp. 543–557). London, UK: Sage

Edensor, T. (2000). Staging tourism – Tourists as performers. *Annals of Tourism Research, 27*(2), 322–344.

Edensor, T. (2001). Performing tourism, staging tourism: (Re)producing tourist space and practice. *Tourist Studies, 1*(1), 59–81.

Ek, R. (2006). Media studies, geographical imaginations and relational space. In J. Falkheimer & A. Jansson (Eds.), *Geographies of communication. The spatial turn in media studies* (pp. 45–66). Göteborg, Sweden: Nordicom.

Fornäs, J. (2014). Mediatization of popular culture. In K. Lundby (Ed.), *Mediatization of communication* (pp. 483–504). Berlin, Germany: Walter de Gruyer.

Franklin, A., & Crang, M. (2001). The trouble with tourism and travel theory? *Tourist Studies, 1*(1), 5–22.

Fry, R. W. (2017). *Performing Nashville music tourism and country music's main street.* London, UK: Palgrave Macmillan.

Gibson, C., & Connell, J. (2005). *Music and tourism: On the road again.* Clevedon, UK: Channel View Publications.

Goffman, E. (1959). *The presentation of self in everyday life.* New York, NY: Double Day.

Goffman, E. (1982). The interaction order. American Sociological Association, 1982 Presidential Address. *American Sociological Review, 48*(1), 1–17.

Hepp, A., & Hasebrink, U. (2014). Human interaction and communicative figurations: The transformation of mediatized cultures and societies. In K. Lundby (Ed.), *Mediatization of communication: Handbooks of communication science* (pp. 249–272). Berlin, Germany: De Gruyter Mouton.

Hultman, J., & Hall, M. (2011). Tourism place-making. Governance of locality in Sweden. *Annals of Tourism Research, 39*(2), 547–570.

Jenkins, O. (2003). Photography and travel brochures: The circle of representation. *Tourism Geographies, 5*(3), 305–329.

Keeling, D. (2011). Iconic landscapes through popular music: The lyrical links of song and music. *Focus on Geography, 54*, 113–125.

Kusenbach, M. (2018). Go-alongs. In U. Flick (Eds.), *The SAGE handbook of qualitative data collection* (pp. 344–361). London, UK: SAGE Publications Ltd.

Lashua, B., Spracklen, K., & Long, P. (2014). Introduction to the special issue: Music and tourism. *Tourist Studies, 14*(1), 3–9.

Lew, A. (2017). Tourism planning and place making: Place-making or placemaking? *Tourism Geographies, 39*(3), 448–466.

Long, P. (2014). Popular music, psychogeography, place identity and tourism: The case of Sheffield. *Tourist Studies, 14*(1), 48–65.

Månsson, M. (2015). *Mediatized tourism: The convergence of media and tourism performances.* Diss. Lund: Media-Tryck, Lund University, Sweden.

Mauss, M. (2000) [1950]. *The gift. The form and reason for exchange in archaic society* (2nd Ed.). New York, NY: W.W. Norton & Company Inc.

Murdoch, J. (2006). *Post-structuralist geography. A guide to relational space.* London, UK: Sage Publications Ltd.

Pierce, J., & Lawhon, M. (2015). Walking as method: Toward methodological forthrightness and comparability in urban geographical research. *The Professional Geographer, 67*(4), 655–662.

Reader, I. (1993). Introduction. In I. Reader & T. Walter (Eds.), *Pilgrimage in popular culture* (pp. 1–28). London, UK: The MacMillian Press Ltd.

Reijnders, S. (2011). Stalking the count: Dracula, fandom and tourism. *Annals of Tourism Research, 38*(1), 231–248.

Sheller, M., & Urry, J. (2004). Places to play, places in play. In M. Sheller & J. Urry (Eds.), *Tourism mobilities. Places to play, places in play* (pp. 1–10). London, UK: Routledge.

Sound Diplomacy and ProColombia, (2018). *Music is the new gastronomy: White paper on music and tourism – Your guide to connecting music and tourism, and making the most out of it.* London, UK: Sound Diplomacy.

Tedlock, B. (2000). Ethnography and ethnographic representation. In N. Denzin & Y. Lincoln (Eds.), *Handbook of qualitative research* (pp. 455–486). Thousand Oaks, CA: Sage Publications Inc.

Urry, J. (1990). *The tourist gaze. Leisure and travel in contemporary societies.* London, UK: SAGE Publications.

Urry, J., & Larsen, J. (2011). *The tourist gaze 3.0.* London, UK: Sage.

van Gennep, A. (1960) [1909]. *Rites of passage.* Chicago, IL: University of Chicago Press.

Xie, P. F., Osumare, H., & Ibrahim, A. (2007). Gazing the hood: Hip-hop as Tourism Attraction. *Tourism Management, 28*, 452–460.

26

TOURISTS' FILMIC REPRESENTATIONS ON YOUTUBE

A case study analysis of two mediatized visits to the Mursi in Ethiopia

Tom Sintobin and Anke Tonnaer

Introduction

Much primitivist tourism is based on the gap between the ahistorical wild Other and the modern civilized Self (Stasch, 2014). In this chapter, we focus on what we see as an emergent practice of (neo-colonial) visual consumption: filmic representations by tourists and semi-professional travellers of their visits to 'primitive' places, which they subsequently post on social media. As case studies, we have selected two YouTube-videos on Ethiopia's Mursi that are, despite obvious differences, similar enough to allow for a systematic comparison. In choosing the Mursi, often presented as an extreme example of cultural alterity because of their characteristic lip plates, we continue where the provocative documentary *Framing the Other* (Kok & Timmers, 2011), which had its focus on Dutch tourists visiting the Mursi, left off. The authors' main goal is to understand the contemporary nature of cultural consumption and production and, in their wake, the power differentials underlying this growing virtual domain of representations. Using theory and analytical approaches from both cultural studies and anthropology, we look at historical tropes employed to frame self and other, and consider the possible consequences, ethical and otherwise, of this type of mediatization of tourist encounters.

Case 1: 'Journey to the Mursi-tribe' (Ethiopia, 2013)

Description of the footage

On August 27, 2014, a Russian traveller, whom we have anonymized, posted a short (5'29) movie in Russian online. It opens with X sitting in a jeep, dressed in khaki and driving through what appears to be a village. The first 37 seconds X explains about the trip. We are going to the 'Mursi-tribe', he says in Russian, that is 'known for being thirsty for blood'. He continues: 'They all have Kalasjnikovs. At times they behave in an aggressive way. We will try to make some nice pictures for a special album on Ethiopia.' The next moment, he looks out of the window at the landscape, characterized by bushes, trees, and hills. The road is curvy, and in the background, we hear 'Solomon Vandy' (2006), composed by James Newton Howard. One minute later, X says: 'We have arrived at the Mursi-tribe' – and in both Russian and the local language: 'Hé, hello!' The camera looks around at the people, houses and cattle. X comments: 'People are gathering together now. They are very pretty.' One of the Mursi men puts on a white t-shirt and the next moment another one is painting X's face. Next, we see a young girl, bare-breasted and with a lip plate, who looks through the window of the car. Simultaneously, a song starts in the background: 'African queen' by 2face Idibia (2006). The movie ends with a series of photo shoots of different Mursi. We see X – naked torso by now like most of the local men – making arrangements for the shooting occasionally touching the Kalashnikovs. This goes on for a minute, after which a text appears on the screen: 'During our visit to the Mursi tribe, everyone remained unharmed. And as a result, we were able to make cool pictures.' The film ends with a series of those pictures.

Eroticizing the other

Mursi are presented as interesting and, in a way, beautiful and seductive, but at the same time wild and dangerous. X is depicted as both a brave and amiable man, who manages to establish close contact with the Mursi on friendly terms. The juxtaposition of the t-shirt scene with the painting scene bears evidence to that: the suggestion is that X is accepted as a warrior, whereas the Mursi man is allowed to dress up like a Russian. Another scene that suggests a successful contact is when the young girl comes close to the car window. The fact that the love song 'African queen' starts at that exact moment lends this quite long shot the meaning of a love interest.

The simplistic narrative of a civilized 'us' and wild 'them' may, however, be complicated when looking in more detail at the power relations that shine through and which are strongly reminiscent of a colonial and orientalist discourse. The camera systematically focuses on the breasts of the Mursi women and on the weapons of the men. The women are, however, eroticized *through* their representation: nudity only becomes a sign of the erotic if it is combined with an obtrusive camera perspective or with an amorous musical score. In a similar way, the men are more than weapon bearers. The camera does not really pay attention to it, but a careful analysis of the shot when X has arrived in the village shows one of the armed men busy with paperwork. He is dressed up in a military outfit, not in local dress, as is the man standing next to him, but these two characters do not get to play a role in the rest of the film. Their presence would, no doubt, have disturbed the radical racialized dichotomy of cultural difference.

At the centre of all scenes are the jeep and the camera, around which people gather (or are made to gather). Significantly, both are examples of technology and symbolically closely associated with practices of hunting and conquests (see Grimes & Venbrux 2010; Sontag, 2008). Furthermore, despite the 'friendly terms' on which the movie's narrative rests, it is clear that inequality and a feeling of white, Russian superiority are fostered. When the Mursi speak, there are no subtitles, which means that their voice is not really meant to be understood but rather a marker of the exotic. X tries to greet them in the local language but gets no reaction whatsoever.

Another scene that could be interpreted as an attempt at communication is the one with the t-shirt and the facial painting. In a sense, this exchange of gifts seals and shows the 'friendship' between guest and host and reduces the differences between the two of them to a certain extent. X gives a t-shirt, comparable to what players of opposing football teams do after the finals, and he gets the Mursi equivalent in return. One of the reasons why Mursi paint each other's bodies is to welcome back a relative after an absence (Mursi.org). In that respect, X is accepted as one of them. X's gift can, however, also be interpreted in an ironical way. The t-shirt – bearing the logo of Russia's leading bank – that is supposed to reduce the difference between white and black is bright white. It is, in fact, the only white thing in the entire film and stands out tremendously and consequently ironically enhances the contrast instead of diminishing it.

'African queen' is a song by 2Face Ibidia, a Nigerian artist. As his name suggests, he belongs to two worlds: he is extremely popular both internationally and in his home country. Moreover, he is a very successful businessman and at the same time leader of an NGO whose motto is 'service to humanity'. He even had a double wedding: a traditional wedding in Nigeria, and a wedding in white in Dubai. The lyrics of 'African queen' reveal the song to be an aubade to his black beloved. Sung by a Nigerian artist, it might be a nice love song. However, in the context of video, the song carries, implicitly, very different connotations, that in view of the white middle-aged producer, become arguably rather obscene, if not imperialistic.

There is a similar issue with regard to the second song, 'Solomon Vandy'. The song is a Christian prayer sung by the African's Children's Choir and Youssou N'Dour. It was part of the soundtrack of the movie *Blood Diamond*, directed by Edward Zwick. *Blood Diamond* tells the story of the atrocious Sierra Leone Civil War, in which rebels and governmental forces fight over the power, and ends with the historic meeting in Kimberley, South Africa, that was supposed to end the trade in blood diamonds. In other words, X picks a song that was part of a movie critical about the way white man deals with Africa, and he uses it to mean 'pure wildness and danger' in connection with an entirely different part of the continent.

Subversive voices

Thus far we have interpreted the tourist (X) as the agent of the discourse and the locals (the Mursi) as its object. However, the scene in which gifts are exchanged allows another reading as well, in which the host is less victimized and granted the role of one of the 'Foucauldian agents' in tourism discourse instead of just that of a 'target' (Cheong & Miller, 2000). In a reaction on a comment on his movie, X claims that the signs on his face were 'simply ornamental', so it appears he is not aware of the potential symbolism attached by Mursi to body paintings. Was he not told, or was it simply not all that important after all? It is possible that he is the subject of Mursi's irony in this scene as well. '[A]esthetic body painting is only practiced by older boys, seeking to attract the attention of the girls and of one another' (Mursi. org). Given his possible love interest for the very young Mursi girl, it may be the case that they are decorating him as 'one of the boys', with all the irony that is present in the decoration of a middle-aged white man as if he were a young warrior. Indeed, as such X may have unwittingly, been ascribed a clownish part in the exchange, seen from the Mursi perspective. Put differently, who is in control of the actual encounter may not be so straightforward to tell in spite of the editorial authority X has in composing the video.

Reception by the online audience

X's movie triggered 181 comments (last checked on 17/8/2016) mostly in Russian, some in English, and was viewed 886.861 times. It resulted in 287 likes and 104 dislikes. They are mostly about the musical score, about the women and girl (people want to know whether he became her lover; some become vulgar in their comments), about the Kalashnikovs, about the history of the lip plates. X is very active in reacting, providing people with extra information. The practice of the lip plates, he explains for instance, stems from colonial times, when men consciously mutilated their women to make them less attractive to slave traders. An explanation that is not only incorrect – the practice of lip plugs or piercings are neither unique to the Mursi (or Africa for that matter) nor limited to females – but also points out how X is reproducing a colonial discourse, since this explanation is as old as the colonial era in that region itself. Furthermore, as David Turton writes, such explanations of the history of lip plates 'as a "form of disfigurement," tell us more about the assumptions and values of those who find it persuasive than they do about the practice which it is supposed to explain' (Turton, 2004, p. 4).

Some people question the authenticity of the village, claiming that it is a set-up for tour-ists only. X readily admits that he was not the first tourist to visit them but insists that it is a real village, 'since ancient times'. One person writes: 'Advice, be nice, these Africans are only playing the "game of the savage" for tourists but others are not'. X reacts only to the last part of the comment: 'you are right that in Africa I should be careful in a wild place but you know in order to take very good photos I am ready for everything' and adds a link to his online photo album. Of all the comments, there is only one that is considering the ethics of his journey, exclaiming in three languages and capital letters: 'we, they are people…. Not animals to be viewed in a zoo'. This critic is one of the few not to get any reaction from X.

Case 2: Diaries of a backpacker, 'The Mursi tribe: The wildest of the wild'

Description of the footage

Daniela Zavala, the Venezuelan/American woman behind the profile 'Diaries of a back-packer', uploaded her six minutes video on July 29, 2011 (and again, in Spanish with some minor differences, the next day). It opens, just as X's film, in a jeep, with Zavala's voice-over explaining about her trip. When we reach the village, we see the reporter for the first time, full profile. She complains that it is very difficult to record there, because people keep both-ering her but promises to do her best. The voice-over then gives information on the Mursi, whose way of life is claimed to have changed little 'for hundreds of years'. Next, we look at a tour guide, who explains about Mursi culture until Zavala's voice-over takes over again to provide the viewer with more information. Meanwhile, the camera gathers examples of what is said in the village. About halfway the video, the 'documentary' aspect is left behind. From then on, Zavala is constantly in the picture herself, in full 'selfie mode': trying to keep Mursi from looking in the lens themselves, having them repeat English and Spanish phrases, having a laugh when one of the local men allegedly tries to buy her for 50 cows, and making fun with a heavily painted elderly woman that makes hissing sounds. After a few seconds, a vaguely audible song starts, the screen blackens, and her brand, 'Diaries of a backpacker', complete with logo, appears.

Conquering the heart of darkness?

Zavala's video relies heavily on the trope of the 'almost first visitors'; she explicitly refers to 'the very few tourists who make it to this very remote region'. Curiously enough, her own video seems to contradict her, as one of the very first persons the camera registers when entering the village, is already a white man. Indeed, the large number of parked 4WD's that are visible in one instance suggests the presence of more visitors. It also contrasts with her companionship, a tour guide, introduced as 'Tariku W/Aregay, founder of Yama Tours'; in fact, this makes clear that she most probably booked this trip through an official travel agency (www.yamatoursethiopia.com).

Nevertheless, Zavala incessantly stresses the adventurousness of her journey. First, there is the way the camera is held. The shaking camera while the car is driving, for instance, suggests that the road is bad, which creates the impression that this woman is leaving the official track and heading for new, uncharted territory. The frequent panning of the camera copies the way Zavala is exploring this site; the unsteady camera suggests movement and dynamism but also a chaotic and therefore 'authentic', unframed, immediate experience. Second, her choice of words is telling. 'The Mursi tribe', Zavala explains, 'lives in one of the most isolated regions of Ethiopia [...] They are considered to be the most aggressive tribe and are known for their unique definition of beauty' – the camera focuses on a gun and on a lip plate. While the car is driving, she comments upon the landscape: arid, secluded, so 'hostile to life that it is inconceivable that humans could actually live here'. The implications are clear: what she expects to find in this 'no man's land', as she calls it, will hardly qualify as human. Just like X, she is travelling to the heart of darkness. After her arrival, the voice-over comments: 'as soon as we get out of the car we are surrounded and almost attacked by people'; later on, she refers to what is happening as 'the situation'. The Mursi are 'aggressive' but Zavala is brave: a young woman who ventures among these 'wild' people on her own, assisted only by a local guide and does not even blink an eye when she explains to us that the Mursi man who wants to marry her has many scars because he has taken the lives of many enemies. Zavala is a domineering observer, another effect of the panning movement of the camera, which puts her in the position of the dominant focalizer and quite literally in the centre. In one instance, when she has switched to a full 'selfie mode', holding her camera in front of her face, she even tells the Mursi to leave: 'I'm working', she says, 'You cannot be in the camera. I have to be in the camera because I am the reporter'. In other words: there can only be one watch(wo)man, and all the others are objects of her gaze. Her gaze controls this representation, as exemplified by a scene in which a young girl, partly hidden behind a tree, looks back for a few seconds, but then, clearly embarrassed, looks down again: Zavala monopolizes the gaze. The same goes for her speech. She is a particularly dominant narrator, talking almost all the time, 'live' or as a voice-over; she tells us what to see and what to think of it. Very few other voices are allowed to feature in her monologue.

Subversive voices

In her description of the video, Zavala claims that 'to my surprise at the end the[y] embrace me as one of them.' The confrontation with the elderly lady at the end of the film is clearly supposed to prove that: 'Maybe it scares other tourists [...] but I won't have that and a war of tickles starts between grandma and me as a way to deeply bond with the Mursi tribe'. The choice of the word 'grandma' suggests that Zavala belongs to the family. However, the scene has a grim element to it as well. It is difficult to pinpoint what causes this bleaker impression, but the sharp contrast between the giggling Zavala and the attitude of the older woman, who only hisses but never loses her seriousness, plays an important part in it. To the attentive viewer, Zavala's interpretation of this 'war of tickles' is doubtful. For one, it is remarkable that the scene is cut off: it does not end with an embrace or a hug or a picture together that would seal it as an act of friendship but with the appearance of Zavala's brand name and logo. Secondly, the Mursi woman is not necessarily engaged in a tickling fight. All the time she is holding a lip plate in her hands and seems to want to insert it into Zavala's mouth. Bystanders are watching the two women and at least one of them is laughing. Taking these observations into consideration, it could be argued that this is a different power play, in which Zavala has become an object of observation herself. It is significant that this scene is not filmed from Zavala's point of view since she loses control of her camera during the encounter.

If our estimate is right that the woman suggests that Zavala wear a lip plate, the evident question is what such a lip plate symbolizes. As LaTosky (2004) has shown, Mursi girls get their lips cut when they reach puberty, 'giving them a new identity' because she makes the 'passage from girlhood to womanhood' (p. 385). She quotes earlier research that claimed that 'bridewealth consists, ideally, of thirty-eight head of cattle [...] and is often agreed upon before a girl has her lip cut' (p. 386). This helps to understand what is going on in Zavala's film, since just before the old Mursi woman approaches her with the lip plate, one of the young warriors has offered 50 cows for her. What is going on here is then, perhaps, not a 'war of tickles', but a rite of passage, mock or not, that Zavala fails to recognize as such. Girls who do not have their lip cut are referred to as 'Ngidini', a member of another ethnic group in the area of which the Mursi 'do not think very highly [...] because they have very few cattle' (LaTosky, 2004, p. 384). Not having your lip cut or not wearing your lip plate has important social consequences for a woman: 'she lacks the graces associated with womanhood, namely to be calm, quiet, hard working, and above all, proud' (LaTosky, 2004, p. 386) and gets punished for it. The fact that the elderly Mursi lady approaches Zavala with a lip plate, well knowing that she did not have her lower lip cut, gets an entirely different meaning in this respect: she is not inviting Zavala to become a worthy member of the group but exposes her unworthiness. Some of the old lady's movements could even be interpreted as attempts to hit her.

Reception by the online audience

The English version of the video was watched 488.039 times (last check 4/6/2019), receiving 651 likes and 228 dislikes, and 344 comments, mostly in English, some in Spanish. Some people comment upon Zavala's outward appearances in different tones – from pornographic and sexist to respectful – or talk about the history of the ritual of lip plates. Many comments, however, are very critical towards Zavala and her video itself. She is not well informed, it is said, and naïve in thinking that she can 'go to these peoples [*sic*] homeland and after a few days you think you are welcomed as one of them'. Others are surprised that Zavala believes that what she sees is authentic – 'It is obvious that this tribe is used to see tourists everyday', someone states, and quite a few people remark that the little language course that Zavala is giving to the Mursi was suspiciously successful: they repeat her with little accent. Her claim that she is in a remote and dangerous area is debunked as a lie because she was 'never more than 10 kilometers away' from the 'police paid by the central government' and the Mursi 'didn't seem that wild. It's not like they took you hostage or anything'. People call her a 'Tourist', 'retarded', 'racist', and 'rude': 'You can't just go to someone's home and act all bitchy – no humility whatsoever, and she calls herself a backpacker. The purpose of backpacking is to actually LEARN [*sic*] something, and humble yourself.' Several commentators refer to scientific literature to make their point. 'May I suggest Edward Said's work on the Other? It will help you better reflect on your own role and the ethics of backpacking.' And someone quotes from Claude Lévi-Strauss: 'A barbarian is someone who believes in barbarianism.'

Zavala reacts twice on this surge of comments. She explains that she depends on her own funding and does not have a crew who does the job of documenting for her. She does not pretend to 'know it all' and makes these movies in the hope that 'through these videos those who watch them can be transported to these amazing places and understand better these cultures that are usually misunderstood or feared.' People who dislike her videos have the choice 'to watch millions of YouTube [*sic*] channels so don't watch mine if it brings the worst of you.'

Two tourist productions: a comparison

Despite the many obvious differences, both videos have a lot in common. Firstly, they both show a 'traveller at work': X while he is taking pictures, Zavala while she is making a video blog. This means that there are at least two cameras and one other person present. This second filmmaker, however, does not feature in the video itself – he or she remains 'undramatized'. Neither of the videos has a list of credits, so it remains unclear whether there were additional people involved in its production, such as other filmmakers, editors, or sound engineers.

Secondly, both videos are clearly 'composed'; that is, the material is not just raw but underwent a process of narrativization. Both journeys move from 'dark' to 'light' and from hostility/fear to a supposed bonding/friendship. Both videos also assume, to a certain extent, a documentary function: they offer some information on the Mursi to outsiders, and both X and Zavala take on an expert role. In that sense, the videos might be associated with the typical *National Geographic*-style documentary. However, the use of two cameras has an effect that the viewer gets to see a kind of 'making-of' or 'behind the scenes': a genre that used to be nothing more than an extra but in postmodern times rapidly gained importance and sometimes even became the main product. Paradoxically, it may be argued that as a result the illusion of authenticity is considerably enhanced: the viewer gets the impression that he or she is seeing 'the truth', the 'raw reality', instead of a mediated product that requires a 'willing suspension of disbelief' (cf. Bruner & Kirshenblatt-Gimblett, 1994).

Both videos have, as a third common feature, a large and active audience, thanks to the (in most countries) freely accessible social medium YouTube. Both producers benefit from it, since their hidden agenda is a commercial one: X is promoting trips he organizes and Zavala tries to build a career as a respected travel journalist. Clearly, two case studies are not sufficient to allow us to draw any conclusions about different reactions on videos in English or in Russian – or by mainly Anglo-Saxon or Russian audiences – but one pattern that does seem to arise from our comparison is that the ethical aspect of this kind of video is much more important with regard to the former than to the latter. Put differently, whereas X gets only one remark about the ethical dimension of his video, Zavala is condemned in the strongest terms numerous times about her approach. If this observation is right, one explanation might be that in the Anglo-Saxon world, a public voice on and awareness of issues of racism might be more pronounced.

Conclusions

So what are we to make of these films and its creators? And, more importantly, what does it mean for the kind of questions we are asking in relation to tourism in the virtual realm? The three points just mentioned raise the issue whether these filmmakers should be considered professionals or amateurs. Moreover, how can we distinguish between these two categories and does it matter? The quality of the images is no longer a distinctive trait, since hi-tech cameras, manuals, and software are currently available to anyone at more or less affordable prices. Furthermore, social media facilitate in the size of the audience and the reach of the channel, further blurring the distinction between professional outreach and audiences connected to amateur productions. The last difference might be located in a form of 'institutionalization', i.e., whether someone is getting paid for his or her job by an official 'institution', like a production house, a TV channel, a newspaper, and so on. Zavala and X clearly are not, but does it matter, especially in view of that many YouTube channels nowadays present a lucrative model of revenue?

On the one hand, many 'official institutions' have nowadays become nothing more than a name that hides the homemade patchwork that is behind it. On the other, it could be argued that X and Zavala are part of a new group of self-made players in the field of tourist image making machinery. Their films are semi-professional productions, aimed at a large audience, published through a global channel. This undermines the classical view on tourist photographic practices as purely acts of consumption of existing images. Indeed, these 'tourists' are not just consumers but also producers – one might use the term 'prosumer' for them (Ritzer & Jurgenson, 2010). It also shows that this practice shares many characteristics with what we formerly recognized as belonging to the domain of 'professionals'.

Precisely because of the increasingly blurred boundaries between amateur and professional filmmakers, further questions on the ethical aspects of these cultural productions can be raised. In the representation of Ethiopia, and especially the Mursi, in the two videos discussed, we see the particular imagery and aesthetic conventions of the marketing mill of the tourism industry that emphasize cultural alterity through a neo-colonial frame of the (ig)noble savage. This is 'the hermeneutic circle of representation' (Jenkins, 2003) that gains new momentum in the virtual domain, further strengthening the dichotomous imagination of the world so typical for tourism (cf. Salazar & Graburn, 2014). At the same time, new actors and Internet bulwarks, such as YouTube, introduce new forms of critique and discussion (or lack thereof) regarding the representation of others. Put differently, ethical codes of conduct to which professional offline productions have to comply seem absent or arbitrarily followed in the virtual realm. Indeed, in the past few years, YouTube guidelines and policy of censorship have been subject of debate because of their apparent arbitrariness in evaluating and censoring the content of videos put online. At the time of writing this chapter, the video by X is no longer online, and it remains a guess whether this may be related to the racist, erotic undertones of the content, or whether he removed it himself. Remarkably, the English version of Zavala's video has now a minimum age constraint attached to; the Spanish version of the video, entitled *La tribu Mursi* instead of 'The Wildest of the Wild', is still easily accessible. In short, these generally opaque rearrangements urge further study into this seemingly self-governing domain.

Such ethical considerations should obviously also be paralleled in the study of these mediatized tourism practices. What may be considered as public and available for critical scrutiny, and when should we, as scholars, apply consent procedures for the analytical use of material posted online? In writing this chapter, we have closely followed the movement and public media life of the selected videos. As the film by the Russian tourist producer was taken offline, we have decided to anonymize his personal details. For the other video, we have taken its public nature including all associated sites of this producing tourist-traveller as open for study. Surely, our aim has not been to remedy touristic hermeneutics but rather to point to a lively domain of cross-cultural representations that have not yet received the scholastic attention it deserves.

The comments show that the majority of viewers learned about the Mursi for the first time through YouTube. Now surely, the Mursi and other so-called primitive 'hosts' are far from naïve in the tourism arena as others have shown (e.g., Abbink, 2000; Kok & Timmers, 2011; Turton, 2004) and as we observed too in between the lines of the video narratives of X and Zavala. Yet, the lack of opportunity to speak back to, let alone control the image production and accompanying copyright that proliferates and reiterates colonial tropes online, does raise the issue of socio-economic inequality and access to digital means to influence the discourse. It could be argued that forms of virtual media through which anyone can become producer enable neo-colonial representations that not only anonymize and, to an extent, dehumanize the cultural Other but also assist to neglect the design of proper filmmaker-filmed agreements, including basic rules of consent relating to filming and using footage in semi-commercial productions.

In line with the focus of this volume, through the critical analysis of the narrativization of two YouTube videos made by self-proclaimed traveller-explorers, we have focused on a type of convergence of mediatized products that not only affects the actual practices of tourists but also their ability to author destinations and its peoples. In so doing, we have shown that the interrelations between specific 'selfie-oriented' media content and touristic encounters may foster particular infrastructures of socio-cultural exchanges that in our view echo colonial discourses and customs. Finally, this type of convergence also points to a necessary convergence of analytical approaches, i.e., the need to transcend disciplines to both develop means for interpreting these tourist products within cultural and historicized contexts and accordingly form an academic rejoinder for those 21st-century skills in which the majority of tourists are increasingly so well versed.

References

Abbink J. (2000). Tourism and its discontents. Suri-tourist encounters in Southern Ethiopia. *Social Anthropology, 8*(1), 1–17.

'Body Painting'. (n.d.) Retrieved from: http://www.mursi.org/introducing-the-mursi/Body%20 Decoration/Body%20Painting.

Bruner, E., & Kirshenblatt-Gimblett, B. (1994). Maasai on the lawn: Tourist realism in East Africa. *Cultural Anthropology, 9*(4), 435–470.

Cheong, S. M., & Miller, M. (2000). Power and tourism. A Foucauldian observation. *Annals of Tourism Research, 27*, 371–390.

Grimes, R., & Venbrux, E. (2010). Shooting the dead. In D. Gross & C. Scheikardt (Eds.), *Die Realität des Todes: Zum gegenwärtigen Wandel von Totenbildern und Erinnerungskulturen* (pp. 63–75). Frankfurt, Germany: Campus Verlag.

Jenkins O. (2003) Photography and travel brochures: The circle of representation. *Tourism Geographies, 5*(3), 305–328.

Kok, I., & Timmers, W. (Directors) (2011). *Framing the other* (Documentary film). Copper View productions.

LaTosky, S. (2004). Reflections on the lip-plates of Mursi women as a source of stigma and self-esteem. In I. Strecker & J. Lydall (Eds.), *The perils of face: Essays on cultural contact, respect and self-esteem in Southern Ethiopia* (pp. 382–397). Münster, Germany: Lit verlag.

Ritzer, G., & Jurgenson, N. (2010). Production, consumption, prosumption: The nature of capitalism in the age of digital 'prosumer'. *Journal of Consumer Culture, 10*(1), 13–26.

Salazar, N., & Graburn, N. (2014). *Tourism imaginaries: Anthropological approaches*. New York, NY: Berghahn.

Sontag, S. (2008). *On photography*. London, UK: Penguin Books.

Stasch, R. (2014). Primitivist tourism and romantic individualism: On the values in exotic stereotypy about cultural others. *Anthropology Theory, 14*(2), 191–214.

Turton, D. (2004) Lip-plates and 'the people who take photographs': Uneasy encounters between Mursi and tourists in Southern Ethiopia. *Anthropology Today, 20*(3), 3–8.

Zavala, D. (2011). Retrieved from: https://www.youtube.com/watch?v=SeMdiGNzSWI.

PART IV

Tourists as media producers

27

STAR GAZING

The nexus and disparity between the media, tourism, and cultural heritage in Ireland

Aine Mc Adam

Introduction

From a sociological perspective, this chapter examines the interconnectedness of media, tourism, and cultural heritage. Within the contemporary media landscape, tourism, and in turn cultural heritage, have become ever more reliant on media platforms for not only promotion but for sustainability. Tourism depends heavily on the visual, or the 'gaze' to use Urry's apropos phrase (1990), from a Foucauldian perspective that 'gaze' can have a powerful effect on the tourist experience when enhanced or constructed by the media. Tourists arrive at a destination with preconceived expectations of pleasure and experiences that are distinctly different from their everyday life, and these preconceived expectations are typically significantly based on mediatization. The out of the ordinary preconceived experiences have already been presented to them by the mass media through carefully constructed, crafted, and 'endlessly reproduced' objectification of tourist imagery (ibid.). The Irish tourist industry is evolving and while fandom or fan tourism is not a novel concept in Ireland, the methods of media presentations of sites has altered in recent years. Ireland is home to *Star Wars* and *Game of Thrones* filming locations and these sites have become the new 'pilgrimage' destinations for their followers. As Howells and Negreiros (2012) so aptly phrase it, 'we live in a visual world' pointing out that in the age of social media, the use of visual imagery has become increasingly important as a form of self-expression, identity, and belonging. In the world of fan tourism, this visual imagery is used to place the fan in the same place as their heroes and demonstrate their dedication to other fans. This mediatization can blur the boundaries between the physical and imaginary places (Jensen & Waade, 2009), media not only reproduce reality... they produce events of their own... it is becoming increasingly difficult to distinguish between actual and mediated experiences (Jensen, 2010). This form of tourism or 'post tourism' as scholars such as Ritzer and Liska (1997) or Larsen and Urry (2011) refer to it, provides the visitor with a new form of experience rather the conventional cultural heritage of these particular places. Promotion of these movie and television sites is not exclusively through private media companies or social media, the Irish state run and fund successful[1] media campaigns. Therefore, this chapter will address the question, if the marketing for these places concentrates on fantasy, is there room for the established heritage?

What is heritage?

> The past is everywhere. All around us lie features which, like ourselves and our thoughts, have recognisable antecedents. Relics, histories, memories suffuse human experience. Each particular trace of the past ultimately perishes, but collectively they are immortal. Whether it's celebrated or rejected, attended to or ignored, the past is omnipresent.
>
> *(Lowenthal, 2015, p. 1)*

Lowenthal's argument is that heritage is enduring, it is tied to the past, but due to its fabrication, it is also linked to the present and connected to the future. While most would regard heritage as about the past, it is rather a reflection of the present (Graham, Ashworth, & Tunbridge, 2000, p. 2) state 'people in the present are the creators of heritage, and not merely passive receivers or transmitters of it, as the present creates the heritage it requires and manages it for a range of contemporary purposes'. Heritage is an interweaving of time and space, where the past, the present, and the future converge in particular places to 'imply certain immediate and longer-term futures but not others' (Shiels, 2013, p. 39). Heritage is a social construct, it is connected to the past, but as a concept, it constitutes a semblance of human interpretation of what has come before. Heritage is the selected portions of the past purposely chosen for contemporary purposes, whether economic, cultural, political, or social, chosen to bequeath to the future, the worth of which 'rests less in their intrinsic merit than in a complex array of contemporary values, demands and even moralities' (Ashworth & Graham, 2005, p. 7). Although initially the term heritage will conjure up preconceived impressions of its meaning, heritage as a concept is extremely complex.

'Definitions of heritage are notoriously non-specific and therefore flexible in the way they can be interpreted' (Kockel, 1994, p. 1). However, heritage has been defined as 'property that is or may be inherited; and inheritance' in addition it further states heritage is valued objects and/or historic buildings and 'cultural traditions that have been passed down from previous generations' also 'denoting or relating to things of special architectural, historical, or natural value that are preserved for the nation' (Balmer & Chen, 2016, p. 1). This situates heritage broadly as physical places or material culture, with an emphasis on inheritance and conservation. By defining heritage in terms of the physical is to reify the concept so as to fit it into contemporary structures of meaning. To focus on the material minimalises the importance of the intangible i.e. folklore. This definition sets 'the nation' in a prominent position, which allows for nationalistic identity construction (Hobsbawm & Ranger, 1992). How heritage has been defined has expanded beyond the old definitions of inheritance, it now includes virtually everything imaginable, however intangible and unverifiable (Gillis, 1997) and heritage's lure now overshadows other methods of retrieval, such as history, tradition, memory, or myth (Lowenthal, 1998).

McCrone (1995, pp. 1–2) contends that heritage has its roots in the restructuring of the world economy, 'is a thoroughly modern concept belonging to the final quarter of the twentieth century' and that heritage concepts are related to the more recent societal changes connected to colonial and post-colonial experiences. Its meaning is often taken as self-evident, yet the word does not so much represent a precise concept but a vague comprehension of a sentiment. The ambiguity of the meaning of heritage has led authors like Russell to describe the term as 'better understood for its psychological resonance than precise meaning' (1997, p. 72). Consequently, there is the potential for a range of alternative ways to relate to, give meaning to, and understand the significance of heritage objects, sites, and practices.

While most assume that heritage and history are one and the same, in reality how heritage is defined, and what it means is dependent on a number of factors, such as, but not limited to, location, ethnicity, political perspective, socio-economic background, and even sexuality. For many, the word heritage is synonymous with history. A definition of history at its most basic level is the study of the past, and the transmission of that knowledge in the present. Yet, historians frequently criticise how the recreation of the past is represented through reconstructions in the present, in places such as folk parks and museums. These historians have made concerted efforts to distance themselves from what they may characterise as bad history (Harrison, 2010). As Lowenthal so aptly states 'heritage should not be confused with history. History seeks to convince by truth but succumbs to falsehood. Heritage exaggerates and omits, candidly invents and frankly forgets, and thrives on ignorance and error' (1998, p. 7). The truth Lowenthal refers to is the historical narrative rather than an investigation of all influential factors and is restricted by the current presentation.

The focus of Lowenthal's argument is that heritage and history are distinct ways of knowing the past. Both, however, are culturally constructed, with their respective constructs continually changing over time. Consequently, they have their own individual histories of how the past is retrieved and represented in contemporary society. While heritage may be constructed, people are drawn to monuments, artefacts, and sites, as these provide a physical link to the accomplishments of their ancestors. People are attracted to heritage as it is often characterised as a key identity component of a social group (Bessiere, 1998), and as an idea, heritage represents the nation's identity to the world, beyond history (Balmer & Chen, 2016). Heritage is a unique multifaceted concept as it is inclusive of, the tangible, the intangible, and the metaphysical. Therefore, when heritage is presented by the media as a tourist location, it becomes limited by physical construction. Butler et al. argue that heritage tourism should be viewed as a phenomenon which is based on tourist motivations and perceptions rather than specific site attributes (Butler, Poria, & Airey, 2001).

The disparity: tradition heritage site versus a mediatized construction. Irish cultural heritage promotion, protection, and funding

Contemporary society is a consumer society, where the commodification of place prioritises the historical value and where constructed reality overshadows the established accepted heritage. Visitor experience at many heritage sites is constructed in such a way as to make it as entertaining as possible to increase visitor numbers, in the words of Bryman (2004), it has become 'disneyfied'. But heritage, Smith argues, is not a physical thing, building or object, but what takes place at these sites 'heritage....is a cultural process that engages with the present, and the sites themselves are cultural tools that can facilitate, but are not necessarily vital for, this process' (2006, p. 44). The narrative presented at heritage sites is dependent on a variety of conditions and motivations, for example, some heritage centres will use a construction of the past to promote a contemporary political position, or at sites like *Skellig Michael* where the narrative has become embroiled with a fantasy/sci-fi construction compromising the United Nations Educational, Scientific and Cultural Organization (UNESCO) world heritage site. It is precisely this medialisation of Ireland as a fan tourist destination that fuelled the Irish government to pass legislation, namely, section 481 film subsidy, which provides tax breaks for international media companies to use Ireland as a film location. These tax breaks consist of either 32% on expenditure, or, 80% on total film production cost, or, €70 million (Revenue, Irish Tax and Customs, 2019). *Skellig*, thus, resides in a juxtaposition between a place as a world heritage site and as a location in a Sci-fi movie. For it is at sites like *Skellig*, that heritage has become the runner up in the convergence between the popular imagination and the tourist gaze (Crouch et al., 2005; Urry, 2000). Thus, exacerbating the competing constructs of history and the mediatized landscape of heritage in Ireland.

Irish cultural heritage encompasses a diverse, alluring and fascinating assemblage of features, both tangible and intangible. Ireland has comprehensive national policies in place to protect, conserve, and promote these heritage artefacts, such as the Heritage Act 1995, National Monuments Amendment Act 1994 and Historic Monuments Act 1999, as well as having obligations to UNESCO to safeguard the world heritage sites situated in the country. Two immensely successful tourist campaigns promoting Irish cultural heritage are currently running concurrently in Ireland, the *Wild Atlantic Way* and *Ireland's Ancient East* (Tourism Ireland, 2018). These pre-mapped routeways provide the visitor with easily accessible information on designated places of interest and the ideal locations to experience the scenery.

Skellig Michael is included as one of the places of significance along the trail, interestingly included in this information is how to get a framed picture for the island asking if the visitor if they would like '"Skellig Michael" on your wall?' (2018), therefore, allowing the visitor to take home a 'memory'; however, this particular image is taken from exotic angle from the air over the sea, not a generally accessible position for the tourist (Slater, 2003). Selling of pictures or postcards of the 'ideal' view of tourist sites is not new, as long as there have been tourist sites there have been visual reproductions of these places, 'the aestheticisation of production and consumption which is a feature of developed economies in late modernity has long been at work in the 'welcome industry" (Urry & Lash, 1994, p. 111). As opposed to the Tourism Ireland campaign, no reference to *Star Wars* is included in the Wild Atlantic Way promotions where focus is predominately on the vistas and Irish traditional heritage sites.

The visual and the gaze

The very nature of tourism involves imagining, fantasising, or daydreaming and the anticipation of the novel experience, thus, once an image is placed in the visitor's mind, it becomes a very powerful motivational tool. As previously mentioned, Tourism Ireland currently have a very successful tourism campaign to promote and sell *Skellig Michael* as the film location for *Star Wars*, this campaign relies heavily on the media and visual representations of the island to the globe. While fan tourists are already flocking in their thousands to the site, the media promotion has utilised everything from panoramic vistas to aerial photography to images captured from space (emphasising the galaxy far away concept from the movie). All of these visual images are constructed in such a way as to lure visitors, just as in the movie, *Skellig* is presented as exotic, picturesque, and the ideal location for solitude and reflection. In a Foucauldian sense, the idea of a constructed visual or powerful 'gaze' (Urry, 1990) can be applied to the tourist experience. Urry (1990) further argues that this kind of tourist gaze derives from the expectations of visual pleasure and experiences beyond the familiar. These constructed expectations are continuously and 'endlessly reproduced' by mass media and are objectified in tourist imagery. However, Urry's argument attributes the construction of the gaze to the tourists themselves when in reality the production, creation, and control of the gaze are executed by powerful authorities in the tourist industry, who have the support and funding of government bodies, who use media and social media to their advantage. This from a Foucauldian perspective situates the mass media as ubiquitous in terms of power, for Foucault power is not static it cannot be obtained and retained, it is 'produced from one moment to the next, at every point, or rather in every relation from one to another' (1980, p. 93).

UNESCO World heritage site Skellig Michael (*Sceilg Mhichil*)

The most fantastic and impossible rock in the world: Skellig Michael…where in south west gales, the spray knocks stones out of the lighthouse keeper's house…the Skelligs are pinnacled, crocketed, spired, arched, caverned, minareted; and these gothic extravagances are not curiosities of the islands: they are the islands… The rest of the cathedral may be under the sea…An incredible, impossible, mad place… it is part of our dream world.

(Shaw, 1972, p. 941)

Skellig Michael is a small Island, not much more than a large rock in the Atlantic Ocean, off County Kerry, in the South West of Ireland. It was founded in the 6th century by early Christian monks (Figure 27.1). Originally, a place to experience a sojourn of silent solitude[2] and to make the inhabitants feel closer to God, the extreme inhospitality of the island forced the monks to abandon the settlement in the 13th century.[3] The site itself consists of six *Clochán*[4] beehive cells, two oratories, several stone crosses and slabs, a hermitage at the south peak, as well as a later church dated to the 12th century (Edwards, 1996, p. 118). Evidence indicates that no more than 12 monks and 1 Abbot inhabited the site at any one time. Life on the island would have been very hard and the monks would probably have existed on a diet of fish, shellfish, seabirds, and their eggs, any other supplies, such as grains, would have been transported from the mainland.

In 1996, *Skellig* was designated a UNESCO World heritage site through criteria (iii) and (iv). Owing to its distinctive and unparalleled cultural significance, the committee describe it as having 'exceptional universal value' and as a 'unique example of an early religious settlement' (World Heritage Committee, 1997, p. 68). In the agreement between the Irish state and UNESCO, the site must be preserved because of its 'remarkable environment' and 'its ability to illustrate, as no other site can, the extremes of Christian monasticism, characterising much of North Africa, near East and Europe' (ibid.).

Figure 27.1   Skellig Michael from the Sky
Source: Irish Air Corps, 2019.

The nexus: historical monastic site or *Ahch* island, Luke Skywalker's place of exile?

Skellig Michael has been a tourist attraction since procuring its World Heritage status; however, it is only in the last four years, since the island first featured in the closing scenes of *The Force Awakens* (2015), that tourist numbers have increased beyond the permitted allocation from UNESCO (Figure 27.2). While the media promotion is dominated by the *Star Wars* connection to the island, the pre-existing heritage literature and marketing, as well as the more recent campaigns guaranteeing that some tourists are visiting for the heritage. In addition, the information in the visitor's centre, as well as on the island itself, is focused on the monastic settlement. Ensuring that even those that make the journey as fan tourists will be exposed to the cultural heritage of the place, as one visitor notes 'I really wasn't sure what to expect on this excursion. I had seen the island on *Star Wars* and had heard we should try to get tickets if we could' (respondent 6) while another stated

> before there was the force, there were the monks and it was fabulous, I loved trekking to Skelling [sic] Michael. It is a world heritage site. Monks used to live on the top of the island and were there until the Vikings invaded.

> *(respondent 8)*

However, it is clear from the response of one visitor that there is not a universal nexus between the fan tourists and heritage visitors

> people are flocking from all over the world because of *Star Wars*, not because it is a UNESCO heritage site. Therefore, every Tom, Dick or Yoda nerd is trying to get on

Figure 27.2 The Monastic settlement & *Star Wars* film location Sky
Source: Irish Air Corps, 2019.

the island. Here's what you need to know – this is nature at its best and worst. People have died trying to make this climb to see where some *Star Wars* character was practicing with her light saber[sic].

(respondent 5)

Arguably, *Skellig Michael* has once again become a pilgrimage site for the faithful, however, these pilgrims aren't visiting to be closer to the Christian God of the past. Their 'religious' journey is a fan experience expedition, to *Ahch island*. These fan tourists want to experience this place from an embodiment perspective,[5] by standing in the spot where Rey met Luke Skywalker for the first time at the end of *The Force awakens* (2015) or visit the place that *Star Wars* has depicted as the location of the first Jedi temple and the tree library of the sacred Jedi texts, in *The Last Jedi* (2017). These purposeful descriptions thus portray the island as the most sacred of all places in Jedi history, a religious site and a holy place. Tourism as a religious experience or a pilgrimage has been discussed and subscribed to by different scholars (such as Brown, 2015; Davidson & Spearritt, 2000; Fullagar et al., 2012; Graburn, 1989; Lickorish & Jenkins, 1997; Okamoto, 2015). While these visitors are all aware that this is a fictional story, it is about escapism, just as all tourism is, Urry and Larsen (2011) maintains that film landscapes are so pervasive for tourists, that the destination itself becomes 'fantasylands or media worlds'. Similarly, Buchmann et al. (2010) argue there are matches between geographical and 'imagined' locations, this link in the case of *Skellig* and *Ahch island* is discernible. One visitor to the island describes it as an 'otherworldly place' and a second as the 'most impressive island on Earth….dangerous, high winds but makes life worth living, it is a testimony to the human spirit' while another stated 'you come here for *Star Wars* but then you get captured by the sacredness of the island' yet another said 'it's a lot more than a *Star Wars* set'(respondent 4), and finally one of the visitors disclosed that 'we were unaware of the *Star Wars* connection until after we began researching our visit. Due to the movie increasing traffic to the island' (respondent 7) (Tripadvisor, 2018). Tourism Ireland launched and continues to run a very successful campaign to sell what it refers to this as 'screen tourism' stating that 'Tourism Ireland will continue to take every opportunity to maximise the *Star Wars* connection with Ireland' (Tourism Ireland, 2018). In the promotion of any tourist destination, the goal is to make the place seem exotic, different from the norm or even a once in a lifetime experience. The construction of the exotic is done by creating a sense of mystery, a uniqueness, and a lure to the 'other', thus establishing its own sense of self (Slater, 2003).

Evidently, the attitude of visitors highlights the nexus; however, it is their very presence that puts the site at risk. The number of tourists permitted by UNESCO has been exceeded, possibly risking the loss of Skellig's World heritage status if this trend continues. *An Taisce*, the national trust, recently expressed concern 'a continuous rise in numbers of people on what is a fragile and unique site. The numbers are far in excess of the management plan presented to UNESCO in 2008' (Lucey, 2018). Official visitor numbers increased from 11,100 in 2014 to 16,755 in 2017, the island first appeared on screen in *The Force Awakens* in 2015 (The Office of Public Works, 2018), and the most updated data shows visitor numbers reached 16,792 (Brouder, 2019).

Conclusions

Heritage is a complex and subjective concept. Within the discipline of sociology, the interdependency between individuals, society, and its institutions is examined. Heritage, the media, and tourism are prime examples of these institutions. How society perceives their heritage is dependent on a number of factors. While heritage is the transmission of accepted beliefs of what came before, it is socially constructed. History differs from heritage in that it is the telling or the studying of the past, however, like heritage, it is constructed, Walter Benjamin once stated 'history is written by the victors', demonstrating its fragility. That being said, heritage can have global acceptance based on strictly set out criteria from organisations such as UNESCO. Tourism and heritage are interconnected, in that many countries sell their heritage as the principle visitation draw, Ireland is no different in this regard with two highly successful media campaigns running currently. Ireland has and will continue to be presented to the consumer/tourist as a place of aesthetic beauty, and media will persist in selling the 'gaze', as the visual is about presenting an idealised, picturesque, and unique landscape. Control or the showcasing of the visual is in the hands of those in power and authority, this control enhances power for these authorities, as the presentation of the 'gaze' is delivered from their perspective. However, this form of visual lure is not isolated to heritage tourism, the same heritage sites are recognised by fan tourists as places that embody and characterise their favourite movie or television series. Ireland has been used as a location for many films in the past, in part due to the government's generous tax breaks. Production companies bring employment and revenue into the country, it is unsurprising that the government not only support but actively encourages this system. UNESCO site *Skellig Michael* is currently being promoted for its connection to the *Star Wars* movies of 2015 and 2017 as well as its cultural heritage, therefore demonstrating the discrepancy between the actual and mediatized heritage. This reinvention of the place is at the possible detriment to the site with increased visitor numbers reportedly causing damage to the fragile island. Bearing in mind that heritage is a construct, who is to say that one history is more important than another, after all heritage transcends time and place. If we accept that heritage is fluid and evolving, we must then allow for and be aware of the *Star Wars* phenomenon. Would accepting this premise mean that the *Star Wars* narrative could become the prevailing heritage, subsequently overshadowing the early Christian story? As with other chapters in this book, the premise of this chapter was to give an example of mediatization and the creation of tourist places, but in this case from a sociological lens.

Notes

1 Evinced through the increase in visitor numbers to areas concentrated on in these campaigns.
2 'Christian monasticism had its conceptual roots in the belief that union with God could best be attained by withdrawal from civilization into harsh and isolated regions' (Horn, White Marshall, & Rourke, 1990, p. 1).
3 'In the thirteenth century, living conditions on the Atlantic islands of Ireland degenerated to such a degree that year-round occupancy of the island probably became impossible. A general climatic deterioration, linked to a southern shift of the circumpolar vortex, began around 1,200, and, as a result, the polar ice cap expanded. Colder weather and the increasing frequency and severity of sea storms appear to have forced the monks to withdraw to a site on the mainland on Ballinskelligs Bay, near Waterville, County Kerry' (Horn et al., 1990, p. 10).
4 *Clochán* is a type of building constructed with dry stones in a corbelled method (Aalen, 1964).
5 Many critiques have argued in opposition to the privileging of gazing or observation as the sole or primary characteristic of touristic behaviour. Shifting away from a visual focus has involved a redirection of attention to the body and the senses; as 'the tourist 'doing tourism' (Crouch & Desforges, 2003; Crouch et al., 2001, p. 254). Similarly, Coleman and Crang (2002, p. 7) argue 'highlights a more dynamic sense of embodied and performed as well as visualised engagement with places and tourist activities'.

References

Aalen, F. (1964). Clochans as transhumance dwellings in the Dingle peninsula, Co. Kerry. *The Journal of the Royal Society of Antiquaries of Ireland, 91*(1), 39–45.

Balmer, J. M., & Chen, W. (2016). Corporate heritage tourism brand attractiveness and national identity. *Journal of Product and Brand Management, 25*(3), 223–238.

Bessiere, J. (1998). Local development and heritage: Traditional Food and Cuisine as Tourist Attractions in Rural areas. *Sociologia Ruralis, 38*(1), 21–34.

Brouder, S. (2019, 01 17). *Concern over soaring visitor numbers on Skellig Michael.* Retrieved from: Independent.ie: https://www.independent.ie/life/travel/travel-news/concern-over-soaring-visitor-numbers-on-skellig-michael-37715839.html

Brown, L. (2015). The female tourist experience in Egypt as an Islamic destination. *Annals of Tourism Research, 63*, 12–22.

Bryman, A. (2004). *Disneyization of society.* London, UK: Sage.

Buchmann, A., Moore, K., & Fisher, D. (2010). Experiencing film tourism: Authenticity & fellowship. *Annals of Tourism Research, 37*(1), 229–248.

Butler, R., Poria, Y., & Airey, D. (2001). Clarifying heritage tourism. *Annals of Tourism Research, 28*(4), 1047–1049.

Coleman, S., & Crang, M. (2002). *Tourism: Between places and performance.* Oxford: Berghahn.

Crouch, D. (2005). Places and us: Embodied lay geographies in leisure and tourism. *Leisure Studies, 1*(3), 63–76.

Crouch, D., & Desforges, L. (2003). The sensuous in the tourist encounter: Introduction: The power of the Body in Tourist Studies. *Tourist Studies, 1*(3), 5–22.

Crouch, D., Aronsson, L., & Wahlstrom, L. (2001). Tourist encounters. *Tourist Studies, 1*, 253–270.

Davidson, J., & Spearritt, P. (2000). *Holiday business: Tourism in Australia since 1870.* Melbourne: Melbourne University Press.

Edwards, N. (1996). *The archaeology of early medieval Ireland.* London, UK: BT Batsford Ltd.

Fullagar, S., Markwell, K., & Wilson, E. (2012). *Slow tourism: Experiences and mobilities.* Bristol: Channel View.

Foucault, M. (1980). *The history of sexuality* (Volume 1 Ed.). New York, NY: Vintage Press.

Graburn, N. (1989). The sacred journey. In V. Smith (Ed.), *Hosts and guests: The anthropology of tourism* (pp. 21–36). Philadelphia: University of Pennsylvania Press.

Harrison, R. (2010). *What is heritage?* Manchester, UK: Manchester University Press.

Horn, W., White Marshall, J., & Rourke, G. D. (1990). *The forgotten hermitage of Skellig Michael.* Berkeley: University of California Press.

Howell, R., & Negreiros, J. (2012). *Visual culture* (2nd Ed.). Cambridge, UK: Polity.

Jensen, J. L. (2010). On line tourism: Just like being there? In B. T. Knudsen & A. M. Waade (Eds.), *Re-investing authenticity: Tourism, place and emotions* (pp. 213–225). Bristol, UK: Channel View.

Jensen, J. L., & Waade, A. M. (2009). *Medier og Turisme.* Arhus, Denmark: Academia.

Kockel, U. (1994). Culture, tourism and development: A view from the periphery. In U. Kockel (Ed.), *Culture, tourism and development: The case of Ireland* (pp. 1–14). Liverpool: Liverpool University Press.

Larsen, J., & Urry, J. (2011). Gazing and performing. *Environment and Planning D: Society and Space, 29*(6), 1110–1125.

Lickorish, L. J., & Jenkins, C. L. (1997). *An introduction to tourism.* Oxford: Butterworth-Heinemann.

Lowenthal, D. (1998). Fabricating heritage. *History and Memory, 10*(1), 5–24.

Lowenthal, D. (2015). *The past as a foreign country-revisited.* Cambridge, UK: Cambridge University Press.

Lucey, A. (2018, 07 15). *The Irish Times.* Retrieved September 25, 2018 from: https://www.irishtimes.com/news/environment/concern-for-fragile-skellig-michael-as-visitor-numbers-rise-1.3565829

Okamoto, T. (2015). Otaku tourism and the anime pilgrimage phenomenon in Japan. *Japan Forum, 27*(1), 12–36.

Revenue, Irish Tax and Customs. (February 04, 2019). *Office of the Revenue Commissioners.* Retrieved from Revenue.ie: https://www.revenue.ie/en/companies-and-charities/reliefs-and-exemptions/film-relief/index.aspx

Ritzer, G., & Liska, A. (1997). McDisneyization and 'post tourism': Complementary perspectives on contemporary tourism. In J. Urry & C. Rojek (Eds.), *Touring cultures: Transformations of travel and theory* (pp. 96–109). London, UK: Routledge.

Russell, J. (1997). Towards a more inclusive, vital model of heritage: An Australian perspective. *International Journal of Heritage Studies, 3*(1), 71–80.

Shaw, G. B. (1972). *Collected letters [of] Bernard Shaw, 1898–1910.* (D. H. Laurence, Ed.) London, UK: M. Reinhardt.

Shiels, R. (2013). *Spatial questions.* London, UK: Sage Publications Ltd.

Slater, E. (2003). Constructing an exotic 'stroll' through Irish cultural heritage: The Iran islands heritage centre. In M. Croni & B. O'Connor (Eds.), *Irish tourism: Image, culture and identity* (pp. 104–121). Clevedon, UK: Channel View Publications.

The Office of Public Works. (2018, September 25). *OPW.* Retrieved September 25, 2018, from: https://opw.ie/en/heritage/.

Tourism Ireland. (2018). *Global marketing, star wars.* Retrieved October 04, 2018, from: https://www.tourismireland.com/Marketing/Marketing-Highlights/Marketing-Catalogue/Star-Wars.

Tripadvisor. (2018). *Skellig Michael, Star Wars reviews.* Retrieved October 04, 2018, from: https://www.tripadvisor.com/Attraction_Review-g186610-d618321-Reviews-or10-Skellig_Michael-County_Kerry.html.

Urry, J. (1990). *The tourist gaze: Leisure and tourism in contemporary societies.* London, UK: Sage.

Urry, J. (2000). *Consuming places.* London, UK: Routledge.

Urry, J., & Larsen, J. (2011). *The Ttourist Ggaze 3.0.* London, UK: Sage Publications Ltd.

Urry, J., & Lash, S. (1994). *Economies of signs and space.* London, UK: Sage.

Wild Atlantic Way. (2018). Retrieved October 11, 2018, from: https://www.thewildatlanticway.com/skellig-michael.html.

World Heritage Committee. (1997, January). *UNESCO convention concerning the world cultural and natural heritage.* Retrieved September 21, 2018, from: http://unesdoc.unesco.org/images/0011/001121/112194Eo.pdf.

28

COMMEMORATING POPULAR MEDIA HERITAGE

From shrines of fandom to sites of memory

Christian Hviid Mortensen

Introduction

Media tourism, broadly defined as media-themed attractions relating to either media text(s) or media technologies, has become prevalent in recent years (Reijnders, 2011; Waade, 2013). Popular media offerings attract people with an interest in exploring the context further, thus creating a demand for opportunities to delve deeper into a given media experience. Many of these attractions are commercial operations utilizing the hype surrounding successful media outlets to make a profit by providing additional spin-off experiences such as *Game of Thrones: The Touring Exhibition*. These offerings all follow the logic of the experience economy (Pine & Gilmore, 2011). In this chapter, I want to address another subset of media tourist attractions that have taken a markedly different approach. This approach follows a heritage logic, and while turning a profit by providing experiences for visitors might be an ancillary ambition, the primary ambition is to preserve a particular location or collection of objects. While official museums dedicated to media-related collections have proliferated since the 1970s (Mortensen, 2017), here I will focus on unofficial initiatives taken by individuals or communities of enthusiasts. The private collection of wealthy aristocrats or businessmen often formed the basis of what would later become public museums (Imprey & MacGregor, 1985). In a parallel fashion, within popular culture, there are private fan collectors with sizeable and significant collections relating to diverse media phenomena. Over time, some of these collectors have decided to open their collection to the public, effectively transforming the collection from a private shrine of fandom into a site of memory for collective commemoration. Within our late-modern society, the agency of public history has been distributed among novel actors and the traditional knowledge monopoly and expert authority of academia and museums has thus been challenged. Now a growing number of diverse actors engage in history-producing activities such as collecting objects, re-enacting historic events and archival research (De Groot, 2016; Gillis, 1994; Huyssen, 2003; Nielsen, 2010; Samuel, 2012; Urry, 1995). Laypeople are no longer just consuming history, but often also producing it, in effect becoming history *prosumers* (Toffler, 1980). This shift can be seen as part of an emerging convergence culture with ever more complex relations between top-down authoritative producers of culture and bottom-up participatory culture (Jenkins, 2006).

The process of mediatization, whereby the media has become integral to our late modern societies (Couldry & Hepp, 2013), has also profoundly changed the phenomenon of tourism (Månsson, 2011). New sites and attractions in relation to particular media products, e.g. movies and TV, have become destinations, and established destinations have become cast in a new light, e.g. with specific media-themed walking tours. Fans, in general, are more invested in the object of their fandom than the average consumers, and therefore often take a more active and productive role, where they gather and share knowledge or produce new related cultural products to be enjoyed by other fans (Jenkins, 1992; Sullivan, 2013). Thus, the audience engagement with media texts exists on a continuum from passive consumption, over the more engaged roles of enthusiast and fan, to petty producers (or prosumers) who turn their fan activity into a profession by creating products and experiences for the fan community (Sullivan, 2013).

The cultural memories of a particular community are bound up on the objects, monuments, and other forms of material externalizations that the community regards as cultural heritage and recognizes as the carriers of these memories (Assmann, 2008). Therefore, heritage is not something we have but something we do (Smith, 2006). Cultural heritage is the product of a heritagization process, whereby something is extracted physically or conceptually from its environment in order to be preserved and function as an artefact of cultural heritage (Desvallées & Mairesse, 2010). This process has traditionally been subject to a hierarchical organization, when cultural heritage was only defined by official experts. However, this hierarchy is being demolished as the heritagization process is de-traditionalized, when other unofficial actors, such as enthusiasts who possess a lay expertise, claim the right to define cultural heritage (Urry, 1995).

Therefore, this chapter is about the heritagization activities, such as establishing collections of objects or creating sites of memory, which goes on within communities of media fans outside established cultural institutions such as museums. The changes following from mediatization can take the form of *extension, substitution, amalgamation,* and *accommodation* (Schulz, 2004). The notion of heritage is clearly *extended* when including popular media heritage, hitherto ignored by established heritage institutions. The result is *amalgamated* collections and displays containing both media phenomena, as well as ephemera and objects which can be connected to these phenomena. Conversely, heritagization is also an *appropriation* of media objects for a purpose to which they were not intended, thereby *extending* their social life as heritage collectables. Therefore, this change process is bidirectional. Such fan-driven heritagization processes take place in relation to many media phenomena. In Denmark, for example, we have a museum for both printing and radio/TV technology driven by volunteer enthusiasts. Here, my illustrative case study will be the Danish film franchise Olsen-banden [The Olsen Gang] (1968–1998), which have recently inspired fans to establish two separate attractions in Denmark.[1]

The chapter will continue with an introduction to The Olsen Gang as a media phenomenon, before introducing Nora's concept *sites of memory [lieux de mémoire]* for further analyzing the two attractions in question (Nora, 1989). I argue that these sites differ from the places of imagination usually related to media tourism, which is why I prefer Nora's original concept rather than Reijnders' adaption of it (Reijnders, 2011). Then follows a section on the special German connection to The Olsen Gang. The chapter will conclude with a short discussion of whether such fan-driven sites of memory constitute an anti-hegemonic form of heritage in relation to the official heritage sector.

The Olsen Gang: media phenomenon and object of fandom

In the film series about The Olsen Gang, we follow the endeavours of the three gang members – petty criminals on the hunt for their final big score – where their intricate plans always involve an ingenious use of household items with humorous results. The series are among the most popular films in the history of Danish film. Despite being 40 years old, four of the films are still on the top ten of the most viewed domestic films since 1976, where they occupy the three top spots (Alexandersen, 2013). The first 13 films were produced from 1968 to 1981, with an epilogue for the series produced in 1998. The films have also become a transmedia phenomenon with several spin-offs in the form of two feature length animation films in 2010 and 2013; a musical in 2008; a prequel TV series in 1999, where we follow the childhood of the main characters and a cameo appearance in the Donald Duck comics, where they assist in breaking into the vault of Scrooge McDuck (Egmont Serieforlaget, 2010). The Olsen Gang has become such an entrenched part of Danish popular culture, that they are now included in the questions on the Danish citizenship test (Brandsen, 2016a, 2016b).

Even so, The Olsen Gang has not yet left noteworthy traces in the heritage collections of the official Danish museums with few registered objects (Kulturarvsstyrelsen, 2004). Recently, however, The Olsen Gang has made several temporary appearances in museums. In March 2016, the Viborg Museum opened the first exhibition on The Olsen Gang in Denmark *Skidegodt, Egon. Olsen-banden på museum [Fucking Great, Egon! The Olsen Gang goes to the museum]*. The exhibition became a huge success attracting 60,000 visitors – three times the yearly number of visitors to this provincial museum. The success of the exhibition in Viborg also attracted the attention of the copyright-holder Nordisk Film. They decided to remake their small film museum at their production facilities in Copenhagen and dedicate an entire floor to The Olsen Gang. This is now a stand-alone attraction with guided tours called *Olsen-banden vender hjem [The Olsen Gang returning home]*. Until then, Nordisk Film had not considered the objects and ephemera surrounding the production of The Olsen Gang films as something to be preserved. For them, the important cultural artefacts have always been the films themselves. In order to acquire artefacts and ephemera for the new exhibition they had to buy back, some of what they had previously let go, from the community of fan collectors at a considerable cost.

Sites of memory vs. places of imagination

The exhibition at Nordisk Film is the first site of memory for The Olsen Gang with a more permanent character as the previous attempts has been temporary. *Sites of memory [liuex de mémoire]* are places of current significance where the collective memory is condensed (Nora, 1989). They are basically re-articulations of some past in the present. These sites do not necessarily have to be of a physical-material nature. They could also be of a more discursive nature. For example, the popular Danish TV series *Matador [Monopoly] (1978–1981)* could be just as significant a site of memory as Dybbøl Banke the site of the infamous defeat of the Danish army by the Germans in 1864 (Nielsen, 2010). *Matador* depicts the life and times of different citizens in the fictional provincial town Korsbæk from 1929 to 1947. The series is a modern, anno circa 1980, representation and interpretation of a particular segment of the past, and therefore essentially a site of memory from the beginning. In contrast, The Olsen Gang films were originally produced as contemporary portraits of Danish society and only become sites of memory when they are articulated again in the present. These articulations function as sites of memory in a double capacity since they evoke memories of both The Olsen Gang as a fictive media phenomenon and the 1970s Denmark as a historic period.

In his analysis of media tourism, Reijnders adapts Nora's concept of *sites of memory* to characterize locations of media productions as *places of imagination*, where visitors can recall experiencing a fictive media event by associating it with a material symbolic reference such as a particular place or object (Reijnders, 2011). Where Nora's concept designate how authentic places can anchor cultural memories, media tourism consists of a reverse process, where media fans try to establish material and geographical reference points for their media-induced memories. Thereby, the intangible and mediated memories become tangible and un-mediated, whereby they can be appropriated and consumed. This appropriation can take many forms such as visiting a site, purchasing an object, recreating a scene, or taking a selfie at a site.

The sites that I will consider here are different from the media tourism sites explored by Reijnders in that they are not locations from The Olsen Gang narrative universe. Instead they are more or less arbitrary placements for collections and artefacts related to The Olsen Gang that also function as a site of memory for a historic period.

Benny's car and a private Olsen Gang collection

The private citizen Kasper Christensen is an avid fan of The Olsen Gang and over the years has created a sizeable collection of artefacts with relation to the franchise. The collection consists of derivative media products (e.g. VHS and DVD editions, posters, press photos, press clippings, and biographies of the actors), objects that have featured in the films, or look like those in the films (e.g. retro china service and furniture, clothes matching film costumes). Kasper is a car enthusiast and trained auto mechanic. Therefore, the iconic car that Benny drives, which is a recurrent element in the films, was the obvious centrepiece of his collection. At first, he could not locate any of the actual cars from the films but had to make do with one of the same model and year – a Chevrolet Bel Air 1959. The car was fully functional and covered in black foil, so Kasper added rust-red splotches to replicate the battered look of the car known from the films. The imitation rust, however, resulted in a rather non-authentic look that counteracted rather than supported the replica. In contrast, when Kasper finally located and acquired one of the original cars from the films, it was completely disassembled, and he had to assemble it anew. Now the car appears with all the patina of time creating a much more authentic look. It is paramount to Kasper that the car keeps its original engine as the heart and soul of the artefact. In fact, he has decided to retire the car to the garage should the engine no longer run reliably.

In addition to the personal satisfaction in renovating the car, Kasper's aim with showcasing it is to make people smile, stop for a moment and go back in time, maybe put on a film with The Olsen Gang and contemplate the artwork that they are. In order to reinforce this effect, Kasper dresses up as Benny, while family and friends dresses up as the other well-known characters from the films. They do not act out the roles, however.

According to Kasper, he receives visitors from both Denmark and Germany to see his collection almost every day, 20–30 visitors each week. Therefore, he plans to establish a more accommodating display of his collection, which currently is just stored in a single room, where it is difficult to appreciate all the objects. The idea is to display the collection in a small house, with retro-style interiors imitating the 1970s decorations from the films. Kasper is humble about his display and does not want to call it a 'museum'. According to him, the museum is the one at Nordisk Film in Valby, Copenhagen.

Even if he does not use the term *heritage*, Kasper obviously considers Benny's car as a cultural heritage artefact that should be preserved and put on display for the benefit of others. He has spent a lot of resources, time and money, on restoring the car, and many would probably consider him 'bonkers'. However, he knows that among fans of The Olsen Gang, his actions are considered reasonable as within this fan, community Kasper's actions increase his cultural capital. Kasper has acquired a collection of objects with a considerable amount of cultural value within the fan community, and by applying his special skillset as an auto mechanic to renovate Benny's car, he has created a unique cultural artefact to be enjoyed by other fans. In addition to the cultural capital within the fan community, that the possession of a desirable artefact bestows upon its owner, the artefact also represents an economic value in the market among fans for the selling and buying of such artefacts.

While a collection is an economic asset with regard to resell value of the artefacts, it could also become an asset as an attraction for visitors, something Kasper has realized and intend to capitalize on in the future by providing ancillary services such as selling coffee and pancakes. By turning his private collection into a public attraction, Kasper effectively has transformed his personal shrine of fandom to a collective site of memory. The future of this site of memory is uncertain, however, as Kasper is just one person, lacking the organizational framework and resources of official heritage institutions. Further, he does not abide by the traditional heritage and museum ethos, where the ambition is to preserve a collection for all posterity in the interest of society (ICOM. International Council of Museums, 2017). Rather, Kasper feels a close personal connection to his collection, in particular Benny's car, which he plans to destroy before he leaves this World.

The Yellow Mansion and The Olsen Gang fan club

The actual fan community surrounding The Olsen Gang is the official fan club. It was established in 2013, when the founder and current chairman, Rune Clausen, realised that an official Danish forum for fans of the franchise did not exist and decided to found said forum. The fan club also arrange film screenings, signings, and location tours (Olsen Banden Fan Klub DK, 2013). Fans are even equipped with screenshots so they can reconstruct where the camera was placed and where the actors stood. This is a self-organised kind of Do It Yourself and non-commercial form of media tourism. The Yellow Mansion was a commando post of the Copenhagen freight terminal built in 1909, which featured in *Olsenbanden på sporet [The Olsen Gang on track]* (1975). The commando post was decommissioned in 2010 and in 2014 it was destined to be demolished. This came to the attention of a fan with connections in the railway community, who contacted the German fan club, who in turn contacted the Danish fan club. Then the two fan clubs mounted a joint-action funding campaign to preserve The Yellow Mansion as an artefact of cultural heritage (Rune Clausen, personal communication, April, 2017).

Initially, Rune invited home the manager from the railway authorities responsible for the building, in order to convince him of preserving the building because of its cultural significance. The meeting was staged with signatory elements from The Olsen Gang in order to foreground this significance:

> Then we just sat there, the four of us from the fan club, and we used every trick we could think of, right? We had butter cake, like in the films, we had cold beers and red raspberry sodas, like in several Olsen Gang films, right? We've made bubble and squeak like in film number 8 [...] He [the railway authority representative] just sat there chuckling

and said: 'You're crazy, dudes!', right? But he enjoyed it, and he thought it was cool, that we were so enthusiastic about it [...] He went back and consulted his people, and then we got one year to remove the building, and we got it for free, if we handled the removal ourselves.

The project was conducted in collaboration with the preservation society Gedser Remise, a group of railway enthusiasts who specializes in the preservation, renovation and exhibition of train carriages and other railway equipment, and The Yellow Mansion now stands on their premises in the town of Gedser. The crucial seal of approval, that also secured the viability of the project, came with a donation of DKK 1 million from the A.P. Møller Foundation (Rune Clausen, personal communication, April, 2017).

The success of the project, according to Rune, was dependent on the work of volunteer craftsmen and micro-sponsors who donated their time and materials, and without whom the budget would probably have been doubled. The Yellow Mansion might not be the most obvious choice of building for a site of memory for The Olsen Gang as it is only featured in a single film, while the main entrance to the prison in Vridsløse are a fixture of the films and therefore presumably a better candidate. However, one must also be pragmatic with regard to what it is feasible to preserve. As Rune states:

> The buildings featured in the films are mostly larger apartment or office buildings. You cannot move such buildings. It is physically impossible! Here we have a building, which it is actually possible to move. Which is even listed as worth preserving by the Agency for Cultural Heritage. Then it seems obvious, that when you have something which both have film history and railway history, moreover which is movable, then we should strive to preserve it.

The Yellow Mansion measures five by five meters and has two stories. The ground floor is dedicated to the remarkable feat of moving and restoring the mansion with photos and video documenting the process as well as a large poster listing all the micro sponsors (Bevarings-foreningen Gedser Remise, 2017). According to Rune, the ambition with the first floor was to recreate the atmosphere from the film:

> That is to say, we paint the room in the same colors, the same woodwork. We get tele-phones, lamps, control panels and knickknack on the walls similar to the film, so you recreate the atmosphere from it, so people will say: My goodness, yes! That's how it was.

Recreating an atmospheric experience in such a way is a form of gesticulating atmospheres (Albertsen, 2012). However, for the uninitiated who are not familiar with *The Olsen Gang on track*, it is an empty gesture. The memorial force of a site of memory is dependent on the decoding capabilities of a given subject (Hall, 1999). Likewise, it also takes a certain fan knowledge to appreciate and value the objects in Kasper's collection. If the reference is unknown to the subject, it will preclude the person from experiencing any memorial force.

The Yellow Mansion was inaugurated on September 9, 2017, an event which attracted 800 visitors (Bevaringsforeningen Gedser Remise, 2017). The addition of the mansion has raised the number of visitors to Gedser Remise in general with over 10,000 visitors in 2018, and there are now plans to rebuild a side building in connection with the mansion, which was not possible to move and use it to house a café in order to capitalize further on this influx of tourists (Pedersen, 2019).

Figure 28.1   The Yellow Mansion as it stands in Gedser
Source: Mortensen, 2018.

The German connection

Both Kasper's collection and The Yellow Mansion are located in the southernmost region of Denmark on the Islands Møn and Falster, respectively, accessible from Germany by ferry, and in the future by a new tunnel under Femern Belt. Proof of the cultural significance of The Olsen Gang for Germans are numerous: the German fan club predates the one in Denmark and was instrumental in providing artefacts to the exhibition in Viborg (Olsenbandenfanklub Deutschland, 2016). Further, the fan club was engaged in the funding campaign and have now produced a documentary film about The Yellow Mansion project (Olsenbandenfanklub Deutschland, 2018). Also, The Olsen Gang is still frequently the topic for community theatre productions in Germany.

It was primarily in the former East German Republic Deutsche Demokratische Republik, however, that The Olsen Gang was popular entertainment (Mitteldeutscher Rundfunk, 2010). The films passed the strict censorship of the communist regime and was one of the few glimpses of life in the West that the east Germans was allowed to see. In addition, the East Germans could sympathize with the element of social critique in the films, where the targets of the gang are often wealthy and corrupt people in power. The current popularity of The Olsen Gang in Germany can therefore be seen as an expression of *Ostalgie* – a nostalgia for aspects of daily life and culture in the former DDR (Berdahl, 2010). If this is the case, The Olsen Gang sites of memory will not be the first example of capitalizing on the ostalgic sentiments of tourists with a commercial memorialization of life in the DDR (Cliver, 2014). Like other forms of nostalgic retro culture, Ostalgie runs the risk of becoming just an ahistorical and aestheticized assembling of past objects, thus reducing the past to limited set of stereotypes. Keightly and Pickering coin the term *retrotyping* for these kinds of reductive forms of history (Keightley & Pickering, 2012). A particular incoding limits the field of possible decodings (Hall, 1999). For example, when Kasper wants to establish a retro environment in which to experience his collection, it invites a nostalgic decoding.

Conclusions

The two sites of memory presented here clearly represent an *extension* of the traditional conception of cultural heritage within heritage institutions. The cultural memory has become mediatized to the extent that commemorating media phenomena, such as The Olsen Gang, resonate well with many people. This is not a unidirectional process of mediatization however, as it is simultaneously a heritagization process governed by a heritage logic.

Heritage is a narrative device that produce or format differences and identities by foregrounding certain qualities of an object or site. In this way, the value of an object or site increases as differences and identities are the primary resources of enrichment economies (Boltanski & Esquerre, 2016). The Olsen Gang is clearly an effective narrative for ascribing value to objects. It is unlikely that the railway heritage would have been enough to secure funding and save The Yellow Mansion. The Olsen Gang dimension further inspired many micro sponsors and enabled the fan club to stage the meeting with the railway authorities in order to affect a positive outcome. This value also translates to attraction power drawing tourists, from Germany in particular, to the new sites. If this trend continues, it would be an interesting display of regional soft power, as both sites are located in an outlying region of Denmark with few attractions (Ooi, 2016). But with the random placement of the sites on locations without reference to The Olsen Gang, it becomes clear that it is the collection and the mansion as artefact that brings value to the place, and they can be relocated without diminishing the memorial force. In contrast, the place is essential for the exhibitions at the Nordisk Film production facilities as they can be branded as the *home* of The Olsen Gang.

Benny's car is unique in this regard as it is not place-bound. This makes the car a mobile site of memory that can become part of different settings such as auto shows or it can become a site of memory when encountered in traffic and ordinary bystanders are suddenly confronted with their memories of both The Olsen Gang and a past with a vastly different view on gas consumption.

In the case of The Olsen Gang, the official heritage sector has not showed due diligence in their collection policy. When they finally take a temporary interest in the media phenomenon, they were entirely dependent on the preservation efforts of fan collectors that could provide them with significant objects. These fans collected objects without regard to prevalent notions of heritage. In this sense, such collections constitute an antihegemonic form of heritage or what Urry calls detraditionalization (Urry, 1995). However, existing outside the heritage sector, these collections are not subject to the same legal requirements and do not have the usual safeguards for long-term preservation. Therefore, their status as a collective site of memory become untenable. Securing this heritage will require new collaborations between fans and established heritage institutions. It is crucial that such collaboration is based on a mutual respect and understanding. Even if the goal is shared, however, it can prove challenging to reconcile the rule bound expertise of heritage professionals with the emotional investment of fans.

Note

1 A slightly different version of this analysis has previously been published in Danish (Mortensen, 2018). The analysis is based on fieldwork at the sites in question and interviews with key informants: a private fan collector, the chairman of the Danish Olsen Gang Fan Club, and the person in charge of the exhibitions at Nordisk Film.

References

Albertsen, N. (2012). *Gesturing atmospheres.* Presented at the Ambiances in action / Ambiences en acte(s) – International Congress on Ambiances, Montreal, Canada.

Alexandersen, H. (2013). Titanic og Egon Olsen urokkelige på toppen. Retrieved 14 December 2016, from Danmarks Statistik website: http://www.dst.dk/da/Statistik/bagtal/2013/2013-05-28-titanic-og-egon-olsen-urokkelige-paa-toppen.aspx.

Assmann, A. (2008). Canon and archive. In A. Erll & A. Nünning (Eds.), *Cultural memory studies* (pp. 97–107). Berlin, Germany: Walter de Gruyter.

Berdahl, D. (2010). '(N)Ostalgie' for the present: Memory, longing, and East German things. *Ethnos, 64*(2), 192–211.

Bevaringsforeningen Gedser Remise. (2017). Det Gule Palæ. Retrieved 24 June 2019, from Gedserremise.dk website: http://www.gedserremise.dk/det-gule-pale.html

Boltanski, L., & Esquerre, A. (2016). The economic life of things. Commodities, collectibles, assets. *New Left Review, 98*(Mar–Apr), 31–54.

Brandsen, M. (2016a, June 8). Her er den nye prøve: Kan du blive statsborger? [News site]. Retrieved 19 November 2017, from nyheder.tv2.dk website: http://nyheder.tv2.dk/politik/2016-06-08-her-er-den-nye-proeve-kan-du-blive-dansk-statsborger.

Brandsen, M. (2016b, December 1). Her er den nye indfødsretsprøve: Kan du få statsborgerskab? [News site]. Retrieved 19 November 2017, from nyheder.tv2.dk website: http://nyheder.tv2.dk/politik/2016-12-01-her-er-den-nye-indfoedsretsproeve-kan-du-faa-statsborgerskab.

Cliver, G. (2014). Ostalgie revisited: The musealization of Halle-Neustadt. *German Studies Review, 37*(3), 615–636.

Couldry, N., & Hepp, A. (2013). Editorial. Conceptualizing mediatization: Contexts, traditions, arguments. *Communication Theory, 23*, 191–202.

De Groot, J. (2016). *Consuming history. Historians and heritage in contemporary popular culture.* London, UK & New York, NY: Routledge.

Desvallées, A., & Mairesse, F. (2010). *Key Concepts in Museology.* Paris, France: Armand Colin.

Egmont Serieforlaget. (2010). *Anders And & Co., 62*(41), n.p.

Gillis, J. R. (1994). Introduction. In J. R. Gillis (Ed.), *Commemorations. The politics of national identity* (pp. 3–26). Princeton, NJ: Princeton University Press.

Hall, S. (1999). Encoding, decoding. In S. During (Ed.), *The cultural studies reader* (pp. 33–44). Abingdon, UK: Routledge.

Huyssen, A. (2003). *Urban palimpsests and the politics of memory.* Stanford, CA: Stanford University Press.

ICOM. International Council of Museums. (2017). *ICOM Code of Ethics for Museums.* Paris, France: ICOM.

Imprey, O., & MacGregor, A. (Eds.). (1985). *The origins of museums, the cabinet of curiosities in sixteenth- and seventeenth-century Europe.* Oxford, UK: Clarendon.

Jenkins, H. (1992). *Textual poachers. Television fans & participatory culture.* New York, NY: Routledge.

Jenkins, H. (2006). *Convergence culture. Where old and new media collide.* New York: New York University Press.

Keightley, E., & Pickering, M. (2012). *The mnemonic imagination. Remembering as creative practice.* Basingstoke, UK: Palgrave Macmillan.

Kulturarvsstyrelsen. (2004). Museernes Samlinger. Retrieved 24 June 2019, from Museernes Samlinger website: https://www.kulturarv.dk/mussam/Forside.action.

Månsson, M. (2011). Mediatized tourism. *Annals of Tourism Research, 38*(4), 1634–1652.

Mitteldeutscher Rundfunk. (2010). MDR Zeitreise. Mein Leben. Meine Geschichte [News site]. Retrieved 8 December 2018, from www.mdr.de website: https://www.mdr.de/zeitreise/ddr/kultum-olsenbande-in-der-ddr100.html.

Mortensen, C. H. (2017). The legacy of mediatization. When media became cultural heritage. In S. Hjarvard, A. Hepp, G. Bolin, & O. Driessens (Eds.), *Dynamics of mediatization: Understanding cultural and social change* (pp. 271–291). Basingstoke, UK: Palgrave Macmillan.

Mortensen, C. H. (2018). "Skidegodt, Egon!". Olsenbanden som erindringssted. In E. G. Jensen & A. S. Sørensen (Eds.), *Tilblivelser. Aktuel Kulturanalyse* (pp. 127–148). Odense: University of Southern Denmark Press.

Nielsen, N. K. (2010). *Historiens forvandlinger – historiebrug fra monumenter til oplevelsesøkonomi* (1. udgave). Aarhus: Aarhus Universitetsforlag.

Nora, P. (1989). Between memory and history: Les Lieux de Memoire. *Representations, 26* (Special issue: Memory and Counter-Memory), 7–24.

Olsen Banden Fan Klub DK. (2013). Skide godt Egón [Community]. Retrieved 24 June 2019, from Olsenbandenfanklub.dk website: http://olsenbandenfanklub.dk/.

Olsenbandenfanklub Deutschland. (2016). Viborg Museum: Skide godt, Egon! Olsen-Banden på museum [Community]. Retrieved 24 June 2019, from Olsenbandenfanklub Deutschland website: https://www.olsenbandenfanclub.de/news/2016-02-10_viborg-museum.php.

Olsenbandenfanklub Deutschland. (2018). Fanclub-Dokumentarfilm: »Det Gule Palæ« - Eine Rettungsgeschichte [Community]. Retrieved 24 June 2019, from Olsenbandenfanklub Deutschland website: https://www.olsenbandenfanclub.de/news/2018-05-31_doku-stellwerk.php.

Ooi, C. S. (2016). Soft power, tourism. In J. Jafari & H. Xiao (Eds.), *Encyclopedia of tourism*, n.p., Cham, Switzerland: Springer.

Pedersen, K. M. (2019). Gedser på vej mod ny turistmagnet [News site]. Retrieved 24 June 2019, from Folketidende.dk website: https://folketidende.dk/Guldborgsund/Gedser-paa-vej-mod-ny-turistmagnet/artikel/504689q.

Pine, J., & Gilmore, J. (2011). *The experience economy.* Boston, MA: Harvard Business Review Press.

Reijnders, S. (2011). *Places of the imagination. Media, tourism, culture.* Farnham and Burlington, UK: Ashgate.

Samuel, R. (2012). *Theatres of memory: Past and present in contemporary culture* (2nd Revised). London, UK: Verso Books.

Schulz, W. (2004). Reconstructing mediatization as an analytical concept. *European Journal of Communication, 19*(1), 87–101.

Smith, L. (2006). *The uses of heritage.* New York, NY: Routledge.

Sullivan, J. L. (2013) *Media audiences. Effects, users, institutions, and power.* Los Angeles, CA: SAGE Publications.

Toffler, A. (1980). *The third wave* (1.). New York, NY: Morow.

Urry, J. (1995). How societies remember the past. *The Sociological Review, 43*, 45–65.

Waade, A. M. (2013). *Wallanderland. Medieturisme og skandinavisk tv-krimi.* Aalborg: Aalborg Universitetsforlag.

29

MEDIA TOURISM, CULINARY CULTURES, AND EMBODIED FAN EXPERIENCE

Visiting *Hannibal*'s Florence

Rebecca Williams

Introduction

This chapter focuses on a case study of a trip made to Florence in Italy which was inspired by fandom of the television series *Hannibal* (NBC, 2013–2015). Its approach is informed by the discipline of Fan Studies, building upon work on media and fan tourism, and the disciplinary background of Tourism Studies, especially work that explores the importance of food and drink to the tourist experience. In so doing, it contributes to research on the material cultures of fandom to better understand the links between media tourism, particularly fan tourism, and culinary consumption. It also highlights the importance of convergence and participatory cultures in understanding how 'tourists are both consumers and producers of media products, making them highly influential in the reproduction of space' (Månsson, 2011, p. 1635).

Hannibal, the television series, was designed by creator Bryan Fuller to tell the story of author Thomas Harris' fictional serial killer and cannibal Hannibal Lecter (Mads Mikkelsen) before his incarceration. Lecter had previously appeared in Harris' novels *Red Dragon* (1981), *The Silence of the Lambs* (1988), *Hannibal* (1999), and *Hannibal Rising* (2006) and in cinematic adaptations of each, most famously being depicted by Anthony Hopkins in Jonathan Demme's Oscar-winning movie *The Silence of the Lambs* (1991). Conceived as a form of 'remix', Fuller's *Hannibal*

> Reimagine[d] the source material by altering key aspects of the original books, including diversifying the cast; focusing on character development and motivation; and establishing a signature lush, beautiful, and sophisticated style for the program.
>
> *(McCracken & Faucette, 2015)*

Whilst the show failed to attract large audiences, it impressed critics with its dark subject matter, moral ambiguity, and aesthetics, and attracted a dedicated fanbase called Fannibals who engaged in a range of discursive, affective, and textual practices. One of the most unique modes of engagement, however, has been *Hannibal* fans' embracing of the opportunity to cook and consume the food depicted on-screen, which remains relatively unusual in terms of contemporary media fan practice. The show textually offered a 'visually excessive representation of food preparation and consumption' (Fuchs, 2015, p. 107), using food to highlight 'thematic concerns surrounding the transgression of borders' (Fuchs, 2015, p. 99) between categories such as good and evil, love and hate, and man and God. Accordingly, fans became interested in recreating the dishes seen on-screen and acted as culinary producers of these meals, a practice encouraged by the show's food stylist Janice Poon (2013) who published a related blog and cookbook *Feeding Hannibal* (2016). Given such textual representations and attempts to sustain 'fantasies of 'entering' into the cult text' (Hills, 2002, p. 151) by recreating and eating what the characters did, it is perhaps unsurprising that fan tourism based on the series may involve culinary consumption or engagement in the type of gourmand experience seen in the series, and the sharing of information about key locations that were used.

Studies of Food Tourism have discussed how 'food is directly or indirectly connected with specific destinations; it encourages tourists to taste and experience a region's cuisine. More importantly, researchers indicate that food can be used as a means of marketing and branding a tourism destination' (Lin, Pearson, & Liping, 2011, p. 31). However, whilst Florence is well represented via culinary branding (Chrzan, 2006), there are no official media tours in Florence. The visitor must trace their own steps through the city whilst seeking out culinary experiences that best replicate the practices and tastes of Hannibal or consult fan-produced resources such as online blogs or posts on social media platforms such as Tumblr, reflecting the broader tendency towards sharing touristic information online as 'tourists' reviews, comments, and perceptions of destinations can now be spread instantly and globally to friends and far beyond one's own network by social media posts' (Månsson, 2011, p. 1639). Drawing on an account of a fan trip to Florence, the chapter considers how the fan-tourist engages in extra-textual practices related to culinary consumption, arguing that this offers opportunities to form and maintain affective connections to a fan object and, in the case of fan tourism, inhabit important spaces as well as undertaking physical experiences that act on the body. The 'consumption of food and drink related to a fan object offer avenues for connection, participation and pleasure' (Williams, 2015, para. 2), demonstrating how 'the repeated spotlighting and excessive visualization of food may spur viewers to transform textual traces into lived experience' (Fuchs, 2015, p. 108). Engaging in this way in locations related to beloved media texts therefore offers the uniquely situated and physical opportunity to move one's fandom from the textual into the bodily *and* the spatial.

Tourism studies, media studies, and fan studies

Many academic studies have focused on the links between tourism and fandom, exploring how fans undertake forms of 'pilgrimage' (King, 1993; Porter, 1999) to sites associated with their favourite fan objects. Visiting locations used in the filming of television series or films (Brooker, 2005; Couldry, 2007; Hills, 2002), or associated with favourite celebrities, allows fans to gain a closer attachment to favourite texts and visiting important sites allows fans the 'opportunity to relocate in place a profound sense of belonging which has otherwise shifted into the textual space of media consumption' (Sandvoss, 2005, p. 64).

Equally, Tourism Studies has long analysed the intersections between tourist activity and food cultures (see Everett, 2016), highlighting how 'every tourist is a voyeuring gourmand' (Lacy & Douglass, 2002, p. 8) who can be encouraged to visit specific locations due to their 'unique culinary attractions' (Cohen & Avieli, 2004, p. 758). As Parasecoli notes,

'Symbolically, economically, and materially, tourists consume and ingest the communities they visit' (2008, pp. 128–129). Whilst food is now an integral part of the experience for many travellers, the specific culinary tourist has been defined as one who undertakes 'tourism trips during which the purchase or consumption of regional foods (including beverages), or the observation and study of food production (from agriculture to cooking schools) represent a significant motivation or activity' (Ignatov & Smith, 2006, p. 238). The impetus for such trips is the opportunity to experience the cuisine itself since it is

> about individuals exploring foods new to them as well as using food to explore new cultures and ways of being. It is about the experiencing of food in a mode that is out of the ordinary, that steps outside the normal routine to notice difference and the power of food to represent and negotiate that difference.
>
> *(Long, 2004, p. 20)*

My interest here lies, however, in exploring the implications when the motivation for a trip is inspired by fandom of a specific text, and when the opportunity for experiencing the culinary is intrinsically linked to both the fan object and the location being visited, a form of mediatized tourism that has, yet, not been fully explored. This omission is particularly striking since Fan Studies have analysed a range of other material practices, and how objects such as 'sports memorabilia, music collectibles, and theatrical props all constitute meaningful bridges between the abstract semiotics of the screen and the lived, tactile experience of audiences' (Rehak, 2014, para. 1.3). As this chapter argues, such 'meaningful bridges' between texts and audiences' experiences can also be engendered by material practices linked to the culinary.

When prior studies have discussed food or drink, it has tended to be as a minor element of the broader topic under exploration; in, for example, Gray's dismissal of the existence of a Dominos 'Gotham City pizza' to promote the *Batman* movie *The Dark Knight* as an 'unincorporated paratext' as one that 'contribute[s] nothing meaningful to the text or its narrative, storyworld, characters or style' (2010, p. 210). One of the only sustained studies of fandom and food is Magladry's research into fan cookbooks which offer 'an embodied experience of media consumption which further blurs the boundaries between producer and consumer, and politics and pleasure' (2018, p. 112). Of particular relevance is her proposal that cooking recipes from fantasy texts such as *Game of Thrones* function as a form of culinary tourism since

> What it means for food to be "authentic" to either Westeros or [countries like] Italy is based on the consumer's conceptualization of these places: therefore, creation of either place through food presents no distinction between "fictional" or "real" places as they are equally constructed and dependent on mediation.
>
> *(Magladry, 2018, p. 118)*

Whilst Magladry focuses on these imaginative and mediated forms of culinary tourism, other work has noted, albeit in passing, the importance of food and drink to the *physical* fannish experience of 'being there'. Norris' discussion of fans of the Japanese anime *Kiki's Delivery Service,* and their trips to the Ross Bakery Inn and Ross Village Bakery in Tasmania, notes how they were able to 'reinforce [...] a positive link between the Ross Bakery and *Kiki'* (2013, para. 9.2) whilst Reijnders (2013) draws attention to the importance of food to the media/fan experience in tourism based on the Kurt Wallander novels and television series in

Ystad, Sweden where a local *konditori* offers a special 'Wallander cake' (2013, p. 40). Reijnders refers to the concept of *'lieu d'imagination'* (places of imagination) and draws attention to

> the distinctively 'material' character of the tourist's performances on the locations that were studied. Tourists attempt to call up their world of imagination, by eating certain cakes at the location, by sitting on certain chairs, or by drinking certain drinks.
>
> *(2013, p. 51)*

In so doing, he highlights how tourism and food and drink intersect for the media tourist, allowing them to 'consume their own imagination' (Reijnders, 2013, p. 40).

These examples begin to demonstrate the linkages between media texts, fandom and food and drink and indicate the need to broaden our understandings of these in relation to media tourism. As the chapter argues, via its autoethnographically informed approach, studies of media tourism and fan-tourism should pay attention to the range of culinary practices that fans engage in, and that their use of cuisine and other consumable goods whilst visiting important sites is worthy of further attention.

Autoethnography in fan studies and tourist studies

Given the lack of prior research into the links between tourism, fandom and food, there is little methodological precedent to follow. This chapter utilises a broadly autoethnographic approach to elucidate how fan tourism and food cultures operate together to offer a distinctive form of experience and practice. This is appropriate since this research is based on the experiences of the researcher, and their relationship to a fan object, locations associated with that object and the culinary cultures that surround both. Given the prominence of researchers writing from the perspective of 'aca-fans' or 'scholar-fans' (Hills, 2002), autoethnography has often been employed within Fan Studies to explore 'the challenges and contradictions that shape the research experience as the individual moves within and between academic and, [...] fan-audience modes of engagement' (Monaco, 2010, online).

Autoethnography draws reflectively on a researcher's experiences to 'describe and critique cultural beliefs, practices, and experiences' (Adams, Holman Jones, & Ellis, 2014, p. 1) and produce 'stories about the self told through the lens of culture' (Adams et al., 2014, p. 1). This method offers opportunities to develop fan studies research into more embodied accounts that deal not only with the discursive practices of fandom (e.g. the constructs and constraints of identity) but with what it means when people actually take up these discursive practices and really live through them (Evans & Stasi, 2014, p. 15).

The autoethnographic account of my fan trip to Florence thus provides a 'narrative account[...] of what it means to take up [fannish] subject positions and use them to create a sense of self as a lived experience' (Evans & Stasi, 2014, p. 15). When undertaking this research, I have attempted to bear in mind my own background and position-taking as the researcher, following how 'autoethnography asks the person undertaking it to question their self-account constantly, opening the 'subjective' and the intimately personal up to the cultural contexts in which it is formed and experienced' (Hills, 2002, p. 43). In so doing, I have tried to be aware of my own class, educational background, and other key factors, as well as the taste formations that contribute to my fandom of Hannibal and impacted upon how I traversed the cityscapes of Florence.

However, the method has been criticised since the 'the creation of accounts that use self-disclosed and self-interpreted individual experience expressed through first person writing and reporting as data source' means that "questions emerge such as how to ensure rigor, or verify the evidence' (Driessen & Jones, 2016, p. 71). More specifically, autoethnography has often been dismissed in Tourism Studies which has assumed 'that in order to be authoritative it is necessary to adopt the passive, third-person voice, which distances the writer physically, psychologically and ideologically from his or her subject, and minimizes the self' (Morgan & Pritchard, 2005, p. 34). However, its strength in studying fandom and, by extension, moments of fan-led media tourism, lies in how it makes 'visible the reflexive voice of the writer while foregrounding the theoretical implications of objective/subjective dualisms through examinations of the relationship between "academic" and "fan" selves' (Monaco, 2010). This chapter thus follows the tradition of autoethnography by 'using my experience to engage with existing literature on a topic of personal significance – in the hope of furthering understanding in that field' (Holmes, 2016, p. 195). Eschewing the notion of a cohesive stable sense of personal and fan selves, it highlights my status 'as academic, fan and subject' as a 'fractured and fragmented character' where 'the real and fictional may be blurred to evocatively elicit variations of fan affect, performativity, and intensities' and the 'fluid, interchangeable, shifting facets of an ethnographically imagined self' (Sturm, 2015, p. 215) are presented.

Food tours and culinary culture in Florence

Whilst set in Baltimore and the surrounding areas and filmed in and around the Canadian city of Toronto, *Hannibal* set six of its third season episodes primarily in Florence. This narrative arc – also present in slightly different forms in the novel and film versions – sees Hannibal on the run from the FBI and masquerading as the Curator of the Capponi Library, Dr Roman Fell, who he has previously killed. Prompted by my fandom, I undertook a four-night trip to Florence at Easter in 2016, combining visits to Hannibal locations with broader tourist sight-seeing. The series was filmed at several of Florence's iconic sights including the Santa Maria Novella, the Duomo, the Uffizi Gallery, and the Palazzo Medici Riccardi and we visited those, and others, during our trip (Williams, 2019).

Given *Hannibal*'s emphasis on food cultures, we also undertook two cuisine-based tours; one walking tour around Florence's hidden food destinations and one coach trip around the Chianti wine region of Tuscany (although only the former is discussed in-depth here). These indicated our desire to not only visit where *Hannibal* had been filmed but to try to emulate part of the lifestyle that the character would have been immersed in whilst living in the city. For example, we considered attending artistic and cultural events such as opera or performances of classical music whilst in Florence, given Hannibal's canonical interest in these forms of culture and his status as an aesthete. Whilst we were unable to undertake these activities due to time constraints, we nonetheless visited sites around the city, often with the gaze of both the traditional visitor and the media fan-tourist. However, we also viewed such sites through the lens of how the character of Hannibal would respond to these, sometimes playfully tailoring our own behaviour and tastes to allow inhabitation of the spaces seen in the text of the series on our own trip.

The primary culinary tour that we undertook was a walking tour of the city of Florence itself, operated by Walks of Italy; whilst meant to be a small group tour, my husband and I ended up being the only people signed up for the walk which meant that we had three hours of personal touring with our guide. Beginning outside the Santa Croce church, the walking tour included wine tasting and canapes in a local wine shop, a visit to the San Lorenzo Market for traditional Tuscan food, the option to try the Florentine culinary speciality of tripe, and ice-cream at an Italian gelateria. Our second tour involved a coach trip with MyTour Tuscany with 50 other tourists around the Tuscan countryside, stopping at two vineyards to sample Chianti wine and local produce. Whilst *Hannibal* the series was never filmed in these locations, we undertook the tour given the famous reference in the movie adaptation of Thomas Harris' *Silence of the Lambs* to Hannibal having eaten the liver of a census taker 'with some fava beans and a nice Chianti'. Whilst the locations themselves were not used in *Hannibal*, there were thematic links that led us to undertake the tour, speaking both to the transmedia potential of the Lecterverse and to the desire to engage in touristic activity that moves beyond 'a single media product [...] and its influence on tourists in isolation because content in film, novels and other media products continuously intertwine' (Månsson, 2011, p. 1635).

Our willingness to experience the 'authentic' cuisine of Tuscany via the first tour positions us as what Cohen (1979) calls the 'experiential (and experimental and existential) modes' of tourism. In relation to the culinary, this refers to those who 'show a marked interest in local dishes and food habits. [...] They may taste local foods, out of curiosity rather than in quest of enjoyment' (Cohen & Avieli, 2004, p. 774). Indeed, part of our rationale for undertaking this tour was that it would take us 'off the beaten track' of Florence's food scene and allow us to access places we would not otherwise have found or felt comfortable negotiating. However, our willingness to experiment and experience local foods had its limit when neither of us was willing to try the local tripe speciality. Thus, despite our fannish interest in *Hannibal* and our desire to engage thematically with the adventurous culinary tastes of the character, we were fundamentally 'reluctant to expose [ourselves] to the local cuisine and [...] opt[ed] to eat local food in the sheltered environment of tourism-oriented establishments' (Cohen & Avieli, 2004, p. 774). Our quest to follow in Hannibal's culinary footsteps and the character's enthusiasm for all forms of the edible (including offal, tripe, and blood) was fundamentally limited by our own hesitancy, our continued patronage of restaurants and bars near the main tourist sites of Florence, and our subsequent reintegration into the 'recreational (and diversionary) modes' of tourism (Cohen, 1979).

Such experiences demonstrate how both traditional culinary tourism and that linked to media or fan tourism have limits in terms of what the visitor will consider 'acceptable'. This may be linked to issues of food safety or, in some cases, repulsion (Cohen & Avieli, 2004) but, in our instance of *Hannibal*-related gastronomy, also spoke to the threshold of where our fan/touristic attempts to walk in the character's footsteps lay. Despite our efforts to recreate his cavalier gourmand attitude, our own engrained tourist practices led us back to more typical and what may be considered more 'inauthentic' forms of culinary engagement. Indeed, many of the debates around tourism, both generally and within fan studies, have focused on the concept of 'authenticity', for example, questioning whether the tourist is experiencing a 'true' or 'real' encounter with local food and drink (Beer, 2008). Equally, Fan Studies have often focused on how fans operate forms of distinction by privileging certain types of location over others (e.g. those that are more difficult to find, those that are not included on official tours) or judging the authenticity of locations based on 'physical differences, an atmosphere that takes one out of the [fictional world], or moral concerns around commercial exploitation' (Norris, 2013, para. 10.18).

However, our judgements of our *Hannibal* media tourism were more concerned with the balance between the authenticity of the *culinary experience* and the authenticity of the fannish experience of inhabiting, and attempting to quasi-literally ingest, the textual. When visiting more tourist-oriented restaurants, we still often attempted to reflect our fan interest in the series; in one restaurant located near the Palazzo Vecchio, we ordered Pappardelle made with wild boar in an effort to replicate a meal eaten by the character Jack Crawford in the Florence-set episode 'Contorno' (3.5). The restaurant itself may have been rendered inauthentic via its appeal to a tourist crowd and its location. However, our attempt to 'consume [our] own imagination' (Reijnders, 2013, p. 40) via a meal similar to the Pappardelle alla Lepre consumed on screen offered the authenticity of an experience linked to the narrative and material world of *Hannibal*, rather than the physical location of Florence itself. In seeking out related foodstuffs, and other experiences of trying to locate filming locations, it thus falls to the *Hannibal* fan-tourist to seek out locations for themselves, drawing on their own knowledge of the scenes in the series, any existing familiarity with the city of Florence, and information shared online by fellow fans, encouraging participation in forms of mediatized tourism via engagement in participatory cultures.

Indeed, despite our visit not being part of communal fan activity or undertaken with a fan group, the participatory culture around *Hannibal* was instrumental in the planning of the trip to Florence. Fannibals have identified key locations and shared photographs and stories from their own visits to these whilst one site, *Fangirl Quest* undertakes 'sceneframing', the practice of visiting 'filming locations of […] favourite shows and movies, and try[ing] to align a screenshot from that film with the original background used in the film' (FangirlQuest, no date: para. 1). This was used to identify the main *Hannibal* locations, demonstrating how participatory cultures have 'relatively low barriers to artistic expression and civic engagement, […] and some type of information mentorship whereby what is known by the most experienced is passed along to novices' (Jenkins, Puroshotma, Clinton, Weigel, & Robison, 2005, p. 3). Whilst many of the Florentine locations used in the film and TV versions of *Hannibal* were self-evident (such as the Duomo, Ponte Vecchio, and the Palazzo Vecchio), others required the knowledge of fellow fans for identification, highlighting the importance of 'participatory knowledge cultures in which people work together to collectively classify, organize and build information' (Delwiche & Jacobs Henderson, 2013, p. 3).

However, even when drawing on the pre-existing knowledge of transmedia locations available within participatory cultures, debates may still occur when trying to accurately map these and ensure that one is visiting, and performing their fandom, in the right place. Our experiences drew on shared knowledge but still had 'to be (skilfully) worked at via the discovery of hidden information' (Hills, 2002, p. 148), albeit information that can then be circulated in turn to other fans. Equally, use of participatory networks before we departed allowed us to discover that specific locations did not exist; most notably, the iconic Florentine grocer Vera Dahl 1926 which is first mentioned in Harris' novel of *Hannibal* and visited in the series. Such failures to live up to expectation have been noted in previous work on fan tourism when fans' encounter places that feel 'inauthentic' or 'ordinary' in comparison to mediated versions or their own imaginations (Norris, 2013). However, in the case of a transmedia tourism location such as Florence, where different layers of meaning linked to the various versions of the story are activated and drawn on differently, such experiences were viewed less as disappointments and more as interesting discoveries in our exploration and mapping of the multiple mediated Florence(s) of the Lecterverse.

Conclusions

One way for scholars in both Fan Studies and studies of media tourism to further their understandings of media, and mediatized, tourism is to pay greater attention to the links between such tourism, contemporary media fandom, and the consumption of food and drink related to fan objects. Following the disciplinary lead of Tourism Studies, turning attention to how culinary and fan tourism intersect also enables better understanding of the practices that media and fan-tourists engage in when visiting important locations and how fans function as cultural producers of meaning and knowledge. Strengthening our understanding and knowledge of how fans engage in place-specific and situated culinary practices allows us to trace how such experiences may lack traditional forms of authenticity in terms of their links to locations and the cuisine that is on offer (for example, eating at a tourist restaurant rather than one frequented by locals), but may still offer opportunities for consuming food and drink that offers an authentic auratic connection to a text, and enables a connection to the narrative world via material practices (e.g. eating a recipe that had been seen on-screen).

However, when fan-tourists visit key sites and seek to engage in culinary practices, they often turn to the existing information and guidance on the locations, from both official sources (if they exist) and the fragments of knowledge gleaned from fans online which also frames expectations of one's own trip to Florence. Indeed, acts of fan tourism are always already 'mediated by the testimonies, amateur and professional, of previous fans' (Brooker, 2005, p. 27). Much as the various incarnations of Hannibal Lecter's Florence across the novel, movie, and series offered multiple layers or versions of expectation and experience, so too did the knowledge and images of other fans which we consulted before our trip. Rather than this proving detrimental, however, or leading to a situation where the 'reality' of our own trip failed to live up to the expectations engendered by knowledge of other fan experiences, it can instead be read as another example of the transmediality of *Hannibal*'s Florence; this time a form of transmediality or 'mediatized tourism' (Månsson, 2011) read through the lens of the participatory culture that surrounds the series, rather than the multiple iterations that exist in different texts and forms of media (i.e. literature, film, television).

Finally, whilst this chapter draws on a specific case study, future work exploring a range of cases beyond *Hannibal* can enable research into the links between media texts and their surrounding fandoms to better understand how the searching out of related food is 'about creating an affective response and whetting viewers' appetites' (Fuchs, 2015, p. 107) and to contribute to our knowledge of fan and media tourists' material practices. Wider research into how fan-tourists draw on participatory cultures and networks to seek out and share information about fannish culinary sites, practices, and experiences across a range of texts and mediated objects offers scholars in Media Studies, Fan Studies, and Tourism Studies new ways to consider 'tourists' agency in media production, an important process to highlight given that tourists' media consumption/production has an impact on tourist spaces' (Månsson, 2011, p. 1647). Furthermore, such wider study has the potential to further our knowledge and understanding of the relationship between media/mediatized tourism, fandom, and culinary cultures and to highlight how each of these is intimately linked to forms of participatory knowledge, performance, and authenticity.

References

Adams, T. E., Holman Jones, S., & Ellis, C. (2014). *Autoethnography: Understanding qualitative research*. Oxford, UK: Oxford University Press.

Beer, S. (2008). Authenticity and food experience – commercial and academic perspectives. *Journal of Foodservice, 19*, 153–163.

Brooker, W. (2005). The *Blade Runner* experience: Pilgrimage and liminal space. In W. Brooker (Ed.), *The Blade Runner experience: The legacy of a science fiction classic* (pp. 11–30). London, UK: Wallflower.

Chrzan, J. (2006). Why Tuscany is the new Provence: Rituals of sacred self transformation through food tourism, imagined traditions, and performance of class identity. *Appetite, 47*, 388.

Cohen, E. (1979). A phenomenology of tourist experiences. *Sociology, 13*, 179–201.

Cohen, E., & Avieli, N. (2004). Food in tourism: Attraction and impediment. *Annals of Tourism Research, 31*, 755–778.

Couldry, N. (2007). On the set of *The Sopranos*: "Inside" a fan's construction of nearness. In J. Gray, C. Sandvoss, & C. L. Harrington (Eds.), *Fandom: Identities and communities in a mediated world* (pp. 245–270). New York: Routledge.

Delwiche, A., & Jacobs Henderson, J. (2012). Introduction: What is participatory culture? In A. Delwiche, & J. Jacobs Henderson (Eds.), *The participatory cultures handbook* (pp. 3–9). New York: Routledge.

Driessen, S., & Jones, B. (2016). Love me for a reason: An Auto-ethnographic account of Boyzone fandom. *IASPM@Journal*, 6, 68–84. Retrieved from: http://www.iaspmjournal.net/index.php/IASPM_Journal/chapter/view/770 (accessed 10 August 2018).

Evans, A., & Stasi, M. (2014). Desperately seeking methodology: New directions in fan studies research. *Participations: Journal of Audience and Reception Studies, 11*, 4–23. Retrieved from: http://www.participations.org/Volume%2011/Issue%202/2.pdf (accessed 10 August 2018).

Everett, S. (2016). *Food and drink tourism: Principles and practice*. London, UK: Sage.

Fuchs, M. (2015). Cooking with Hannibal: Food, liminality and monstrosity in *Hannibal*. *European Journal of American Culture, 34*, 97–112.

Gray, J. (2010). *Show sold separately: Promos, spoilers and other media paratexts*. New York, NY: New York University Press.

Hills, M. (2002). *Fan cultures*. London, UK: Routledge.

Holmes, S. (2016). Between feminism and anorexia: An autoethnography. *International Journal of Cultural Studies, 19*, 193–207.

Ignatov, E., & Smith, S. (2006). Segmenting Canadian culinary tourists. *Current Issues in Tourism, 9*, 235–255.

Jenkins, H., Puroshotma, R., Clinton, K., Weigel, M., & Robison, A. J. (2005). 'Confronting the challenges of participatory culture: Media education for the 21st century,' *New Media Literacies*. Retrieved from: http://www.newmedialiteracies.org/wp-content/uploads/pdfs/NMLWhite Paper.pdf.

King, C. (1993). His truth goes marching on: Elvis Presley and the pilgrimage to Graceland. In I. Reader & T. Walter (Eds.), *Pilgrimage in popular culture* (pp. 92–104). London, UK: Macmillan.

Lacy, J., & Douglass, W. (2002). Beyond authenticity: The meaning and uses of cultural tourism. *Tourist Studies, 2*, 5–21.

Lin, Y., Pearson, T. E., & Liping, A. C. (2011). Food as a form of destination identity: A tourism destination brand perspective. *Tourism and Hospitality Research, 11*, 30–48.

Long, L. (Ed.) (2004), *Culinary tourism: Exploring the other through food*. Lexington: University of Kentucky Press.

Magladry, M. (2018). Eat your favourite TV show: Politics and play in fan cooking. *Continuum: Journal of Media & Cultural Studies, 32*(2), 111–120.

Månsson, M. (2011). Mediatized tourism. *Annals of Tourism Research, 38*(4), 1634–1652.

McCracken, A., & Faucette, B. (2015). Branding *Hannibal*: When quality TV viewers and social media fans converge. *Antenna*. Retrieved from: http://blog.commarts.wisc.edu/2015/08/24/branding-hannibal-when-quality-tv-viewers-and-social-media-fans-converge/ (accessed 18 February 2018).

Monaco, J. (2010). Memory work, autoethnography and the construction of a fan-ethnography. *Participations: Journal of Audience and Reception Studies, 7*. Retrieved from: http://www.participations.org/Volume%207/Issue%201/monaco.htm (accessed 11 February 2018).

Morgan, N., & Pritchard, A. (2005). On souvenirs and metonymy: Narratives of memory, metaphor and materiality. *Tourist Studies, 5*, 29–53.

Norris, C. (2013). A Japanese media pilgrimage to a Tasmanian bakery. *Transformative Works and Cultures, 14*, doi: 10.3983/twc.2013.0470 (accessed 17 February 2018).

Parasecoli, F. (2008). *Bite me: Food in popular culture*. Oxford: Berg.

Poon, J. (2013). *Feeding Hannibal Blog*. Retrieved from: http://janicepoonart.blogspot.co.uk/ (accessed 16 February 2018).

Poon, J. (2016). *Feeding Hannibal: A connoisseur's cookbook*. London, UK: Titan Books.

Porter, J. E. (1999). To boldly go: *Star Trek* convention attendance as pilgrimage. In J. E. Porter & D. L. McLaren (Eds.), *Star trek and sacred ground: Explorations of star trek, religion, and American culture* (pp. 245–270). Albany, NY: SUNY Press.

Rehak, B. (2014). Materiality and object-oriented fandom. *Transformative Works and Cultures, 16*. Retrieved from: http://journal.transformativeworks.org/index.php/twc/chapter/view/622/450 (accessed 12 February 2018).

Reijnders, S. (2013). *Places of the imagination: Media, tourism, culture*. London, UK: Palgrave.

Sandvoss, C. (2005). *Fans: The mirror of consumption*. Cambridge, UK: Polity Press.

Sturm, D. (2015). Playing with the autoethnographical: Performing and re-presenting the fan's voice. *Cultural Studies ↔ Critical Methodologies, 15*, 213–223.

Williams, R. (2015). Cooking with Hannibal: Food, fandom and participation. In, *In Media Res*. Retrieved from: http://mediacommons.futureofthebook.org/imr/2015/09/23/cooking-hannibal-food-fandom-participation (accessed 13 August 2018).

Williams, R. (2019). Funko *Hannibal* in Florence: Fan tourism, participatory culture, and para-textual-spatio play. *Jomec Journal: Journalism, Media & Cultural Studies, 14*, 71–90.

30
SCENE HUNTING FOR ANIME LOCATIONS
Otaku tourism in Cool Japan

Antonio Loriguillo-López

Introduction

Japanese commercial animation has become one of the most popular forms of audio-visual entertainment, widely accepted in markets around the world. This is a compelling argument for its promotion as an ambassador of Japanese culture worldwide through the Cool Japan Policy endorsed by the Ministry of Economy, Trade and Industry. Concurrently with the growing numbers of international visitors in recent years (28,691,073 in 2017; JNTO, 2018), anime-related tourism has given rise to several 'meccas' around Japan for both foreign and Japanese fans. Some of these sites, such as the increasingly popular otaku districts of Akihabara and Ikebukuro or the 'anime venues' (Denison, 2010) established by the holders of profitable intellectual property (like the Ghibli Museum in Mitaka or Sanrio Puroland in Tama) have benefitted from being easily accessible day trips for visitors to Tokyo. In other cases, however, places traditionally left out of conventional tourist itineraries have turned into high-profile destinations thanks to the spontaneous development of *seichijunrei*, fan pilgrimages to the places that inspired the background settings for popular anime series.

The *seichijunrei* phenomenon represents a very interesting opportunity to promote one of Asia's rising tourist destinations, a fact that the Japan National Tourism Organization has channeled by composing guidebooks and websites dedicated to these appealing destinations for fans around the world (JNTO, 2014). Despite the fact that the first records of these pilgrimages date from the 1990s, the term *seichijunrei* has only begun to receive media attention since 2008. In the past decade, much research has focused on these pilgrimages as paradigmatic examples of mediatized tourism in their capacity to connect the prolific Japanese pop culture industry (Yamamura, 2018) to the paratextual activities of consumption of its products by fans fully integrated into the era of media convergence (Okamoto, 2015), and to the cultural promotion policies driven both by the Japanese government and by local communities (Seaton & Yamamura, 2015).

Anime pilgrimages generate eloquent communicative expressions. In addition to their productive synergies with media-induced tourism (Tung, Lee, & Hudson, 2017) and to the anime-inspired social interactions with typical otaku geographical hubs, *seichijunrei* reflect the degree of media literacy of its practitioners. This is evident in the increasingly popular practice of *butaitanbou* or scene hunting for the real locations on which an anime production was based. The conversion of certain towns into pilgrimage sites has given rise to the production of comparative photomontages combining photographs of the original locations and still frames from their animated version. For these creations, anime otaku need not only to test their ability to gather information about the location scouting of the animation studios but also to demonstrate both their skills in composition and the use of postproduction software to manipulate their own pictures and, of course, their capacity to disseminate the results of their expeditions through the *butaitanbou* communities.

The purpose of this chapter is to observe the connections between this kind of tourism and the otaku subculture in the context of media literacy. The activities of the anime otaku, far from being marginal, may well represent a type of paradigmatic example of competency in the age of media convergence both for their growing global visibility and their symbiotic relationship with the importance of visual culture and communication in the new millennium. Therefore, examining their proactive sightseeing practices from the perspective of media literacy is a good first step towards incorporating this booming prosumer profile into the rich academic debate about the effects of digital culture in tourism.

Specifically, I propose an approach to the implicit know-how of the practice of *butaitanbou* in the case of the tourism generated as a result of the series *Suzumiya Haruhi no Yūutsu*, one of the most popular anime titles of the 2000s, which has given rise to a pilgrimage fever to various locations in Hyōgo Prefecture (Kansai). This case study deals with aspects that demonstrate the digital literacy of otaku in a media environment as eminently multifarious as the media mix. *Butaitanbou* highlights two main features of the otaku subculture noted by the influential postmodern cultural philosopher Azuma (2009): the thorough search for information about the context of fiction universes, and the pleasure derived from processing that information intellectually, mainly in the form of a database. A comparison between the indicators of media competence established by The United Nations Educational, Scientific and Cultural Organization (UNESCO) and the individual and collective skills demonstrated by otaku communities in their wide range of social activities reveals that otaku have a consistently high level of performance in advanced skills associated with 'MIL', or Media and Information Literacy. MIL competencies encompass the skills and needs of people across the world facing the increase of information and media circulation in the Information Age. The concept and its purposes have been widely supported by UNESCO and media educators (Kress, 2003) in order to develop the abilities needed enable media and information literate societies in the 21st century.

Literature review: the paratextual activities of otaku

MIL in Japan

Although Japan is one of the world's biggest production and distribution centres of technological and audio-visual entertainment, the media skills of its society do not reflect the innovative capacity of its industries. Over a decade ago, Liversidge noted the limited implementation of media literacy initiatives in Japan (2005), and studies since then have pointed to the initial absence of technology in the design of the national curricula (Arke, 2012). In the last decade, the Japanese educational system (Takabayashi, 2015) and the public media corporation, NHK (Kodaira & Watanabe, 2013), have developed an informal media education program that has brought the skills level of its youngest participants up to that of their Canadian and British[1] counterparts through coordinated actions that assess the degree of media literacy of Japanese citizens (Mizukoshi, 2017, p. 38). However, as Mizukoshi concludes, we should not allow this data to mislead us because running in parallel with the rise of devices such as smartphones is what he calls an 'invisible illiteracy', directly related to the steep decline in the use of personal computers (2017, p. 38).

This reality seems to contradict both the results obtained by Japan in the usual indications of the population's immersion in the Information Society – 99.1% of the country with access to a bandwidth connection, 94% of users with mobile phones with a 3G connection (Oishi, Takada, Yamakoshi, Nakamura, & Kamino, 2012), and more than 83% of Internet use among the population as a whole (Media Innovation Lab, 2017, p. 6) – and the skills exhibited by the most active and recognizable subculture of contemporary societies: otaku.

A portrait of otaku

The term 'otaku' refers to individuals between 18 and 40 years of age who are immersed in the subculture associated with anime, video games, science fiction, merchandising and other related interests, a variety of what in the English-speaking world has come to be known as nerds. Otaku subculture is paradigmatic of what Jenkins defines as 'collective intelligence' or 'grassroots convergence' (Jenkins, 2006, p. 155). As Azuma – a leading figure in the determination of otaku as inhabitants of modern societies – suggests, one of the most prominent qualities of otaku is their capacity for compiling the building blocks of anime's aesthetic codes – paraphilias, deliberately cheesy elements, the use of irony among fans – in the form of databases, either for individual enjoyment as an evaluative guide of the latest titles or as a repository of expressive and narrative resources that serve as the basis for their own derivative works.

The activities that make up the bulk of what Jenkins, Itō and boyd refer to as 'participatory culture' (2017) have a whole series of protocols defined in the otaku subculture. The successive ethnographic studies of the habits and customs of dōjin scene[2] give special attention to the processes of reappropriation of intellectual properties (Arai & Kinukawa, 2013) and the underlying resilience, inspired by the counterculture ethics of 'Do It Yourself', displayed by fans in the production of derivative works (generally of an erotic or comic nature) of mainstream titles. These actions fall under the current features of the 'free labor' (Jenkins, Ford & Green, 2013, pp. 53–64) happening within online communities of fans. The status of otaku individuals inside their own hierarchies (Hills, 2002) has a lot to do with the display of technological knowledge in the production of audio-visual, non-commercial user-generated contents and their spreading through different platforms.

Anime tourism

Anime tourism is paradigmatic of the transformations in Japanese society in recent years towards a form of tourism focusing on the consumer tastes of minority groups. In demographic terms, it is associated with a market segment that is one of the fastest growing and most attractive to the tourist industry: adults who have completed their formal education but have not yet settled down to have a family. This segment, which has expanded in recent years thanks to the decline in the number of people marrying before the age of 40, is characterized by greater economic and social freedom to choose both travel Companions and destinations. It is especially common among Japanese 20-somethings to travel with friends (Funck, 2013), and in the otaku lifestyle, these friends are often enthusiasts of popular culture niches who have connected online, an increasingly common alternative to face-to-face friendships. It is no accident that pilgrimages to anime locations are presented as an evolution of the *ofukai*, meetings that bring together members of photo or online communities that share specific interests.

This population segment's travel activity usually consists of trips of no more than two or three nights concentrated around public holidays (Funck, 2013) and are commonly typified by *seichijunrei* day trips. This time constraint has been facilitated by the development and extension of the high-speed transport networks that link many of the country's main centres, enabling sporadic detours off the main circuits and the flourishing of niche tourism. As noted above, in contrast with the most popular destinations of their demographic segment (the regions of Hokkaido, Shizuoka, Nagano and Okinawa), otaku tourists favour urban centres whose main economic activity is not tourism.

Anime has been the subject of pilgrimages thanks to both its aesthetic evolution and the structural changes in the online world. On the one hand, there is the noteworthy emergence of the *nichijo-kei* (or daily life anime) since the middle of the last decade. This subgenre is associated with the works of studios such as Kyoto Animation, whose detailed rendering of everyday life spaces and meticulous animation in the adaptations of popular franchises such as *Lucky Star*, *K-On!* or *Uchōten Kazoku*, have become the model for fans and artists. On the other hand, online game groups, blogs and social networking sites help otaku to spread both their emotional responses to and precise information about their favourite entertainment. Thus, anime tourism appears as an extension of the emotional networking that makes full use of individual information technologies, in contrast with conventional travel agency tourism focusing on mainstream destinations. These mediatized destinations are not associated with top-rating TV series, but with cult shows that are highly appreciated by a very small proportion of the masses.

Methodology

The sample used for this study was drawn from the contributions on the *Suzumiya Haruhi* series location featured on two of the most active, longest running, English-language websites dealing with *butaitanbou*: Vito's *Like a Fish in Water* (2018) and Hattsu's *Anime Journeys* (2018). The latter is one of the most accomplished repositories of *butaitanbou*, with regular updates identifying the real spaces where many of the most popular anime are set. The former, run by a regular practitioner since 2012 and a full-fledged member of the local *butaitanbou* community, is one of the most through spaces of reflection on the themes of transit, place, and culture in anime, with weekly updates identifying the real locations where many current anime productions supposedly take place, as well as information on recent pilgrimages. Regarding our case study, *Like a Fish in Water*, has featured 58 posts that mention photographic expeditions to *Suzumiya Haruhi* locations between 2013 and 2018, while *Anime Journeys* features 19 posts over the period 2014–2018.

To determine the media competence implicit in the activity of the *butaitanbou*, the MIL indicators proposed by UNESCO were used. According to its latest report (2013), media literacy can be identified in two main dimensions: environmental factors (that is, media industry, state policies of media education, and the actions of civil society) and individual competence (use, critical thinking, and communication skills in relation to the media). Although designed to frame the priority fields of action for the policy building of educational agencies, its skills matrix is useful for an initial analysis of the level of media literacy in the otaku subculture based on *butaitanbou* (Table 30.1).

Table 30.1 Components of the matrix of individual skills and their descriptors in relation to Media and Information Literacy

Access and retrieval	*Understanding and evaluation*	*Creating and sharing*
Definition and articulation of a need for information	Understanding	Creation of knowledge and creative expression
Search and location	Assessment (also providers)	Ethical effective communication Participating in societal-public activities as active citizen
Access (also to providers)	Evaluation (also providers)	
Retrieval and holding/ storage	Organization	Monitoring influence and use (also providers)

Source: UNESCO, 2013.

Scene hunting *Suzumiya Haruhi*

Nishinomiya and the Suzumiya Haruhi *impact*

The most well-known pilgrimage in the Kansai region is inspired by the settings of the *Suzumiya Haruhi* franchise. The self-consciousness and allusions to popular subculture references that characterize manganime were already present in good measure in the 11 volumes of the light novel saga, which to date has sold over one million copies in Japan and a total of 16.5 million copies in 15 countries. *Suzumiya Haruhi no Yūutsu* (2006, 2009) is the animated television adaptation. The series intertwines mundane situations with genre parody, some of which would only be recognizable to an otaku audience (in particular, the influence of the *nichijo-kei*).

As the centrepiece of the whole *Suzumiya Haruhi* media mix, the animated adaptation of *Suzumiya Haruhi* provides a realistic background to the novel. The locations of the two seasons of the series and its only film adaptation clearly evoke a number of places of the city of Nishinomiya, just 15 kilometres away from Kobe and Osaka. The busiest route for this anime pilgrimage covers the saga's main settings and is often divided into two parts. The first is the route that runs from the Kōyōen train station to the Hyogo Kenritsu Nishinomiyakita high school evokes the setting where the adventures of Haruhi and her Companions usually take place, and the second is the route around the Nishinomiya-Kitaguchi station, which includes the Hirota Shrine, the Café Cream, the pond, and Nagato's apartment. In just an hour and a half, any visitor can see the most emblematic settings for the extracurricular activities of the SOS Brigade, the school club created by Haruhi.

Two takes on the *Suzumiya Haruhi butaitanbou*

Anime Journeys

Hattsu's *Anime Journeys* represents the quintessential *butaitanbou* blog. In each one of the 19 posts related to his own *Haruhi Suzumiya* pilgrimages over the period 2014–2018, Hattsu offers frame versus shot analogies of different corners of Nishinomiya featured in the series. These locations range from the aforementioned ones to many anonymous pedestrian overpasses and train crosses carefully picked for its correspondence with their fictional counterparts. The shot-for-shot recreations are extremely accomplished. Even when the real location from the frame of reference is hard to spot – blurred background, close-up of a character – the shot of the real location achieves its purpose.

The pictures taken by Hattsu meet the criteria of any *butaitanbou* material as they let a comfortable grasp of the spaces photographed due to their proper morphological traits: clear sharpness, accurate scale, and perspective and well-lit scenes even in indoors. In terms of composition, the faithfulness of the recreation can be appreciated in some of the shots which recreate unusual takes (e.g. Dutch angles, off-centred compositions). The frames are accompanied by captions where the title of the episode of the frame (with the precise minute and second within its running time) and a brief plot contextualization can be read. For their part, the captions of the actual shots include useful instructions about addresses, bus or train stops and also feature sometimes links to other of *Suzumiya Haruhi*-related posts.

The straightforward web layout (under Google's publishing service Blogger) features two useful navigation tools: a chronological blog archive and a tag cloud that groups together all the keywords used in the posts. All entries are conveniently labelled to optimize hyper textual navigation so that all posts referring to specific regions or series are encoded and conveniently classified for consultation. Each post concludes with an embedded Google Map that records each one of the steps taken in Hattsu's *butaitanbou* session in the case others were interested in reproducing it. Finally, each post ends with links to the official sites (when able), and a special shout-out to previous explorers who went through the experience before and shared it publicly.

Like a Fish in Water

In contrast with Hattsu's blog, Vito's *Like a Fish in Water* related posts (typically under the header 'Weekly Review of Transit, Place and Culture in Anime') diverge from the regular activities of *butaitanbou* practitioners in order to encompass a wide array of other pressing issues regarding *Suzumiya Haruhi* pilgrimages (2018). Vito's contributions take the shape of a watching list, as they are divided by series and chapters that Vito himself is watching at the moment of the writing. His updates rarely include a side-by-side comparison with shots taken by himself (although he also does that in his less frequent 'Pilgrimage to…' posts). Instead, they mainly feature many stills from different anime series whose real-life locations are included in captions (mostly accompanied by links to their official website or the English version of their respective Wikipedia entries). Interestingly, Vito also includes under the header that signals the title and episode links to the official SVoD platforms which are legally streaming each series. The stills of each episode appear in chronological order and, along with the informative captions, are complemented by a brief contextualization of the story so far in the show.

Two interesting features in terms of MIL competence distinguish *Like a Fish in Water*. The first is a 'Media and General Interest' section in which Vito comments on recent publications on press and television regarding his interests on urbanism, tourism, and popular culture featured. The second is the inclusion of a section entitled 'Fan Pilgrimage Update' in each post. In these parts, Vito not only summarizes recent posts and pilgrimage trends in the *butaitanbou* community happening in Japanese forums (Butaitanbou Archive, 2018) but also includes URL links to the personal blogs or the Twitter accounts of those other practitioners involved. The interface is also accompanied by links to an Really Simple Syndication feed of the blog, and to his personal accounts in Twitter, Vimeo, and Flickr, all of them positioned within a narrow panel along with a 'Recent Posts' section, a spreadable 'Archive', and an email subscription tool (2018).

The overall tone is more self-conscious and informative and, beyond the valuable function as a *butaitanbou* repository, Vito also delves into approaches typical of cultural ethnography in his reports of the annual *butaitanbou* summits since 2014 and in his interviews to key personalities of the *butaitanbou* scene such as Moriwaki (curator of the Kyoto Film Archive) or Okamoto (leading university scholar in content tourism in Japan; Like a Fish in Water, 2015). In such discussions (Like a Fish in Water, 2014) , Vito even explains the technical specs of his kit using jargon about camera and lenses and other questions such as depth-of-field, ISO, chromatic aberration, tone curves, and colour profiles.

Discussion: otaku media literacy through *butaitanbou*

The results reveal that some of the otaku's media literacy skills are at advanced proficiency level on the UNESCO scale (2013, p. 60), demonstrating a very good level of knowledge and skills acquired in the practice and learning of MIL. This advanced level is defined by abilities displayed by *butaitanbou* practitioners:

- 'Formulate [...] information and media (content) needs into concrete strategies and plans to search for and access information from diverse sources using relevant and where necessary diverse tools in a systematic, explicit and efficient manner, and retrieve existing information for further utilization.' Both sources resort to Japanese online practitioners and digital tools of communication (Google Maps, Blogger, social networking communities) in order to fetch and store *butaitanbou* information.
- '[I]nterpret, compare, critically evaluate, authenticate and hold synthesized information and media content, appreciating work of author(s), and media and information providers within the context of sustainable development of society, organization or community.' Shared by both sites upon appreciating the labour of *butaitanbou* pioneers, but especially notable in the case the argumentative, rich-in-sources (interviews, media links) posts featured at *Like a Fish in Water*.
- 'Combine information and media content for creation and production of new knowledge considering socio-cultural aspects of the target audiences and then communicate and distribute in various appropriate formats and tools for multiple applications in a participatory, legal, ethical and efficient manner, as well as monitor influence and impact made.' The creation of new knowledge is evident in both cases, as is the civic responsibility and ethical concerns deduced from the respect to personal privacy (erasing licence plate numbers and the faces of passers-by) and the proper quotations to third party information and contents.

As if they were script supervisors for a motion picture, anime fans endeavour to seek an artificial continuity between the actual location and its animated version. The preparation of comparative photomontages[3] between images of the original locations and the frames of the animated version and the preparation of maps of the area for and by fans are activities that allow us to glimpse two typical features of the otaku (sub)culture identified by Azuma: the interest in seeking out information on the context of the fictional worlds and the pleasure of processing that information intellectually, usually recoded in the form of databases for further use.

Beyond the egotism evident in the practice of tourist photography in the age of selfie and video-taking (Dinhopl, & Gretzel, 2016a, 2016b) and social networks (Marder, Archer-Brown, Colliander, & Lambert, 2018), the graphic documents shared by *butaitanbou* practitioners are aimed at exchanging knowledge. This type of activity is also redefining how the variables on which the authenticity and identity value of these tourist destinations are formed. Perhaps the most distinctive feature of the *butaitanbou* boom is the fact that it is exclusively fandom driven. Unlike spaces controlled by owners and producers, otaku tourism constitutes an element in the panoply of paratextual activities developed by anime fans themselves. There is also no need for corporate tourist events (Getz, 2008), because it is often a self-sufficient community of fans operating autonomously not only in the organization of commemorative festivals but also in the creation of maps and guides recommending the best routes and times to visit the locations.

Conclusions

In terms of MIL, the competencies demonstrated in the practice of *butaitanbou* go far beyond those developed in the regulated curricula of educational institutions (Pereira, Pinto, & Pereira, 2012). This case study underscores the fact that MIL skills are often developed in contexts outside the education and family environments that are the preferred focus of activity for educommunication experts. For the cultural anthropologist Condry (2013), the activities of the otaku, reflecting their love for anime, is proof of its 'soul' as a cultural product, and they are also indicative of the power of extracurricular motivations for the development of digital literacy.

As we have seen, otaku forms of expression like *butaitanbou* reflect a profile of active users, expert locators of verified information, and distributors of cores of new knowledge on anime for other fans. These qualities are supported by a creative capacity based on informative erudition and the mastery of the techniques of digital audio-visual formats. As we have seen, the panoply of activities of this community of fans transcends the notion of the virtual domain as the only symbolic space to exercise their limited power, bringing in tourism as an essential activity for a community of fans dedicated to a representative cultural manifestation of 21st-century digital culture, and a great example of mediatization.

Acknowledgements

This work was supported by the Universitat Jaume I for the research project *Análisis de identidades en la era de la posverdad. Generación de contenidos audiovisuales para una Educomunicación crítica* under Grant code 18I390.01/1, *El diseño narratológico en videojuegos: una propuesta de estructuras, estilos y elementos de creación narrative de influencia postclásica* under Grant code 18I369.01/1; and the Valencian Community and the European Social Fund through the post-doctoral scholarship program, under grant number APOSTD/2019/068.

Notes

1 According to the OECD (2017), the percentage of Japan's adult population with higher education is the second highest in the world (50.5%).

2 Groups of fans who self-published manga (dōjinshi), video games (dōjin soft), and music (dōjin music), among others cultural products made for other fans.

3 Similar cases were studied in contexts such as user-generated journalism (Day Good, 2013), Google Earth forums (Robinson, 2014), and other social media platforms (García-Palomares, Gutiérrez, & Mínguez, 2015); and in other regions such as Hong Kong (Lo McKercher, Lo, Cheung, & Law, 2011), Peru (Stepchenkova & Zhan, 2013), or Catalonia (Marine-Roig, Martin-Fuentes & Daries-Ramon, 2017).

References

Anime Journeys. (2018). Haruhi Suzumiya. Retrieved May 28, 2019, from: http://mikehattsu.blog-spot.com/search/label/haruhi%20suzumiya.

Arai, Y., & Kinukawa, S. (2013). Copyright infringement as user innovation. *Journal of Cultural Economics, 38*(2), 131–144.

Arke, E. T. (2012). *Media literacy: History, progress, and future hopes.* New York, NY: Oxford University Press.

Azuma, H. (2009). *Otaku: Japan's database animals.* Minneapolis: University of Minnesota Press.

Butaitanbou Archive. (2018). 舞台探訪アーカイブとは. Retrieved May 28, 2019, from: http://leg work.g.hatena.ne.jp/.

Condry, I. (2013). *The Soul of anime: collaborative creativity and Japan's media success story.* Durham, NC: Duke University Press.

Day Good, K. (2013). Why we travel: Picturing global mobility in user-generated travel journalism. *Media, Culture & Society, 35*(3), 295–313.

Denison, R. (2010). Anime tourism: Discursive construction and reception of the Studio Ghibli Art Museum. *Japan Forum, 22*(3–4), 545–563.

Dinhopl, A., & Gretzel, U. (2016a). Selfie-taking as touristic looking. *Annals of Tourism Research, 57*, 126–139.

Dinhopl, A., & Gretzel, U. (2016b). Conceptualizing tourist videography. *Information, Technology & Tourism, 15*, 395–410.

Funck, C. (2013). Domestic tourism and its social background. In C. Funck & M. Cooper (Eds.), *Japanese tourism. Spaces, places and structures* (pp. 62–95). Oxford, UK: Berghahn.

García-Palomares, J. C., Gutiérrez, J., & Mínguez, C. (2015). Identification of tourist hot spots based on social networks: A comparative analysis of European metropolises using photo-sharing services and GIS. *Applied Geography, 63*, 408–417.

Getz, D. (2008). Event tourism: Definition, evolution, and research. *Tourism Management, 29*(3), 403–428.

Hills, M. (2002). *Fan cultures.* London, UK: Routledge.

Jenkins, H. (2006). *Fans, bloggers and gamers: Exploring participatory culture.* New York: NYU Press.

Jenkins, H., Ford, S., & Green, J. (2013). *Spreadable media: Creating value and meaning in a networked culture.* New York: NYU Press.

Jenkins, H., Itō, M., & Boyd, D. (2017). *Participatory culture in a networked era: A conversation on youth, learning, commerce, and politics.* Cambridge, UK: Polity.

JNTO. (2014). Japan Anime Map. Retrieved May 28, 2019, from: http://www.jnto.go.jp/eng/anime-map/ANIMEmap_front.pdf.

JNTO. (2018). Japan tourism statistics – Trends in visitor arrivals to Japan. Retrieved May 28, 2019, from: https://statistics.jnto.go.jp/en/graph/#graph--inbound--travelers--transition.

Kodaira, S. I., & Watanabe, S. (2013). *The diversifying media environment of Japanese classrooms and educational content.* Tokyo, Japan: NHK Broadcasting Culture Research Institute.

Kress, G. (2003). *Literacy in the new media age.* London, UK: Routledge.

Like a Fish in Water. (2014). Animation, urbanism and *Tamako Market*: A discussion with Moriwaki Kiyotaka. Retrieved May 28, 2019, from: http://likeafishinwater.com/2014/11/27/animation-urbanism-and-tamako-market-a-discussion-with-moriwaki-kiyotaka/

Like a Fish in Water. (2015). Contents tourism discussion with Okamoto Takeshi. Retrieved May 28, 2019, from: http://likeafishinwater.com/2015/11/14/contents-tourism-discussion-with-okamoto-takeshi/

Like a Fish in Water. (2018). Haruhi Suzumiya. Retrieved May 28, 2019, from: http://likeafishinwater.com/?s=suzumiya+haruhi

Liversidge, G. (2005). Media literacy: An unknown concept in Japan? *Otsuma Women's University, annual report. Humanities and Social Sciences, 37*, 374–362.

Lo, I. S., McKercher, B., Lo, A., Cheung, C., & Law, R. (2011). Tourism and online photography. *Tourist Management, 32*, 725–731.

Marder, B., Archer-Brown, C., Colliander, J., & Lambert, A. (2018). Vacation posts on Facebook: A model for incidental vicarious travel consumption. *Journal of Travel Research, 58*(6), 1–20.

Marine-Roig, E., Martin-Fuentes, E., & Daries-Ramon, N. (2017). User-generated social media events in tourism. *Sustainability, 9*, 1–23.

Media Innovation Lab. (2017). *Information media trends in Japan.* Tokyo, Japan: Dentsu.

Mizukoshi, S. (2017). Media literacy and digital storytelling in contemporary Japan. Retrieved May 28, 2019, from: https://iias.asia/the-newsletter/article/media-literacy-digital-storytelling-contemporary-japan.

OECD. (2017). *Education at a glance 2017.* Paris, France: OECD Publishing.

Oishi, Y., Takada, Y., Yamakoshi, S., Nakamura, Y., & Kamino, A. (2012). *Mapping digital media.* London, UK: Open Society Foundations.

Okamoto, T. (2015). Otaku tourism and the anime pilgrimage phenomenon in Japan. *Japan Forum, 27*(1), 12–36.

Pereira, S., Pinto, M., & Pereira, L. (2012). Resources for media literacy: Mediating the research on children and media. *Comunicar, 20*(39), 91–99.

Robinson, P. (2014). Emediating the tourist gaze: Memory, emotion and choreography of the digital photograph. *Information, Technology & Tourism, 14*, 177–196.

Seaton, P., & Yamamura, T. (2015). Japanese popular culture and contents tourism – Introduction. *Japan Forum, 27*(1), 1–11.

Stepchenkova, S., & Zhan, F. (2013). Visual destination images of Peru: Comparative content analysis of DMO and user-generated photography. *Tourist Management, 36*, 590–601.

Takabayashi, T. (2015). Media use as an element of self-directed learning: The learning strategies and media-related behaviors of Japanese University Students. *IJEMT International Journal for Educational Media and Technology, 99*(1), 80–82.

Tung, V. W. S., Lee, S., & Hudson, S. (2017). The potential of anime for destination marketing: Fantasies, otaku, and the kidult segment. *Current Issues in Tourism, 22*(12), 1423–1436.

UNESCO. Communication and Information Sector, & UNESCO Institute for Statistics. (2013). *Global media and information literacy (MIL): Assessment framework: country readiness and competencies.* Paris, France: United Nations Educational.

Yamamura, T. (2018). Pop culture contents and historical heritage: The case of heritage revitalization through "contents tourism" in Shiroishi city. *Contemporary Japan, 30*(2), 144–163.

31

BEHIND-THE-(MUSEUM)SCENES

Fan-curated exhibitions as tourist attractions

Philipp Dominik Keidl

Introduction

Museum exhibitions have been a popular part of the *Star Wars* franchise since the 1990s. Primarily drawing on Lucasfilm Archives' collections, these exhibitions are generally produced in collaboration with established heritage and educational institutions, and tour international museums and science centres (Bartolomé Herrera & Keidl, 2017). However, in addition to Lucasfilm-sponsored exhibitions, fans have founded their own museums. In such 'DIY museums' (Baker, 2015), fans have the sole curatorial agency to determine the exhibitions' content and form. Although fan curators do not necessarily have professional training, they perform similar work to that of trained curators, as they are responsible for the care, development, study, enhancement, and management of collections as well as the curation of exhibitions and publications (Ruge, 2007). Fans' curatorial agency is what distinguishes fan-run from professional museums that do not necessarily share fans' aims and objectives (Brandellero, van der Hoeven, & Janssen, 2015). While Lucasfilm exhibitions emphasize the theatrical releases of the franchise and present George Lucas as the singular auteur of the *Star Wars* universe (Bartolomé Herrera & Keidl, 2017), fan-run museums put more focus on merchandise and ephemera, fans' relationship to the franchise, and also occasionally address other popular media texts beyond *Star Wars*.

Fan museums' curatorial imprint differs from exhibition to exhibition. Consider the following three US-American examples. In Arkansas, The Galaxy Connection is as much about *Star Wars* as it is about the curator's own religious beliefs. As the project's website explains, the exhibition features 'the owner's [...] personal collection of toys, props and life size characters from *Star Wars* [...] while tying in his own personal story' (The Galaxy Connection, n.d.). In New York, the curator of Brett's Toy Museum 'wanted to display [his collection as an art object] in a way that symbolized the creative (and marketing) genius of George Lucas' (About Brett, n.d.).[1] The motivation for the Star Toys Museum, on the other hand, was 'to get recognition and draw the attention of other collectors [since] no [collector] ever invited people into their home to see the stuff, which was one of the things [Atkinson] thought was important' (Jensen, 1999, para. 11). Despite these different curatorial approaches, however, these museums have in common that they have become destinations for tourists.

Drawing on fan studies, this chapter argues that fan-curated museums represent a new form of museum and tourist attraction—even if they cannot compete in size, scale, budget, and visitor numbers with Lucasfilm or other industry-run exhibitions and sites. Fan-curated museums offer a distinct experience from filming locations, studio tours, themed environments, conventions, and professionally produced museum exhibitions, all of which have become a central part of contemporary tourist industries (Brooker, 2007; Couldry, 2007; Larsen, 2015; Lee, 2012; Norris, 2013; Porter, 1999; Roesch, 2009). Crucially, instead of primarily drawing fans into immersive and affective encounters with their objects of fandom, the experience of fan-produced works, and therefore fandom itself, becomes the main interest and attraction. As such, this chapter will contribute to research on fan and media tourism, albeit from a different perspective. Instead of theorizing fans as consumers of media tourism, it examines how fans actively produce and run their own attractions, targeting members of their community as well as tourists in general. As this chapter demonstrates, their primary appeal is the encounter with, and experience of fandom and fan productivity itself. Direct connections to a media text, a shooting location, or production materials are only secondary. Consequently, fan-curated museums represent a shift from film-induced tourism (Beeton, 2005) to fan-induced tourism. Whereas film-induced tourism describes the motivation to travel to a certain destination inspired by films, fan-induced tourism emphasizes the experiences and encounter with fans and their works themselves.

The primary case study will be the fan-run museum Stars of the Galaxy (hereafter SOTG) in the city of Mönchengladbach in Germany. SOTG presents a vast array of merchandise predominantly from the *Star Wars* franchise and a few original props from other science-fiction productions (Digital Tour Guide, n.d.). The project started in 2005, when five *Star Wars* fans transformed a former cinema into a public exhibition space for their collections. Opened under the name StarconstruX-Hauptquartier, the exhibition showcased approximately 30 dioramas with more than 300 four-inches action figures, 70 life-size movie figures and original and replica movie props (StarconstruX Headquarters, n.d.). In 2011, the exhibition moved into a vacated former indoor swimming pool facility. By transforming the former changing rooms and showers into approximately 400 square metres of exhibition space, the renamed Filmfiguren Ausstellung (Film Figures Exhibition) provided more exhibition space for further displays and enough room to organize workshops, host special events, and establish a museum shop. Since then, the curators and volunteers have further expanded the museum. Most notably, the actual pool now functions as a multi-level exhibition space. This spatial increase to a total of 1,000 square metres followed a thematic expansion, and yet another renaming to Stars of the Galaxy in 2017. Although the vast majority of the museum remains dedicated to merchandise, SOTG now also presents original costumes from productions other than *Star Wars* in a section named 'Hall of Fame' (Digital Tour Guide, n.d.).

The analysis of SOTG is primarily based on the museum's website, German news coverage, and visitor reviews on social media such as YouTube and Facebook. Moreover, I conducted onsite research in 2015, when the exhibition was still called Filmfiguren Ausstellung. The chapter focuses on media coverage for two reasons. First, my onsite research took place before SOTG's expansion, and my knowledge of the current form and focus of the exhibition is built on the museum's website and press and fan coverage. Second, as I noticed early on in my research, SOTG, like other fan-run museums, has been very careful in the presentation of their exhibition in order to circumvent claims of copyright infringement. This is evident in the naming of the exhibitions, which avoid any direct reference to *Star Wars* or any of their characters, arguably because of Lucasfilm's history of restrictive policies and legal actions regarding fan productions (Brooker, 2002). Indeed, most fan-curated *Star Wars* museums refer in their title to abstract terms such as 'galaxy' or focus on the toys on display to avoid copyright infringement. In the case of SOTG, this type of caution is even evident in interviews, when curator Thomas Manglitz carefully reminds interviewers that they are not at a *Star Wars* museum but an exhibition of merchandise that also includes *Star Wars* toys (Die Pierre M. Krause Show, 2017).

Manglitz generally avoids questions about the museum's expenses or revenues. He claims that the museum does not count visitor numbers and hesitates to give information about the costs that go into running SOTG, which therefore limits a quantitative assessment of the museum's success among fans and tourists. One of his strategies to dodge budget-related questions is to do a nondescript 'currency transfer' into 'monthly pocket money', leaving it to the viewers to calculate how many months they would have to save to afford the acquisition (Vorhees82, 2015). The information he provides about partnerships remains vague. In interviews, he mentions that there had been an initial contact with Disney after it purchased the rights to the franchise. He also indicates that Hasbro provides toys for the large dioramas, and that the museum, in turn, provides the toy manufacturer with dioramas for fairs and other exhibition events. But the details behind these contacts are not further explained (Thiele, 2014; Vorhees82, 2015). Consequently, this chapter builds its argument with the information the curators feel comfortable sharing with a wider public through the media, as well as visitor reactions that are available online.

Fan-induced tourism

Fan scholars have addressed the intersection of media and tourism from various perspectives. Hills proposes the idea of 'cult geographies' to discuss fans' emotional attachment to sites associated with popular media texts. *'Cult geographies'*, he explains, 'are diegetic and pro-filmic spaces (and "real" spaces associated with cult icons) which cult fans take as the basis for material, touristic practices' (2002, p. 110). Reijnders highlights the importance of imagination in his conceptualization of media tourism sites as 'places of the imagination' that function 'for certain groups within society [...] as material-symbolic references to a common imaginary world' (2011, p. 14). Other fan scholars theorize such travels as 'fan pilgrimages', describing media-themed travel destinations as liminal spaces outside everyday life during which fans similarly negotiate and play with the dialectic and discrepancy of real-world locations and their (fictional) representation (Brooker, 2005; Larsen, 2015; Zubernis & Larsen, 2018). Fan practices associated with tourism include the discovery of filming locations and the sharing of knowledge thereof, re-enactments of scenes, vernacular memory practices such as leaving one's name somewhere at the destination, as well as interaction with others visitors (Alderman, 2002; Kim, 2010; Waysdorf & Reijnders, 2017; Williams, 2017). Such practices often take centre stage in fans' negotiation of questions regarding authenticity, commercialism, identity, and the relationships among fans, anti-fans, and general tourist of those sites associated with an object of fandom (Williams, 2017). Moreover, fan communities and practices have formed around places such as themed environments like Disney parks or the Hard Rock Café franchise (Geraghty, 2014; Williams, 2018). Importantly, fans often travel to such destinations several times. Waysdorf and Reijnders conceptualise such repeat visits as 'fan homecoming', which describes a 'return visit to a familiar fandom-related place' that shapes individual fan identities and fan communities in general (2019, p. 52).

Fan tourism falls into the broader category of film-induced tourism, a concept developed by tourism scholar Beeton (2005) to examine destinations that attract a broader mix of travellers to sites connected to particular media works. Film-induced tourism refers to activities such as individually planned excursions to original filming locations, participation in commercial tours of centres built after initial filming took place, visitation to sites where a film is set but was not actually filmed, outings to places such as film studio tours and theme parks, as well as armchair travelling by watching travel programs on television (Beeton, 2005).[2] Museums and museum-style environments are also part of film-induced tourism. Most prominently, the wide range of dedicated film and media museums around the world attract visitors through their diverse range of exhibitions, central location, and architecture. For example, the Deutsche Kinemathek—Museum für Film und Fernsehen, the Australian Centre for the Moving Image, and the EYE Filmmuseum have been part of urban renewal and tourism projects with their locations at Potsdamer Platz in Berlin, Federation Square in Melbourne, and Amsterdam-Noord, respectively. Moreover, museum-style displays have also been integrated into studio tours (e.g. Warner Bros. Studio Tour) or themed environments such as Planet Hollywood (Trope, 2011).

SOTG falls under the broader category of film-induced tourism, due to its focus on *Star Wars* and other science-fiction productions. However, fan-curated museums represent a form of fan-induced tourism, which can be understood as a subcategory of media tourism. Fan-induced tourism offers a way of thinking about tourism motivated by the promise to see fan productions and hear about film and media history, as well as the history of (their) fandom, from the point of view of fan curators. Of course, interaction with other fans has always been part of the pleasure of fan tourism (Waysdorf & Reijnders, 2019; Zubernis & Larsen, 2018), and fans themselves have opened their own tours and stores nearby cult geographies (Erzen, 2011). Yet, in these instances, fan interaction occurs at sites outside of fans' everyday life and practices, or at ephemeral places such as conventions, where the making of fan productions is less the focus of attention than the finished works fans present on such occasions. Attractions such as original filming locations, themed environments or conventions offer individual or communal engagement with, and often immersion into, a fictional diegesis. In turn, fan-curated museums offer a more direct and physical access to the everyday lives of fans and their practices. Although some fan-run museums are built on sites were media texts were filmed (e.g. Mad Max 2 Museum, Silverton, Australia), or at places with direct connections to celebrities (e.g. Laurel and Hardy Museum, Ulverston, UK), many fan-run museums like SOTG have not such a direct connection. For instance, none of the *Star Wars* films or series was produced in Mönchengladbach, no famous talent involved in the franchise has ties to the city, and it has never hosted important official events such as national premieres other than fan-made films (Jedrychowski, 2017). Further, no merchandise is produced in or nearby the city. The curatorial team established Mönchengladbach as a landmark for *Star Wars* fans from scratch, and without even presenting noteworthy original production materials from the franchise's films or series.[3]

Hence, the experiences that SOTG offers are vastly different from Lucasfilm-produced exhibitions. Lucasfilm exhibitions draw in visitors with original production materials but lack any display of merchandise that would redirect attention away from the films and George Lucas's role as an auteur. In turn, SOTG primarily displays mass-produced merchandise, with only few original production materials, most of which are not from the *Star Wars* franchise. Some of the most valuable *Star Wars* items in SOTG are action figure prototypes sponsored by Hasbro, and limited-edition replicas of lightsabers. These differences are best exemplified by a set of dioramas on display in SOTG that depict scenes from the pre-production process of the first trilogy. Since SOTG cannot offer original production materials, and copyright restrictions and licensing fees prohibit the use of behind-the-scenes footage to document how the films were made, the fan curators make do with these limitations by offering highly detailed miniature depictions of post-production moments. For example, one diorama shows special effects artists working on the battle scene on Hoth from *The Empire Strikes Back* (1980) and includes miniature reproductions of models, concept art, and other work materials. If Lucasfilm exhibitions marvel fans with auratic originals, SOTG convinces them with the industrious originality of its curators and a fan perspective on the franchise and its fandom.

The interest in fan-curated museums as a fan project and demonstrations of fandom becomes evident when we take a look at the media coverage and visitor reviews of SOTG. They reveal that the prospect of learning and engaging with the *Star Wars* franchise is not the sole motivation for a visit. Rather, the fact that the museum is run by fans without any formal curatorial training and no evident or apparent economic interest seems to be one of the biggest selling points for SOTG. In contrast, media coverage and visitor reviews of the official Lucasfilm exhibition Star Wars: Identities, which was on display in Germany in Cologne and Munich in 2015 and 2016, do not address how the exhibition was made and only occasionally mention the name of the curators and their intentions behind it. Instead, the focus is on the opportunity to see original production materials, gain background information on the making of, and the chance to create your own *Star Wars* character (Die FILM Seite, 2015; Nerdkultur, 2016; Westphal, 2015). In other words, Star Wars: Identities is presented as an exhibition for fans but not as a site founded and run by fans for their peers such as SOTG.

Stars of the Galaxy

Several fans are involved in running SOTG, but Manglitz usually appears in interviews as the spokesperson for the museum and has therefore become SOTG's public face. Manglitz's own fan biography has become central in the reporting on the project, with one recurring question being when and how he became a *Star Wars* fan, and what motivated him and his colleagues to open the museum. Indeed, the work involved in running SOTG and the quality of the displays are often foregrounded in visitor reviews and exhibition reports. Consider the nearly hour-long video exhibition tour and interview with Manglitz, by vlogger Vorhees82, which repeatedly accentuates how much love and detail went into making the exhibition. When he talks about the Mos Eisley diorama, which includes approximately 300 figures, he explains how 'every little' detail is acknowledged and how much 'heart and soul' went into the space, stressing that one can see the extent of fandom that motivates such a project. Noticeably, the majority of questions for Manglitz are not about his detailed knowledge of the franchise, but rather concern the background of the museum and its everyday practices, such as visitor numbers, the number of figures on display or the decision to use a swimming pool as an exhibition venue (Vorhees82, 2015). Thus, the video indicates how the fan curators' labour shapes the experience of the exhibition and demonstrates the important role in the presentation of SOTG as an attraction for visitors.

Space and time are themes that are particularly relevant in the framing of SOTG as a fan-run institution. In the case of space, Manglitz answers the recurring question of what motivated him and his peers to begin the museum project by explaining how he and his peers suffered from a problem that he considers to be common among fans: too little space at home (FANwerk, 2016; RTL West 2015). Indeed, Woo argues that fans constantly face 'constraints of materiality', and must therefore '(inter alia) research, curate, organize, clean, repair, move, and dispose of objects' to manage the 'domestic object-world' of their fandom (2014, paras. 1.5 & 1.4). SOTG is the result of these restrictions, as well as the transition of these practices from the private into the public sphere. In terms of time, reports recognize the time that goes into the production of detailed dioramas that range in size but can take up several square metres in the exhibition. For instance, the local television station CityVision focuses in their coverage on one of the museum's curators, showing the time-intensive and detail-oriented work of preparing and placing the figures in the dioramas (Schommer, 2014). Similarly, reports addressing a self-made Star Destroyer LEGO model and life-size set replica of inside the spaceship, both donated to the museums by fans who did not have enough space in their homes anymore, foreground how, where, and for how long fans have worked on building their versions of the spaceship rather than an explanation of its meaning in the films (GROBI.TV, 2015). By providing insights into fandom and fan practices, SOTG establishes an aura of authenticity, both for the objects on display and the knowledge and skills of the fans who made them.

The themes of time and space are also apparent in questions relating to SOTG's exhibition venue. Almost all media coverage of SOTG refers to the museum's location. In their tours, Manglitz and his peers repeatedly explain which parts of the former indoor swimming pool were transformed into exhibition space (GROBI.TV, 2014; Schommer, 2014). Moreover, Manglitz has shown in reports the gradual transformation of the swimming pool into a multi-level exhibition space (GROBI.TV, 2015; Vorhees82, 2015). Such tours were not exclusively reserved for reporters, as fan reviews indicate that visitors could enjoy a tour of the unfinished space while it was still in the making (Volkerc, 2013). Therefore, the experience of the repurposed space becomes as much an attraction for visitors as the actual objects on display.

This accent on space and time represents an interesting shift of emphasis from the making of a film to behind-the-scene insights into the making of a museum. If official *Star Wars* exhibitions provide a polished look at the production histories of the franchise's storyworld, SOTG distinguishes itself by openly showing what it takes to run a museum. The unusual re-making of an indoor swimming pool into a museum, combined with the industriousness productivity of fans displayed within its walls, frames the museum as an ongoing 'work-in-progress'. When I visited the exhibition in 2015, its 'work-in-progress' character could be felt in the construction noise that echoed from the downstairs pool into the finished exhibition space upstairs. It was audible proof that the fans were not only running the museum from behind the counter but were still actively working on finishing it in their free time. As such, the visibility and audibility of the renovation of the building represented a unique aspect and experience of the museum space.

This 'work-in-progress'-character is also referenced by Manglitz (GROBI.TV, 2015) and by reviewers when they describe their engagement with the richness of the objects on display. As one visitor remarks, one has to go several times, as there is always something new to discover (Rip, 2018). Such comments indicate that the exhibition itself is not a finished and polished product but like other fans productions, such as fan fiction, a continuing and collaborative process. Before displays enter the museum, for example, the curators post pictures of smaller dioramas in progress on their Facebook page (Stars of the Galaxy—Photos, n.d.). Further, when discussing the transformation of the swimming pool into an exhibition space, Manglitz explains how mass-produced figures were customized and landscapes created with the help of such mundane objects as paper plates (Vorhees82, 2015). Thus, a visit to the exhibition includes the promise of finding details from the films depicted in the dioramas but also the enjoyment of figuring out how and with what materials the displays were built.

SOTG's authenticity as a fan project is also evident in accounts of the tours given by the curatorial team, which highlight the knowledge and enthusiasm of the curatorial team. Such encounters contribute to SOTG's self-understanding and reputation as a project 'by fans for fans' (Jovanovic, 2017; RP Online, 2017). However, the success of the museum is described by other reviewers as dependent on fan support beyond the core curatorial team. One reviewer urges readers to visit the exhibition, arguing that fans are actually contributing to the survival of the exhibition (Dyckers, 2014) Without any direct funding from the state or other heritage bodies, the survival of the museum depends not only on the motivation of the curators but also on the support of the fan community.

In sum, such examples demonstrate how fan-curated exhibitions and their visitors foreground the labour that goes into the making and running of these museums. Similarly, visitors put more emphasis in their reviews on fan labour than the media texts and merchandise on display, thereby framing it as the main attraction of the museums. This, then, is in stark contrast to reports and reviews of commercial or industry-produced exhibitions that focus on the display of original costumes and the kinds of activities these exhibitions offer for fans. Hence, while official *Star Wars* exhibitions fall under the category of film-induced tourism—visitors attend the exhibition to engage with the films and their production—SOTG is an example of fan-induced tourism: visits to attractions that are motivated by engagement with other fans and their work in specific local contexts.

Conclusions

SOTG is only one of many museums founded and run by fans that exist all over the world. Besides museums dedicated to the *Star Wars* franchise, similar projects have been dedicated to media texts and characters such as Superman (USA), Batman (Thailand) and Doctor Who (England), Other focus on specific objects, such as media-themed merchandise, such as the Toy and Action Figure Museum (USA). In some cases, exhibitions develop from fan projects into professionally curated institutions, such as the Ava Gardner Museum (USA), which has been managed by the Ava Gardner Museum Foundation after its initiators passed away (True Fans and Collectors for Life, n.d.). Despite differences in their curatorial foci, these projects have in common that they are shaped by fans' curatorial agency. Fan-run museums present exhibition narratives without sacrificing fandom as the core of their curatorial imprint. Unlike professional museums, topics such as collecting, art and religion emphasize rather than overshadow fan curators' interpretations of their objects, as they are mostly free from the need to justify their collection and programming to funding agencies. Thus, the fact that fans are running a museum is one reason why people are visiting these places. Fan-run museums are clearly built around the curators' affective relationship to their exhibitions and objects of fandom, as well as the experience of fan practices (e.g. diorama building) and skills in museum management that require knowledge of cultural production, exhibition, and public relations.

The promotional potential of fan-run museums for the media industries becomes evident in Lucasfilm's endorsement of Rancho Obi-Wan in California, which houses Stephen J. Sansweet's *Star Wars* collection. The museum follows the example of Sansweet's many publications and frames merchandise an integral factor in making *Star Wars* a pop cultural phenomenon (Sansweet, 1992, 2009). However, although Rancho-Obi-Wan is advertised as 'the world's largest privately-owned Star Wars collection' (History, n.d.), it is somehow a hybrid between fan-run and industry museum. As a former Lucasfilm-employee, Sansweet has the endorsement of the company and the project's board of directors includes in addition to prominent *Star Wars* fans (and occasional Lucasfilm-collaborators) a current Lucasfilm employee responsible for fan relations (Board of Directors, n.d.). Hence, the museum can make explicit reference to one of the franchise's most well-known characters. Its focus on merchandise and memorabilia complements official Lucasfilm-produced exhibitions that neglect mass-produced objects as well as fan works in general. More importantly, however, Rancho Obi-Wan is a valuable addition to Lucasfilm as it commemorates and advertises the franchise in line with the company's policies without compromising the notion that it is run by fans for fans and therefore a celebration of fandom in itself.

Fan studies and the idea of fan-induced tourism help to further theorize fans as makers of tourism and makers of tourist destinations rather than to think of them as sole participants in touristic activities and cultural producers as tourists, respectively. The phenomenon of fan-run museums in particular redirects questions of fan productivity during travels to fans' making of tourist destinations as one, if not the main, attraction. They indicate that fans and general tourists may desire to travel to sites with connections to their object of fandom, but locations that promise encounters of other fans and their works in their local environments also mobilize them. Fan-run museums are not only a sign of the popularity of a media text or star but also reflect the increasing presence and integration of fandom into mainstream culture and the potential for media industries to benefit from fan curators' free labour in curating and therefore automatically advertising media texts and personalities. Thus, fan studies and fan-induced tourism are also crucial for understanding a shift from an interest in media production to curiosity in local media reception practices, as well as the growing presence—and potential mergers—of amateur and DIY initiatives alongside industry and official heritage bodies in contemporary museum and tourism culture.

Acknowledgements

The project was funded by the Faculty of Fine Arts and the School of Graduate Studies at Concordia University, Montreal. The article preparation was supported by the DFG-Graduiertenkolleg 2279 "Configurations of Film" and the European Regional Development Fund/Mobilitas Pluss programme [MOBJD349].

Notes

1 Brett's Toy Museum has closed its website while this essay was finalized.
2 Aden refers to armchair travelling as "symbolic pilgrimage" to describe imaginative journeys of fans through the immersion into a narrative without actual physical travels (Aden, 1999).
3 The only original objects from any *Star Wars* film are parts of the set design for the desert planet Tatooine from George Lucas's *A New Hope* from 1977 (Digital Tour Guide, n.d.).

References

About Brett (n.d.). Retrieved May 28, 2018, from: http://brettstoymuseum.com/about_brett.
Aden, Roger C. (1999). *Popular stories and promised lands: Fan cultures and symbolic pilgrimages.* Tuscaloosa, AL: University of Alabama Press.
Alderman, D. H. (2002). Writing on the Graceland wall: On the importance of authorship in pilgrimage landscapes. *Tourism Recreation Research, 27*(2), 27–33.
Baker, S. (2015). Identifying do-it-yourself places of popular music preservation. In S. Baker (Ed.), *Preserving popular music heritage: Do-it-yourself, do-it-together* (pp. 1–16). New York, NY: Routledge.
Bartolomé Herrera, B., & Keidl, P. D. (2017). How Star Wars became museological: Transmedia storytelling in the exhibition space. In S. Guynes & D. Hassler-Forest (Eds.), *Star Wars and the history of transmedia storytelling* (pp. 155–168). Amsterdam, The Netherlands: Amsterdam University Press.
Beeton, S. (2005). *Film-induced tourism.* Clevedon, UK: Channel View Publications.
Board of Directors (n.d.). Retrieved June 20, 2019, from: https://ranchoobiwan.org/board/ (Accessed on April 30, 2020).

Brandellero, A, van der Hoeven, A., & Janssen S. (2015). Valuing popular music heritage: Exploring amateur and fan-based preservation practices in museums and archives in the Netherlands. In S. Baker (Ed.), *Preserving popular music heritage: Do-it-yourself, do-it-together* (pp. 31–45). New York, NY: Routledge.

Brooker, W. (2002). *Using the force: Creativity, community and Star Wars fans.* London, UK and New York, NY: Continuum.

Brooker, W. (2005). The Blade Runner experience: Pilgrimage and liminal space. In W. Brooker (Ed.), *The Blade Runner experience: The legacy of a science fiction classic* (pp. 11–30). London, UK: Wallflower Press.

Brooker, W. (2007). A sort of homecoming: Fan viewing and symbolic pilgrimage. In J. Gray, C. Sandvoss, & C. L. Harrington (Eds.), *Fandom: Identities and communities in a mediated world* (pp. 149–164). New York: New York University Press.

Couldry, N. (2007). On the set of The Sopranos: 'Inside' a fan's construction of nearness. In J. Gray, C. Sandvoss, & C. L. Harrington (Eds.), *Fandom: Identities and communities in a mediated world* (pp. 139–148). New York: New York University Press.

Die FILM Seite (2015, May 28). Star Wars identities Odysseum Köln [video file]. Retrieved from: https://www.youtube.com/watch?v=HMHz3GrYUMc.

Die Pierre M. Krause Show (2017, November 28). Die Pierre M. Krause Show | Folge 526 | Mirja Boes [Video file]. Retrieved from: https://www.youtube.com/watch?v=Pp6oF3Z9on0&feature=youtu.be.

Digital Tour Guide (n.d.). Retrieved June 20, 2019, from: https://spark.adobe.com/page/ZyTHVTqW9G0CJ/.

Dyckers, R. (2014, September 1). In *Facebook* [Museum page]. Retrieved February 8, 2018, from: https://www.facebook.com/pg/starsofthegalaxy/reviews/?ref=page_internal.

Erzen, T. (2011). The vampire capital of the world: Commerce and enchantment in Forks, Washington. In M. Parke & N. Wilson (Eds.), *Theorizing twilight: Critical essays on what's at stake in a post-vampire world* (pp. 11–24). Jefferson, MO: McFarland.

FANwerk (2016, December 1). Interview zur Filmfiguren Ausstellung & Saber Con | FANwerk [Video file]. Retrieved from: https://www.youtube.com/watch?v=Ws5rwJrES-o.

Geraghty, L. (2014). It's not all about the music: Online fan communities and collecting Hard Rock Café pins. *Transformative Works and Cultures, 16.* doi: 10.3983/twc.2014.0492.

GROBI.TV (2014, July 10). Hier ist das Paradies für Star Wars Fans—die Filmfigurenausstellung in Mönchengladbach [Video file]. Retrieved from: https://www.youtube.com/watch?v=biULxYqQBEc.

GROBI.TV (2015, December 30). Wir waren wieder in der Filmfigurenausstellung in Mönchengladbach [Video file]. Retrieved from: https://www.youtube.com/watch?v=UySJSMqjz48.

Hills, M. (2002). *Fan cultures.* London, UK: Routledge.

History (n.d.). Retrieved June 20, 2019, from: http://www.ranchoobiwan.org/about/history/.

Jedrychowski, N. (2017, August 28). Star-Wars-Nebendarsteller hautnah. *RP Online.* Retrieved from: https://rp-online.de/nrw/staedte/moenchengladbach/star-wars-nebendarsteller-hautnah_aid-20819979.

Jensen, B. (1999, April 13). Charmed life. *Baltimore City Paper.* Retrieved July 2, 2019, from: http://www.readersadvice.com/museum/charm1.html.

Jovanovic, L. (2017, May 14). In *Facebook* [Museum page). Retrieved February 8, 2018, from: https://www.facebook.com/pg/starsofgalaxy/reviews/?ref=page_internal.

Kim, S. (2010). Extraordinary experience: Re-enacting and photographing at screen-tourism location. *Tourism and Hospitality Planning and Development, 7*(1), 59–75.

Larsen, K. (2015). (Re)Claiming Harry Potter fan pilgrimage sites. In L. S. Brenner (Ed.), *Playing Harry Potter: Essays and interviews on fandom and performance* (pp. 38–54). Jefferson, MO: McFarland.

Lee, C. (2012). Have magic, will travel: Tourism and Harry Potter's United (magical) Kingdom. *Tourist Studies, 12*(1), 52–69.

Nerdkultur (2016, June 9). Original-Requisiten in Deutschland | Star Wars Identities [Video file]. Retrieved from: https://www.youtube.com/watch?v=ETclkzVpV9I.

Norris, C. (2013). A Japanese media pilgrimage to a Tasmanian bakery. *Transformative Works and Cultures, 14.* doi: 10.3983/twc.2013.0470.

Porter, J. E. (1999). To boldly go: Star Trek convention attendance as pilgrimage. In J. E. Porter & D. L. McLaren (Eds.), *Star Trek and sacred ground: Explorations of Star Trek, religion, and American culture* (pp. 245–270). Albany, NY: SUNY Press.

Reijnders, S, (2011). *Places of the imagination: Media, tourism, culture.* Farnham, UK: Ashgate Publishing.

Rip, M. (2018, January 21). In *Facebook* [Museum page]. Retrieved February 8, 2018, from: https://www.facebook.com/pg/starsofthegalaxy/reviews/?ref=page_internal.

Roesch, S. (2009). *The experiences of film location tourists.* Bristol, UK: Channel View Publications.

RP Online—Wir sind NRW (2017, October 13). Mönchengladbach: Hier können Fans Star Wars hautnah erleben [Video file]. Retrieved from: https://www.youtube.com/watch?v=GoljaOfwYnc.

RTL WEST zu Besuch in der Filmfiguren Ausstellung [Video file]. In Facebook [Museum page]. Retrieved April 28, 2020, from: https://www.facebook.com/watch/?v=974911602555160.

Ruge, A. (Ed.). (2007). *Frame of reference for museum professions in Europe (preliminary edition 2007).* (n.p.): ICTOP.

Sansweet, Stephen J. (1992). *Star Wars: From concept to screen to collectible.* San Francisco, CA: Chronicle Books.

Sansweet, Stephen J. (2009). *Star Wars: 1000 collectibles.* New York, NY: Harry N. Abrams.

Schommer, A. (2014, February 13). Filmfigurenausstellung, CityVision TV Bericht [Video file]. Retrieved from: https://www.youtube.com/watch?v=I9h7vpbs__M.

StarconstruX Headquarters (n.d.). Retrieved May 28, 2018, from: http://www.starconstrux.de/hauptquatier/headquater.htm.

Stars of the Galaxy—Photos (n.d.). In *Facebook* [Museum page]. Retrieved May 28, 2018, from: https://www.facebook.com/pg/starsofthegalaxy/photos/?ref=page_internal.

The Galaxy Connection (n.d.). Retrieved May 28, 2018, from: http://www.thegalaxyconnection.com.

Thiele, M. (2014, September). Filmfigurenausstellung: Paradies für *Star Wars*-Fans. *Hindenburger: Die Stadtzeitschrift für Mönchengladbach und Rheydt.* Retrieved from: https://issuu.com/hindenburger-mg/docs/hindenburger_09-2014.

Trope, A. (2011). *Stardust monuments: The saving and selling of Hollywood.* Hanover, NH: Dartmouth College Press.

True Fans and Collectors for Life (n.d.). Retrieved June 20, 2019, from: https://avagardner.org/history.html.

Volkerc (2013, October 8). Filmfiguren Ausstellung Mönchengladbach [Forum post]. Retrieved February 8, 2018, from: forum.starwars-figuren.com/forum/topic.asp?TOPIC_ID=584.

Vorhees82 (2015, August 1). Follow me around – Filmfiguren Ausstellung in Mönchengladbach [Video file]. Retrieved from: https://www.youtube.com/watch?v=gfXmaYuCA04&t=2544s.

Waysdorf, A., & Reijnders, S. (2017). The role of imagination in the film tourist experience: The case of Game of Thrones. *Participations: Journal of Audience & Reception Studies, 14*(1), 170–191.

Waysdorf, A., & Reijnders, S. (2019). Fan homecoming: Analyzing the role of place in long-term fandom of The Prisoner. *Popular Communication: The International Journal of Media and Culture, 17*(1), 50–65. doi: 10.1080/15405702.2018.1524146.

Westphal, H. (2015, May 20). Star-Wars-Ausstellung in Köln: Am Freitag startet Star Wars Identities im Kölner Odysseum, *Kölnische Rundschau.* Retrieved from: https://www.rundschau-online.de/region/koeln/star-wars-ausstellung-in-koeln-am-freitag--startet--star-wars-identities--im-koelner-odysseum-1283678.

Williams, R. (2017). Fan tourism and pilgrimage. In M. A. Click & S. Scott (Eds.), *The Routledge Companion to media fandom* (pp. 98–106). London, UK and New York, NY: Routledge.

Williams, R. (2018). Replacing maelstrom: Theme park fandom, place, and the Disney brand. In R. Williams (Ed.), *Everybody hurts: Transitions, endings, and resurrections in fan cultures* (pp. 167–180). Iowa City: University of Iowa Press.

Woo, B. (2014). A pragmatics of things: Materiality and constraint in fan practices. *Transformative Works and Cultures, 16.* doi: 10.3983/twc.2014.0495.

Zubernis, L., & Larsen, K. (2018). Make space for us! Fandom in the real world. In P. Booth (Ed.), *A Companion to media fandom and fan studies* (pp. 145–159). Hoboken, NJ: John Wiley and Sons, Inc.

32

CREATIVE FANDOMS AND THE MEDIATIZED SACRED SITES

Kyungjae Jang and Takayoshi Yamamura

Introduction

This chapter illustrates the process and structures of creating "sacred sites" of pop culture in the Internet age via what we term "creative fandom" (fan communities with active creativity). In the Internet and social media era, individuals who were solely recipients have started providing and sharing information with a large number of people online (Ma & Alhabash, 2017). While some of them create original information or creative works, other fans create information, images, and derivative works based on existing mediatized works. As both recipients and creators, these creative fans play powerful roles in the mediatization of sites and the creation of sacred sites of pop culture. They share their culture across physical boundaries and produce and reproduce images of media-related works in various parts of the globe regardless of geographical limitations.

The hallmark of the new Internet era is that space that is not directly related to popular culture works is made into sacred sites by the works' fans. The fandom of the previous era mainly focused on visiting the place where the actual work had been staged, or the background of the specific work (Beeton, 2005). In addition to the spread of the Internet, fans began to assign meaning to places that were not directly related to creative works (Beeton, Yamamura, & Seaton, 2013). For example, Kidlington, a small town in the UK, and the University of Sydney's main campus were mistaken as Harry Potter filming locations by Chinese tourists, based on wrong information provided by their tour guide. Beeton (2005) described this as a case of mistaken identities, but in reality, this was done intentionally by the tour companies and guides. Another popular example of this is Japanese virtual idols. Fans of *Idolmaster* and *Love Live!* often grant meaning to places that have only their names in common with certain characters in the two multimedia games, even if there is no other similarity or connection (Seaton, Yamamura, Sugawa-Shimada, & Jang, 2017). Such a phenomenon is related to the increase in online communication. Communication through the Internet has begun to create new meanings in a bottom-up rather than the conventional top-down method, often described as user generated content (UGC). With pop culture as material, fans have started numerous communication initiatives on the Internet. This communication fostered the creation of many sacred sites.

To illustrate the dynamism of the construction process of sacred sites, this chapter presents and analyzes two cases of fan-created sacred sites of pop culture. These sites are unrelated to the original pop culture works and have had fans identify and assign meaning to them. The first case study involves Japanese game/animation *The Idolmaster* (2005) fans visiting Yayoiken Restaurant in Takatsuki city, Osaka, Japan, and the second case study focuses on the fan-made sacred site of the Japanese anime film *Kiki's Delivery Service* (1989) in Tasmania, Australia. Through these two cases, the chapter shows the process through which sites that were unrelated to the original film/works become sacred sites by having fans assign special meaning and value.

What we want to highlight in this chapter is not only how a place gets a new meaning due to media works but also how fandom contributes to placemaking. Particularly, we want to focus on the role of social media and how it enhances fans' placemaking.

Creative fandoms

Fans are typically consumers of popular culture. Fans pay for and buy their favorite works as a commodity. Fans also engage in symbolic production through the consumption of popular culture. Fiske (1992) argues that this not only occurs in fan culture but also in popular culture. "Semiotic productivity is characteristic of a popular culture as a whole. It has been argued that there is a great deal of empirical evidence to support this hypothesis" (Fiske, 1992, p. 37). Fans contribute to the spread of works based on their passion for their favorite works. Lee (2011) refers to the process of fans creating subtitles for foreign works as participatory media fandom in the global mediascape. Lee insists that the Internet has encouraged the spread of these activities. These activities such as scanning and uploading works could constitute copyright infringement, but they indicate the fans' enthusiasm and motivation. Fans represent themselves in the resistance against protecting the value of a work and not just as a consumer (Hills, 2002). Ironically, the company that made the media work sometimes becomes the resistant opponent citing intellectual property rights or other legal arguments.

A special passion among the fans is often expressed in the form of spin-off or fanfic (fan fiction) of the original works. Over 100 years ago, literary fans started enacting scenes from "Sherlock Holmes" on stage. Today, the *Star Wars* series (1977-on-going) demonstrates how many fans create a fictional universe beyond the one described in the film and share the creation with others. The concept of the "fannish place" was proposed by Toy (2017) to describe the process of how fans assign sacred meaning to places they consider to be related to creative works, regardless of the frequency and importance of the work. And the Internet plays an important role in this process, as it speeds up the reach of its distribution. Recently, the Internet has expressed various meanings of a place by using coordinates of space and actively communicating the derived meaning.

In media studies, Jenkins (2006) and Steinberg (2012) have discussed the concept of convergence by focusing on consumer participation, media mix, and so on. However, they basically analyzed this concept" from a media study perspective and offered very few descriptions of tourism phenomena" (Graburn & Yamamura, 2020, p. 6). On the other hand, based on their frameworks, Norris (2013) presented a research article on the process of media pilgrimage and media scaffold related to tourist sites. Norris referred to the concept of "media tourism" and discussed the creativity of fans as media tourists. He also discussed the process of editing original media content by fan tourists. The transformative fandom that Norris suggests has a lot in common with the creative fandom here. However, it can be said that creative fandom can cover a wider range because it contains more diverse images.

Transformative fandom discusses the relationship between fandom and a place based on the similarity of images. Creative fandom, on the other hand, focuses on placemaking, which can be connected with fandom more easily.

As Ishimori and Yamamura (2009) pointed out, nearly anybody can produce and consume information and images of tourist destinations and share it through social networking sites (SNS) as prosumers. In other words, as Graburn and Yamamura (2020) mentioned, "anybody can be a creator of mediascape and tourist/tourism imageries." This situation "cannot be analyzed and sufficiently understood using previous approaches of tourism studies" such as the framework of "hosts and guests." It is therefore necessary and important to focus on "the formation process of a new community through communication" that utilizes media content that goes "beyond position and nationalities" (Graburn and Yamamura, 2020, p. 7). Yamamura (2020) focused on the formation process of these communities by creativities (fandom who act creatively) and named them "creative fandom."

Mediatized sacred sites and social media

Creative fans may create their own celebration venues. Regardless of the method of expression and genre, such as writing (literature), painting (cartoon or animation), photography, video (film), and so on, various places become mediatized sacred sites for fans (Seaton et al., 2017).

During their pilgrimage, pilgrims; particularly fans; that consider their sacred sites to be intersections of reality and imagined worlds (Reijnders, 2011). However, they are more than that: "they are co-constructed and ritualized sites of interpretation that emphasize the agency and emotional attachments of fans" (Toy 2017, p. 252). Buchmann, Moore, and Fisher (2010) also described this process in detail, showing how "secular" fans anticipate and create pilgrimage-like journeys to the sites of their longing. On the other hand, fan making sacred sites are related to the use of the Internet, especially social media. Visiting a mediatized place by fans is a form of media consumption that gives the place a new meaning, as Sandvoss (2005) insists 'opportunity to relocate in place a profound sense of belonging which has otherwise shifted into the textual space of media consumption' (Sandvoss 2005, p. 64). Such visits may serve as a pilgrimage by fans (Magasic, 2016). What is important in such personal pilgrimage is to indicate/mark that the place as sacred sites to fans (Williams, 2017). However, fans have limitations to mark physical places. In this process, uploading photos to social media is an important part of a marking. Particularly, Jansson (2018, p. 102) emphasizes the role played by photos that are uploaded on social media as:

> Tourists, like people in general, are to an increasing extent (co-)producers of media text that can be spread and discussed far beyond the closed circles of traditional family albums. Instagram images, for example, can be geo-tagged and immediately commented upon, which in turn contributes to the cultural (re)coding of tourism places and practices.

Sometimes, religious forms also appear in the process of fans' marking their sacred sites. Ryosuke Okamoto insists it is a new form of personal religion in the modern world where secularization proceeds (Okamoto, 2015). Of course, seemingly religious expression does not necessarily imply religion. As McCloud (2003) argued, pop culture fans may perform a seemingly religious action because it is the most effective way of expressing their emotion and not because they consider their fandom to be a religion. In other words, such actions by fans are self-projections that accidentally become kind of canonical. Nevertheless, religiosity in pop culture-related rituals cannot be totally ruled out. Fundamentally, people pursue what they rely on. With the spread of spirituality that substitutes institutional religion, pop culture plays a role in the worship of people. Thus, researchers need to avoid both overestimating and underestimating the meaning of pop culture. Pop culture does not merely serve as an effective means of self-expression, as McCloud argued, but pop culture-based rituals are also a new type of religion (Bickerdike, 2016; Davidsen, 2013). It is important to note that these activities, whether religious or not, can make pop culture sacred. This means that even though these activities did not have any religious significance, they seemed to have a religious meaning especially through the recent increase and reproduction through the Internet, especially social media.

These phenomena, as Beeton et al. (2013, p. 150) pointed out, can be regarded as "two parallel processes" of mediatization "occurring in tourist sites." First, the physical sites "become media sites in which meanings are spread" from the creative fans who have converted a pre-existing site into a tourist site. Second, the "act of visiting those sites is mediatized by the recording and dissemination of experiences via media (such as blogs)" in this "age of smart phones and high-speed Wi-Fi" (Beeton et al., 2013, p. 150).

Case studies

The Idolmaster fans' pilgrimage to Yayoiken Restaurant in Takatsuki city, Osaka

The Idolmaster (officially stylized *THE iDOLM@STER*) is a Japanese idol training game produced by Namco in 2005, eventually evolving into a media franchise with the release of TV animation series and other works. The game's objective is for the player, acting as an entertainment producer (usually called P) in a production office, to train female idol characters and turn them into stars—in other words, to become an 'idol master'. The game includes five productions and a total of 300 idol characters. After its introduction as an arcade game, a home console version was released in 2007 and a portable version in 2009; however, initial interest in the game was relatively low. It finally gained significant popularity with the release of the 2011 TV animation series. The mobile game released in the same year, *The Idolmaster Cinderella Girls*, ultimately surpassed 5 million users, and the overall media franchise of *The Idolmaster* has grown to 10 billion yen (100 million US dollars). The major newspaper *Nihon Keizai Shimbun* describes it as the *'Idolmaster economic zone'*.

As a game, *The Idolmaster* reflects Japan's unique idol culture. Whereas idols in other countries, such as South Korea and the United States, make their debut in perfect form after a fierce selection process and years of practice, especially since the 2000s, Japanese idols make their debut as imperfect amateurs in both singing and dancing. Fans support them while they watch them grow. In other words, Japan's idol fandom system is a nurturing one (Hamano, 2012). This form originated with and is typified by AKB48, now the leading idol group in Japan. In 2005, the same year that *The Idolmaster* was first released, AKB48 began performing in the streets and small theaters under the image of 'idols you can meet'. The group was initially composed of students with no celebrity experience to effectively depict the process of each member's growth as an idol; it is now an icon of Japanese pop culture and imagination in the 2000s (Uno, 2008). *The Idolmaster* is a virtual representation of this culture.

A unique place created by a fandom through social media can be seen in the pilgrimage of *Idolmaster* fans to the restaurant Yayoiken in the city of Takatsuki every year on 25 March. Run by Plenas, Japan's leading restaurant franchise, Yayoiken is a chain providing *teishoku* (traditional Japanese meals of rice, soup, and side dishes). Originally operated under the name of Meshiya-dong, the chain was renamed Yayoiken in 2004, after a restaurant run by the Plenas founder's grandfather. The restaurant has no real connection with *The Idolmaster*, but since 2008, it has developed into a sacred place for fans.

One of *The Idolmaster's* characters, a member of 765 Production, is named Yayoi Takatsuki. The official website reports that she is 13 years old, born on 25 March (PROJECT IM@S, n.d.). Her hometown is in Saitama Prefecture; she has no particular link with Takatsuki city, which is in Osaka Prefecture. However, around 2008, fans of her character discovered that there was a Yayoiken branch in Takatsuki, Osaka. Because 'ken' means 'house' in Japanese, these fans considered the Takatsuki Yayoiken to be 'the home of Takatsuki Yayoi'. But whereas local information such as this is typically confined to the local area, knowledge of the Takatsuki Yayoiken spread throughout fans of *The Idolmaster* on a national scale. In an article on the pilgrimage to Takatsuki's Yayoiken, J Town (2017), a news agency in Japan, indicated that social media, especially Twitter, was instrumental in making the Takatsuki Yayoiken sacred to Yayoi fans. Social media platforms enabled fans to attach meaning to the place and spread it in a grassroots way. Among those Twitter posts still remaining in 2019, the first reference to Takatsuki's Yayoiken as sacred for Yayoi fans appears in February 2008 (ug3, 2008). However, no posts show that fans visited the restaurant to celebrate the character's birthday on 25 March of that year. The first birthday visit post appeared in 2009, and in 2010 and 2011, the number of visiting fans gradually increased with the accumulation of information on Twitter.

Beginning in 2012, the number of fans who visited the Takatsuki Yayoiken on 25 March increased sharply. This is likely a result of the success of the 2011 TV animation version of *The Idolmaster*, in addition to the fact that Japan's social media usage rate exceeded 50% with the spread of smartphones in 2012 (ICT Research & Consulting, 2013). Many fans waited in line because there was no open seating, and some took pictures of Yayoi figures with their food (maniaxch, 2012; marina3000turbo, 2012). Today, the fan pilgrimage to Takatsuki's Yayoiken on 25 March has become a part of the fandom culture. In 2014, enthusiastic fans drove and parked a car called 'Ithasha', decorated with depictions of Yayoi, in front of the restaurant; in 2015, more than 20 fans were still waiting in line even after 9 pm.

The mediatized place created and spread by *Idolmaster* fans on social media has now grown more formal. Since 2015, the Takatsuki Yayoiken has posted a sign saying 'thank you for visiting' on 25 March; beginning in 2018, visitors were presented with a commemorative card showing the date. In 2016, the city of Takatsuki presented Yayoi character goods to those who paid the city a hometown tax (a Japanese tax system that allows urban taxpayers to contribute to rural areas through income and residence taxes). Despite the fact that it has no concrete connection to *The Idolmaster*, participants in the pilgrimage to the Yayoiken only continue to increase year by year.

In the next section, as a general example, we will look at how fans make a place sacred with similar images of media works.

The sacredness of mistaken identities: case of the fan-made sacred site of Kiki's Delivery Service in Tasmania

Kiki's Delivery Service is a 1989 Japanese anime film based on the title of the 1985 novel by Eiko Kadono. The film was written, produced, and directed by Hayao Miyazaki and was successful worldwide (Clements & McCarthy, 2015, p. 430). It still gathers many admirers throughout the world as one of the most famous anime by studio Ghibli, directed by

Miyazaki. As Miyazaki officially mentioned, the location model of the film was actually the city of Visby, Gotland, Sweden (Takahata, Miyazaki, & Kotabe, 2014). However, the fans have identified their own destination not in Sweden but in Tasmania, Australia, just because the atmosphere of a bakery in Tasmania is similar to the one in the anime (Hooper, 2015). A local bakery named, "Ross Village Bakery," has become a tourist destination as a sacred site of the anime *Kiki's Delivery Service*, and many fans believe that it is a model of the bakery in the anime. In the field of film-induced tourism studies, this kind of destination is often referred to as "mistaken identity" which means "film tourism to places where the filming is only believed to have taken place" (Beeton, 2005, p. 10). It is clear how a site can obtain special meaning and sacredness through the imagination and creativity of fans, even if the site is irrelevant to the original work and has mistaken identities.

According to Sugawa (2015), the concept of mistaken identity was started "by one Japanese tourist who traveled to Australia and incidentally visited the town of Ross, Tasmania" (p. 120). "He found Ross Village Bakery to have a bread oven and an attic room very similar to those in the anime and wrote about it in a Social media" (p. 120). In addition, he informed the owners (husband and wife) of the bakery that their bread oven and attic room were very similar to those in the anime. Though they had never watched the Japanese anime, they watched the film and perceived that their bakery was similar to the one in the anime. Consequently, they converted the attic room to a guest room inn (Hooper, 2015).

Many fan tourists have since stayed and brought related goods such as figures and posters to the room. In addition, they have written messages and impressions on the notebooks placed in the room, and many of them express the deep emotion of re-experiencing the narrative world of the anime that they love. In this way, the discourse of mistaken identity that Ross Village bakery is a model of the anime *Kiki's delivery service* has been widely spread through SNS especially among young people who have an Australian working holiday visa (Yamamura, 2018, personal communication. 10 May). Many fans believe this bakery is a sacred site and fans from Japan and other countries especially from East Asian countries visit the bakery.

Norris (2013) referred to the practice of fan tourism as "the media pilgrimage." He further stated that, "fans bridge their ordinary reality to the special world of media, and the media scaffold, where Kiki becomes a way to interpret the world around them" (para. 0.1) through the media pilgrimage. Moreover, he pointed out that media pilgrimage is "a creative process where fans transform existing elements of popular culture through their physical surroundings to express changes in culture, geography, and identity" (para. 11.4).

The town and the bakery have no connection to the anime film. Nevertheless, this place has been assigned the meaning or narrative quality of *Kiki's Delivery Service* by the anime's fans and the bakery's owner and has become a popular tourist destination (Figure 32.1). In other words, while the site may bear mistaken identity, the fact that the imagination and creativity of fans (including the owners as fans) has created the sacred site is significant, and fans' experiences during their stay are authentic and sacred. Therefore, this is also an interesting example of the tourism imagery and sacred site constructed by creative fandom. In addition, we can find similar cases of the sacred sites of Hayao Miyazaki's films all over the world, which became sacred owing mistaken identities by creative fandom and are unrelated to the film's creation (e.g. Jiufen, Taiwan, as a site of a 2001 film *Spirited Away*, Uluṟu-Kata Tjuṯa National Park, Australia, as a site of a 1984 film *Nausicaä of the Valley of the Wind*, etc.). It is also interesting that fans of the anime films directed by Miyazaki created these kinds of imagined sacred sites even though they were unrelated to the original work and attract media pilgrimages due to mistaken identities. This in turn shows the power of mediatization, a core concept for any studies of this phenomenon.

Figure 32.1   Ross Village Bakery, the fan-made sacred site of Kiki's Delivery Service, Tasmania, Australia

Source: Takayoshi Yamamura.

Conclusions

This chapter showed two cases of fan-created sacred sites of pop culture. In particular, the chapter illuminated how the value of the mediatized place diffuses and reproduces in terms of tourism anthropology, and media studies. These places are unrelated to the original pop culture works and yet fans identified and provided significances to these sites. Both cases entail the process of assigning meaning and converting a pre-existing site into a sacred site of media contents. Then, special meanings are assigned to these sights converting a pre-existing site into a sacred site of the contents. In particular, the Internet and social media have made it possible for fans to share specific values and to concentrate these values in one place. In both cases in this chapter, sites that had nothing to do with creative works were assigned new meanings through communication on the Internet. It is possible that not only the location and the stage of a specific work but also the relationship between creative works and a particular space can be formed more diversely.

In this process, social media plays a role in strengthening trivial associations. Places that are not directly related to the media works are given value and spread through social media. Visitors upload all their actions at those places in real time, and these contents are spread further to give a deeper meaning to the place. This can be said to be a process in which social media gives value to a place. In other words, the place becomes both mediatized and social mediatized. These places serve as sacred sites for fans only, and the significance of the sacred sites is shared through the Internet and social media sites after the visit. This can be said to be the mediatization process of sacred sites of pop culture and also the process of building tourism imagery.

Now the cross-border movement of people, goods, and information has increased dramatically, and the media content is consumed more and more transnationally. Therefore, the transnational mediatization of sites by creative fandom is expected to occur more frequently. This chapter provides a new perspective on media and tourism in that it shows that fans can be the active actor of place-making, not a passive recipient of the media.

References

Beeton, S. (2005). *Film-induced tourism.* Clevedon, UK: Channel View.

Beeton, S., Yamamura, T., & Seaton, P. (2013). The mediatization of culture: Japanese contents tourism and pop culture. In J. Lester & C. Scarles (Eds.), *Mediating the tourist experience: From brochures to virtual encounters* (pp. 139–154). Surrey, UK: Ashgate Publishing.

Bickerdike, J. O. (2016). *The secular religion of fandom:Ppop culture pilgrim.* London, UK: Sage.

Buchmann, A., Moore, K., & Fisher, D. (2010) Experiencing film tourism: Authenticity & Fellowship. *Annals of Tourism Research, 37*(1), 229–248.

Clements, J., & MacCarthy, H. (2015). *The anime encyclopedia, third edition: A century of Japanese animation.* Berkeley, CA: Stone Bridge Press.

Davidsen, M. A. (2013). Fiction-based religion: Conceptualising a new category against history-based religion and fandom. *Culture and Religion, 14*(4), 378–395.

Fiske, J. (1992). The cultural economy of fandom. In L. Lewis (Ed.), *The adoring audience: Fan culture and popular media* (pp. 30–49). London, UK: Routledge.

Graburn, N. & Yamamura, T. (2020). Contents tourism: background, context, and future. *Journal of Tourism and Cultural Change, 18*(1), 12-26.

Hamano, S. (2012). *Maeda Atsuko wa Kirisuto o koeta: < Shūkyō > toshiteno AKB 48.* Tokyo, Japan: Chikumashobō.

Hills, M. (2002). *Fan cultures.* London, UK: Routledge.

Hooper, F. (13 March, 2015). Kiki's delivery service: Japanese anime fans flock to Tasmania's Ross bakery to see little witch's room, *ABC News.* Retrieved from: https://www.abc.net.au/news/2015-03-13/tasmanias-ross-bakery-continues-to-attract-japanese-anime-fans/6308038.

ICT Research & Consulting (2013). 2013-nen SNS riyō dōkō ni kansuru chōsa. Retrieved from: https://ictr.co.jp/report/20130530000039.html.

Ishimori, S., & Yamamura, T., (2009). Jouhou shakai ni okeru kankou kakumei: bunmeishiteki ni mita kankou no grōbaru tornedo. *JACIC Jouhou, 94/24*(2), 5–17.

Jansson, A. (2018). Rethinking post-tourism in the age of social media. *Annals of Tourism Research, 69,* 101–110.

Jenkins, H. (2006). *Convergence culture: Where old and new media collide.* New York, NY: New York University Press.

J-townnet. (2017, March 27). Takatsuki-shi ya yoi noki e no junrei no rekishi tsuini shinbun mo toriageta... 3 Tsuki 25-nichi no `Yayoi-chan panikku' ga kakudai suru made. Retrieved from: https://j-town.net/tokyo/news/localnews/241099.html?p=all

Lee, H. K. (2011). Participatory media fandom: A case study of anime fansubbing. *Media, Culture & Society, 33*(8), 1131–1147.

Ma, M., & Alhabash, S. (2017). A tale of four platforms: Motivations and uses of Facebook, Twitter, Instagram, and Snapchat among college students? *Social Media + Society, 3*(1), 1–13.

Magasic, M. (2016). The 'selfie gaze' and 'social media pilgrimage': Two frames for conceptualising the experience of social media using tourists. In A. Inversini & R. Schegg (Eds.), *Information and communication technologies in tourism 2016* (pp. 173–182). Cham, Switzerland: Springer.

maniaxch. (2012, March 25). Son'na wake de oishiku tabete Takatsuki no Yayoiken o demasu 5-nin kurai de raiten shite seki ga nakute hairenakatta gurūpu ga nan kumimo itakedo ten'nai wa semai-kara nan'ninka ni wakarete hairu to ī kamo yo! u~tsuu. [Tweet]. Retrieved from: https://twitter.com/maniaxch/status/183775413422653440.

marina3000turbo. (2012, March 25). Kyō wa takatsukiyayoi no tanjōbi to iu koto de, jimoto no Yayoi-ken nite kikan gentei no bīfusutēki&ebifuraiteishoku. [Tweet]. Retrieved from: https://twitter.com/marina3000turbo/status/183837141791551488.

McCloud, S. (2003). Popular culture fandoms, the boundaries of religious studies, and the project of the self. *Culture and Religion, 4*(2), 187–206.

Norris, C. (2013). A Japanese media pilgrimage to a Tasmanian bakery. *Transformative Works and Cultures, 14*, 2013. Retrieved from: http://journal.transformativeworks.org/index.php/twc/article/view/470/403.

Okamoto, R. (2015). Seichi junrei-seikaiisan kara anime no butai made. Tokyo, Japan: Chūōkōronshinsha.

PROJECT IM@S. (n.d.). THE IDOLM@STER [Yayoi Takatsuki]. Retrieved from: http://www.idolmaster.jp/imas/character/04_yayoi.html.

Reijnders, S. (2011). *Places of the imagination: Media, tourism, culture.* Surrey, Canada: Ashgate.

Sandvoss, C. (2005). *Fans: The mirror of consumption.* Cambridge, UK: Polity Press.

Seaton, P., Yamamura, T., Sugawa-Shimada, A., & Jang, K. (2017). *Contents tourism in Japan: Pilgrimages to "sacred sites" of popular culture.* New York, NY: Cambria Press.

Steinberg, M. (2012). *Anime's media mix: Franchising toys and characters in Japan.* Minneapolis: University of Minnesota Press.

Sugawa, A. (2015). Majo no Takkyūbin: Kaigai ni okeru josei no contents tourism. In T. Okamoto (Ed.), *Contents tourism Kenkyū* (pp. 120–121). Tokyo, Japan: Fukumura Shuppan,

Takahata, I., Miyazaki, H., & Kotabe, Y. (2014). *Maboroshino 'Nagakutsushita no Pippi'.* Tokyo, Japan: Iwanami Shoten.

Toy, J. C. (2017). Constructing the fannish place: Ritual and sacred space in a Sherlock fan pilgrimage. *The Journal of Fandom Studies, 5*(3), 251–266.

ug3 (2008, February 24). Kuso ~u tanoshi-sōda. Ichido wa Takatsuki no Yayoi noki ni junrei ni itte mitai monoda. [Tweet]. Retrieved from: https://twitter.com/ug3/status/750789872

Uno, T. (2008). *Zero-nendai no sōzō-ryoku.* Tokyo, Japan: Hayakawa Publishing.

Urry, J., & Larsen. J. (2011). *The tourist gaza 3.0.* London, UK: Sage.

Williams, R. (2017). Fan tourism and pilgrimage. In M. A. Click, & S. Scott (Eds.), *The Routledge Companion to media fandom* (pp. 98–106). New York and London: Routledge.

Yamamura, T. (2020). Contents tourism and creative fandom: the formation process of creative fandom and its transnational expansion in a mixed-media age. *Journal of Tourism and Cultural Change, 18*(1), 12-26.

33

I CAME, I SAW, I SELFIED

Travelling in the age of Instagram

Ana Oliveira Garner

Introduction

This chapter sheds light on how Instagram is impacting on tourism and how tourism is submitting to the logic of social media. This is done under the framework of mediatization, which refers to a 'long-lasting process, whereby social and cultural institutions and modes of interaction are changed as a consequence of the growth of the media's influence' (Hjarvard, 2008, p. 114). Whilst Media Studies tend to focus on the effects that some media have on society, mediatization broadens this perspective by seeking to comprehend the 'consequences of media's embedding in everyday life' (Couldry & Hepp, 2013, p. 195). This concept is interested in describing changes, be they within the media and communication institutions, or in culture and in society due to the emergence and development of different types of media. Furthermore, within the field of Cultural Studies, I also discuss how this cultural practice of travelling with Instagram connects to power relations in society.

Smartphones, high speed Internet, and social media are technological affordances that influence many aspects of our everyday lives, including the way we travel. The invention of digital cameras has changed the way we take and interact with travel photos, as these enable people to immediately check and retake photos as many times as they want. With the proliferation of good mobile phone cameras, and apps that enable the easy retouching of images, non-professional photographers are now also able to produce high-quality images. Additionally, where previously travel photos were put into an album and shown mostly to friends and family, nowadays these images have become public, as they can be more widely shared via the Internet.

Some authors (Pan, Lee, & Tsai, 2014; Urry, 1990) have argued that tourism is fundamentally related to visual experiences and that changes in travelling practices could be related to the way people gaze at objects and places. This is why it is important to understand how this gaze is changing. The fact that we are now able to take many pictures wherever we go and upload them instantly, even whilst travelling, has had an impact on the ways in which people experience and interact with the locations they visit. It also means that both touristic sites, and tourists, have needed to adapt.

Travellers' Instagram practices are often inserted in the context of other representations of places found in official media like websites and advertisements, and in peer platforms, such as in blogs, forums, and social media. Månsson (2011) refers to this process as mediatized tourism in which 'media products accordingly converge and float around in people's awareness without demarcation in an ongoing circle of references' (2011, p. 1635). The author advocated the need to use the convergence concept to investigate how tourism is affected by changes in media usage. To do this, researchers should concentrate their efforts on the media content created by tourists.

One way to do this is by exploring how people are adapting their travel photo albums to the social media era. This chapter does that, by asking: How do tourists' narratives on Instagram impact on the practices of tourism? To find out, a combination of three methods was used. In the first, 15 semi-structured interviews were conducted with Instagram users from different countries. These participants were found via an open call on Facebook to those interested in talking about their social media practices. The first participants then indicated further interested acquaintances. Secondly, a participant observation of Instagram profiles, posts, and hashtags was undertaken, by following ten of the participants' Instagram accounts, with their knowledge and consent, and up to 120 posts were then analysed. All the participants had photos of touristic sites in their feeds, even though the initial study did not intend to focus on tourism but asked how people visually narrate their lives on social media, which therefore enabled the comparison of travel photographs to other kinds of posts. This study considers a post on Instagram to be not only a photo but the combination of an image and any other elements that may be present. Therefore, the following aspects of the posts were observed: the kind of image according to its content; and whether the images had been altered through filters; captions; hashtags; comments; and likes. The third step was to make a compilation and analysis of popular media articles that covered stories related to selfies and tourism.

Travelling with Instagram

The interviews made it possible to describe the tourist experience of travelling with the intention of posting on Instagram. When taking photos, most participants preferred to use some kind of self-technique such as selfie-sticks, tripods, and timer functions. This is partly because they were often too embarrassed to ask people to take their photos but mainly because they like having the control over the photos and the possibility to retake as many photos as necessary to achieve their aim. With regard to the kind of spaces they like to photograph whilst travelling, people quickly acknowledged places like restaurants, streets, famous buildings, markets, and natural landscapes. They were more reluctant to admit to museums, means of transportation, religious places, and places such as war memorials and cemeteries. Taking photos was mentioned as a way to interact with the place or event when it is boring, or not so exciting, and to have fun during long journeys or in hotel rooms. When asked about what kind of things they do to get a nice image, all participants mentioned that they pay attention to aspects like angle, light, and background. All participants had a favourite pose and particular photo face, or at least some ideas about what position/smile/side make them look more attractive in the photos.

Regarding the sharing of travel photos on Instagram, half of the participants declared that they treat all, or most, of their photos before posting. These were the participants who posted more often, and they prefer to edit separate features rather than applying the app's pre-set filters. The cited edits were increasing brightness; changing the saturation to make more the image look more colourful; changing the contrast; cutting the photo; removing blemishes; and changing facial features. Additionally, all participants demonstrated that they reflect on and have some kind of self-regulation of their use of the app. Some are careful not to post too often, others established a maximum number of photos to post about the same event. Some declared they avoid posting selfies or try not to post too many. They are also careful to avoid certain kinds of posts. The female participants stated that they were cautious about posting pictures of themselves in bikinis, or other images that could potentially be seen as vulgar, such as where their breasts have been emphasized. Participants also found strategies to repeatedly post, without the risk of seeming to over post in their feeds. One method is to use the Stories feature, which does not appear in the home screen of their followers and disappears after 24 hours. Another is to keep posting after the trip is over, with hashtags that make clear they are no longer travelling, such as #throwback, and by adding captions that state that they miss the place in the picture.

The observation showed that captions, not always present in other posts, are a frequent occurrence in travel photos. These captions tell a micro story, that mix facts with opinions. The most common types of captions are a brief explanation of, or basic details about, the place itself; a personal story that happened at that location; and impressions, emotions, or feelings that the place provoked in the user. Emojis are often used to enrich the text, by adding an extra visual layer that can quickly convey an emotion and add more meaning to the words. Another element that is not always present in other kinds of posts, but is frequent in travel posts, is hashtags. They are mostly used when there is a desire to reach out to the wider world, because anybody searching for that hashtag can also potentially see the image. All travel photos in the sample contained hashtags, and they occurred in larger numbers than in other types of posts. Most hashtags refer to: the country, the city, the name of the building or place shown in the image, or some kind of generic hashtag like: #trip, #travel, #wonderlust, #beautifuldestinations, #travelphotography, #instatravel. They could also refer to a feeling or an impression (#fun, #iloveit, #beautiful). This is an indication that when posting travel photos, users desire to talk to a broader public, beyond their followers.

We can perceive that the process of posting involves a series of actions: going through the photos to select which ones to post; editing the photos; writing the caption and hashtags; and then interacting with the comments. All of this is time consuming and when asked if these took up sightseeing time, participants stated that they preferred to do these on occasions such as when they are back in hotel rooms, siting in cafes or restaurants, or waiting in queues.

Instagram as travel inspiration

Another fundamental element of Instagram is the likes and comments from other users. As one participant confessed, she needs to register her travels because it 'is a way for my family to travel a bit with us as well. Family and friends said: 'cool, show more. It seems we are travelling with you'. People ask us to take photos so they can see' (participant 11, pers. comment, 13.01.2018). In return, followers leave likes and comments. So, when tourists post their photos whilst travelling, they take other people on the trip with them, by sharing the moment. The participant observation revealed that the comments are usually positive, brief, and filled with exclamations and emojis. They can also come in the form of funny statements or questions. This interaction in travel photos is significant, because the study showed that some people are using Instagram to plan their trips. Participants shared that they search for geotags and hashtags of places to which they are interested in going. They also said that strangers posted questions, such as if they recommended the restaurant in the photograph. As with other consumer-generated content online, such as travel blogs and forums, tips and suggestions from other travellers are considered more authentic, reliable, and unbiased than paid or official publicity (Mack, Blose, & Pan, 2008; Xiang & Gretzel, 2010). All the participants admitted to having gone to a particular place when travelling, with the specific goal of taking nice photos, having been inspired by something they saw on Instagram. Some reported following hashtags of places to which they wish to visit. Travel posts not only help them to decide on places to explore on a trip, but they are also a source of creative inspiration, by indicating what type of angles, details, and poses other people use. Furthermore, online search engines show that there are several online lists of the most 'Instagrammable' locations in a city, which compile the best places to take Instagram photos determined either by those with the most photos posted on Instagram, or by opinions of places that an individual feels are good for producing nice photos to post.

Based on this research, I argue that a consequence of looking for travel inspiration on Instagram is the promotion of places and events that are outside the main touristic experiences. These places are not necessarily touristic but are good backdrops for photos. To better illustrate this tendency, I will describe two cases that were researched based on my in loco observation and review of media articles. The first one is in Hong Kong, where recently tourists have become interested in visiting residential buildings, called housing estates (South China Morning Post, 2018). These are not touristic attractions but are tall and old buildings that, arguably, would not normally be considered aesthetically pleasant. However, the volume of windows and their bright colours or geographical patterns make them a perfect background for cool urban photos (as shown on Figure 33.1). Six out of 13 places in a list of best Instagram places in Hong Kong, by online magazine Timeout (Parkes, 2018) are in housing estates, whilst another four are also not touristic attractions. Some are far from traditional touristic areas, and they therefore encourage people to visit less touristic parts of the city.

Figure 33.1   Instagram post of walls painted in rainbow colours. Photo taken in Choi Hung housing estate in Hong Kong

Source: Ana Oliveira Garner, 2018.

The second example happens in South Korea, where tourists are renting hanboks to take photos and post online (Seoul magazine, 2016). Hanboks are Korean traditional clothes that until a few years ago were seen as something from the past, but which are now trending among young people. Social media is being held responsible for the revival of the tradition, which has led to an increase in the number of rental shops in 2015. In the traditional palaces in Seoul, it is possible to see a number of tourists dressed up and taking photographs. A search for #hanbok on Instagram reveals thousands of pictures, including many of these with dressed tourists in traditional places.

Touristic places in the age of Instagram

Analysis of popular media revealed that some places are trying to capitalize on Instagram practices, by encouraging the taking of pictures, and their subsequent uploading online. Matchar (2017) listed some examples of places that have adapted due to the influence of Instagram. Restaurants are creating more colourful and beautiful looking food, brighter spaces and are using white dishes and darker tables, all to facilitate the production of better photos to attract more customers. Commercial establishments are including elements made for Instagram, such as displays with giant accessories and funny furniture. Museums have found within social media a way to motivate people to visit and see their exhibitions. Indeed, some museums have been created with the specific purpose providing scenarios for pictures, such as the Art in Island Museum in Manila; the Trick Eye Museum in Hong Kong; and the Museum of Ice Cream in Los Angeles. Even cities, like Penang in Malaysia, are commissioning public art and murals that offer good photographic backgrounds. The author called this redesign of cultural spaces, the Instagramization of the world (Matchar, 2017).

Figure 33.2   Instagram post with a photo taken at the Yick Cheong residential building in Hong Kong

Source: Ana Oliveira Garner, 2018.

Whilst cases like this may increase business opportunities, some underprepared places can suffer consequences related to the disturbances caused by the increase in tourists. According to the South China Morning Post (2018), residents of the housing estates in Hong Kong who have seen an influx of Instagram visitors, complain about issues such as privacy, safety, and damage to the facilities. In the Yick Cheong Building, a photography ban was imposed, and signs were posted stating that the buildings are private property, and that no photos or videos should be taken without prior approval. However, these are largely ignored and tourists and locals continue to go there to take pictures (see Figure 33.2). In addition, some places like museums, sporting venues, and even cities are banning self-sticks or the practice of selfies, alleging they are a hazard or a nuisance (Lekach & Ciechalski, 2017).

Discussion

Analysing the observations within the larger context of neoliberalism, we can say that social media users are inserting themselves within the capitalist notion of consumption. Their posts are a way to be consumed by others, and this also serves the economic objectives of online platforms. Schwarz (2010) stated that users produce and consume content for the cultural gain, which is the acquisition of social status and ties, such as likes, friends, followers, and admirers. The better a person masters the advertising-inspired images, the more social capital they gain. The logic of social media in general is that of a branding tool, in which individuals mediate their personas and emerge as self-marketers. By narrating their trips on Instagram, users are working on their own personal brand, and on what kind of people and tourists they want to be seen as, whilst at the same time producing locations.

Because these narrations are inserted in social media, they are also subjected to some of their negative characteristics. One widespread criticism is the assumption that people engage less with places because they are more concerned with taking pictures than actually seeing and observing the scene. This is a particular worry for museums, and for those who hold the idea that a work of art needs to be viewed and admired appropriately. A study (Soares & Storm, 2018) found that participants who took pictures of paintings had poorer memories of it than those who had just observed. This has been used to support the idea that people should take fewer photographs when visiting cultural and touristic sites. There is no doubt that humans often offload our memories to external devices, such as cameras. Looking at the pictures after the visit is finished could help with the remembering process. However, research showed that due to a large volume of digital photos, and a lack of organization, many people do not actually frequently view or reminisce about these (Bowen & Petrelli, 2011). Nevertheless, due to social media practices, people spend a long time with the photos prior to posting, and this may contribute to the formation of better memories. Yet, remembering is only one aspect of the touristic experience. These arguments imply that touristic sites and works of art are better appreciated only if more details about them can be remembered later. Some works of art seem to beg for interaction and their meaning is often increased by interaction from viewers, which can include the act of taking selfies. There are voices claiming that Instagram has the potential to bring new dimensions to art. Suess and Budge (2018) stated that 'banning photography on the basis that it interferes with the visitor's experience could be seen as cultural elitism; expressing a view that art can only be appreciated in an orthodox manner' (2018, para. 8). The authors argued that museum visitors can use Instagram as part of their aesthetic experience, connecting to the exhibitions in a way that 'they can control and is meaningful to them' (2018, para. 12). Budge and Burness (2018) found that visitors engage with artworks when taking photos, and exercise their authority and agency, by sharing their experiences on Instagram. The authors stated that engagement, which is central to learning, occurs in the physical space of the museum and also around and through the images that visitors share on Instagram. The platform, thus, can be a way to account for and document their experience of visiting a museum.

Another criticism relates to the common complaint that photos posted on social media are inauthentic, because they depict ideal situations. These voices claim these images are flattering and forged, since people choose only to show what they want others to see, hiding any unpleasant aspects. Indeed, travel photos on Instagram follow the tendency in social media of only sharing positive things. Most captions and images are associated to feelings like happiness and wonder. Common problems that often occur during trips are shown less often or are even shown in a positive light. However, this does not mean that posts on social media are inauthentic. For Stylianou-Lambert (2012), the camera encourages performances and tourists 'spend considerable amounts of time striking poses that are related to their image of self, their dreams and achievements; poses that indicate how they want to be seen, with whom, and what they are proud to have seen' (2012, p. 1822). By posting these carefully planned photographs, tourists create stories about their travels. Goodman (2007) affirmed that tourism exists in the interplay between places and stories, and that these stories help in making sense of both, the travelling, and ourselves. For McCabe & Stokoe (2004) through these stories, travellers build an identity of being a particular type of tourist. So, on Instagram, people actively build stories that not only contribute to their identities but also add to the visual imaginary of the destinations.

This implies a more active role by tourists in the construction of the visual discourses about a place. Previous studies have noticed that tourists tend to replicate the images they constantly see in postcards, guidebooks, media, and promotional materials. Urry (1990) stated that these images help develop the tourist gaze by stating what is extraordinary and worth seeing. For the author, tourists seek to escape from their everyday lives and mundane activities, by searching for the extraordinary. This search is socially constructed by images that guide and frame the touristic experience. This creates a circle of representation in which tourists seek what has been photographed and published in tourism media, in order to photograph it for themselves. However, with the development of technological affordances, tourists can now also be seen as performers actively involved in the building of visualities, and so, more than only cultural consumers, they become cultural producers (Haldrup & Larsen, 2010). In the same vein, Stylianou-Lambert (2012) stated that, despite some social and technological constraints, people actively build their narratives, meanings, and identities and, by doing that, they are at the same time both reproducers and producers of images. As we can see, by producing their own images, with good quality and by telling their travel stories online, tourists are not only passive consumers of place representations circulated in the media but are also producers of representations, further circulating media content.

Within the perspective of Cultural Studies, we should ask how this can impact on power relations. On one hand, whilst many people know in theory that media images are often retouched and manipulated, social media enables them to do this for themselves. When people try to make good photographs, they experiment with light, angles and may additionally apply filters and retouch features. These are precisely some of the same techniques that are embedded within mainstream images. Social media and associated apps make available a series of techniques for manipulating photographs, which were previously restricted to professionals. A lot of these apps are free and easy to use and do not require any need for expert knowledge on image manipulation. By interacting and participating in the creation of media content, they are exercising their agency. Users also benefit by submitting themselves to the media logic, and by learning how to gain more followers and likes. Furthermore, those look-ing for places to visit gain insights that are more independent than those of official tourist organizations, whilst at the same time not being restricted by the limits of word of mouth recommendations.

On the other hand, users produce content for free, which is then utilized by social media for commercial interests. In this way, users serve the economic interests of these platforms, but they do not always acknowledge this, making the power of the social media more diffi-cult to unmask. Another aspect of who wins and loses with this practice is how even though Instagrammers may gain more likes and followers, in the end it is the tourist industries that are the big beneficiaries. This is because, as we saw, most places are still being represented in a good light. This means free advertising, with the added advantage of being perceived as more trustworthy. Additionally, there are those who lose with this practice, such as those residents of places that are not suitable for tourism or are not prepared for an unexpected influx of visitors.

Conclusions

This chapter sought to comprehend how visual stories on Instagram connect to spaces and how the practice of posting affects the way we travel. By travelling with the intention of posting photos, 'tourists' relation to their visual recording equipment and social media au-dience becomes an integral component of the tourist experience' (Dinhopl & Gretzel, 2016, p. 136). Exploring how people use Instagram to visually narrate their travel experiences is important because it also plays an important role in shaping narratives associated with places. Photos have always been connected to travel, but now some people seem to find on Insta-gram their reasons to travel. As one participant stated: 'Ok, a trip is to have fun, but for me the main goal of the trip is to take photos' (participant 11, pers. comment, 13.01.2018). A study (Philip, 2017) revealed that for 40% of millennials, 'how instagrammable the holiday will be' is the number one factor for choosing a travel destination. Instagram has also become a place to make sense of, and give meaning to, the photos taken during the trip: to share them with friends, family, and even strangers, whilst telling stories that help them perform being a certain kind of traveller.

The findings in this study point to the need to redefine the touristic experience and the tourist gaze, in order to adapt these to the current context of social media. After all, 'one of the principal consequences of the mediatization of society is the constitution of a shared ex-periential world, a world that is regulated by media logic' (Hjarvard, 2008, p. 129). Destina-tion management organizations have traditionally dominated the propagation of the desired images for destinations, but because of social media, these organizations need to compete with user-generated content that are themselves influencing tourists' decision-making be-haviour (Akehurst, 2009). One consequence of this is that the power of the media to shape the tourist gaze will continue to fade (Urry & Larsen, 2011). Månsson (2011) stated that a shift of control is taking place: 'from primary control by producers of tourist spaces in traditional marketing, to now including tourists in mediatized tourism' (2011, p. 1648). It is important to reiterate that, as in official photos, these places are still usually portrayed at their best. Therefore, these kinds of posts hardly impact negatively on destinations, since their representations are usually positive. However, by doing this, users are exercising their agency and are becoming a part of the production of tourism discourses, in which photos are a primary mechanism, with the possibility of producing new locations.

This chapter contributed to Cultural Studies' aims of understanding and trying to change dominance in capitalist societies, by discussing how Instagram travel photos can contribute to tourists as location creators, whilst reducing the power of traditional tourism institutions to dictate a locations' visual imaginary. It also contributed to the Media Studies field, by describing how mediatization occurs in tourism, more specifically how Social Media's embedding in everyday life affects travelling practices and what are the consequences for both for tourists and locations. However, diverse aspects of the mediatization of tourism need to continue being investigated. As Månsson (2011) warned, the circuit of tourism 'will be less predictable and more multifaceted in the future. Thus, the convergence of media and tourism consumption is like opening Pandora's Box, with layers and layers of new challenges and/or possibilities' (2011, p. 1649). Certainly, visual consumption has become one of the dominant ways in which societies interact with their environments (Crawshaw & Urry, 1997) and this is why we need to better understand how people are using Instagram in their travels, in order to fully comprehend the implications of travelling in the age of Instagram.

References

Akehurst, G. (2009). User generated content: The use of blogs for tourism organisations and tourism consumers. *Service Business, 3*(1), 51–61.

Bowen, S., & Petrelli, D. (2011). Remembering today tomorrow: Exploring the human-centred design of digital mementos. *International Journal of Human – Computer Studies, 69*(5), 324–337.

Budge, K., & Burness, A. (2018). Museum objects and instagram: Agency and communication in digital engagement. *Continuum, 32*(2), 137–150.

Couldry, N., & Hepp, A. (2013). Conceptualizing mediatization: Contexts, traditions, arguments. *Communication Theory, 23*(3), 191–202.

Crawshaw, C., & Urry, J. (1997). Tourism and the photographic eye. In C. Rojek & J. Urry (Eds.), *Touring cultures* (pp. 176–209). London, UK: Routledge.

Dinhopl, A., & Gretzel, U. (2016). Selfie-taking as touristic looking. *Annals of Tourism Research, 57*, 126–139.

Goodman, E. (2007). Destination services: Tourist media and networked places. *UCB iSchool Report 2007-004.*

Haldrup, M., & Larsen, J. (2010). *Tourism, performance and the everyday: Consuming the orient.* London, UK: Routledge.

Hjarvard, S. (2008). The mediatization of society. A theory of the media as agents of social and cultural change. *Nordicom Review, 29*(2), 105–134.

Lekach, S., & Ciechalski, S. (2017). Don't even think about bringing your selfie stick to these tourist destinations. Retrieved June 08, 2019 from: https://mashable.com/2017/07/29/selfie-sticks-banned-travel-tourist-destinations/#ZXDFErApiZqL.

Mack, R. W., Blose, J. E., & Pan, B. (2008). Believe it or not: Credibility of blogs in tourism. *Journal of Vacation Marketing, 14*(2), 133–144.

Månsson, M. (2011). Mediatized tourism. *Annals of Tourism Research, 38*(4), 1634–1652.

Matchar, E. (2017). How instagram is changing the way we design cultural spaces. Retrieved June 08, 2019 from: https://www.smithsonianmag.com/innovation/how-instagram-changing-way-we-design-cultural-spaces-180967071/.

Mccabe, S., & Stokoe, E. H. (2004). Place and identity in tourists' accounts. *Annals of Tourism Research, 31*(3), 601–622. doi: 10.1016/j.annals.2004.01.005

Pan, S., Lee, J., & Tsai, H. (2014). Travel photos: Motivations, image dimensions, and affective qualities of places. *Tourism Management, 40*, 59–69. doi: 10.1016/j.tourman.2013.05.007

Parkes, D. (2018). The top 13 best places to instagram in hong kong. Retrieved June 08, 2019 from: https://www.timeout.com/hong-kong/things-to-do/best-places-to-instagram-in-hong-kong.

Philip. (2017). Two fifths of millennials choose their holiday destination based on how 'Instagrammable' the holiday pics will be. Retrieved June 08, 2019 from: https://www.schofields.ltd.uk/blog/5123/two-fifths-of-millennials-choose-their-holiday-destination-based-on-how-instagrammable-the-holiday-pics-will-be/.

Schwarz, O. (2010). On friendship, boobs and the logic of the catalogue: Online self-portraits as a means for the exchange of capital. *Convergence, 16*(2), 163.

Seoul magazine. (2016, Oct 12). Hanbok: Not just yesterday's wear. *SEOUL Magazine*. Retrieved June 06, 2019 from: http://magazine.seoulselection.com/2016/10/12/hanbok-not-just-yesterdays-wear/.

Soares, J. S., & Storm, B. C. (2018). Forget in a flash: A further investigation of the photo-taking-impairment effect. *Journal of Applied Research in Memory and Cognition, 7*(1), 154–160.

South China Morning Post. (2018). Photo ban at one of hong kong's most unconventional tourist hotspots. Retrieved June 06, 2019 from: https://www.scmp.com/photos/hong-kong/2131892/photo-ban-one-hong-kongs-most-unconventional-tourist-hotspots?page=2.

Stylianou-Lambert, T. (2012). Tourists with cameras: Reproducing or producing?: Reproducing or producing? *Annals of Tourism Research, 39*(4), 1817–1838.

Suess, A., & Budge, K. (2018). Instagram is changing the way we experience art, and that's a good thing. Retrieved June 06, 2019 from: http://theconversation.com/instagram-is-changing-the-way-we-experience-art-and-thats-a-good-thing-90232.

Urry, J. (1990). *The tourist gaze*. London, UK: Sage.

Urry, J., & Larsen, J. (2011). *The tourist gaze 3.0*. London, UK: Sage.

Xiang, Z., & Gretzel, U. (2010). Role of social media in online travel information search. *Tourism Management, 31*(2), 179–188.

34

THE MEDIATIZATION OF SHERLOCK HOLMES

Autoethnographic observations on literary and film tourism

Annæ Buchmann

Introduction

This research article analyses the significant changes in how tourists engage with and contribute to the phenomenon of Sherlock Holmes tourism. It conceptualises the nature of experiences gained by tourists and gives insight into the role media plays in influencing their journeys. Understanding changes in visitor motivation, expectation, and behaviour allows us insights into the changing nature of fandom and fan engagement and furthermore has implications for attraction management. For this, the study examines 'links between tourism and communications as part of the social fabric in different parts of the world' (Buckley, 2010, p. 318) in the case of tourists seeking a Sherlock Holmes experience, by exploring literature and film about how media influences tourism choices, at key locations in London, United Kingdom, and Meiringen, Switzerland.

Research approach and method

This exploratory study into tourists' Sherlock Holmes' ('Holmes' from here on) journeys applies an autoethnographic method based on personal travel and personal communication, as well as observation at selected locations. Autoethnography has emerged as an accepted method over recent years (e.g., Sparkes, 2000). There are varied definitions as well as different styles of autoethnography in action. Here, autoethnography is seen as being a personalised account that drawings on 'the experience of the author/ researcher for the purposes of extending sociological understanding' (Sparkes, 2000, p. 21). It is a process in which the researcher chooses to make explicit use of 'positionality, involvements, and experiences as an integral part' (Cloke, Crang, & Goodwin, 1999, p. 333) instead of hiding these facets in academic construction (Bochner, 2000). This approach 'utilizes data about self and its context to gain an understanding of the connectivity between self and others within the same context' (Wambura Ngunjiri, Hernandez, & Chang, 2010, p. 2). This study seeks to allow a deeper understanding of the Sherlock Holmes-phenomenon by investigating an embodied experience and building on the 'aca-fans' or 'scholar-fans' perspective (Hills, 2002). The benefit is that practises and experiences are described more richly through researcher's experiences (cf. Voneche, 2001) with the added benefit that reflecting on the experience looking back to the original experience may introduce 'a measure of coherence and continuity that was not available at the original moment of experience' (Bochner, 2000, p. 270).

Practically, I began by constructing a thematically focused autobiographical timeline of the events relating to my experiences. I then expanded this timeline using a narrative structure, alternating between the autoethnography, my personal notes, and my timeline. One challenge was to remember the physical and emotional circumstances of my site visits and to reflect on these memories with systematic sociological introspection. This was a way to see how my story related to wider issues in media-influenced tourism and the broader tourism world. This is a great value of autoethnography as it methodologically appropriate for researching lived experience in, for example, complex tourism settings (Buckley, 2012). As indicated above, it meant experiencing the destination as a tourist while also reflecting on one's self and others in a move 'from participant observation to the observation of participation' (Tedlock, 1991, p. 69). The data for this analysis is derived from 21 documented days of location visits over several years including repeat visits to key locations such as Baker Street, North Gower Street, and Barts in London. I kept notes of most observations though across various media, including paper-based and electronic notebooks, audio recordings, and photographs.

In order to enhance the credibility and authenticity of the results, this study includes data beyond my ethnographic experiences with official canon material, audio-visual recordings, promotional material, and online sources including newspaper articles, publicly accessible message board postings, and blog entries.

Limitations of an autobiographic approach

Still, autoethnography remains a contested research approach and its academic rigor has been questioned (see Sparkes, 2000). Anderson (2006, p. 378) proposed five requirements to enhance the reliability and validity that I fulfil: I was the primary participant in the investigation and my experiences inform this study; have reflected on the emergence of 'aca-fan' literature; drew from further sources to reflect on and position my self-dialogue, and sought to expand the theoretical understanding of the mediatized tourism phenomenon. In fact I found myself drawn to 'structurally complex narratives, stories told in a temporal framework that rotates between past and present reflecting the nonlinear process of memory work—the curve of time' (Bochner, 2000, p. 270), which can be challenging at times, especially as my notes written on site were abbreviated. And I experienced diverse, complex, and often contradictory feelings that could lead to changes in my behaviour or attitude.

In assessing the benefits and challenges in choosing an autoethnographic approach, I opted to further aid enhance interpretation by also analysing official canons (defined as the original written work by Sir Arthur Doyle published between 1887 and 1927), audio-visual recordings, promotional material, and online sources including newspaper articles, publicly accessible message-board postings, and blog entries. These two sources of data reveal similarities and differences within the fandom(s) and the sites they visited and experienced.

The Sherlock Holmes phenomenon

The foundations of the Sherlock Holmes phenomenon were laid in 1887 when Sir Arthur Canon Doyle introduced Holmes' character to the world, in what would soon expand to 56 short stories and four novels. The character and adventures of the Great Detective quickly became popular and stage productions begun as early as 1899, and a first film was made in 1900 (*Sherlock Holmes Baffled*). By the 1940s, informal groups had evolved into full blown societies like the Baker Street Irregulars (BSI Society) which is just one of around 250 societies worldwide today. These societies' activities included holding formal dinners, writing journals, celebrating traditions, and experiencing the inevitable growing pains of a subculture, evolving in an interplay of original literary text, pastiche, and mediatized versions. Many fans decided to see Holmes as a real person and labelled this phenomenon *The Game*. Sherlock Holmes may be 'the most prolific screen character in the history of cinema' (Redmond, 1993). Castleman (2012) counts 254 interpretations, while DeWaal (1974) lists over 25,000 related productions and products. And as one of the oldest continuous franchises in existence, it stands to reason that Holmes has a correspondingly extensive fandom. 'Long before the possibilities of today's mediated world, he was one of the first characters to massively, irrevocably, step off the page and into the world' (Klimchynskaya, 2014). Thus, Sir Arthur Conan Doyle's stories about Holmes are believed to be the first works of fiction to have created an organised and active fandom that made its presence known all over the world.

One of my first Sherlock Holmes' connections was forged in Edinburgh, Scotland: it is here where Doyle lived and wrote most of his stories, and it is here that some of my earliest memories begin. The Scottish capital and its dark-faced buildings soon came to symbolise the Victorian era to me, which I saw as a complex agglomeration of an age of discovery and scientific inquiry. For me, Doyle was very much associated with these times and developments. Over time, I noticed how the Sherlock Holmes phenomenon grew with further stage plays, movies and exhibitions, to the emergence of Baker Street pubs even in Boulder, United States, and downtown Melbourne, Australia.

And there has always been a strong Holmes presence in London. This is somewhat surprising as Doyle wrote most of the original stories in Edinburgh, and in fact had only a very limited knowledge of London at the start of his writing and used a Post Office map for guidance for the first stories (Werner, 2014). Still, the Great Detective soon became synonymous with Victorian London and with it his supposed home address *221 Baker Street*. Soon, a nearby Baker Street office experienced a flurry of fan mail starting in the 1940s that was being dealt with by a specially appointed secretary (Pollard-Gott, 2010). It was not for many years that the official Holmes museum was finally assigned the famous address '221b Baker Street' in a disrupt of the actual official street numbering. Of course, further London landmarks also found their way into Doyle's work and increased the range of possible tourism locations. Today, fans today can observe diverse sign markers such as plaques, statues, and museums commemorating Doyle and Holmes (cf. Laing and Frost, 2012) all over London, and worldwide.

The original Reichenbach Falls

A major location for Sherlock Holmes' enthusiasts is situated in Switzerland, where Holmes' tourism tentatively started in the 1940s in the small town of Meiringen.

I have visited the Meiringen region several times throughout my life, at first with family and later with friends. I also visited Lucens, though I did not enter the Castle when I was there in the mid-1990s. I still have vivid memories of observing people in Meiringen dressed in iconic short cloaks and deerstalker hats with smoking pipes. It is particularly those pipes who I seem to remember, and while I recall them as (most likely incorrectly, as they are supposed to be calabash) Meerschaum pipes, I do remember the surprisingly sweet smell of tobacco, that was quite different from the more stingy cigarette smoke I was more familiar with at the time. I also remember a Doyle-reading circle in the late afternoon in the foyer of our hotel, which I quietly observed from a distance. I remember marvelling at this group of grown-ups, who found such strong and communal interest in books and travelled as a group.

The reason one of the most iconic locations of the literary canon is in Switzerland, is most likely that Doyle was an early adapter of skiing. He knew the area well and used the setting to see Holmes and Professor Moriarty (seemingly) plunge to their deaths together down the local Reichenbach waterfall in 'The Final Problem' in December 1893. The public reactions to the detective's death were impressive:

> British society dressed in mourning. Black armbands were worn to commemorate the great detective's passing. People cancelled their subscriptions to The Strand (the newspaper that then published the Holmes stories), but not before sending piles of angry letters. Even more piles of pleas and petitions arrived on Doyle's doorstep. Obituaries appeared in newspapers.
>
> *(Klimchynskaya, 2014)*

The 'death of the world's first consulting detective was taken up by the wire services and reported all over the world as front-page news' (Duane, 2014). In 1903, the so-called 'hiatus' concluded when Doyle finally relented and admitted in 'The Empty House' that Holmes had survived his fall. Still, the Reichenbach Falls had become an attraction, though early descriptions are rare. Either way, Doyle has been accredited for his significant contribution to Swiss tourism 'through his writings and lectures, and together with his great friend Henry Lunn' (Bechtel, 2014). In 1986, the Sherlock Holmes Society of London visited the Falls in an expedition that saw participants being 'assigned roles from the canon... dressed in Victorian garb... [who] then climbed the famous cliffs to stage Holmes and Moriarty's struggle. The trip drew the attention of newspapers the world over' (Klimchynskaya, 2014). Such 'pilgrimages, continue, and it is a somewhat common picture to see people traveling in Victorian attire to the Reichenbach Falls, especially around significant dates such as anniversaries' (Sherlock Holmes Museum staff, personal communication, 2014). The authenticity of the places is marked by plaques at the waterfall and in the nearby town of Meiringen. There is a Holmes statue made by the same sculptor who created the one outside the London 'Baker Street' tube station, and a local museum that includes a mock-up of the sitting-room of 221b Baker Street (similar to the Museum in 221b Baker Street and the Sherlock Holmes pub in Northumberland Street, both in London). Another museum, the 'Château de Lucens' claims to host the most authentic recreation of Holmes's drawing room at 221b Baker Street. The windows, fireplace, and walls of the sitting room in the Lucens museum were used in the 1951 'Festival of Britain' exhibition, while the furniture and fittings belonged to Arthur Conan Doyle (Lucens Tourisme, 2014). Still, more tourists visit the Meiringen Museum than the Lucens one.

I remember my first trip to the Reichenbach Falls. It was early spring at the time of our visit, or at least it seemed so: I remember seeing my breath, and ice in shaded areas of the river. It was difficult going and slippery, and some of the adults proceeded with a near-comical caution. The Falls themselves seemed brooding and hostile. I remember taking pictures on an old-fashioned camera but not developing or seeing those photographs afterwards. Now I wish I could remember more of those early visits, but my visual memory has been overlaid with media images of the Falls. Still, over the coming years I would return to Meiringen, eating at the Sherlock Holmes Hotel, buying postcards featuring Doyle on skis, the Reichenbach train or, really, any available Holmes souvenirs, and of course, revisiting the Reichenbach Falls. Over time I noticed some changes, most significantly, an increase in Sherlock Holmes-related signage and merchandise.

And while the Holmes' franchise kept developing, the domineering narrative continued to be that of a middle-aged man living in Victorian London. Then the Game changed.

The BBC Sherlock phenomenon

In 2009, the BBC commissioned a remake of Sherlock Holmes to Steven Moffat and Mark Gatiss ('BBC Sherlock' from here on) set in contemporary London. The portrayal of a young Sherlock Holmes by Benedict Cumberbatch alongside his Doctor John Watson (Martin Freeman) was complimented by a generous use of iconic urban London landmarks (2010, 2012, 2014 & 2017) and has been described as a 'visual love letter'.

I am fascinated with how the creators of BBC Sherlock, Gatiss and Moffat are playing with the literary canon. For example, the 'Criterion' is the restaurant where Sherlock and John are introduced to each other, and nowadays even features a brass plaque commemorating this occasion (Figures 34.1 and 34.2). However, in the BBC version, John Watson and Mike Stamford meet more casually and discuss flatmate options in a public park—while sipping from coffee take away cups branded with 'Criterion'. It becomes a popular activity on newsgroups and bulletin boards to identify such 'canon nods'.

BBC Sherlock soon became a phenomenon with considerable reach as the third series became the UK's most watched drama series since 2001 (Jones, 2014). The show was a massive seller with global success (Hill, 2014) and sold to over 200 territories (BBC, 2014). New fans emerged, book sales of the original stories increased and literary, but especially film, location tourism increased. However, while the original canon and newest adaptation share locations, one crucial location has been re-located from its Swiss waterfall. Instead, the iconic name is retained in the story but attributed to BBC Sherlock's recovery of a Turner painting of the Reichenbach Falls, resulting in the press headline 'Reichenbach Hero Sherlock Holmes'. The BBC adaptation then places the final confrontation between Moriarty and Holmes on the rooftop of The Royal Hospital of St Bartholomew, also known as Barts.

Figure 34.1 Criterion Restaurant, London, UK
Source: Buchmann, 2014.

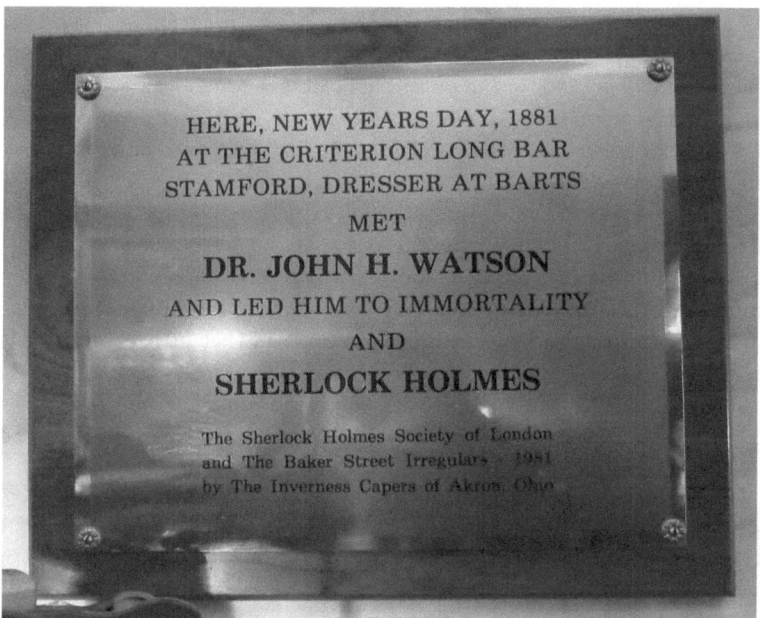

HERE, NEW YEARS DAY, 1881
AT THE CRITERION LONG BAR
STAMFORD, DRESSER AT BARTS

MET

DR. JOHN H. WATSON
AND LED HIM TO IMMORTALITY

AND

SHERLOCK HOLMES

The Sherlock Holmes Society of London
and The Baker Street Irregulars 1981
by The Inverness Capers of Akron, Ohio

Figure 34.2   Sherlock Holmes-Plaque in the Criterion Restaurant, London, UK
Source: Buchmann, 2014.

I remember being positively intrigued by the visual narratives and smart twists in BBC Sherlock but genuinely puzzled by the cliff-hanger of its sixth episode. Here, Sherlock falls to his death in the middle of London! But Sidney Paget's 1893 illustration of Holmes and Moriarty at Reichenbach Falls was legendary and had been copied hundreds of times in audio-visual media. Furthermore, for over 100 years, people had been traveling in Victorian attire to the original Reichenbach Falls in Switzerland, especially around significant dates such as anniversaries.

BBC Sherlock had initially interpreted several locations as supporting characters, including the hospital. It is here that the initial meeting between Holmes and Watson takes place. Later on, Mycroft Holmes describes Barts as his brother's 'home away from home'. And in the season 2 finale, it is on the roof of Barts that the final confrontation and Holmes jump is set, allowing for an impressive backdrop to the cliff-hanger scene. The choice and change of location see BBC Sherlock retained in his beloved London with glimpses of the city visible throughout the scene. Incidentally, the final minutes show John breaking down over Sherlock's lifeless body on the pavement below, only for the last camera angle to reveal a very much alive Holmes observing his friend at the burial site. It was this revelation that would drive a significant public debate over the following two years hiatus, when message boards and even newspapers sought to figure out how the detective could have survived such a fall. Either way, it did not take long for tourism to that site to grow.

At several locations, and especially on the Baker Street's stand-in North Gower Street, I notice organised as well as self-organised tours by individuals and groups, often sporting fans in BBC costume or merchandised shirts; and many carrying smart phones and/or tablets, which they use to re-play film clips. I have taken to consciously pass Barts over the last week. Each time and at different hours I observe individuals who, and more often groups which, can be identified by their BBC Sherlock costume consisting of a sweeping great coat, preferably with a blue scarf. There are now messages on the hospital's walls and windows, which have become places for fan messages. I stand in front of one of the iconic English red phone boxes and am surprised to realise the number of notes and drawings clearly referencing Sherlock with popular comments such as 'I believe in miracles' and 'Come back to me. John'—notes clarifying what fans expect from series three. Tellingly, the TV writers acknowledged their awareness of these fan activities.

There is no official signage yet besides the fan-made ones, and Brown (2012) comments on how such 'an informal shrine at this site is a splendid idea, and very much in keeping with the long tradition of pretending Holmes is a real man, rather than a fiction'. It seems that the iconic status of Barts has been well established within BBC Sherlock's canon as well as in fanon, meaning fandom-expanded canon, and that its tourism will continue to grow; indeed, more and more media outlets report on the new Sherlock locations (cf. Metro, 2013). When the BBC 2016 Sherlock Special 'The Abominable Bride' was set in a dream-like atmosphere in Victorian times, the key confrontation between Holmes and Moriarty was relocated once again a waterfall deep in the mountains, suggesting a reference to the original Reichenbach Falls. However, the pilgrimage to the urban Reichenbach 2.0 location remains strong. So how is it possible for the audience to be so accepting of the geographical re-location of such an iconic literary and visual setting and deeming the added location to be a true landmark?

Analysis

There has been a long tradition of devotion to the Great Detective and creative output ranging from pastiches to re-enactment of fans. It seems that both the Holmes and BBC Sherlock fandom are close and yet differ in their canonical reference (Doyle canon versus the expanded BBC Sherlock canon), their iconography and sign reading (for example, regarding the *Criterion*). The fans speak the same language and yet have different codes; the new fandom shares the source of the historical text and understanding but now has 'tumblr extensions' (a popular social media site). In particular, the online behaviour of BBC Sherlock fans is showing active engagement on message board sites, YouTube (with fan-made trailers and compilations), tumblr (with its vast visual outputs) (compare with Couldry (2000, p. 70) who mentions the importance of surrounding 'vast textual field'), Twitter (where real life sightings of cast and crew are relayed instantly), and in fanfiction featuring slash pairings of Sherlock and John ('Johnlock'), and thus unlike most pastiches written in the style of the original canon by earlier fans. In short, the new breed of 'Sherlockians' displays a variety of 'fan-ish' behaviour, and their strong online presence shows the marked evolution of a fandom that has developed from its literary roots into a mediatized world.

Ever since MacCannell (1976) popularised the notion of 'authenticity' as a putative motive for tourist activity and experience, controversy over the applicability and usefulness of the term has continued. And yet its dominance in travel narratives has ensured that it maintains a central concept in understanding the tourist experience (Li, 2000; Reisinger & Steiner, 2006; Steiner & Reisinger 2006; Wang, 1999). Reisinger and Steiner (2006) have argued that scholars should abandon both the concept and term 'object authenticity' and recommended the use of the term 'existential authenticity' instead. Wang (1999) had argued that 'existential authenticity, unlike [the] object-related version, can often have nothing to do with the issue of whether toured objects are real' (Wang, 1999, p. 359). This study shows that both literary and mediatized tourists remain concerned with 'authentic' depictions that are 'loyal' to the original. Even for mediatized tourists, the seeking out of the exact spot is crucial and is unmistakably identified by freeze-framing the television show recording on smart phones and tablets. Often, these virtual sign markers are included in a photograph of the toured location. Thus, physical places and social settings (as demonstrated in shared markers and in social media) remain pivotal to understanding tourists' sense of what is and is not an authentic experience. In this, it will involve returning the 'place'—which includes its meanings as a physical, spatially 'fixed' location understood in terms of its physical geography—to the tourist experience. However, the contradiction lies in the fact that two destinations claim to present the same attraction. And yet, Sherlock Holmes, at least in this case, navigates this apparently complex pursuit of the authentic in a way that reveals its anchor in some basic social, moral, and spiritual concerns. Tourists navigate the multi-layered construction of the 'real locations' they wish to experience—from the fictional books, through the mediatized version of them, through the mix of geographical places, and fictional places. In this time of virtuality, the embodiment of the individual proves crucial; from the point of view of the tourists, their socially constructed understandings, and their direct, empirical encounters merge when they stand on the doorsteps of 221b Baker Street or lie on Barts footpath.

As shown, at times it is possible to distinguish 'sub-fandoms' at locations; for example, the more traditional fans favour the iconography of deerstalker hat and Inverness cape and smoking a large curved calabash pipe, as institutionalised by early theatre plays (starting with William Gillette in 1899). However, BBC Sherlock-fans favour dark coats in reference to the character's iconic Belstaff coat, and the accompanying blue scarf. There are also differences in displayed behaviour and discussions of the locations: the more traditional Holmes fans seem to display affirmative behaviour (through discussing books) while modern fans seek transformational experiences (through roleplaying). And mediatization is demonstrated when The Barbican's *Hamlet* sessions became the fastest ever selling tickets because they featured Benedict Cumberbatch and attracted a non-traditional audience. Ultimately, it could be argued that the BBC fandom draws from a greater diversity of locations including those based on the originally described locations by Doyle (e.g., Criterion, The Strand, Barts); institutions like 221b Baker Street and the museum(s); the filming locations of BBC Sherlock; and cast and crew-related sites. It seems the allure of Sherlock Holmes has never been stronger, and even with a very mixed reception to BBC Sherlock season 4 (2017), fans remain eager to explore literary and film locations.

Conclusions

This study sought to conceptualise the nature of experiences gained by Sherlock Holmes tourists and to gain insight into the role of media as sign markers for tourists' choices. In doing so, it focused on tourism related to a pivotal scene and location in the original literary Sherlock Holmes canon (1887–1927) and contrasted it with the recent BBC adaptation of 'Sherlock' (2010–2017). The observations show how the mediatization and with it the fan tourism experience has changed. In observing visitors in several locations, I found evidence for demonstrated principles of tourism development, with suggestions of a sector on the verge of formalizing and the significant supportive role of social media. The evidence also showed that society has moved on from literary and film tourism to mediatized tourism; a shift poignantly described by Månsson (2011) in an early visionary article reframing film tourism research.

The autoethnographic approach has shown transformation of the self and society as reflected in observable changes in the phenomenon of Sherlock Holmes tourism relating to visited location, choice of clothing and accessories, as well as reading and interpretation of the original and extended text. And while the story has also been my story, it is not just my story as it is produced and consumed within the constraints of the wider culture. In tracing the changes within Sherlock Holmes tourism over several years, I describe a changing world. It has been said that 'every generation gets the Holmes they deserve'. In the latest BBC interpretation, the character is no longer a hero by Victorian standards but instead represents a modernised rendering of heroic characteristics; more specifically, the character might be seen as a neoliberal ideal and an instrument of reassertion of Britishness—the new cool indeed.

> Sherlock offers the viewer a chance to experience modern day London. Just as Conan Doyle is credited to have described the streets of London so well as to make the city a living, breathing character, so, too, do Moffat, Gatiss, and the production team bring London to life on screen in Sherlock. Various locations, including historically and culturally significant ones such as Buckingham Palace also lend to this sense of being let in on British culture.
>
> *(Stein & Busse, 2012, p. 60)*

I am partially disquieted with the re-writing of Sherlock Holmes history and profess to a certain confusion stemming from the displacement of the Reichenbach Falls. The modern re-interpretation of Sherlock is anything but coincidental and part of a bigger project of nation building (cf. Buchmann & Frost, 2011). It demonstrates the power audio-visual media has to re-write history, like in Scotland, where a William Wallace Monument controversially showed a Mel Gibson likeness following the popularity of his 'Braveheart' movie (1995). Similarly, Tolkien's England barely capitalises on its Tolkien association despite Birmingham being home to several key attractions of the 'Lord of the Rings', including the Hobbiton Mill, and the Two Towers. However, for the time being, New Zealand claims to be the 'Home of Middle-earth' and is strengthened by the great match of its market image with a fictional one (Buchmann, Moore, & Fisher, 2010). And similar to the perceived match between the images of New Zealand and Lord of the Rings, there is a now a match between Britain and Sherlock, re-affirming each other. The role of landscape and setting is high, as Doyle's context had always been meaningful. The writer made conscious connection, which allows us to put his literature into context. Now new context is created, and the BBC Sherlock phenomenon requires discussions of power, power of image formation, and promotion; as well as the role of social media and its overstated ability to break hegemonic discourse. In fact, none of the original destinations has a marketing budget rivalling the reach of BBC Sherlock's production; consequently, finding it hard to showcase their location and grow tourism interest.

Personally, I found my own concepts of sense making, history, and place challenged though the dominating feeling was the joy of re-discovery, and an increasing interest in showcasing the complexity of modern identity, image formation and negotiation of it. From a practical view, BBC Sherlock tourism was mostly ignored. Until 2014, VisitBritain's website only referred to locations of the Hollywood blockbuster movies with Robert Downey Jr. (2009, 2011) which did not attract much tourist interest. Some information on BBC Sherlock appeared on VisitLondon's (2014) website while the location-rich city of Cardiff as of 2020 is still not officially promoting this link. Either way, the initial lack of official tourism-related information led to thriving informal tourism fuelled by social media-based communication. I have observed the influence the mediatized world has on our travel behaviour resulting in a cautionary tale of the thin line of rejuvenation and hype, image formation, and stability. Today, both dedicated fans and coincidental tourists visit literary locations near Meiringen.

Some tourists carry Sherlock Holmes-related books and portray a sense of scholarship. But it is, as yet, rare to encounter fans of BBC Sherlock (Sherlock Holmes Museum staff, personal communication, 2014). The managerial implications show that influential factors outside the control of Meiringen and Switzerland are re-shaping Sherlock Holmes tourism. There is a sense of ownership of the character due to the long-standing association with the author and the vivid descriptions in his work; however, the tourism sector may need to re-package its experience by pushing the original canon more as well as acknowledging fanon. This study shows how Sherlock Holmes continuous to be adapted and modernised, all while delivering examples of mediatization: 'The Game is Afoot!'

References

Anderson, L. (2006). Analytic autoethnography. *Journal of Contemporary Ethnography, 35*(4), 373–395.

BBC (2014). Sherlock in five languages – BBC worldwide showcase. Retrieved from: http://www.youtube.com/watch?v=jNzeuzYPiLc (accessed 14 September 2014).

Bechtel, D. (2014). Sherlock Holmes success no mystery. Retrieved from: http://www.swissinfo.ch/eng/sherlock-holmes-success-no-mystery/12820 (accessed 14 September 2014).

Bochner, A. P. (2000). Criteria against ourselves. *Qualitative Inquiry, 6*(2), 266–272.

Brown, J. (2012). A phone box shrine to Sherlock At St Barts. Retrieved from: http://londonist.com/2012/11/a-phone-box-shrine-to-sherlock-at-st-barts.php (accessed 6 January 2014).

Buchmann, A., & Frost, W. (2011). Wizards everywhere? Film tourism and the imagining of national identity in New Zealand. In Frew, E. & White, L. (Eds.), *Tourism and national identities: An international perspective* (pp. 73–85). Abingdon, UK: Routledge.

Buchmann, A., Moore K., & Fisher D. (2010) Experiencing film tourism: Authenticity & Fellowship. *Annals of Tourism Research, 37*, 229–248.

Buckley, R. (2010). Communications in adventure tour products. Health and safety in rafting and kayaking. *Annals of Tourism Research, 37*, 315–332.

Buckley, R. (2012). Rush as a key motivation in skilled adventure tourism: Resolving the risk recreation paradox. *Tourism Management, 33*(40), 961–970.

Castleman, A. (2012). Sleuthing Sherlock: On the trail of the most filmed character ever. Retrieved from: http://shine.yahoo.com/visit-britain-us/sleuthing-sherlock-trail-world-most-filmed-character-175725835.html (accessed 6 January 2014).

Cloke, P., & Crang, P., & Goodwin, M. (1999). *Introducing human geographies*. London, UK: Arnold Press.

Couldry, N. (2000). *Inside culture: Re-imagining the method of cultural studies*. London, UK, Thousand Oaks, CA, and New Delhi, India: Sage.

DeWaal, R. B. (1974). *The universal Sherlock Holmes*. Boston, MA: New York Graphic Society

Duane, D. (2014) The affair of the black armbands. Retrieved from: http://dianeduane.com/outofambit/2012/01/17/the-affair-of-the-black-armbands/ (accessed 6 January 2014).

Hill, L. (2014). How 'Sherlock' made Holmes sexy again. Retrieved from: http://www.rollingstone.com/movies/news/how-sherlock-made-holmes-sexy-again-20140124#ixzz2s4QlStgj (accessed 6 January 2014).

Hills, M. (2002). *Fan cultures*. London, UK: Routledge.

Jones, P. (2014). Sherlock is most watched BBC drama series for over a decade. Retrieved from: http://www.radiotimes.com/news/2014-01-22/sherlock-is-most-watched-drama-series-for-over-a-decade (accessed 14 September 2014).

Klimchynskaya, A. (2014). Sherlock Holmes: The original fandom. Retrieved from: http://www.denofgeek.us/tv/sherlock-holmes/232593/sherlock-holmes-the-original-fandom (accessed 6 January 2014).

Laing, J., & Frost, W. (2012). *Books and travel: Inspiration, quests and transformation*. Bristol, UK: Channel View.

Li, Y. (2000). Geographical consciousness and tourism experience. *Annals of Tourism Research, 27*(4), 863–883.

Lucens Tourisme (2014). Musée Sherlock Holmes. Retrieved from:http://www.lucens.ch/TOUR/Sherlock.html (accessed 14 September 2014).

MacCannell, D. (1976). *The tourist: A new theory of the leisure class.* New York, NY: Schocken Books.

Månsson, M. (2011). Mediatised tourism. *Annals of Tourism Research, 38*(4), 1634–1652.

Metro (October 22, 2013) Sherlock: Top 10 London locations. *Metro Daily London.*

Pollard-Gott, L. (2010). *The fictional 100: Ranking the most influential characters in world literature and legend.* New York, IN: IUniverse.

Redmond, C. (1993). *A Sherlock Holmes handbook.* Toronto, Canada: Dundurn Press.

Reisinger, Y., & Steiner, C. (2006). Reconceptualising object authenticity. *Annals of Tourism Research, 33*(1), 65–86.

Sparkes, A. (2000). Autoethnography and narratives of self: Reflections on criteria in action. *Sociology of Sport Journal, 17,* 21–43.

Stein, L. E., & Busse, C. (2012). *Sherlock and transmedia fandom.* Jefferson, NC, and London, UK: McFarland.

Steiner, C. J., & Reisinger, Y. (2006) Understanding existential authenticity. *Annals of Tourism Research, 33*(2), 299–318.

Tedlock, B. (1991). From participant observation to the observation of participation: The emergence of narrative ethnography. *Journal of Anthropological Research, 47*(1), 69–94.

VisitLondon (2014). Sherlock Holmes London Itinerary. Retrieved from:http://www.visitlondon.com/things-to-do/sightseeing/one-day-itineraries/sherlock-holmes-london (accessed 14 September 2014).

Voneche, J. (2001). Identity and narrative in Piaget's autobiographies. In J. Brockmeier & D. Carbaugh (Eds.), *Narrative and identity: Studies in autoethnography, self, and culture* (pp. 219–246). Amsterdam, Netherlands: John Benjamins.

Wambura Ngunjiri, F., Hernandez, K., & Chang, H. (2010). Living autoethnography: Connecting life and research. *Journal of Research Practice, 6*(1), 1–17.

Wang, N. (1999). Rethinking authenticity in tourism experience. *Annals of Tourism Research, 26*(2), 49–370.

Werner, A. (2014). *Sherlock Holmes: The man who never lived and will never die.* London, UK: Museum of London.

35

CULTURAL INTIMACY OF FANS/TRAVELLERS

Popular culture and the politics of classification

Andreja Trdina, Barbara Pavlaković and Maja Turnšek

Introduction

As pop cultural tourism is primarily a collectively driven phenomenon, characterized by motives of social belongingness (Gyimóthy, Lundberg, Lindström, Lexhagen, & Larson, 2015), fan cultures have long been an important focus of research on the interrelationships between media and tourism. The theoretical advances put forward so far have acknowledged the practical and emotional engagements of fans as tourists, the role of authentic relationships within the community of fellow travellers and respective fan validation of experience as key elements of pop cultural tourism (Buchmann, Moore, & Fisher, 2010; King, 1993). This contribution aims to extend the research on fan cultures in tourism to Herzfeld's (2005) concept of cultural intimacy which has been in these debates so far overlooked. We investigate cultural intimacy as a specific aspect of fellowship or common sociality that fans of low legitimacy genres engage in vis-à-vis outsiders and that is being manifested in external embarrassment or discomfort of sharing one's fandom with non-fans in particular, and ground our discussion within politics of classification and taste distinctions (Bennett et al., 2009; Bourdieu, 1984/2010). Based on qualitative and quantitative empirical research, we consider illustrative cases of performing cultural intimacy and the role cultural capital plays in structuring this sensibility among fans. In this way, our study contributes to fan culture debates by presenting a more socially sensitive perspective on cultural consumption, dealing with socially inflected ways in which the fans engaged with, and make sense of, the TV text and their media-induced travels. We thus start from the disciplinary perspective of cultural sociology which draws on Bourdieu's heritage (particularly on his conception of relational organization of the social) while also revising significantly his main arguments regarding cultural taste, practices of cultural consumption, and social power (Bennett, 2006; Savage & Gayo, 2011; Warde, Wright, & Gayo-Cal, 2007).

We focus the study on the case of 'der Bergdoktor' (a popular Austrian-German TV series 2008–on-going) phenomenon in Slovenia and its induced tourism. Drawing on focus groups and survey among Slovenian Bergdoktor fans, we aim to demonstrate how cultural intimacy is articulated in many situational cases of feeling uncomfortable with the outsiders/non-fans, indicating that the sense of fans' enjoyment may be also a source of external embarrassment. We argue that such manifestations of cultural intimacy can be regarded as a reflection of perceived cultural hierarchies and value discourses surrounding popular culture. We suggest that fans as travellers acknowledge the effects of politics of classification and taste distinctions and consequently feel the fear of being ridiculed. What is more, we seek to demonstrate that cultural intimacy is not equally distributed among fans, rather it is conditioned by their cultural capital or social position. Cultural theories seem to be in considerable distress in dealing with where to place the material in relation to the symbolic. According to Reckwitz (2002), the social has been redefined as the cultural in many theoretical branches, furthering the perspective that orderliness of social order is a result of symbolic structures. Our main goal is, conversely, to place the significance of the 'material' back into the debates on culture and in this way offer sociological account on fan culture. Our contribution thus proposes a subtler understanding of the practices of pop cultural tourists with special regard to taste hierarchies and social power.

Hierarchies of popular culture: cultural legitimacy and distinction reconsidered

Recent studies suggest that Bourdieu's contention on the tension between high and low culture proved to be considerably muted today. Not only that Bourdieu underestimated the capacity of broadcasting to cut across cultural fields and complicate the relations between audiences, as argued by Bennett et al. (2009), the relationship between high culture and popular culture itself is undergoing a profound transformation as well. Specifically, 'high-pop' (Collins, 2002) – in terms of many forms of popularization of elite tastes like superstar opera singers, commercial theatres, TV literary adaptations, etc. – is blurring the distinction between elite culture of refinement and popular culture of amusement. Additionally, exclusive highbrow snobbery has also been contested, most prominently through cultural omnivorousness thesis (Peterson, 1992). As Frow (2002, p. 70) suggested, in recent years, it has come to 'a convergence ... between the mass-marketing of high-cultural rarity and the aestheticization (rarefication) of certain form of popular or commercial culture'. This challenged the legitimacy of validated distinction between high and low tastes and the lingering cultural authority of high culture as the legitimate culture.

Nonetheless, literature on social organization of cultural consumption (relationship between social stratification and cultural consumption) focused so far mainly on high cultural formats, such as music, reading, and visual arts. Consequently, socio-cultural divisions articulated in the field of television consumption have long been almost entirely neglected in studies on cultural consumption. Jontes (2014) attributes the lack of consideration of television consumption in relation to the class and/or cultural capital to the nature of television viewing, which is commonly bounded to the private sphere and, consequently, not so much related to the public display of status. TV, considered as a bearer of a mass culture, has long served as a negative point of reference in relation to which other cultural practices register their cultural superiority. Many authors (Garnham, 1993; Lahire, 2008) also discuss the impact of television in flattening out the relationships between classes and culture so that class distinctions are no longer so easily identifiable in terms of sharply differentiated tastes. Garnham (1993, p. 188) argues that what we appear to see with television is 'a breaking down of the class-specific patterns of consumption that Bourdieu identifies elsewhere'. Divisions within the television audience tend to be fluid and overlapping, as argued by Bennett (2006), and so they are less sharply drawn than in other cultural fields. However, this does not diminish their significance. Contrary to idea that the link between class and patterns of TV preferences and practices is non-existent or weak, Kraaykamp, van Eijck, Ultee, and van

Rees (2007) confirm that media preferences are highly differentiated and formed into clear taste repertoires according to age, gender, and class. Even though these cultural practices are typically undertaken at home, and thus not directly visible as status marker, status effects are significant, because preferences for TV programs become regular conversation topics and their popularity differs between social strata (ibid.). In his UK study, Bennett (2006) also demonstrated that practices of distinction are operative within the space of broadcast television, showing that occupational class and level of education play important roles in stratifying television audiences along traditional high/low line. Bennett (2006) provided empirical evidence of three levels of genre legitimacy: high legitimacy genres (news, documentary, arts, new drama etc.), medium legitimacy genres (comedy/sitcoms, sports, films, cookery etc.), and low legitimacy genres (quizzes, chat shows, soap operas, reality TV). This implies that the division between high and low is actually operative within each cultural field, including pop culture, which takes over as a primary field in the practices of distinction and thus forms its own cultural hierarchies (the difference between a high popular culture with higher legitimacy and a low or commercial mass culture with lower legitimacy). In this way, today 'legitimate culture … extends beyond specialist expertise and appreciation of canonical figures within any one field (which may have been a concomitant of 'snob culture') and involves a more omnivorous orientation' (Bennett et al., 2009, p. 172).

Because of its accessibility and absence of exclusivity, television may not actually play a central role in forming and maintaining cultural hierarchies today. Nevertheless, it is not a 'no-brow' distinction-free zone, but noticeably differentiated space of consumption. It could be argued that there are more complex and subtle distinction patterns at work in today's highly fractured cultural order, and that cultural differences as class distinctions persist, only the way in which they are articulated is changing significantly. Culture (and cultural taste) as a source of exclusion therefore retains its central place in the reproduction of social inequalities. Social groups reproduce their status also by monopolizing distinct travel practices, travel aspirations, and destination preferences (Ahmad, 2014). With the rise of modern global tourism and democratization of tourist practices, the distinction has shifted from 'what' to 'how' as status pursuits have become grounded more in the modes and attitudes towards travelling than in travel as such. Nevertheless, the distinctive ways in which different social classes consume tourism today is still vastly under-researched, as argued by Mowforth and Munt (2003). Especially, as we will show below, both the conceptualizations and empirical research into ways that popular culture tourism is considered as a source of both external, class related, embarrassment, and a source of common sociality amongst the fans – what Herzfeld (2005) defines as cultural intimacy – is still an uncharted terrain.

'Der Bergdoktor' mania in Slovenia

Der Bergdoktor, a German-Austrian drama TV series (2008–), gained a considerable cross-generational audience in Slovenia when first aired in summer 2016 on a local commercial TV channel. TV series achieved record breaking ratings when it was first aired in an otherwise increasingly high-choice media environment. It was the one that captivated the most viewers in the summer period among domestic and foreign series for the last ten years. In the target group from 18 to 54 years, it was mostly watched by viewers with a university degree over 45 years of age; in average, the TV series was watched by almost every other Slovenian, i.e. 39% of viewers who watched live TV – every other Slovenian woman and every third Slovenian man. In general, the highest ratings were in the age group above the age of 55 (Kotnik & Cah, 2016). These numbers are impressive especially in the context of pervasive global trend of TV audience fragmentation. High audience ratings prompted an increased tourist interest from Slovenia in filming location of Wilder Kaiser region Tyrol (comprising four villages Elmau, Going, Scheffau, and Söll) in years 2016 and 2017 with offerings of organized commercial travels by Slovenian tourist agencies.

Der Bergdoktor TV series focuses on idealization of the Alpine way of living, a bucolic return to the tradition, family values, and nature. Through the portrayal of enchanted village-life of order, diligence, and humility, episodes focus on dramatic and emotional stories about private life and professional work of the main character, doctor Martin Gruber, offering audiences distinct melodramatic pleasures. Due to its simple, recognizable and recurring storylines, predictable endings, simple and stereotypical characters TV series could be characterized as an easy, one-layered TV drama and thus, in line with Bennet's (2006) evidence on internal differentiation/hierarchization of TV field, considered as a TV text with lower legitimacy.

Fan dis-identifications: the cases of performing cultural intimacy

Herzfeld (2005, p. 3) defines cultural intimacy as 'the recognition of those aspects of a cultural identity that are considered a source of external embarrassment but that nevertheless provide insiders with their assurance of common sociality'. In the first part of our analysis, we identify concrete performances of cultural intimacy fans as tourists engage in when talking about their viewing of series or their trip to Bergdoktor's Tyrol. For this purpose, we have conducted three in-depth focus groups. The participants were Bergdoktor fans who have travelled either on their own or via organized travel-agency package to Austrian Tyrol region, the location of the TV series. The focus group participants of different age, gender, and place of residence were mobilized using a convenient sampling technique: via posting a request to participation on the mailing list of the students of the Faculty of Tourism, University of Maribor and via snowballing amongst the researchers' friends and acquaintances.

'One may be called upon to identify oneself – to characterize oneself, to locate oneself vis-à-vis known others, to situate oneself in a narrative, to place oneself in a category – in any number of different context', argue Brubaker and Cooper (2000, p. 14). Here, we are concerned with the reverse process of dis-identification – with ways fan identity is being in particular circumstances purposely refused by strategically displacing fan's interest, thus distancing oneself from fan's identity. While all focus group participants had no reservations stating they enjoy der Bergdoktor, some were much more careful in proclaiming themselves proudly as fans, albeit they all joined the Tyrol fan trip. Furthermore, similar to Ang's (1985) findings in canonical study Watching Dallas, many of our informants who admittedly enjoyed watching the series felt the need to be apologetic about their viewing pleasure:

> 'It's a bit for fun of course. Just for relaxation it is nice to watch it.'
>
> (Participant N. G., age 18)

> 'Well, surely it's better to watch such shows than some kind of quarels or arguing.'
>
> (Participant J. G., age 48)

Comparable to Ang's Dallas viewers, our informants seem to have a strong awareness that in the dominant cultural hierarchy of value theirs is perceived as a lowly pleasure. What is more, we documented particular, very situational, instances of fan dis-identifications that we discovered in fans' testimonials about narrating their trip to others and that indicate diverse efforts of denying or evading fan identity – efforts not being recognized as a part of der Bergdoktor's fan community. We understand these fan dis-identifications as a result of the workings of cultural intimacy. One of such cases, recorded among focus group participants, was not telling co-workers why you need a day off when going to the Bergdoktor trip, in this way not discussing what feels to be a weakness before a 'foreign' audience (Herzfeld, 2005). Whereas if going on vacation to Greece or skiing in Alps, they would presumably gladly share the information with the outsiders as well. Another significant example illustrating the elusive effects of cultural intimacy is intentionally reframing your travel experiences of Bergdoktor's Tyrol to those who do not watch or like the series. For instance, one participant testified of such tactful self-presentation vis-à-vis non-fans, talking deliberately about good local food only, leaving out references to the series and her 'spiritual' experiences at the site:

> *Instead of impressions from the series, I tried to emphasise the local food there. In fact, you can always attract others with something else.*
>
> *(Participant M. B., age 53)*

Likewise, when talking about reasons to join the trip to Tyrol, one informant tried to relativize his interest beyond the TV series itself:

> *I would not go just for him [the main character] and the series /.../, there are all those ski resorts there.*
>
> *(Participant J. G., age 48)*

Our empirical data then suggests that cultural intimacy vis-à-vis outsiders is exposed as operating mechanism in the form of induced shame/presupposed embarrassment that is being reflected in apologetic interpretations of fans' viewing pleasures and careful accounts of their trip experiences. Instead of understanding such situationally specific fan dis-identifications (efforts of denying, evading, or distancing from fan identity) as the absence of a fan identity, it seems more appropriate to consider them as its constitutive part. Namely, these are actively produced by fandom itself and are primarily a product of feeling inadequate or inferior in relation to non-fans according to the taste hierarchies as perceived in everyday life. We should therefore grasp performative ways in which the fan identifications are vigorously blurred in more detail and investigate conditions in which they are concealed or that put them into effect.

Cultural capital and (un)easy awareness of supposedly 'lowly' pleasures

To investigate the role of cultural capital in structuring cultural intimacy among fans, we conducted a survey amongst the visitors of the 'Bergdoktor in Ljubljana' event. Media company PRO Plus organized an event for the Bergdoktor TV series fans, inviting the main actor Hans Sigl (playing the main character Martin Gruber) to Ljubljana to a public gathering with his Slovenian fans. The event was held on March 13, 2017, and was reportedly visited by about 5,000 people (Simončič, 2017). We managed to collect 140 answers to a survey amongst the visitors of the event. The sample confirms the predominant older female audience of the series: 85% were females, 45% were retirees, 79% had high school education or less, and average age was 50.

Herzfeld (2005) defines the cultural intimacy as 'the sharing of known and recognizable traits that not only define insiderhood but are also felt to be disapproved by powerful outsiders' (p. 132). According to that, cultural intimacy was operationalized and measured with a Likert-type scale on three indicators intended to gauge the level of discomfort fans feel in relation to presenting themselves as fans (of low legitimacy TV series) to non-fans: (a) 'It sometimes feels awkward to admit that I like the series to people who do not watch it', (b) 'I often find myself embarrassed by those who do not watch the series', and (c) 'I would proudly show a photo of me and the main actor to everyone' (reverse statement). As seen from the mean values and standard deviation, the analysed sample of fans felt relatively comfortable in admitting that they are fans of the TV series. For example, 75% of the interviewees agreed or strongly agreed with the statement 'I would proudly show a photo of me and the main actor to everyone.' This was expected since these were the fans that were willing to attend the event and, in this way, publicly proclaim their fandom. Although all three statements were selected to measure the concept equally, factor analysis (principal component analysis) showed that only first two statements belong to the same component (Component Matrix: 0,904 and 0,913). Compared to the third (reverse) statement (Component Matrix: −0,284), formulated in more general way (replacing non-fans with 'everyone'), the first two describe the level of discomfort of sharing one's fandom with non-fans in particular, hence including the relational/boundary-making aspect vis-à-vis outsiders as the concept of cultural intimacy presupposes. We thus included only the first two statements into index of cultural intimacy, measured as an average value on both indicators. This was done for the purpose of regression analysis whereby cultural intimacy was taken to be the dependent variable and the cultural capital indicators, measured by formal education, extent of travelling and effort made to get to the event, as the independent variables (model summary: Adjusted R square: 0,039; St. error of the estimate: 1,46. ANOVA F: 2,206; Sig.: 0,073). We were interested particularly to what extent experiencing cultural intimacy is affected by fan's social position (as measured by cultural capital indicators).

The results show that in cases of two cultural capital indicators: the education level and number of travels abroad, there is an indication of negative correlation between the cultural capital indicator and extent of discomfort felt as the Bergdoktor TV series fan amongst non-fans. The higher the education, the lower the level of discomfort, showing a moderately strong negative correlation (Beta: −0,186, Sig: 0,043). Similarly, yet not significantly (results should thus be taken merely as an indication), the more often a fan travels abroad, the less discomfort s/he feels (Beta: −0,161, Sig: 0,088). It seems that fans in higher class position or with higher cultural capital (those who are more educated and travel abroad more often) are more competent and self-assured as regards revealing their preferences. While fans in lower class position or with lower cultural capital (less educated, less travelling abroad) feel less sovereign in their cultural choices and more aware of the politics of classification, taste hierarchies, and effects of external taste judgements (the taste for commercial mass culture being perceived as inferior taste).

That cultural intimacy (or feeling of embarrassment) is stronger among lower social strata could partly be explained by the omnivore thesis (Peterson, 1992) which empirically demonstrates the transformation of the Bourdieu's model based on duality between high and low culture into a model that distinguishes between omnivore consumers with high cultural capital that are open to diverse cultures, and univore consumers with lower cultural capital that consume distinctly one culture only, especially commercial popular culture.[1] What cultural omnivores have developed is 'an openness to appreciating everything' (Jarness, 2015, p. 66), in this way it seems plausible that higher strata fans do not feel so much discomfort vis-à-vis non-fans when it comes to watching, among other TV content, also a low-legitimacy TV text. Status for them is gained by knowing about and participating in different cultural forms, including those regarded low-brow. They do not any more appropriate exclusively a fixed set of high-brow forms of culture (e.g. classical music, opera) and they do not despise all things low-brow either (e.g. country music, sitcoms). Upper social classes today show broad, eclectic, and hybrid tastes and in their cultural consumption cross high and low culture. In their study, Bennett et al. (2009) note that it is the quantity of television viewing that is structured by class mainly and, especially, that it is not that important what people are watching, rather how. People appropriate cultural goods differently, so we could argue that what distinguishes Bergdoktor fans with higher cultural capital from fans with lower cultural capital is the different mode of consumption of the same text. Fans with higher cultural capital also engage in what Ang (2007) calls ironic pleasure, a mode of viewing that is informed by a more intellectually distancing, a superior subject position which allows having pleasure in the series while simultaneously expressing a confident knowingness about its supposedly low quality. The same could be said for the consumption of the Tyrol, as how Bergdoktor fans consume and engage with destination could vary. This suggests that distinction today is predominantly achieved through embodiment (how fans/tourists consume) and not in its objective form (what fans/tourists consume), since a given text or destination can be appropriated in many different ways. Structural position of fans/tourists (their cultural capital or social class) sets the limits for acquisition of different cultural dispositions and cultural codes that systematically and in different ways affect the process of decoding TV text or consuming a destination.

On the other side, lower social strata are more limited in their consumption according to omnivorousness thesis and seem to feel more strongly the burdens of their univore cultural choices, as they mostly participate only in one type of culture (commercial pop culture in particular). Cultural intimacy presupposes the awareness of power in society, which strengthens the feeling of intimidation. It provides 'familiarity with the bases of power that may at one moment assure the disenfranchised a degree of creative irreverence and at the next moment reinforce the effectiveness of intimidation' (Herzfeld, 2005, p. 3). Fans in lower social positions/with lower cultural capital seem to be well aware that their preference for Der Bergdoktor and associated forms of fandom expression, such as travelling to the destination, has in the broader culture a lesser respectability factor and are thus fairly cautious when associating themselves with choices that are routinely condemned and despised amongst non-viewers and non-fans. In this respect, it seems that fans in lower social positions predict the effects of external taste judgements and overall recognize implications of cultural tastes in social organization of power. Consequently, they express higher level of cultural intimacy in order to diminish the power of effects of the perceived level of cultural legitimacy of their TV preferences or travel choices and to reduce the degree of cultural shame. As an audience at a greatest distance from cultural legitimacy they seem to experience cultural intimacy more intensely. The performances of cultural intimacy are then primarily determined by their awareness of existing taste hierarchies and politics of classifications and are testifying on how cultural taste is closely integrated into the complex interaction of forms of social and cultural power.

Conclusions

Bennett et al. (2009, p. 22) argue that audiences are able to engage with texts by stepping outside the limiting frames arising from their own social position, suggesting a much greater fluidity of reading and interpretative contexts than Bourdieu's approach allows. Yet, our study suggests that social position somewhat still matters when reading texts and engaging in associated forms of fandom expression. Our contribution provided empirical examples of cultural intimacy as a specific aspect of fans engagement in the case of 'der Bergdoktor' phenomenon (a popular TV series and its induced tourism) in Slovenia. We demonstrate how cultural intimacy is articulated and manifested in many situational cases of feeling uncomfortable with the non-fans indicating that the sense of fans' enjoyment of pop cultural text and related destination visit may be also a source of external embarrassment. Our results show that cultural intimacy is experienced differently among fans and is conditioned by their cultural capital/social position; therefore, fans should not be considered as homogenous entity. We conclude that it is a reflection of perceived cultural hierarchies and value discourses surrounding particular genres of popular culture as supposedly 'lower' culture and in this respect also an implication of boundary work (distinction making) through media consumption and related tourism practices in general. This indicates how cultural consumption is implicated in the organization and exercise of social power.

Advocating Gans's (1999) idea of aesthetic pluralism, we believe that within the context of more and more globalizing tastes local vernacular aesthetics, as reflected in Der Bergodktor, occupies a notable position as one of the equally valuable taste cultures in specific locality. Our primary aim was then to critique the ideological work of distinction supporting the perceived hierarchies of value regarding taste cultures. What our study testifies is that fans themselves are quite aware of the organization of cultural hierarchies and their effects. La Rochefoucauld (in Bourdieu, 1984/2010, p. 255) maintains that our pride is more offended by attacks on our taste than on our opinions. This could be explained if we understand that patterns of cultural taste are closely integrated into the complex interaction of cultural and social power. Differences in tastes are then socially functional in that they establish cultural and social hierarchies, thereby not only marking but also creating and maintaining social differences even when regarding tourism destinations, since destinations referring to supposedly 'lower' popular culture are perceived as a choice of supposedly 'lower' taste. By discovering local television cultures, this contribution then also addresses the issue of historically constructed cultural hierarchies and markers of cultural legitimacy that transfer from media field to tourism field. In this regard, it offers intervention into ongoing debates about the changing dynamics of taste while investigating increasingly unsettling and complex relations between (popular) culture and class, as reflected in intersections of media consumption and tourism practices.

Note

1 The omnivore thesis marks a historical shift in distinction making. Yet, it does not explicitly deny homology between class and culture; it merely argues that the nature of this relation has been fundamentally transformed – from distinctive snobbism to a more omnivore posture in cultural consumption of higher social strata. For criticisms of the empiricism of approach, see Warde et al. (2007) and for later corrections of the concept of a cultural omnivore also Savage and Gayo (2011).

References

Ahmad, R. (2014). Habitus, capital, and patterns of taste in tourism consumption: A study of western tourism consumers in India. *Journal of Hospitality & Tourism Research, 38*, 487–505.

Ang, I. (1985). *Watching Dallas: Soap opera and the melodramatic imagination*. London, UK: Methuen.

Ang, I. (2007). Television fictions around the world: Melodrama and irony in global perspective. *Critical Studies in Television, 2*, 18–30.

Bennett, T. (2006). Distinction on the box: Cultural capital and the social space of broadcasting. *Cultural Trends, 15*, 1–21. Retrieved from: http://www.uws.edu.au/__data/assets/pdf_file/0006/185865/Bennett_DistinctionOnTheBox_ICS_Pre-Print_Final.pdf.

Bennett, T., Savage, M., Silva, E., Warde, A., Gayo-Cal, M., & Wright, D. (2009). *Culture, class, distinction*. New York, NY: Routledge.

Bourdieu, P. (1984/2010). *Distinction: A social critique of the judgement of taste.* London, UK & New York, NY: Routledge.

Brubaker, R., & Cooper, F. (2000). Beyond 'identity'. *Theory and Society, 29*, 1–47.

Buchmann, A., Moore, K., & Fisher, D. (2010). Experiencing film tourism: Authenticity and fellowship. *Annals of Tourism Research, 37*, 229–248.

Collins, J. (Ed.). (2002). *High-pop: Making culture into popular entertainment.* Oxford, UK: Blackwell Publishers.

Frow, J. (2002). Signature and brand. In J. Collins (Ed.), *High-pop: Making culture into popular entertainment* (pp. 56–74). Oxford, UK: Blackwell Publishers.

Gans, H. J. (1999). *Popular culture and high culture: An analysis and evaluation of taste.* New York, NY: Basic Books.

Garnham, N. (1993). Bourdieu, the cultural arbitrary and television. In C. Calhoun, E. LiPuma, & M. Postone (Eds.), *Bourdieu: Critical perspectives* (pp. 178–192). Chicago, IL: University of Chicago Press.

Gyimóthy, S., Lundberg, C., Lindström, K. N. Lexhagen, M., & Larson, M. (2015). Popculture tourism: A research manifesto. *Tourism Research Frontiers: Beyond the Boundaries of Knowledge, Tourism Social Science Series, 20*, 13–26.

Herzfeld, M. (2005). *Cultural intimacy: Social poetics in the nation-state.* New York, NY: Routledge.

Jarness, V. (2015). Modes of consumption: From 'what' to 'how' in cultural stratification research. *Poetics, 53*, 65–79.

Jontes, D. (2014). Televizijski okus, omnivornost in kulturni kapital. In B. Luthar (Ed.), *Kultura in razred* (pp. 105–123). Ljubljana: Fakulteta za družbene vede.

King, C. (1993). His truth goes marching on: Elvis presley and the pilgrimage to graceland. In I. Reader & T. Walter (Eds.), *Pilgrimage in popular culture* (pp. 92–104). London, UK: Palgrave Macmillan.

Kotnik, B., & Cah, K. (2016, August 30). On je Bog, ki ga vsi potrebujejo. *Delo.* Retrieved from http://www.delo.si/prosti-cas/on-je-bog-ki-ga-vsi-potrebujejo.html.

Kraaykamp, G., van Eijck, K., Ultee W., & van Rees, K. (2007). Status and media use in the Netherlands: Do partners affect media tastes? *Poetics, 35*, 132–151.

Lahire, B. (2008). The individual and the mixing of genres: Cultural dissonance and self-distinction. *Poetics, 36*, 166–188.

Mowforth, M., & Munt, I. (2003). *Tourism and sustainability: New tourism in the third world.* London, UK & New York, NY: Routledge.

Peterson, R. (1992). Understanding audience segmentation: From elite and mass to omnivore and univore. *Poetics, 21*, 257–282.

Reckwitz, A. (2002). The status of the 'material' in theories of culture: From social structure to artefacts. *Journal for the Theory of Social Behaviour, 32*, 195–217.

Savage, M., & Gayo, M. (2011). Unravelling the omnivore: A field analysis of contemporary music taste in the United Kingdom. *Poetics, 39*, 337–357.

Simončič, J. (2017, March 16). Nasmejani Hans Sigl je razveselil množico oboževalcev, ki so ga pričakali v Šiški. *24ur.com.* Retrieved from: https://www.24ur.com/ekskluziv/domaca-scena/hans-sigl-prispel-v-slovenijo.html.

Warde, A., Wright, D., & Gayo-Cal, M. (2007). Understanding cultural omnivorousness: Or, the myth of the cultural omnivore. *Cultural Sociology, 1*, 143–164.

PART V

Transmedia tourism

36

THE ROLE OF STORIES IN TRAVEL POSTS TO SOCIAL MEDIA

John Pearce and Gianna Moscardo

Introduction

In 2018, booking.com, a major international tourism business, used a marketing campaign spread across multiple media and social networking platforms, based on the tagline 'These are our stories. What will be yours? Book your next story'. This campaign highlighted two contemporary trends in tourism marketing – an explicit call to tourists to experience a story while they travel, and the dominance of online, especially social, media for customer communication in tourism. Some might argue that one of the defining features of these first decades of the 21st century is the widespread adoption of online social media combined with mobile communication technologies. Tourism is no exception. In tourism, the massive growth in user-generated content (UGC) in travel blogs, reviews, and posts to social media such as Instagram, Facebook, Weibo, and Snapchat has been acknowledged as an important feature of destination image and promotion, and travel decision-making (Amaro, Duarte, & Henriques, 2016; Kim, Lee, Shin, & Yang, 2017). Many destination marketing organizations now regularly use this UGC in their destination marketing strategies. To date, tourism research into this UGC about travel has focused on descriptions of the content and what it means for destination and tourism business image and performance. Very little attention has been paid to the structure of this content or the role that UGC plays for the users. Despite the focus of tourism research on the content of UGC, a small number of studies have noted the value of UGC that is structured as a story (Hsiao, Lu, & Lan, 2013; Moscardo, 2017a; Tussyadiah & Fesenmaier, 2008; Volo, 2010). The importance of stories in social media has been recognized by tourism practitioners but not yet by tourism researchers.

The present chapter therefore focuses on stories in social media and argues that stories are embedded in human thought and action and a better understanding of the roles that stories play in tourist experiences may provide insights both into the nature of tourism as a social phenomenon and its value to those that travel. The chapter seeks to contribute to both tourism and mediatization research by developing a conceptual framework that identifies the various ways in which stories exist in tourist experiences and their reports on those experiences. To develop this conceptual framework, the chapter will synthesize research from psychology and sociology on the roles that stories play in human thought and action with research in tourism on stories, to identify where tourist stories posted to social media exist

349

within the larger tourism system. It will then briefly review work in uses and gratifications theory applied to social media use to describe a set of possible roles that social media travel stories may play for the tourist storyteller. The chapter will then integrate all these areas to present an expanded conceptual model of storytelling in tourism. The chapter will conclude with a discussion of the implications of this framework for tourism research.

Mediatization, stories in social science, and transmedia research

Aristotle argued that a story must be about a challenge or an adventure, must include a description of a set of events triggered by something surprising or unexpected, and must describe how various characters respond to these surprises and each other (Husain, 2002). Polletta, Chen, Gardner, and Motes (2011) describe a narrative as an account of events in their order of occurrence, and a story as a description of how characters respond to events and how the events and the character actions are linked in causal chains to form a plot. The overall aim of a story is to engage, entertain, and change its audience (Moscardo, 2017b). Stories and storytelling are central topics of interest across several areas of psychology, sociology, anthropology, marketing and consumer behaviour, education, planning, and organizational behaviour, making any attempt at an exhaustive literature review impossible. Such a review is not, however, necessary as across all these areas there is consensus about three key story functions:

- Stories are how humans organize and make sense of the information they get from their experiences;
- Stories are one of the key elements of long-term memory and therefore are critical to self-identity; and
- Stories are a core tool for human communication and therefore are critical to social identity and interaction (Goldstein, 2015; Haigh & Hardy, 2011; Kent, 2015; Matlin, 2013; Popova, 2015; Reed, 2013).

Not surprisingly research into stories and storytelling is essentially interdisciplinary and stories have been the subject of research across a number of disciplines and topic areas (Bamberg, 2009). However, as the authors of the present chapter have backgrounds in psychology and sociology as applied to tourism and communication, the overall theoretical tradition of the chapter lies in the overlap between these two disciplines where social psychology merges into sociology and shares an interest in how individuals interact to construct social realities within the context of existing social structures and forces.

This disciplinary background and the overall aims of the chapter fit well with the concept of mediatization which forms the basic framework for this edited volume. Couldry and Hepp (2013) describe mediatization as a concept that attempts to bring together previously disconnected approaches to understanding how various media contribute to and, in turn, are affected by social and cultural realities. They note that previously different disciplinary traditions tended to focus on just one aspect of media and human communication. Thus, sociology tended to look at the processes of production and how these processes reflected social structures. Psychologists tended to explore the impact of different media on audiences. Other disciplines within social science and the humanities examined the nature of the content of media. Mediatization encourages an integration of these elements and a recognition of the extent to which media is itself embedded in social realities. The present chapter seeks to focus on the commonalities in the overlap between social psychology and sociology and integrate individual audience variables with social structures. In addition, this combination of two disciplinary approaches, each assuming a different level of agency, allows for an examination of not just how media usage and content influence tourists but also how tourists use these media to create new social realities. In this latter sense, the chapter is consistent with Månsson's (2011) work introducing mediatization to tourism which incorporated the concept of convergence highlighting the importance of understanding tourist agency in the creation and use of media products in tourism.

Mediatization has become a particularly important concept as the rise of new digital and social media represents a significant change in the nature of media communication. Not only do these media allow for significant audience engagement in media product creation (Månsson, 2011), they are also associated with idea of transmedia. Transmedia refers to the presentation of information and the conduct of communication across multiple media channels simultaneously (Hancox, 2017). This is most obvious in the area of transmedia storytelling where stories can be told across books, films, television, on websites, through interactive online games and in multiple social media channels (Ryan, 2016). While the present chapter focuses on the single channel of social media to explore existing evidence on the reasons tourists might post stories to these platforms, the integrative framework being developed and expanded is relevant for storytelling in tourism across multiple media channels.

Tourism research into stories

According to Rezdy (2018), nearly half (42%) of all Facebook stories posted on timelines and in status updates were about travel experiences. Despite evidence that suggests stories are one of the main ways tourists share their experiences with others and the growing number of examples of tourism organizations using stories in their communications with travellers and increasingly in tourist experience design, there has been little detailed examination of stories in the tourism research literature. Moscardo (2018) notes that in addition to some emerging information on stories in UGC about tourist destinations, there are only two other streams of research into stories in tourism. The first are studies analysing the effectiveness of stories as ways to attract and engage tourist attention in heritage settings such as museums, historic precincts, and national parks (cf. Xu, Cui, Ballantyne, & Packer, 2013) or to support effective communication of conservation messages (cf. Wilson & Desha, 2016). The consistent conclusions from this research stream are that tourists seek and respond positively to the presentation of stories (Moscardo, 2010), and that even when an explicit story is not provided by the official site interpreters, tourists create their own stories from the resources provided (cf. Chronis, 2012). The second stream of research into stories in tourism is that of Woodside and his colleagues (Woodside, 2010; Woodside & Martin, 2015) which examines the use of stories as a form of sense making within tourism. Building upon the research developed by Woodside and his colleagues, Moscardo (2017b, 2018) provides a storytelling framework (see Figure 36.1 for a simplified version) for better understanding the different types of stories within a destination and how these connect to tourist experiences.

Moscardo (2017b, 2018) argues that there are four interrelated dimensions that connect stories to tourism and tourists. Firstly, there is a hierarchy that connects stories at different levels to the tourist's overall life story. As noted earlier, each of us has an autobiographical life story and each trip that we do potentially contributes to that life story. Then, within each trip, there is at least one, often more, destination, each of which is linked to a story in the tourist's memory. Then, within each destination are specific locations, attractions, and activities which are further sources of tourist stories. For example, the first author of the

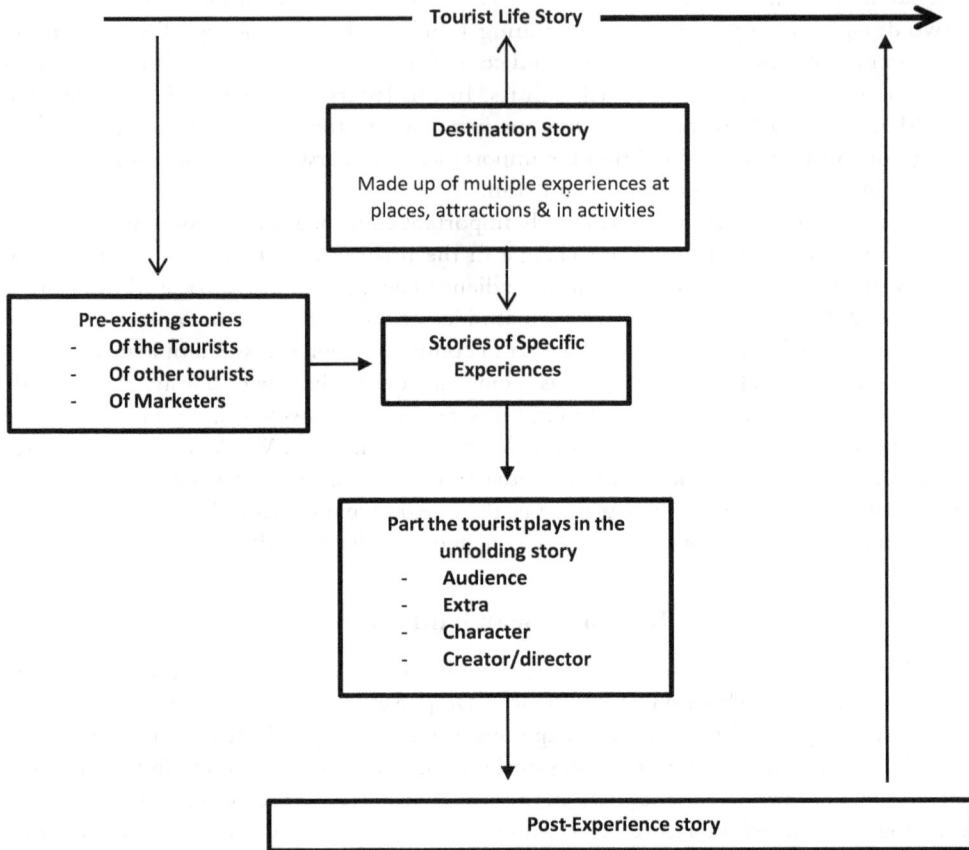

Figure 36.1   A storytelling framework for tourism (adapted from Moscardo, 2018)

present chapter recently spent three weeks travelling in Europe. That trip overall contributes to his life story. Within that trip are several destinations including Lucerne, Venice, Rome, and Paris, each of which can be associated with a story in his memory. Then within each destination, there are specific incidents, for example, visiting St. Marks or the Acqua Alta bookshop in Venice that may also be linked to stories.

The second dimension recognizes that in any single tourist situation, there are multiple stories including:

- stories that tourists bring with them either from their previous experiences or that were told to them by other tourists,
- stories told by staff or destination residents involved in the setting,
- stories about the operating organization, and the stories of the setting or the activity itself.

In our example of the Acqua Alta book shop in Venice, a tourist may arrive with stories others have told them about the place, there are the stories of the people who work there, and the story of the building, stories of the cats that live in the bookshop and that are often photographed by visitors, and the history of the bookshop, how it came to be there and why it is organized the way it is. A third dimension is the role that tourists may play in the stories of the experiences. These include being an audience member, the most common role, an extra or a character in someone else's story, or the creator and/or director of their own new story.

Finally, there is a temporal and relevance dimension. This dimension distinguishes between pre, during, and post experience stories. Pre-experience stories about a setting provide tourists with information to help their planning and serve as advertising or promotion for the setting operators. During the experience or unfolding, stories emerge from the activities, events, and social interactions that people have in the setting. Post-experience stories are used by tourists to organize and share their experiences with others and by setting operators to monitor quality and to feed into setting management and marketing. It is these post-experience stories that are most commonly shared by tourists in their social media posts.

Understanding why people post to social media

Like many other aspects of this framework, there has been little research exploring how and why tourists share these post-experience stories with their social networks across different media channels. The issue of how and why people share information in social media posts has, however, become a major area of communications research. The available evidence confirms that stories are a dominant structure in the material people post or share across all social media platforms (Dayter, 2015; De Fina & Gore 2017; Page, 2012; Rottberg, 2014; West, 2013), with many social media platforms specifically designed to facilitate online storytelling and sharing (Georgakopoulou, 2017a; Page, 2012). There is also evidence that audiences respond positively and engage more with content that is recognized as a story (De Fina, 2016; Utz, 2015). Two areas of research into social media usage can be seen as relevant to the present discussion – research into how people organize their online posts, especially how stories are created and told through social media, and research into motivations for, and benefits of, using social media, especially research into why people post things to social media.

As might be expected much of the work in the first stream on how people structure their social media posts is focused on storytelling and story creation. This work concludes that while the basic elements of stories remain the same in social media, especially blogs (Rottberg, 2014), they are created in new ways and take on new forms in social media. Firstly, social media stories are told in fragments or episodes that are posted both across time and across multiple platforms simultaneously (Dayter, 2015; Gan & Wang, 2015; Ham, Lee, Hayes, & Bae, 2019; Rottberg, 2014; West, 2013). These are sometimes referred to as small or tiny stories (Dayter, 2015; Georgakopoulou 2017a). Secondly, social media stories are collaboratively created between multiple tellers and multiple audiences (Page, 2012). Georgakopoulou (2017a, 2017b) argues that people often post a short statement about an event or their current state which has the potential to become a story if their audience responds positively. When these audiences ask for more details, offer their own stories relevant to the topic, or share other stories linked to the topic, then the original poster can respond with more details and the process of co-creation of the story begins and expands (Georgakopoulou 2017a; Page, 2012; West, 2013). Thirdly, social media stories are very visual both in terms of stories told through the posting of images and in the addition of memes, gifs, and emoticons in comments and in posts (Georgakopoulou 2017a; Page, 2012). Finally, storytelling in social media is influenced by audience response, the affordances of the social media platforms and reasons why the individual decides to post to a social media platform (Page, 2012).

The expected benefits of, or reasons for, using social media is currently the dominant area of research into communication through social media. This work can be seen as falling into two main approaches – sociological analyses expanding on Goffman's presentation of self and psychological approaches expressed through the Uses and Gratifications (UG) framework (Seidman, 2013). Goffman argued that we present ourselves to others by our public performances, which offer insights into our values and attitudes and by managing the information we share with them through how we exhibit material possessions (Hogan, 2010). While some studies have focused on social media posting as performance (cf. Jung, Youn, & McClung, 2007), more recent research has seen it more as curation of exhibitions of digital possessions or artefacts (Hogan, 2010). Zhao and Lindley (2014), for example, found evidence that people do see social media as a way to store important personal mementos and memories suggesting that social media sharing both helps present self-identity to others and also acts as an archive or diary to aid autobiographical memory.

Table 36.1 summarizes the findings of UG research into the expected benefits or motives for using social media. There is consistent support across a range of different platforms offered across a range of countries, for social media usage offering four overlapping categories of benefits – identity development and self-presentation, social interaction, information, and entertainment. Although the work reviewed in Table 36.1 is mostly focused on social media usage in general, research into more specific uses, such as sharing photos on Facebook (Malik, Dhir, & Nieminen, 2016), sharing hyperlinks through Twitter (Holton, Baek, Coddington, & Yaschur, 2014), and posting selfies (Sung, Lee, Kim, & Choi, 2016), generally reveals the same sets of motives and benefits. Similarly, the available research into sharing of travel stories reports the same sets of motives and benefits as those listed in Table 36.1 (Munar & Jacobsen, 2014; Pera, Viglia, & Furlan, 2016).

Table 36.1 Uses and gratifications associated with using social media

Main categories	More specific motives/benefits
Identity development & self-presentation	*Reputational* – present myself to others, build reputation, tell people who I am
	Creative – express artistic needs, be creative
Social interaction	*Personal* – maintain and build social relationships, keep up with friends, support people who matter to me, maintain professional networks
	Collective – build connections to people like me, share problems,
Information	*Personal* – pursue personal interests/hobbies, keep up to date, learn new things, seek help with problems
	Collective – provide information to help others, because everyone else does it, keep up with trends
Entertainment	Fun, relaxation, pass the time, escape pressures, follow celebrities

Sources: Baek, Holton, Harp & Yaschur, 2011; Gan & Wang, 2015; Hwang & Choi, 2016; Jung et al., 2007; McCay-Peet & Quan-Haase, 2016; Quan-Haase & Young, 2010; Seidman, 2013; Sheldon & Bryant, 2016; Utz, 2015; Zhang & Pentina, 2012.

An expanded storytelling framework for tourism

Combining this evidence outlining the ubiquitous role of stories in social media posting and the consistent set of expected benefits, it is possible to suggest a set of roles that posting of travel stories might play for the storytellers. This allows for an expansion of the storytelling framework for tourism to incorporate the perspective of the tourists telling the stories and linking the travel experience more closely to the rest of their lives. Figure 36.2 offers this expansion with additional elements in broken lines. For each of the four main roles that travel storytelling plays, there are additional linkages. Thus, telling stories to build social connections links the tourist's post-experience stories to the larger collective stories about destinations and travel in general. Telling stories to build identity and present that identity to others is the function that links travel stories to a tourist's life story. Informational roles links to travel stories are the primary link between the tourist and the pre-experience stories, and

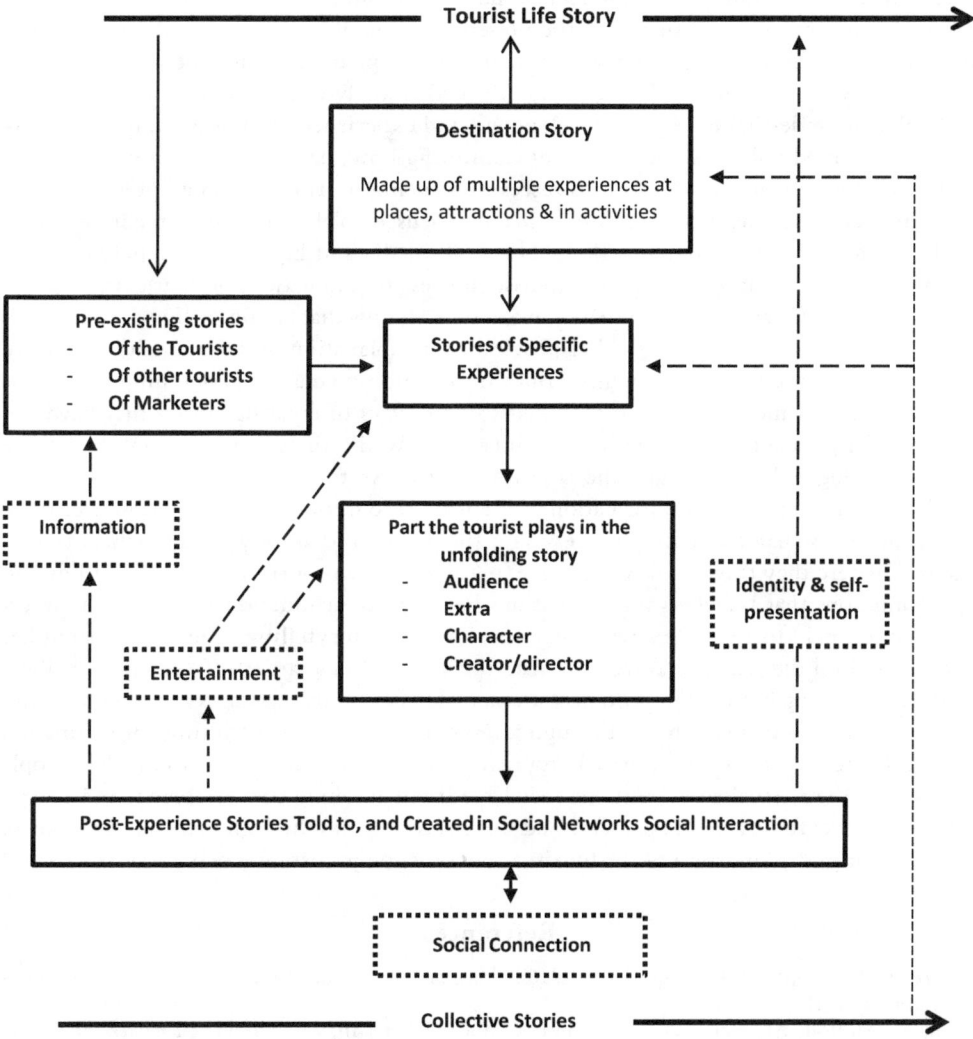

Figure 36.2   An expanded storytelling framework for tourism

entertainment links the tourist to both the type of role they seek in the unfolding experience story and the choices and actions in the experience itself. This is a preliminary and simplified model that leaves out potential links between the collective story and all the other elements in the tourism framework. Such links may be very important. For example, collective stories may influence destination and experience choices. The collective story will also combine multiple travel stories to create new pre-experience stories.

Conclusions

The expanded storytelling framework for tourism emerges from the integration of research in several disciplines, more specifically social psychology and sociology, and uses the construct of stories and storytelling to connect all features of the media communication process. This framework links media creators and audiences to each other and to the content being communicated through various media channels. According to Couldry and Hepp (2013), this is the aim of mediatization, and the present chapter sits within a social constructivist tradition which is emerging as a major element within mediatization. The present chapter and the storytelling framework offer a way to further analyse the role tourists are increasingly playing in destination image development and experience promotion, thus contributing to this edited volume's overall aim of examining how mediatization can be critical in understanding the current and futures forces influencing tourism. As noted previously, the literature review element of the present chapter focused solely on a single medium, social media platforms, but the framework applies to all media and highlights the potential role of transmedia stories in understanding tourist choices and evaluations of their experiences.

Standage (2013) argues that humans have evolved brains that are uniquely wired for both storytellers and social networking thus providing an explanation of the ease with which we have taken to social media storytelling. But this new transmedia world of multiple channels and varied social media platforms is linked to new ways of creating stories that have not yet been fully recognized in the tourism literature. While there is recognition that these new technologies have changed the way the tourism sector communicates with tourists and how tourists find travel information, there is no recognition of the importance of these new collective social storytelling processes for the decisions that individual tourists make in terms of where they travel and what they experience. Much tourism research is still driven by assumptions that travellers make decisions based on their individual needs with almost no attention paid to the collective dimensions of travel storytelling. Tourism research has also focused on the text that travellers write in the social media posts, typically in a single platform, ignoring both the importance of visual elements in travel stories and the possibility that people tell their travel stories through fragments spread across both time, platforms and media. There has also been very little tourism research into the conversations that people have about their travel story posts in social media. If booking.com is correct and people primarily travel to be able to co-create and share travel stories through their social network, then a significant proportion of the tourism phenomenon is currently being ignored.

References

Amaro, S., Duarte, P., & Henriques, C. (2016). Travelers' use of social media. *Annals of Tourism Research, 59*, 1–15.

Baek, K., Holton, A., Harp, D., & Yaschur, C. (2011). The links that bind. *Computers in Human Behavior, 27*(6), 2243–2248.

Bamberg, M. (2009). Identity and narration. In P. Hühn, J. Pier, W. Schmid, & J. Schönert (Eds.), *Handbook of narratology* (pp. 132–143). Berlin, Germany: William De Gruyter.

Chronis, A. (2012). Tourists as story-builders. *Journal of Travel & Tourism Marketing, 29*(5), 444–459.

Couldry, N., & Hepp, A. (2013). Conceptualizing mediatization. *Communication Theory, 23*, 191–202.

Dayter, D. (2015). Small stories and extended narratives on Twitter. *Discourse, Context & Media, 10*, 19–26.

De Fina, A. (2016). Storytelling and audience reactions in social media. *Language in Society, 45*(4), 473–498.

De Fina, A., & Gore, B. T. (2017). Online retellings and the viral transformation of a twitter breakup story. *Narrative Inquiry, 27*(2), 235–260.

Gan, C., & Wang, W. (2015). Uses and gratifications of social media. *Journal of Systems and Information Technology, 17*(4), 351–363.

Georgakopoulou, A. (2017a). Sharing the moment as small stories. *Narrative Inquiry, 27*(2), 311–333.

Georgakopoulou, A. (2017b). Small stories research. In L. Sloan & A. Quan-Haase (Eds.), *The Sage handbook of social media research methods* (pp. 265–282). London, UK: Sage.

Goldstein, E. B. (2015). *Cognitive psychology* (4th Ed.) Stamford, CT: Cengage Learning.

Haigh, C., & Hardy, P. (2011). Tell me a story—A conceptual exploration of storytelling in healthcare education. *Nurse Education Today, 31*(4), 408–411.

Ham, C. D., Lee, J., Hayes, J. L., & Bae, Y. H. (2019). Exploring sharing behaviors across social media platforms. *International Journal of Market Research, 61*(2), 157–177.

Hancox, D. (2017). From subject to collaborator: Transmedia storytelling and social research. *Convergence, 23*(1), 49–60.

Hogan, B. (2010). The presentation of self in the age of social media. *Bulletin of Science, Technology & Society, 30*(6), 377–386.

Holton, A. E., Baek, K., Coddington, M., & Yaschur, C. (2014). Seeking and sharing. *Communication Research Reports, 31*(1), 33–40.

Hsiao, K. L., Lu, H. P., & Lan, W. C. (2013). The influence of the components of storytelling blogs on readers' travel intentions. *Internet Research, 23*(2), 160–182.

Husain, M. (2002). *Ontology and the art of tragedy.* Albany: State University of New York Press.

Hwang, H. S., & Choi, E. K. (2016). Exploring gender differences in motivations for using sina weibo. *KSII Transactions on Internet and Information Systems (TIIS), 10*(3), 1429–1441.

Jung, T., Youn, H., & McClung, S. (2007). Motivations and self-presentation strategies on Korean-based "Cyworld" weblog format personal homepages. *CyberPsychology & Behavior, 10*(1), 24–31.

Kent, M. L. (2015). The power of storytelling in public relations. *Public Relations Review, 41*(4), 480–489.

Kim, S. E., Lee, K. Y., Shin, S. I., & Yang, S. B. (2017). Effects of tourism information quality in social media on destination image formation. *Information & Management, 54*(6), 687–702.

Malik, A., Dhir, A., & Nieminen, M. (2016). Uses and gratifications of digital photo sharing on Facebook. *Telematics and Informatics, 33*(1), 129–138.

Månsson, M. (2011). Mediatized tourism. *Annals of Tourism Research, 38*(4), 1634–1652.

Matlin, M. W. (2013). *Cognition* (8th Ed.) Hoboken, NJ: Wiley.

McCay-Peet, L., & Quan-Haase, A. (2016). A model of social media engagement. In H. O'Brien & P. Cairns (Eds.). *Why engagement matters* (pp. 199–217). Heidelberg, Germany: Springer.

Moscardo, G. (2010). The shaping of tourist experience. In M. Morgan, P. Lugosi, & J. R. B. Ritchie (Eds.), *The tourism and leisure experience* (pp. 43–58). Bristol, UK: Channel View.

Moscardo, G. (2017a). Exploring mindfulness and stories in tourist experiences. *International Journal of Culture, Tourism and Hospitality, 11*(2), 111–124.

Moscardo, G. (2017b). Stories as a tourist experience design tool. In D. R. Fesenmaier & Z. Xiang (Eds.), *Design science in tourism* (pp. 97–124). Basel, Switzerland: Springer International Publishing.

Moscardo, G. (2018). Tourist experience design: A storytelling framework. In L. P. Cai & P. Alaedini (Eds.), *Quality services and experiences in hospitality and tourism* (pp. 93–107). Bingley, UK: Emerald.

Munar, A. M., & Jacobsen, J. K. S. (2014). Motivations for sharing tourism experiences through social media. *Tourism Management, 43*, 46–54.

Page, R. E. (2012). *Stories and social media.* New York, NY: Routledge.

Pera, R., Viglia, G., & Furlan, R. (2016). Who am I? *Journal of Interactive Marketing, 35*, 44–55.

Polletta, F., Chen, P. C. B., Gardner, B. G., & Motes, A. (2011). The sociology of storytelling. *Annual Review of Sociology, 37*, 109–130.

Popova, Y. B. (2015). *Stories, meaning, and experience.* Abingdon, UK: Routledge.

Quan-Haase, A., & Young, A. L. (2010). Uses and gratifications of social media. *Bulletin of Science, Technology & Society, 30*(5), 350–361.

Reed, S. K. (2013). *Cognition.* Belmont, CA: Wadsworth.

Rezdy (2018). *Travel statistics for tour operators.* Retrieved from: https://www.rezdy.com/resource/travel-statistics-for-tour-operators/ (accessed 1 October 2018).

Rottberg, J. W. (2014). *Blogging* (2nd Ed.) Malden, MA: Polity.

Ryan, M. L. (2016). Transmedia narratology and transmedia storytelling. *Artnodes, 18,* 37–46.

Seidman, G. (2013). Self-presentation and belonging on Facebook. *Personality & Individual Differences, 54*(3), 402–407.

Sheldon, P., & Bryant, K. (2016). Instagram. *Computers in Human Behavior, 58,* 89–97.

Standage, T. (2013). *Writing on the wall.* London, UK: Bloomsbury.

Sung, Y., Lee, J. A., Kim, E., & Choi, S. M. (2016). Why we post selfies. *Personality and Individual Differences, 97,* 260–265.

Tussyadiah, I. P., & Fesenmaier, D. R. (2008). Marketing places through first-person stories—An analysis of Pennsylvania roadtripper blog. *Journal of Travel & Tourism Marketing, 25*(3–4), 299–311.

Utz, S. (2015). The function of self-disclosure on social network sites. *Computers in Human Behavior, 45,* 1–10.

Volo, S. (2010). Bloggers' reported tourist experiences. *Journal of Vacation Marketing, 16*(4), 297–311.

West, L. E. (2013). Facebook sharing. *Discourse, Context & Media, 2*(1), 1–13.

Wilson, K., & Desha, C. (2016). Engaging in design activism and communicating cultural significance through contemporary heritage storytelling. *Journal of Cultural Heritage Management and Sustainable Development, 6*(3), 271–286.

Woodside, A. G. (2010). Brand-consumer storytelling theory and research. *Psychology & Marketing, 27*(6), 531–540.

Woodside, A. G., & Martin, D. (2015). Introduction: The tourist gaze 4.0. *International Journal of Tourism Anthropology, 4*(1), 1–12.

Xu, H., Cui, Q., Ballantyne, R., & Packer, J. (2013). Effective environmental interpretation at Chinese natural attractions. *Journal of Sustainable Tourism, 21*(1), 117–133.

Zhang, L., & Pentina, I. (2012). Motivations and usage patterns of Weibo. *Cyberpsychology, Behavior, and Social Networking, 15*(6), 312–317.

Zhao, X., & Lindley, S. E. (2014, April). Curation through use. In *Proceedings of the 32nd annual ACM conference on human factors in computing systems* (pp. 2431–2440). ACM. doi: 2556288.2557291.

37

EVALUATING MULTIPLE PORTRAYALS OF DESTINATION IMAGE

Assessing, categorising, and authenticating visuals on Facebook posted by national tourism organisations

Nicholas Wise

Introduction

We rely on the internet today to find and consume information about destinations. National Tourism Organisations (NTOs) are responsible for implementing national tourism policies, disseminating information and producing (online) content to market and promote tourism opportunities. In addition to how NTOs promote a destination, social media sites such as Facebook, Instagram, Twitter, and TripAdvisor have added another dimension to how we consume destination information. Images communicated to potential consumers through Facebook offer insight into how NTOs want consumers to perceive a destination by highlighting appealing destination attributes. In addition to this, people interact and engage with social media providing user-generated content (UGC) to offer personal experiences and reflections, thus generating an online narrative. While there has been a lot of interest recently in UGC (e.g. Cox, Burgess, Sellitto, & Buultjens, 2009; Easton & Wise, 2015; Lu & Stephenkova, 2014), it is now increasingly important that we consider destination image by assessing content from both NTOs and users. While national promotion has always been the focus of NTOs, they increasingly use social media platforms to reach potential tourists, so assessing what content is communicated by NTOs about a place is important in contemporary tourism research (Hays, Page, & Buhalis, 2013; Xiang & Gretzel, 2009). This chapter contributes insight for scholars working across the areas of tourism studies and communications, especially those focusing on transmedia and how social media content adds to the narrative being communicated.

There are various examples from the tourism literature where scholars analyse Facebook content (e.g. Kim, Kim, & Wise, 2014; Wise & Farzin, 2018; Yoo & Lee, 2015). Facebook helps construct and reinforce particular place imaginaries, and such imaginaries are important when considering destination image. Some countries seek to promote images covering a range of opportunities, whereas others limit the scope of content covered, focusing more on competitive features (Govers, Go, & Kumar, 2007). Regardless, visual images generate an impression of a place to effectively communicate/market/promote unique products or attractions. Visual content thus establishes mental schema (Rose, 2001), linked to destination images—where people associate places with particular images (Baloglu & McCleary, 1999). An example is Facebook cover photos and profile pictures. Pictures (as visuals) meditate images, promote experiences, and highlight unique attractions, and users then add insight and narrative to supplement the photo.

This chapter discusses how the photographs that NTOs post on official Facebook pages can be assessed and categorised. Moreover, this chapter conceptualises how UGC then authenticates the content that those viewing or seeking both objective and subjective insight on a destination, using the cases of Croatia, Serbia, and Slovenia as examples. When we consider mediatization (see Couldry & Hepp, 2013), in the case of this chapter, we see the traditional promotors of tourism who prepare content for social media, and we have the UGC, which blurs the boundaries between traditional promotion and the increasing role of the consumer in contributing to destination image and promotion. Månsson (2011) argues that this positions the tourist as more active in the consumption and production of presenting and promoting destinations. This book argues that media convergence blurs boundaries between media texts and other cultural performances such as tourism. This is true when we consider traditional media such a newspapers that offer a one-way flow of information (see Wise & Mulec, 2012), the increasing influence of social media and UGC now positions the tourist as consumer and promoter (see Wise & Farzin, 2018). Arguably, so to scrutinise assumptions, in this case, transmedia actually occurs on the same platform, but through interactions with different users: the NTO that displays particular content and visuals of the destination, and the users who offer experiential insight. With the NTO and users both creating content, this enhances marketing and communication of a destination because the NTO is using visuals and particular marketing content to attract tourists and the users then enhance or supplement the images presented with subsequent associations, meanings, and stories based on their personal experiences in a destination.

Destination image and social media

With increased opportunities and competition in the international tourism market, NTOs face increased pressure to present unique insight about their country and particular attractions to catch the attention of those planning future travel. Destination images are oftentimes generated through marketing strategies to highlight significant developments, attractions, and tourism resources in a destination (Govers et al., 2007). According to Dwyer and Kim (2003), destination image is an important situational condition, because image, and how people perceive a place, influences the likelihood of receiving tourists. Moreover, tourism images relate to how we come to recognise a place (Baloglu & McCleary, 1999; Wise & Mulec, 2015). Beerli and Martín (2004) present nine dimensions pertinent to how they describe perceived destination images: natural resources, tourist leisure/recreation, natural environment, general infrastructure, culture/history/art, social environment, hotel/self-catering, political/economic factors, and place atmosphere. While these wider dimensions offer a conceptual base, an adapted version, which borrows from and adds to the list (as necessary), are presented in Table 37.1. This creates a base for conducting research and interpreting destination image content posted on social media platforms. Kim and Perdue (2011) note natural environments play a key role in image perceptions. Linked to promoting and advertising what consumers seek (Govers et al., 2007), NTOs and tourism managers

Table 37.1 Dimensions/attributes of destination image

Dimension	Attributes of destination image
Natural resources	Beautiful scenery; Aesthetics; Weather; Protected nature reserves; Fresh water lakes/rivers/waterfalls; Mountains/valleys/gorges; Rugged coastline & sea; Beaches; Islands; National parks; Variety/uniqueness of flora; Variety/uniqueness of fauna
Natural & built environment	Beautiful scenery; Aesthetics; Attractiveness of towns; Attractiveness of cities; Cleanliness; Mass-tourism/overcrowding; Pollution
Leisure, recreation & entertainment	Sport/physical activities; Leisure activities (fish/hunt/camp); Adventure/adrenaline activities; Swimming; Hiking/trekking; Nightlife/parties; Events/festivals; Shopping/fashion
Culture, history, heritage & art	Museums/memorials/monuments; Historical buildings/structures; Music/theatre/drama; Religion; Folklore/local customs; Sport
Tourist infrastructure & marketing	NTO image; General information; Hotels/accommodations; Restaurants/bars
Gastronomy	National cuisine; Food culture/marketplaces; Drink culture; Viticulture; Agriculture
Atmosphere & appeal	Family-oriented; Exotic; Mystic; Relaxing; Stressful; Fun/enjoyable experiences offered; Boring; Pleasant; Attractive/interesting
Political & economic factors	Political stability/political tendencies; Safety/security; Economic development; Prices

(Adapted from Beerli & Martín, 2004, p. 659).

also must consider psychological factors, split between personal (i.e. values, age, motivation) and stimulus (i.e. information sources, previous experiences) factors (Baloglu & McCleary, 1999). Stimulus factors invoke cognitive memories—linked to what past associations people might have of a particular country/place. For instance, the media communicates place images, and research has addressed how past events provoke images (Wise, 2011), positively and negatively—important to consider when looking at places with a recent tragic past (such as former-Yugoslavia countries).

Baloglu and McCleary (1999) assess how destination images form when people have not experienced a destination, but today potential travellers can simply scan through images posted online. In relation to image, the success of a destination is dependent upon how place awareness develops—as this helps shape and create perceptions (Wise & Mulec, 2015). Arguably, perceptions are linked to several destination image attributes, and expected image is shaped by what general destination knowledge and/or awareness tourist have.

Social media plays a significant role in promoting, marketing, and affecting how we consume tourism opportunities and destination images (see Easton & Wise, 2015; Hays et al., 2013; Munar & Jacobsen, 2013). Moreover, tourism research in social media is concerned with the links between image formation and UGC (see Camprubí, Guia, & Comas, 2013; Cox et al., 2009; Easton & Wise, 2015; Lu & Stephenkova, 2014; Sigala, 2011). Within this, there has been a lot of recent research looking at the role of Facebook and tourism (e.g. Hsu, 2012; Wise & Farzin, 2018; Yoo & Lee, 2015). From this literature, Hays et al. (2013) looked directly at how NTOs and destination marketing organisations use social media (Facebook) to reach wider audiences and promote destination image.

When addressing destination image, Facebook helps communicate an image of a place, creating a source of information and appeal—referring to stimulus factors (Baloglu & McCleary; 1999; Kim & Perdue, 2011). Considering how Facebook and other forms of social media contribute to destination image, the process is ever changing and involves numerous attributes that have been identified, considered, and compared (Baloglu & McCleary; 1999; Beerli & Martín, 2004; Wise & Mulec, 2015). Earlier insight discussed by scholars previously addressed mass media and the more traditional form of word-of-mouth (e.g. Fakeye & Crompton, 1991) long before the rise of social media. Today people (referring to those planning travel) put much trust in available sources and seek UGC insight (this is probably most evident in hotel reviews). Munar and Jacobsen (2013) note that tourists today put trust in social media to when making tourism plans and communicate experiences, and Easton and Wise (2015) note that it is important to evaluate the content being communicated based on the subjective nature of UGC. UGC can enhance visuals and marketing because they allow users to bring in their own personal stories, which contributes more to destination image because of the opportunity to reflect and/or critique based on experiences in the places and spaces being communicated. In Facebook, the organisation, or the person, controlling the content will portray some impression of the destination based on what they expect tourists want to consume, whereas users can add content to images. Wise and Farzin (2018) then added to this conceptual discussion by suggesting interaction and UGC helps authenticate content by adding subjective insight.

Because place images can be constructed through photographs and extended tourism textual narratives, researchers need to further evaluate what unique destination attributes are being presented and discussed (Kim et al., 2014). Today, the increasing influence (and popularity) of utilising social media in destination marketing and promotion helps ease communication, contributing to place images, or representations—which are claims about place characteristics (Dittmer, 2010). Because consumption is also increasingly visual (Francesconi, 2014; Rose, 2001), social media platforms such as Facebook allow users (the prospective travellers) to search through a Facebook page by scrolling through content on a pages timeline to get some sense of what to expect or see what others have experienced. Francesconi (2014, pp. 136–137) notes: 'visual and aural interaction creates an appealing atmosphere that invites further exploration.' Therefore, it can be argued that the content in social media discourses help construct visual imaginaries that can influence destination choice.

Assessing and categorising visuals on Facebook

Table 37.2 outlines NTO and website details for the countries included in this chapter. This section and subsequent sections will present what image content each countries NTO engages with and posts on Facebook (cover photos and profile pictures) and presents conceptual directions for assessing UGC that brings in subsequent content to the picture being posted. Photographs posted on Facebook offer researchers content with meanings and contexts embedded within them. Cover photos and profile pictures are depictions of what the NTO generating the site wants people to associate with their country—what they can do, experience, or consume. This is where the dimensions and attributes come in first as a way of categorising content, which is useful for also looking at similarities and differences between how NTOs promote their country. Thus, visuals are content used to communicate a particular place image to appeal to and attract consumers (tourists).

As previously discussed, a framework of destination image is used, by adopting dimensions and attributes from Beerli and Martín's (2004) conceptual framework (Table 37.1) to

Table 37.2 Facebook pages analysed

NTO	Details	Short description (directly quoted from Facebook page)
Croatia Full of Life	Tourist information	*Welcome to the official Facebook page of Croatia! Thank you for connecting with our country*
Serbia Travel	Government organisation, Travel agency	*Official Facebook page of the National Tourism Organisation of Serbia http://www.serbia.travel/*
Feel Slovenia	Organisation, travel agency, tourist information	*We're the official Facebook page for Slovenian Tourism. Bringing you travel ideas to FEEL & SHARE S♥NIA. Find us at www.Slovenia.info or follow @Sloveniainfo on Twitter*

assist with assessing and categorising wider contexts and representations being communicated through three NTO Facebook pages (Table 37.2). Adapting existing frameworks assists with linking content to position patterns of communication (Ellingson, 2011). Researchers conduct content analyses quantitatively or qualitatively, and researchers sort content based upon subject matter, locations, or set dates, or in this case linking visual interpretations to predetermined dimensions and attributes. The purpose of content analyses is to categorise representations found in similarly relevant resources (Ellingson, 2011; Hsu, 2012). Rose (2001, p. 16) suggests researchers consider three points of reference: (1) 'the site(s) of the production or an image,' (2) 'the site of the image itself', and (3) 'the site(s) where it is seen by various audiences.' Such insight from Rose (2001) frames how one platform represents as a (virtual) space for transmedia and different user interactions to occur. In this case, each NTO is presenting images from across their respective country and adding them as profile pictures or cover photos to their respective NTO Facebook page. Table 37.3 outlines what content is available on each Facebook page, showing how each NTO engages with Facebook. In total, from the three Facebook pages, NTOs collectively posted 249 cover photos or profile pictures from the time each page's creation to 1 September 2018. Tables 37.4 and 37.5 present the assessment and categorisation of attribute content for cover photos and profile pictures, respectively. The intention is to show and compare the range of dimensions and attributes each country uses. Attributes are recorded in Tables 37.4 and 37.5 once, despite the breadth of image attributes communicated—for instance, Croatia's NTO posted several photos of mountains rugged coastlines.

As noted above, attributes included in this analysis refer to Beerli and Martín's (2004) conceptual outline, bit given prior knowledge of destinations and emerged interpretations, it was necessary to include additional attributes as necessary. Given the case examples used here in this chapter, additional attributes help align the conceptual examples on destination image in the literature with case-specific knowledge to make sense of particular content communicated on Facebook pages. Cover photos display attractions, experiences, and opportunities in each destination. Cover photos and profile pictures portray particular insight (about a particular place) linked to how the NTO wants to communicate, create, or (re)create an image. After analysing all the images, it was determined that cover photos covered a much wider range in terms of destination image attributes. In profile pictures, Croatia, Serbia, and Slovenia used their respective NTO logo, as companies will often use their logo (Hsu, 2012).

By addressing destination image dimensions and attributes, adapted and amended from Beerli and Martín (2004), such existing models are useful and are guidelines for policy makers and destination managers to consider when deciding what image content to add and to

Table 37.3 Number of profile pictures, cover photos and descriptive content by country

	Croatia	Serbia	Slovenia
Page started	29 May 2009	7 October 2010	16 July 2009
Page likes (to 1 September 2018)	1,702,088	52,290	546,378
# of profile pictures	43	6	8
Most recent Profile image	NTO Logo with Slogan: Croatia Full of Life	NTO Logo	NTO Logo and Impressum: **I Feel Slove**nia
Number of cover photos	101	35	56
Most recent cover photo image	Croatia Feeds: National Football Team (with the caption: Welcome to Croatia *Full of excellent players*)	Vineyard promoting Serbia's wine	Cycling along Lake Jasna near Kranjska Gora (with the caption: Make New Memories)
Content features and other social media links	Timeline, About, Photos, Reviews, Events, Activity, Twitter, Likes, YouTube, Videos, Pinterest, Notes, Discover Croatia	Timeline, About, Photos, Reviews, #lifesyleserbia/ serbiatourism, Videos, Twitter, Events, Likes	Timeline, About, Photos, Videos, Reviews, Twitter, YouTube, I feel Slovenia PicBadges, Likes, Instagram, Pinterest, The Skocjan Caves-8th Wonder of the World, Events, Notes, Top Fans, Contact, Coccole estive
About: page info	Address, Short/Long Description, Company Overview, Mission, Phone, Email, Website	Address, Start Date, Short/ Long Description, Company Overview, Mission, Products, Phone, Email, Website	Address, Start Date, Short/ Long Description, Impressum, Company Overview, Mission, Products, Phone, Email, Website
Language(s)	English	English	English then Slovenian

distinguish themselves from their neighbours and main competitors. Interpretations focusing on how images are communicated differ, as the focus is on using images to attract tourists opposed to managing actual attractions, facilities, or infrastructures. Therefore, the general infrastructure dimension was removed, as was social environment to instead address infrastructure and marketing, because social media platform is used in such cases for destination promotion. One dimension in Beerli and Martín's (2004) initial study was natural environment, here amended to natural and built environment because of the range of landscape features presented in the photographs. Gastronomy was an added dimension that emerged from the images analysed given the focus on attributes such as food, drink, and viticulture. While the dimension of gastronomy and associated attributes could be argued to closely align with other dimensions, it was decided that products of food and drink deserved its own dimension because different parts of the supply chain were presented from the field to consumption. This dimension is useful for other (European) countries to adapt since vineyards, viticulture and wine are part of destination promotion.

Table 37.4 Image content based on cover photos by country

Dimension attribute	Croatia	Serbia	Slovenia
Natural resources			
Beautiful scenery	+	+	+
Aesthetics	+	+	+
Weather	x		
Protected nature reserves	x		
Fresh water lakes/rivers/waterfalls	x	x	x
Mountains/valleys/gorges	x	x	x
Rugged coastline & Sea	x		x
Beaches	x		
Islands	x		x
National parks	x		x
Variety/uniqueness of flora	x	x	x
Variety/uniqueness of fauna	x		
Urban/built environment			
Beautiful scenery	+	+	+
Aesthetics	+	+	+
Attractiveness of towns	x	+	+
Attractiveness of cities	x	x	x
Cleanliness	+	+	+
Mass-tourism/overcrowding	−	−	−
Pollution	−	−	−
Leisure, recreation & entertainment			
Sport/physical activities	x		
Leisure activities (fish/hunt/camp)			
Adventure/adrenaline activities	x	x	x
Swimming	x	x	
Hiking/trekking	x		x
Nightlife/parties		x	x
Events/festivals			x
Shopping/fashion		x	x
Culture, history, heritage & art			
Museums	x	x	
Memorials/monuments/statues	x	x	
Historical buildings/structures	x	x	
Music/theatre/drama	x		x
Religion	x		
Folklore/local customs	x	x	x
Sport	x		

(Continued)

Dimension Attribute	Croatia	Serbia	Slovenia
Tourist infrastructure & marketing			
NTO image	x	x	x
General information	+	+	+
Hotels/accommodations			
Restaurants/bars	x		
Gastronomy			
National cuisine			
Food culture/marketplaces	x		
Drink culture	x	x	x
Viticulture	x	x	x
Agriculture		x	
Atmosphere/ appeal			
Family-oriented	x	x	
Youthful	x		
Exotic	+	+	+
Mystic	+	+	+
Relaxing	x		x
Fun/enjoyable experiences offered	+	+	+
Stressful	−	−	−
Pleasant	+	+	+
Boring	−	−	−
Attractive/interesting	+	+	+
Political & economic factors			
Political stability/tendencies	−	−	−
Safety/security	−	−	−
Economic development	+	+	+
Price	+	+	+

Looking at Tables 37.4 and 37.5, when analysing the photographs, present attributes are marked with an (x). In Table 37.4, it is important to acknowledge that some attributes (i.e. *fun* or *pleasant*) are difficult to determine presence in photos but were widely observed based on impressions given. In such instances, a (+) highlights these attributes as interpreted impressions. Whilst images were overwhelmingly positive, however, to forego negative images, other attributes important to consider when discussing destination image (i.e. *pollution, mass-tourism/overcrowding, political,* or *safety*) were denoted with a (−). Content from cover photos in Table 37.4 relate to all destination image attributes, whereas Table 37.5 includes only attributes covered in profile pictures.

Table 37.5 Image content based on profile pictures by country

Dimension attribute	Croatia	Serbia	Slovenia
Natural resources			
Weather	x		
Fresh water lakes/rivers/waterfalls	x		x
Mountains/valleys/gorges			x
Rugged coastline & sea	x		
Beaches	x		
Islands			x
Variety/uniqueness of fauna	x		
Leisure, recreation & entertainment			
Adventure/adrenaline activities			
Hiking/trekking			
Nightlife/parties			
Events/festivals	x		
Shopping/fashion			
Culture, history, heritage & art			
Memorials/monuments/statues	x		
Historical buildings/structures			
Religion	x		
Folklore/local customs	x	x	
Tourist infrastructure & marketing			
NTO image	x	x	x
General information			x
Restaurants/bars			
Gastronomy			
National cuisine			
Food culture/marketplaces			
Drink culture			
Viticulture	x		
Atmosphere/appeal			
Family-oriented	x		
Youthful	x		
Relaxing	x		
Fun/enjoyable experiences offered	x		

UGC and authenticating

While we can interpret image content above, this section discusses how visuals supplement additional texts, thus adding experiences/reactions to the visuals displayed. Mariani, Felice, & Mura (2016, p. 321) note, 'the internet has deeply transformed the manner in which travellers access information, plan for and book trips, and subsequently share their travel experiences'. Because our consumption is increasingly becoming more-and-more interactive and visual (Francesconi, 2014), social media platforms better allow prospective travellers to search content and get some sense of what to expect (Sigala, 2011; Xiang & Gretzel, 2009) and plays an important role in making travel decisions. Facebook, especially, is used to exhibit experiences, creating a source of information and appeal, which refers to stimulus factors (relating to destination image). Knudsen and Waade (2010) discuss phenomenological experiences as encounters which help produce and seek a new awareness, adding to image content posted. Experiences are further realised through understandings of interpersonal and intrapersonal experiences and interactions (Cohen & Cohen, 2012), further detailed below. Munar and Jacobsen (2013) highlight that tourists today put trust in social media and communicated experiences. This means, when an NTO posts an image to promote a destination, the subsequent UGC can help users and future travellers by reflecting and adding their own insight to give narrative to the image. Such posting can then further create, (re)create, or authenticate place images by posting questions, detailed narratives, descriptions, or photographs that describe or capture unique destination attributes (Wise & Farzin, 2018).

Users typically offer a quick response in the form of a 'reaction', and this links to Cohen and Cohen's, (2012) notion of 'hot' authentication based on other users know or may have experienced based on the image presented or to confirm a comment. We see this in the reactions to the most recent profile pictures and cover photos: Croatia (added 13 July 2018 with 1,265 reactions), Serbia (added 31 August 2018 with 47 reactions), and Slovenia (added 14 August 2018 with 375 reactions). This shows user engagement and involvement in posts. While inquiry posts are common, Cohen and Cohen (2012) suggest user interactions are common in authentic production posts because they inform and help present particular dimensions and attributes. Such posts are more subjective but existential authenticity can emerge from these posts to offer insight into the social dynamics. This helps connect users with similar interests—thus further producing 'hot' authentication.

Given the above points about the use of online content from Facebook, tourists in the past have generally sought altered realities (see McKercher & du Cros, 2002). Thus, the use of social media allows for multiple stories, one that the NTO seeks to promote and the other a hodgepodge of content based on subjective insight and reviews that help position multiple perspectives. On Slovenia's NTO Facebook page, for instance, the content posted by users on the most recent cover photos detail questions and reactions to the picture of people cycling along Lake Jasna near Kranjska Gora posted by the NTO:

> **[User]** Always lovely memories in Slovenia xxt (and added an image of themselves at the site)
>
> **Feel Slovenia** [user name] thanks for sharing and welcome back soon!

> **[User]** i would also like to know where this is
>
> > **[User]** Looks like lake Jasna
> > **Feel Slovenia** dear [user name], it is Lake Jasna near Kranjska Gora.

[User] we did! (and added an image of themselves at the site)

[User] OMG 😎Couldn't wait to see it in real. We're coming to you Slovenia.!! Very soon! 😊

[User] Beautiful 😍 🎞️ ▬ **SI** #ifeelsLOVEnia

[User] Joanne! 😄 Time to book!

[User] C'est fait!

[User] Wo ist das?

> **[User]** Lake Jasna i think

The subsequent UGC here in this example helps authenticate the image posted by the NTO. Now instead of the potential tourist only consuming what the NTO selects to promote the destination, they can reflect and inquire. Similarly, in the case of Croatia, users note their enthusiasm for Croatia (given their recent success in international football competition) and in earlier images ask questions about the places, sites, or attractions depicted in the cover photos. This helps to confirm content, whilst connecting users to share experiences and propose questions. Such user engagement extends the narrative of the image, and positive comments in turn can enhance place image and attractiveness.

Conclusions

Tourists continually seek everyday (or local, or real) experiences opposed to commodified experiences. Such experiences are easily and widely disseminated via social media platforms (Knudsen & Waade, 2010; Munar & Jacobsen, 2013), so to authenticate the content (Cohen & Cohen, 2012; Wise & Farzin, 2018) of transmedia narratives. What tourists often encounter in places is an alternative reality, and as Leite and Graburn (2009, p. 43) argued: 'knowing that everything "on stage" is put there as entertainment, tourists believe that the "real," authentic parts of the world are to be found backstage, hidden from view'. Without the subjective and experiential insight that people expand on through experiences (as they do in authentic productions and authentic encounters), this may explain why comments and reactions can help assure and encourage a potential traveller to visit a particular destination in the country, or it may encourage them to travel to the destination if UGC experiences supplement the more traditional promotion measures (intended by the NTO by posting an image of a particular place or landscape).

The approach presented here offers insight into how visuals have become institution-alised for enhancing destinations image. Today we communicate using visuals as they enhance destination attributes. More evaluation of the UGC will help scholars authenticate images that offer subjective insight and subsequent narratives (based on experiences and/or reactions). This framework has also been applied to teaching and during seminars students evaluate image content and use conceptual frameworks such as the one adopted from Beerli and Martín's (2004) study to understand destination image formation, place marketing and promotion, and consumer insight. Managing how a destination is perceived was once in the control of the NTO, but UGC now allows the tourist to respond and react. Such transmedia narratives can either strengthen a places appeal or it can have an adverse impact if content is overwhelmingly negative. Such content and analysis that looks across NTO and UGC is especially important today among tourism studies scholars focusing on communications. UGC is helping to authenticate meanings and using social media as a platform means transmedia narratives reach a wide audience of future tourists who want to learn about other people's experiences in a destination.

References

Baloglu, S., & McCleary, K. (1999). A model of destination image formation. *Annals of Tourism Research, 26*(4), 868–897.

Beerli, A., & Martín, J. D. (2004). Factors influencing destination image. *Annals of Tourism Research, 31*(3), 657–681.

Camprubí, R., Guia, J., & Comas, J. (2013). The new role of tourists in destination image formation. *Current Issues in Tourism, 16*(2), 203–209.

Cohen, E., & Cohen, S. A. (2012). Authentication: Hot and cool. *Annals of Tourism Research, 39*(3), 1295–1314.

Couldry, N., & Hepp, A. (2013). Conceptualizing mediatization: Contexts, traditions, arguments. *Communication Theory, 23*, 191–202.

Cox, C., Burgess, S., Sellitto, C., & Buultjens, J. (2009). The role of user-generated content in tourists' travel planning behaviour. *Journal of Hospitality Marketing & Management, 18*(8), 743–764.

Dittmer, J. (2010). *Popular culture, geopolitics, and identity.* New York, NY: Rowman and Littlefield Publishers.

Dwyer, L., & Kim, C. (2003). Destination competitiveness: Determinants and indicators. *Current Issues in Tourism, 6*(5), 369–413.

Easton, S., & Wise, N. (2015). Online portrayals of volunteer tourism in Nepal: Exploring the communicated disparities between promotional and user-generated content. *Worldwide Hospitality and Tourism Themes, 7*(2), 141–158.

Ellingson, L. L. (2011). Analysis and representation across the continuum. In N. K. Denzin & Y. S. Lincoln (Eds.), *The SAGE handbook of qualitative research* (pp. 595–610). London, UK: SAGE.

Fakeye, P., & Crompton, J. (1991). Image differences between prospective, first-time, and repeat visitors to the lower Rio Grande Valley. *Journal of Travel Research, 30*(2), 10–16.

Francesconi, S. (2014). *Reading tourism texts: A multimodal analysis.* Toronto, Canada: Channel View Publications.

Govers, R., Go, F. M., & Kumar, K. (2007). Promoting tourism destination image. *Journal of Travel Research, 46*(1), 15–23.

Hays, S., Page, S. J., & Buhalis, D. (2013). Social media as a destination marketing tool: Its use by national tourism organizations. *Current Issues in Tourism, 16*(3), 211–239.

Hsu, Y. L. (2012). Facebook as international emarketing strategy of Taiwan hotels. *International Journal of Hospitality Management, 31*(3), 972–980.

Kim, D., & Perdue, R. R. (2011). The influence of image on destination attractiveness. *Journal of Travel and Tourism Marketing, 28*(3), 225–239.

Kim, S. B., Kim, D. Y., & Wise, K. (2014). The effect of searching and surfing on recognition of destination images on Facebook pages. *Computers in Human Behavior, 30*, 813–823.

Knudsen, B. T., & Waade, A. M. (2010). *Re-investing authenticity: Tourism, place and emotions.* Bristol, UK: Channel View.

Leite, N., & Graburn, N. (2009). Anthropological interventions in tourism studies. In T. Jamal & M. Robinson (Eds.), *The SAGE handbook of tourism studies* (pp. 35–64). London, UK: SAGE.

Lu, W., & Stephenkova, S. (2014). User-generated content as a research mode in tourism and hospitality applications: Topics, methods and software. *Journal of Hospitality Marketing & Management, 24*(2), 119–154.

Månsson, M. (2011) Mediatized tourism. *Annals of Tourism Research, 38*(4), 1634–1652.

Mariani, M. M., Felice, M. D., & Mura, M. (2016). Facebook as a destination marketing tool: Evidence from Italian regional destination management organizations. *Tourism Management, 54*, 321–343.

McKercher, B., & du Cros, H. (2002). *Cultural tourism: The partnership between tourism and cultural heritage management.* New York, NY: The Howarth Hospitality Press.

Munar, A. M., & Jacobsen, J. K. R. (2013). Trust and involvement in tourism social media and web-based travel information sources. *Scandinavian Journal of Hospitality and Tourism, 13*(1), 1–19.

Rose, G. (2001). *Visual methodologies.* London, UK: SAGE.

Sigala, M. (2011). Special issue on web 2.0 in travel and tourism: Empowering and changing the role of travellers. *Computers in Human Behavior, 27*(2), 607–608.

Wise, N. A. (2011). Post-war tourism and the imaginative geographies of Bosnia and Herzegovina, and Croatia. *European Journal of Tourism Research, 4*(1), 5–24.

Wise, N., & Farzin, F. (2018). 'See you in Iran' on Facebook: Assessing 'user-generated authenticity'. In J. Rickly & E. Vidon (Eds.), *Authenticity & tourism: Productive debates, creative discourses* (pp. 33–52). Bingley, UK: Emerald.

Wise, N. A., & Mulec, I. (2012). Headlining Dubrovnik's tourism image: Transitioning representations/narratives of war and heritage preservation, 1991–2010. *Tourism Recreation Research, 37*(1), 57–69.

Wise, N., & Mulec, I. (2015). Aesthetic awareness and spectacle: communicated images of Novi Sad, the Exit Festival and the event venue Petrovaradin Fortress. *Tourism Review International, 19*(4), 193–205.

Xiang, Z., & Gretzel, U. (2009). Role of social media in online travel information search. *Tourism Management, 31*(2), 179–188.

Yoo, K. H., & Lee, W. (2015). Use of Facebook in the US heritage accommodations sector: An exploratory study. *Journal of Heritage Tourism, 10*(2), 191–201.

38

THE DIGITAL TOURIST BUREAU

Challenges and opportunities when transferring to a digital value creation

Sara Leckner and Carl Magnus Olsson

Introduction

Society is becoming increasingly connected through its infrastructure of internet-connected airplanes, trains, buses, and cars, and we are seeing a growth in digitally represented additional information and services in physical spaces such as parks, museums, and shopping centres. This explosion of connectivity between sensors, embedded systems, and mobile technology, has made it increasingly possible to share information and experiences, for example, cross- and transmedially, and is one of the absolute strongest growth areas for digital innovation. For companies and organizations, technological development creates countless of opportunities but also means challenges. One of the main issues is how to transfer a business where the key value is grounded in physical materiality into a digital one, and at the same time maintaining the core value, as well as creating and facilitating new ones with the help of digital properties.

This chapter examines how the increasingly connected society can be approached to create value for the tourist and the tourism industry, guided by the underlying holistic understanding of mediatization. This includes a comprehension of people's increasingly 'mediated' access to reality, using a media-centred perspective where media are seen as drivers of social processes (e.g., Adolf, 2017). In particular, this chapter examines the digital transformation of a tourism organization's hosting of visitors through cross- and transmedia features. Hosting can here, and in a wider sense, be understood as the art of making people feel welcome, related to material and immaterial aspects connected to a destination (see e.g., Hanefors, 2010). The investigated transformation is the journey of the organization from foremost being based on promoting storytelling of real-life interactions with a few expert assistants at a physical tourist bureau, towards becoming a primary digitally-based cross- and transmedia operation, where support of the hosting relies on a larger number of connected media and actors. As neither cross- nor transmedia can be conceived as only textual but exist also as economic and social phenomena – i.e., are conditioned by a complex and historical mesh of forces, such as industrial power struggles, market inertia, empowering of users, etc. (Ibrus & Scolari, 2012) – this wider understanding has been used as a frame to describe a digitalization-driven reorganization in a tourism setting.

The organization used as case is Malmö Tourism (MT). MT started in the year 2000 and is a part of a municipal organization, located in the city of Malmö in the south of Sweden.

MT's overall mission is to get more visitors, events, and meetings to Malmö, in order to increase growth and sustainable employment in the city and the region. According to themselves, MT was the first tourist organization in Sweden to make an extensive effort towards digitalization. In 2016, they made a strategic decision to close their physical bureau in 2017 and to shut down their official website in 2018. Hereafter, they aimed to provide their hosting through digital media, where tourists and potential visitors were already expected to interact, such as popular social networks like Facebook and Twitter. The study has been undertaken through individual semi-structured interviews and a workshop – based on personal construct theory for repertory grid sessions – with tourist assistants and managers at MT in 2016, and followed up in 2018 to capture the actual outcomes of the digital transformation. This chapter contributes with insight and understanding from a novel case that other studies of change may contrast their findings with and learn from. In particular, it contributes to an understanding on how digital transformation shapes the material and immaterial values of an organization, and the strategic and exploratory journey that eventually can lead to more efficient and timelier visitor interaction through cross- and transmedia features. Since the visiting industry is a sector that requires a lot of contact with the customer, this interaction can be crucial for how a company succeed and has been pointed out as an essential part of the future tourism development (e.g., Dielemans, 2008; Forss Karlsson & Jäger, 2014).

Digitization in the visiting industry

In our progressively mediatized society, crossmedia is becoming a strategic endeavour for many companies (Ibrus & Scolari, 2012). Along with the development of networked technologies and mobile media, such strategies are evolving and becoming more complex. While many crossmedia experiences do not include a narrative that connects the different content entities and thus making them transmedial, all transmedia storytelling is crossmedia (Ibrus & Scolari, 2012). In transmedia storytelling, a single story expands autonomously across multiple platforms using different forms of media, compared to in crossmedia where it does not have to expand (Jenkins, 2003). Used in a tourism setting, transmedia create a more efficient marketing tool, where the tourist is not merely buying the tourist product but also buying the stories behind the product (Mossberg, 2007). An essential part of both cross- and transmedia is the increased possibility for users to share information and experiences (Canavilhas, 2018; Jenkins 2003). This creates a more direct relationship between audiences and content producers, facilitating, among other things, media pluralism and empowerment of a diversity of actors (Canavilhas, 2018; Ibrus & Scolari, 2012) but also marketing possibilities such as electronic word of mouth (e-wom), which is of increasing importance for the tourism industry (Jalilvand & Samiei, 2012).

As with crossmedia, transmedia storytelling has emerged as a consistent communicative strategy (Saldre & Torop, 2012). In the tourism industry, the use of these concepts has facilitated new experiences, in particular through social networking services (Korez-Vide, 2017; Paiano, Passiante, Valente, & Mancarella, 2017). Research related to the area of cross- and transmedia in the visiting industry is starting to grow, in particular studies on models and management for audience engagement and visitor meaning-making in relation to digital media and information technology tools (e.g., Ferreira, Alves, & Quico, 2012; Korez-Vide, 2017; Paiano et al., 2017). The use of information technology in the industry is, however, nothing new. Prior research has emphasized that the visiting industry has been at the forefront of interest in and use of digital technology, which has shaped the way it works, facilitating more efficient booking and marketing on a global scale (Buhalis & Law, 2008; Neuhofer, Buhalis, & Ladkin, 2015). At the same time, others have argued that the industry has been quite slow to absorb technological innovations and that new digital possibilities are implemented slowly and with delay (Egger, Gula, & Walcher, 2016). Regardless, there is a consensus that digital technology, including cross- and transmedia opportunities, will change the existing visitor industry and bring many new values. Despite that the adoption of new technologies have had profound effects on organizations and their decision-making for a long time (cf. Huber, 1990), there is no blueprint on how the transition to digitalization should occur. Compare, for example, with the media industry which still struggles with developing successful digital business models (e.g., Nygren, Leckner & Tenor, 2018). Thus, there is a further need to investigate how organizations in the industry strategically can use cross- and transmedia as tools to create value in their hosting of visitors.

Malmö tourism's change towards digitalization: before the reorganization

Hosting and value creation before the digitalization

In 2016, MT's hosting of visitors mainly emanated from personal real-life encounters at a physical tourist bureau, led by primarily four tourist assistants (TAs). These encounters were complemented with both analogue and to some extent digital channels (Table 38.1). Good hosting, according to the TAs, included being *genuine, curious,* and *open* – interested in people and eager to help visitors – as well as *compassionate* and *responsive* – understanding visitor needs and wishes. Effectively using a mix of standard answers based on most commonly asked questions, together with *interpreting* the responses and the visitors' characteristics – e.g., nationality, age, potential interest – provided the input. These were then combined with the TAs' *knowledge* – a set of personal tips based on the TA's experiences, which highlighted the local and unique about Malmö depending on season, weather, and time. Based on these parameters, the TAs could adjust the tip, or 'story', to each individual visitor, aiming to create a *trustworthy* and *authentic* interaction. Furthermore, it was viewed as highly important for the TAs to work *efficiently* and to be able to multitask, for example, by handling several communication channels at the same time and change between front and back office. These values are similar to the five dimensions of service that Hanefors (2010) claims influence on how a tourist perceives quality of hosting; safety, willingness, reliability, tangibility, and understanding. Thus, both storytelling and crossmedia existed before MT's reorganization, but the work process emphasized analogue elements and a top-down workflow.

Of the crossmedia channels used for hosting visitors in 2016 (Table 38.1), real-life encounters over the front office desk were considered the most efficient and valuable due to the personalized face-to face experience. This was greatly emphasized by the TAs during the interviews. Next were interactions through telephone, chat, and Cargo bikes – mobile versions of a tourist bureau used during the summer seasons – due to the channels relatively personal mediation. Other crossmedia channels offered by MT at the time – physical as well as digital (see Table 38.1) – were not emphasized to any extent by the TAs. However, MT management expressed a wish to become more knowledgeable in using digital crossmedia more efficiently, including transmedia features. The TAs' inexperience of digital channels meant that some expectations did not turn out as expected and were perceived as less efficient. For example, the chat channel was expected to generate quick responses but was instead used for

Table 38.1 Pros and cons of the crossmedia used for visitor interaction before the move to digitalization (in 2016).

Channels used	Before digitalization (2016)	
	Pros	Cons
Physical bureau	Personalized face-to-face interactions, avaliable materials (e.g. printed folders), central location (easy to find)	Only a fraction of the tourists that visit, short visits, cost intensive
Cargo bikes	Active, outreaching, mobile, spontaneous meetings, different meetings e.g., also with locals	Seasonal, weather dependent
Info points	Smaller versions of the physical bureau, located at strategic places	
Telephone	Reasonable well used for bookings (but less than before) and follow-up questions	Most outdated channel, static, lesser used by visitors nowadays
E-mail	Good for longer answers, not location dependent (yet, usually answered at the office)	10% of the volumes they would like, interactions take time (the longer answers)
Chat	Location-independent (e.g., TAs can work from home), not dependent on the mood of the TA, extensive opening hours (compared to physical bureau), quick responses of basic questions (although has extended to longer communication), fast feedback from visitors, flexible (can be put aside if attention needed in other channels)	Lesser volumes than expected, can be stressful if unusual questions, have taken some time to get used to
Facebook	More integrated than Instagram, everyone can read the answers	Mostly used for images, mostly used for first contact, TAs not enough experienced yet on how to use efficient
Instagram	Wish to use more	Not used as much
Twitter	Used for info about Malmö that can raise interest or debate. Mostly used for information ads	Not used for customer service
YouTube	Wish to use more	Not used. Do not know how to use cost efficiently (e.g., how to produce own video material)
Webpage	Contains a lot of content	Contains little updated information
Printed material	Handy	Not entirely synced with other information channels (harder to update)
Event calendar	Gathers info at one place	Not in English, too general (both for citizens and visitors), not synced with other channels
Arrival guides	Not used	Not used
TripAdvisor	Not used	Not used

longer communications. Moreover, when more complicated questions were asked in some digital channels, it became stressful for the TAs, since they were not used to handle these in addition to their other (foremost analogue) channels.

Vision of the digitalization

In 2016, MT management made a strategic decision to close their physical tourist bureau from May 2017 and furthermore to shut down their website in early 2018. Instead, they decided to focus the hosting and interaction with visitors on digital, and foremost, social networking channels. One of the managers explained that she got an epiphany one day at the tourist bureau when serval visitors asked for 'The knotted gun' (a renowned sculpture actually called 'Non-violence'). Apparently, it was number three on TripAdvisor's must-see list in Malmö, while MT had never advertised it. MT thus realized that there was a discrepancy between descriptions and visitors' image of Malmö online, compared to the typical tourist visiting MT's tourist bureau. While the four very experienced TAs at the bureau held deep knowledge, the mix of commercial actors, residents, and former visitors of Malmö, were apparently also telling stories of the city online that many visitors were interested in. Additionally, MT had experienced decreasing number of visitors at the bureau (down 8.4% between 2015 and 2016). The bureau was also cost intensive, and foremost reached people that were visiting Malmö for a few hours. Thus, MT decided it was time for a digital strategy. Inspired by Visit Sweden (2015), MT's desired target group was the global traveller, an international group with strong potential to choose Sweden as a destination. This group was highly connected and used social media for planning and during their trips, as well as for sharing their experience. Thus, MT decided it was more efficient and cost-effective to be present in the channels already visited by the desired target group and, with the help of expertise from other actors, tell the story of Malmö, instead of through four tourist assistants at a physical office.

The main challenge to the change was that MT's work process was based on well-established best practices, scrutinized and largely optimized given their TAs' expertise. Any changes had to be considered carefully not to disrupt the work practices of the TAs, even if the type of interaction changed. The overarching goal, thus, was to identify and embrace the changes that emerging technologies were bringing, without losing the personal touch and direct customer contact that the present value of MT's hosting inherently held. While the organization looked forward to these changes at the time of the interviews 2016, they had not finalized their digital strategy by this time. The TAs appeared somewhat insecure about their future roles but expressed trust and hope for the changes. Since no other tourism organization – nationally or internationally – was deemed to have 'cracked the code' (as one of the managers expressed it) of mainly digitally based hosting, there were no established best practices to copy. This made MT's change exploratory and thus of relevance to follow over time.

Malmö tourism's change towards digitalization: the outcome

In 2018, the traditional TAs did not exist anymore. Instead, they were now project coordinators with areas of responsibility, connected to a framework of cross- and transmedia functionalities. The strategy expected all employees (not only the former TAs) to work more closely with hosting visitors.

The outcome of the vision after the digitalization

2018, MT foremost used a crossmedia framework consisting of Facebook, Twitter, Instagram, and YouTube for interacting visitors (Table 38.2). These digital channels were each seen as particularly suitable for specific information, together forming the transmedial narrative of hosting visitors. The framework did not consist of a single top-down told story but a juxtaposition of storytellers and stories having the promotion and marketing of Malmö as a valuable destination for travellers in common. Questions from visitors to MT primarily came through Facebook Messenger (rather than the feed of posts on Facebook). Instagram stories were deemed one of the most direct interactions with visitors and residents, where MT got quick feedback, for example, on what places the visitor wanted to see or hear more about. Similarly, Twitter was also considered a very direct channel for interaction – especially since the space is limited – where the story tended to be more informal, often mischievous and jokingly phrased. Still, however, MT did not believe they used cross- and transmedia to their full potential and thought they could put the channels, and in particular their trans-medial functionalities, to better use. They wished to be more proactive and outreaching in their own storytelling, posting fully researched information as a complement to their current reliance on asking questions as a way to stimulate user interaction. Posted images were mainly taken by other actors, while video was self-produced by MT given that high-quality video-based material has been harder to find. It was furthermore considered more strategic to complement what other actors produced online, rather than to compete, as this gives a broader space for interaction and storytelling. In 2018, the video-based approach was still under development, as producing such material in a cost-effective way remained a challenge.

At the time of the follow-up interviews in the end of 2018, the website was yet to be shut down, however, it was only used as a landing platform, transferring users to MT's other crossmedia outputs. The chat service connected to the site, one of the few digital channels emphasized in 2016, had been closed due to lesser traffic when social media channels become more utilized. Additionally, most of the analogue crossmedia channels used in 2016 were closed, which from a management perspective was natural as part of the strategy to 'go all in' with digitalization (Table 38.2).

Hosting and value creation after the reorganization

During the interviews in 2016, the real-life personal interaction was essential to MT's value creation. 2018, it was still regarded as important but not emphasized in the same way. With digitalization, the personal encounter was considered to be different and more varied. Fore-most, this was due to the different materiality afforded by different channels and how it affected the storytelling, but it also depended on who was greeting the visitor, in terms of telling the story of Malmö, and the type of questions they received. In 2016, the visitors were a fairly homogeneous group of people, asking similar questions. While these questions still came up 2018, the questions were now considered to be more varied, which was perceived as stimulating and a sign of responding to greater needs than the physical bureau managed. MT stressed that digital encounters must have a personal tone and differ from a message perceived as automatically generated or being a default answer, and that audience participation was crucial in enriching the content, thus for the content to be considered transmedia (cf. Canavilhas, 2018). MT did not have more time for visitor interaction 2018 but worked more efficient, for example, by forwarding questions to other actors more suitable to answer, when deemed appropriate.

Table 38.2 Pros and cons of crossmedia used for interaction with visitors after the move to digitalization (2018)

Channels used	After digitalization (2018) Pros	Cons
Physical bureau	Not used	Not used
Cargo bikes	Not used	Not used
Info points	Strategic physical places in the city, meetings in real life	Type of location vary a lot, therefore needs different info materials. Handled by other actors than MT, but visitors may think it is a tourist bureau (and get disappointed when it is not). Still not completely efficient, may close down in a near future
Telephone	Not used	Not used
E-mail	Not used	Not used
Chat	Not used	Not used
Facebook	Reach a much larger number of visitors or potential visitors than a physical bureau. Large engagement from Malmö residents and from visitors (tagging, commenting, etc.) in feeds. Efficient communication through Messenger	Some comments improper (e.g., in relation to the media image of Malmö); easier to criticize when anonymous
Instagram	One of the most positive channels. People proud and happy that MT tags their images. Instagram stories the most direct interaction (because easy to comment/give feedback). Fun for MT staff to experiment with	Could not stand alone as channel, but fulfils its function together with other channels. MT would like to show more breadth of Malmö (now a lot from the city centre), but believe they cannot since they may lose followers
Twitter	Very direct (since limited space to write text). Tone can be more mischievous	Could be even better put to use, could be more proactive and outreaching
YouTube	Wish to use more. (Still) Investigates how	Still not used to any great extent. Still do not know how to use cost efficiently.
Webpage	Transfer platform to other channels	Include only static statistics and information
Printed material	n/a	n/a
Event calendar	Not used	Not used
Arrival guides	n/a	Currently exist, but not active. May be removed as communication channel
TripAdvisor	n/a	May be removed as communication channel

Thus, while MT saw the digital encounter as simply different from their prior ways of working, they did not believe they had lost any core value. Half of the former TAs had however left the organization. While these TAs partly felt it was time to do something else, they partly believed that visitor interaction, and the storytelling, had changed too much.

The use of other actors – residents, traders, former visitors – in the hosting of visitors has thus been an important part of MT's realization and made the transmedia storytelling a more efficient destination marketing tool. Besides tagging, linking and posting other people's content, as well as providing possibilities for more extensive participation, MT had started to work actively with influencers. In 2018, the influencers were 60 persons who got their expenses paid to visit Malmö and communicate their experiences to their followers. MT also had begun to cooperate with 'info mates', local residents with interests or knowledge of particular value to visitors. On a voluntary basis, these info mates answered questions online, sometimes leading to real-life meetups to facilitate rewarding visitor experiences.

Conclusions

The decision to no longer provide real-life interaction in a physical office is a bold move for a municipal organization. Not the least when there is no blueprint to follow on how to realize it successfully. This is, however, not unique for the visiting industry but a challenge facing most industries transferring to digitalization. At the same time, to be at the forefront can benefit the commitment to change and lead to better feedback and better service, in particular through the challenges and failures that follow being an early adopter of new technology and new ways of working. Altogether, MT considers their strategy successful as no major negative issues have surfaced. Although mistakes may be costly, MT argued that change is about having the guts to take the plunge rather than to focus on the monetary aspects, as one simply cannot make large-scale changes if one does not dare to venture into the unknown. It can be argued, however, that the commitment to change could have been different if the organization had been private and/or commercial. MT does not have the burden of being economically successful; hence, it may affect their rather bold attitude. Nonetheless, radical innovation and change inherently implies testing the unexplored and strong support from management is one of the key factors for successful organizational change (e.g., Angelöw, 1991). Throughout the change process, MT has had strong support from their main (municipal) organization, and the management at MT has strived to be supportive of the staff – something which our interviews also confirm. The change has not, however, happened overnight. The reorganization is by MT considered to be work in progress, although the greatest changes have been made. Nor has it been a straight-forward process but included new challenges along the way. Overall, the process has required a new understanding and an open mind, so the materiality afforded by a physical space and face-to-face interactions were not just transferred to the digital, but transformed to take advantage of the unique aspects of media materiality. This included storytelling that was continuous (the overall story) and episodic (the side plots). It has also involved engagement and participation in the storytelling process by external actors, and to keep this involvement sustainable over time, at the same time as the distinct qualities which acting and speaking as a public authority implies, has to be maintained. For the TAs, who previously were the main hosts and narrators, the process has been long and challenging. They believe, however, that altogether the digitalization has resulted in a better working situation. The real-life encounter with visitors is still regarded as important, but not essential, in order to be a good host. The TAs personalized expertise is still needed, but in other ways than before. Today, the former TAs act more as conductors of an orchestra of storytellers rather than as main storytellers themselves.

This fits well with the mediated nature of a cross- and transmedia context. Thus, the gate-keeping of the storytelling now means controlling what already exists, rather than producing and actively influencing the direction of it oneself. The increased involvement of other actors keeps the story changing, and thus possibly also makes it more current, but makes MT taking a more passive role. Yet, some aspects of the physical meetings and real-life interaction of hosting visitors have not been transferable to a digital environment. Those of the TAs who valued and desiderated that type of interaction the most, have quit. Thus, the hosting has to a large extent become a *new* way of telling the story of Malmö.

While MT believe they have lost some visitors, the less digitally mature, they overall think they reach more people, meet the needs of more heterogeneous interests, and have more loyal followers, than before the reorganization. MT argues that critique for closing the physical bureau foremost has surfaced from those that view a physical office as a de facto form of public service, rather than the underlying and actual needs of visitors. However, there are no statistics to support the claim, and it is foremost based on user feedback through their crossmedia channels. A challenge with measuring interaction online is the often taken for granted significance based on the number of clicks, or in this case, those who interact through social media. Those who do not, have lesser possibility to get in touch, and thus to provide potentially negative feedback. This includes also them who use www but not social media, since MT in principal has closed down their website. Although the large majority of the Swedish population uses social networking sites, populations in, for example, other European countries do not to the same extent (Internetstiftelsen, 2018). Overall, the number of visitors in Malmö has increased since the reorganization, but it has increased overall in Sweden during the studied time period (e.g., Tillväxtverket, 2018). Hence, it is hard to clearly measure the impact of the digitalization of MT's hosting. Moreover, future research is suggested to look at the changes from the visitors' perspective, for example, visitors' experiences and attitudes toward digitalized services compared to face-to-face ones, and the value of different channels in terms of pre-visit and on-site hosting.

In summary, for MT, as a part of the visiting industry, which is in turn is a significant service industry, the personal encounter is an important guarantor of authenticity and trust. Research on hosting in the tourism industry claims that fewer face-to-face meetings mean fewer possibilities of hosting, and that technology never can replace the personal face-to-face interaction (Forss Karlsson & Jäger, 2014). What this chapter has shown is that it is possible, at least from the perspective and experience of the organization. Besides on how one chose to work with different materiality and technologies, the success of a digital transformation depends on the ability to change the mindset of the employees, to have the backing of the organization, and to study the pre-conditions but not the least to dare to think outside the box. This fits well with the arguments for understanding mediatization as a long-term process (see Hepp, Hjarvard, & Lundby, 2015). In a time of increased globalization, technological development, and self-sufficient travellers, many organizations will likely embrace similar challenges as described in this chapter, in order to maintain a good hosting.

References

Adolf, M. A. (2017). The identity of mediatization: Theorizinga dynamic field. In O. Driessens, G. Bolin, A. Hepp, & S. Hjarvard (Eds.), *Dynamics of mediatization: Institutional change and everyday transformations in a digital age* (pp. 11–34). Cham, Switzerland: Palgrave Macmillan.

Angelöw, B. (1991). *Det goda förändringsarbetet: om individ och organisation i förändring.* Lund, Sweden: Studentlitteratur.

Buhalis, D., & Law, R. (2008). Progress in information technology and tourism management: 20 years on and 10 years after the Internet – The state of eTourism research. *Tourism Management, 29*(4), 609–623.

Canavilhas, J. (2018). Journalism in the twenty-first century: To be or not to be transmedia? In R. R. Gambarato & G. Alzamora (Eds.), *Exploring transmedia journalism in the digital age* (pp. 1–14). Hershey, PA: Information Science Reference.

Dielemans, J. (2008). *Välkommen till Paradiset*. Stockholm, Sweden: Atlas.

Egger, R., Gula, I., & Walcher, D. (2016). *Open tourism: Open innovation, crowdsourcing and cocreation challenging the tourism industry*. Berlin, Germany: Springer.

Ferreira, S., Alves, A., & Quico, C. (2012). Location based transmedia storytelling: The travelPlot Porto experience design. *Journal of Tourism and Development, 17/18*(4), 95–99.

Forss Karlsson, A., & Jäger, A. M. (2014). *Vad är värdskap? En studie om värdskapets olika sidor inom turismbranschen*. Karlstad, Sweden: Karlstads Business School.

Hanefors, M. (2010). *Värdskap inom turism och resande*. Lund, Sweden: Studentlitteratur.

Hepp, A., Hjarvard, S., & Lundby, K. (2015). Mediatization: Theorizing the interplay between media, culture and society. *Media, Culture & Society, 37*(2), 314–324.

Huber, G. P. (1990). A theory of the effects of advanced information technologies on organizational design, intelligence, and decision making. *Academy of Management Review, 15*(1), 47–71.

Ibrus, I., & Scolari, C. A.(2012). Introduction: Crossmedia innovation?. In I. Ibrus & C. A. Scolari (Eds.), *Crossmedia innovations: Texts, markets and institutions* (pp. 7–22). Frankfurt Am Mann, Germany: Peter Lang.

Internetstiftelsen (2018). *Svenskarna och internet 2018: En årlig studie av svenska folkets internetvanor*. Sociala medier. Retrieved from: https://2018.svenskarnaochinternet.se/sociala-medier/.

Jalilvand, M. R. & Samiei, N. (2012). The impact of electronic word of mouth on a tourism destination choice: Testing the theory of planned behavior (TPB). *Internet Research, 22*(5), 591–612.

Jenkins, H. (2003). Transmedia storytelling. MIT Technology Review. Retrieved from: http://www.technologyreview.com/news/401760/transmedia- storytelling/

Korez-Vide, J. (2017). Storytelling in sustainable tourism management: Challenges and opportunities for Slovenia. *Journal of Advanced Management Science, 5*(5), 380–386.

Mossberg, L. (2007). A marketing approach to the tourist experience. *Scandinavian Journal of Hospitality and Tourism, 7*(1), 59–74.

Neuhofer, B., Buhalis, D., & Ladkin, A. (2015). Smart technologies for personalized experience: A case study in the hospitality domain. *Electron Market, 23*, 243–254.

Nygren, G., Leckner, S., & Tenor, C. (2018). Hyperlocals and legacy media: Media ecologies in transition. *Nordicom Review, 39*(1), 33–49.

Paiano, A. P., Passiante, G., Valente, L., & Mancarella, M. (2017). A hashtag campaign: A critical tool to transmedia storytelling within a digital strategy and its legal informatics issues. A case study. In V. Katsoni, A. Upadhya, & A. Stratigea (Eds.), *Tourism, culture and heritage in a smart economy* (pp. 49–71). Cham, Switzerland: Springer Proceedings in Business and Economics.

Saldre, M., & Torop, P. (2012). Transmedia space. In I. Ibrus & C. A. Scolari (Eds.), *Crossmedia innovations: Texts, markets and institutions* (pp. 25–44). Frankfurt Am Mann, Germany: Peter Lang.

Tillväxtverket (2018). Turismens årsbokslut 2017: *Omsättning, sysselsättning och exportvärde*. Rapport 0252. Stockholm, Sweden: Tillväxtverket.

Visit Sweden (2015). *Besöksnäringens målgruppsguide*. Version 1.0. Retrieved from: http://partner.visitsweden.com/Documents/VisitSweden%20Besöksnäringens_målgruppsguide_2015_VER1.0.pdf.

39

TOURIST INFORMATION SEARCH IN THE AGE OF MEDIATIZATION

Lena Eskilsson, Maria Månsson, Jan Henrik Nilsson,
and Malin Zillinger

Introduction

The digitalisation in society has increased significantly over the last three decades. Today's society is highly dependent on information and communication technology, ICT, for maintaining its basic functions. Information is generally regarded as a very important factor in most value-creating processes, most of which is transmitted by different sorts of media. We live in a society influenced by transmedia where stories, knowledge, and information are shared, developed, and accessed through a range of media platforms simultaneously (Jenkins, 2006). Transmedia implies both analogue and digital platforms. However, digital platforms make it easier to share and reach out with content to different people while also enabling and stimulating user agency, even if it is done in combination with non-digital activities (cf. Freeman & Gambarato, 2018). This development is permeating all parts of the tourist system, where producer-generated and consumer-generated content is increasingly co-present on various online platforms and sites. For example, information is now spread and shared by a range of stakeholders including destination marketing organisations, tourists, or influencers working through social media platforms such as TripAdvisor. Through online platforms, it is now a faster and easier process to gain information about a destination, to book a journey and share its stories and images. In many parts of the world, tourist destinations have been quick to adapt to using these information channels. The number of travel-related homepages and blogs have multiplied, information is increasingly downloaded instead of being printed and booking systems are facilitating access to hotels, attractions, and transport solutions. Tourist information is today available at any time on the Internet as long as users have access to reliable Internet. Tourism – and to be a tourist – has become highly digitalised. Due to the digitalisation, Tourist Information Centres (TICs) are closing down in many countries and cities, not the least in Sweden – the empirical case of this chapter. Instead, new access points for information such as mobile tourism information units and new digital resources like maps and apps replace them. There is a common understanding in Sweden's main urban areas that TICs and other more traditional information channels have lost their relevance for today's tourists. Furthermore, there is a belief among stakeholders within a number of tourist destination organisations that to be digitally connected is something that is valuable for all tourists with more and more information exclusively accessible through digital resources. However, the recent developments pose a series of questions: Has new technology changed tourists information search behaviour, or are traditional channels – such as TICs – still important for tourists when they search for information about destinations and attractions? Are TICs and analogue media really irrelevant and out-of-date just because of the possibilities of transmedia practices?

Mediatization is a concept that is used to analyse the current changes in consumers' media usage on the one hand, and changes in culture and society on the other (cf. Couldry & Hepp, 2013; Hepp & Krotz, 2014; Jansson, 2002). One aspect of mediatization is transmedia (Jansson, 2018). A TIC can, for example, display tourist information on a range of media platforms: their own website, Facebook, Twitter, and Instagram all at the same time. It might be the same information or information that is altered according to the chosen media platform. Moreover, any destination information could be initiated by the TICs themselves or be tourist generated. Transmedia specifically addresses the growing circulation of media content between various media platforms run by different types of organisations and people (cf. Jenkins, 2006). Consumers (in this chapter tourists) are increasingly involved in the production of media content on, for example, different social media platforms (Jansson, 2013). Jansson and Lindell (2015) argue that different kinds of content and news today is spread through multiple channels, which means that media users can access this material from any platform, at home or on the move. Although Jansson and Lindell (2015) focus on news media, the same understanding can be applied to tourism information and tourist practices.

The aim of the chapter is to analyse tourists' information search behaviour and their use of different information channels from a transmedia tourism perspective. Which media platforms do tourists use? Do tourists combine different media platforms, and if so, how? And what are tourists' preferences when it comes to the individual platforms? The empirical case builds on a study of German tourists and their search for tourist information about Sweden (2016–2017). The conceptualisation and analysis build on the contributing researchers' backgrounds in human geography, service studies, and strategic communication.[1]

Targeting German tourist and their information search behaviour

In our study, we investigate both the use of different information channels as well as the reasoning behind tourists' choice of media platforms. Mediatization and transmedia is used as a theoretical frame for the analysis. A mixed methods approach has been applied, as the investigation into the choice of different information channels requires a quantitative approach, while searching for reasons behind information-related actions requires a qualitative approach.

The two methods used consisted of interviews and questionnaires. In total, 136 spontaneous interviews with German tourists were made in the towns of Vimmerby and Ystad during the summers of 2016 and 2017. In most cases, we talked to more than one person during the interview, as many of the Germans appeared in groups, families, or couples. The questionnaires were sent out to Germans via the Newsletter and Facebook page of Visit Sweden's office in Hamburg. In total, 292 respondents completed the questionnaire. Despite the relatively low number, it gave us strong indications on how the German tourists in rural Sweden think when it comes to tourist information. The quantitative results coming out of the questionnaire were qualified by the face-to-face interviews (see Zillinger, Eskilsson, Månsson, & Nilsson, 2018).

Tourist information search behaviour: setting the scene

Tourists' information search is considered to be the first step in a decision-making process, and we know that tourists use a wide range of information channels when planning a trip. These information channels may be categorised into public sources, tourism sources, friends and relatives, and governmental information (Björk & Kauppinen-Räisänen, 2011). Sources like family, friends, and relatives (word-of-mouth) continue to be ranked in the top three for tourists' information search, with guidebooks and Internet also seen as important information channels (Grønflaten, 2009). In other words, despite a range of media platforms with both producer-generated and consumer-generated content, tourists continue to value traditional channels such as, for example, friends and family. This is interesting from a transmedia perspective.

In order to understand tourist information search about destinations and tours, several models have been suggested over the years: Fodness and Murray (1999) give an overview of internal and external quests in tourist information search. The general search models proposed by Gursoy and McCleary (2004) explain information search before decisions are made at home. Bieger and Laesser (2004) highlight factors that have an impact on the information search process, such as type of trip, degree of packaging, choice of destination, and choice of accommodation. Following their analysis, search behaviour and choice of information channels may also depend on the level of familiarity with the destination, and on the general travel experience.

In short, there are several theoretical approaches to tourist information search. However, the rise of the smartphone together with a proliferation of two-way communication via web 2.0 since 2008 has resulted in rapid changes in tourists' information search behaviour, due to transmedia possibilities as part of everyday life (cf. Chapter 2 by Jansson). Tourist information is in constant circulation and can be accessed and shared at any time. Early on travel-related information search was considered as one of the most popular online activities (Ye, Law, Gu, & Chen, 2011). Today, the Internet is an important way of reaching information channels for travellers at all stages of a journey (Bronner & de Hoog, 2016). A range of digital information is available for tourists including material from destination marketing organisations, consumer-generated content on TripAdvisor and other rating sites, as well as social media platforms such as Instagram. This is in line with the transmedia perspective, which also imply tourists' agency in creating content for other tourists (cf. Månsson, 2011). Compared to earlier periods, there are more travel-related material available online, generated by multiple actors, and available with few limitations in time and space.

The real game changer in tourist information is the rise of smartphones, due to its ubiquitous access to information, nearly independent of any restrictions in time and space. There is now an abundance of transmedia possibilities and for the tourists there is a constant flow of information to access and process. Research shows that people have similar search behaviours when it comes to internet-based information, no matter if they use desktops, laptops, or smartphones (Ho, Lin, Yuan, & Chen, 2016). However, smartphones are used for more collaborative forms of information search as information sharing with others. Dickinson et al. (2014) affirm that social practices are undergoing radical transformations due to mobile technology, not the least when it comes to the domain of travel. Mobile technology has made it possible for people to be more flexible when it comes to travel decisions and has thus enabled more ad-hoc decision-making on the way. Such omnipresent admission to information and people is a sign of the transmedia society.

Hitherto, both researchers and tourism marketers have considered tourists to plan their activities at home and then to fulfil their plans at the destination. Tourists have been believed to be rational individuals, choosing their information channels in a determined way, and picking the sources that are most reasonable at the moment of access. Recent research reveals, however, that travel decisions are in fact flexible, as they are subject to both situational and information-related changes (Kah & Lee, 2016). Hence, tourists' information search and decision-making are found to be less rational than previously believed. Transmedia, with its digital possibilities and high mobility, allows people to negotiate their daily decisions with growing fluidity, which grants the possibility for ad-hoc decisions made on the move. We argue that this has substantial bearing on tourists' information search behaviour, as it simplifies access to information in the actual travel phase, thus also contributing to delaying the decision-making process. In addition, Mieli (2017) shows that serendipity plays a major role in determining which information tourists engage with. Serendipity in this case means to be able to take advantage of opportunities that occur when planning and conducting travel. In other words, coincidence plays a greater role in travel-related decision-making processes than previously believed.

Tourist information search behaviour: Lessons from the case study

Tourists use a combination of media platforms

Empirical results from different studies show that digital information channels have become increasingly important for tourists (Vallespin, Molinillo, & Muñoz-Leiva, 2017). Such research simultaneously shows that tourists still tend to use a combination of different information channels (Tan & Chen, 2012). These findings are similar to Snepenger, Meged, Snelling, and Worrall's (1990) results that discussed tourists' combined use of information channels before the use of Internet. Even with the emergence of the Internet and the possibilities of digital transmedia information, there is still a need of a range of media platforms for information, analogue as well as digital. It is important because tourists tend to use a range of media platforms simultaneously. The result from our study confirms that German tourists nearly always use several parallel media platforms for information. In our study, looking at Internet-based material (homepages) is the most favoured way of searching travel information prior to a trip: 28% of all answers are stating the use of homepages. Homepages are followed by traditional channels like guidebooks (16%), own experience (13%), maps (10%), brochures (10%), and information received from TICs in analogue form (6%).

During the actual journey, non-digital channels like guidebooks (19% of all answers), maps (15%), and TICs (10%) become relatively more important. Tourists mention their own experience and word-of-mouth as important information channels both before and during a journey. Channels like travel agents, popular culture, travel magazines, social media, and rating sites rank significantly lower. In our questionnaire, social media was ranking among the least important channels of information (1, 3% of answers for use before and during travel) for this market. The interviews confirm this finding. The low number of social media users is surprising considering the importance attributed to this information channel in the age of transmedia tourism (cf. Ho, Lin, & Chen, 2012). Our data shows that consuming and producing tourist-generated content in social media cannot be considered as standard tourist behaviour for all types of tourists. Having access to all sorts of information does not necessarily mean that all tourists are interested in them, especially not this German traveller segment.

Moreover, while there is an abundance of information available through transmedia that tourists can access to make decisions, many interviewees talked instead about spontaneous decisions while travelling. Such decisions depend on the mood of the family members, the company of travellers and on weather forecasts. On the road, tangible objects like road signs, public maps in city centres, and information signs facilitate spontaneous actions and hence become important information channels. The result of the study shows that the tourists in many cases take advantage of opportunities that occur during travel when confronted with sign markers. Coincidence and flexibility play an important role, and so does the feeling of spontaneity and adventure, as one of the interviewees explains: 'You don't want to know everything; you want to explore things a bit as well'. Another interviewee describes how 'we love it, I think, to be flexible because in this way, you can see so much of a country' (authors' translations). Thus, even if there are plentiful possibilities of accessing information on transmedia platforms at any time, decisions are left on hold on purpose. Tourists' other preferences and feelings have more significance for travel choices than, for example, online information accessed before travelling. Searching information is thus a complex undertaking, where digital media are crucial – but where other factors play a significant part too.

A traditional use of digital platforms

German tourists in Sweden have a rather traditional view on information search, even though homepages are the most common information channel. By traditional view, we mean that this group of German tourists have an interest in engaging with homepages such as destination marketing homepages since they valued for their (perceived) up-to-date information. These sites are also associated with a high degree of trust. This is in contrast to a low interest for social media in general that is associated with a lower trustworthiness since anyone can create and share information. A key part of transmedia tourism is tourists sharing content themselves, but there was a low interest in the group examined for this activity. Moreover, tourists are generally not willing to rely on digital platforms alone, perceiving the system as fragile. The share of respondents using solely digital information channels prior to the journey was only 5%. During the trip, the number goes down to 3.5%. Several interviewees also choose to escape the digital world altogether during their vacation. This corresponds well to the emerging field of research that addresses tourists' needs and desires to stay digitally disconnected while on vacation. Dickinson, Hibbert, and Filimonau (2016) found that some tourists want to escape their daily lives and routines and chose to abandon the digital world altogether while on holiday. Thus, although it is argued that we live in a transmedia society, our results show that German tourists travelling to Sweden do not necessarily follow this belief. There are significant differences between national markets, such as the German and the Swedish (Zillinger et al., 2018). This knowledge is vital when planning information campaigns.

The remaining popularity of guidebooks

Despite many people's conviction that analogue guidebooks are increasingly replaced by digital information, travel guides still sell in record numbers (Peel & Sørensen, 2016). Reading a guidebook typically becomes popular after the choice of destination is made. Guidebooks contribute to giving their readers an overview, to shaping their sense of place, and to giving meaning to the presented places. In doing so, guidebooks have the power to define what an attraction is, and what it is not (Zillinger, 2007). Hence, guidebooks are sign markers and contribute to mediating tourism practices (Mieli, 2017).

In our study, guidebooks are used by 70% of the respondents. Their use is somewhat more important during the trip than prior to travel (16% and 19% of all answers, respectively). Guidebooks are viewed as the most authoritative information channel by a large number of interviewees, as they are supposedly based on non-biased experts' opinions. This contributes to their high level of trustworthiness. The technological development has made guidebooks become valued in new ways by their readers. First, guidebooks are valued for their credibility and stability in relation to digital information channels. Second, while they used to be preferred mainly for their practical information, hedonic values soon take over. Third, after the journey, guidebooks turn into souvenirs and status symbols in the book shelfs at home. In short, guidebooks are preferred over other information channels due to several reasons.

However, guidebooks are not only accessed in the traditional form of a physical book but are increasingly downloaded and retrieved when needed, as seen in Figure 39.1. Thus, the same information channel is now accessible and used in a digital way.

Another interesting finding from our study is the interplay between reading guidebooks and searching online. German tourists tend to first read guidebooks to get an overview, and then turn to the Internet to do a more targeted search. The separate media platforms provide a range of possibilities for accessing information in a different way. This combination of searching modes might be described according to the sandwich model (Mieli, 2017), where information technology and word-of-mouth are used prior to travel for general planning purposes, i.e. choice of destination, length of travel and for booking flights and accommodation. Thereafter, guidebooks are used to get an overview, to plan day-to-day itineraries, to read about history and context, and to find recommendations (Mieli & Zillinger, 2020). During travel, information technology and word-of-mouth provide detailed, up-to-date, and practical information. Metaphorically, information search using guidebooks is central to the process, but chronologically squeezed in between layers of digital modes. This is in line

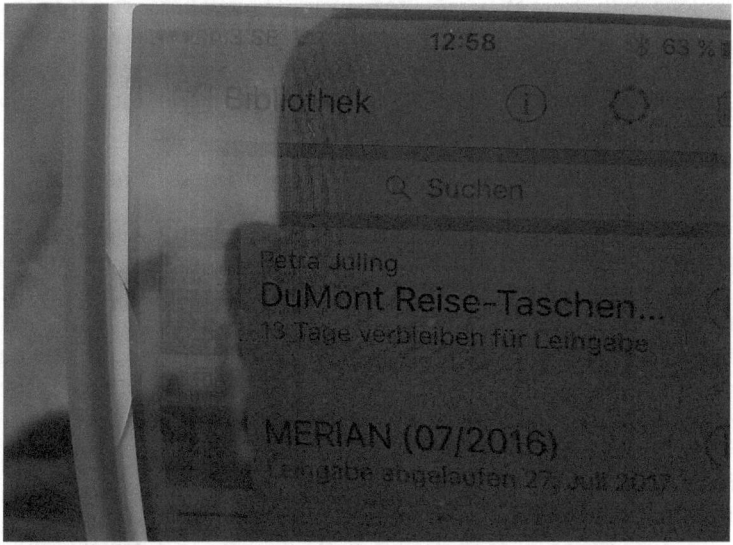

Figure 39.1   Digital guidebook in a German tourist's mobile phone
Source: Maria Månsson 2018.

with a transmedia perspective where tourists use analogue and digital media platforms to access, create, and share information in a range of combinations.

The importance of maps

According to the questionnaire in our study, maps are the third most used travel information channel prior to a trip, second only to webpages and guidebooks. During travel, maps become relatively more important. In general, tourist maps have dual purposes. They guide tourists to find their way, and they promote a destination. Maps can be national, regional, or local; they can be printed, be part of a physical guidebook, or found online since maps are also accessible on many digital platforms. Some tourists prefer the digital ones, whereas other prefer traditional paper maps to get better overview and understanding of the place.

The information provided by maps is especially important for first time visitors and for those being unfamiliar with the destination (Yan & Lee, 2015). Furthermore, tourist maps are particularly vital in urban contexts, for planning activities, and for finding and identifying major tourist attractions (Farias, 2011). They also function as a memento after the holiday, as souvenirs to bring home (Boulaire & Hervet, 2010; Farias, 2011). Rosetto (2012) argues that a map is floating, fluid, and relational in character. Moreover, maps provide opportunities to map activities by showing what people actually have visited (Boulaire & Hervet, 2010). In digital maps, individual maps and travel itineraries can easily be shared afterwards, and they can trigger anticipation before the journey.

In our study, we can see that maps, both printed and digital, play a prominent role for many German tourists. Some interviewees show a large interest in maps and atlases. Many of them consider themselves as having a high degree of cartographic literacy. They read maps to get an idea of what to find, to plan the route, and to get inspired. One interviewee explains that coming to Ystad 'was actually very spontaneous. We looked at the map and turned left […] by the end, we also wanted to visit *Ales Stenar*' (authors' translation). Some interviewees spend hours reading maps. When travelling, maps inside guidebooks are important information channels along with street atlases and road maps. Digital maps are not only used to plan the itinerary but also as a means to find sites and attractions. The use of digital maps is in line with the logistical turn that Jansson address in Chapter 2, regarding transmedia tourism. A map from a transmedia perspective encompass so much more than just a depiction of a place.

The importance of personal information and service

In a study of visitors in Australia, de Ascaniis, Gretzel, & Mistilis, (2012) found that almost 50% of all questions asked in TICs concerned the geography of the destination and included requests for routes and tangible maps. This led to the conclusion that despite digital availability of this information, human interaction was of importance, not only for the visitors to take decisions but also to improve decisions. The staff helped them to choose within the possibilities they had already filtered out. Lyu and Lee (2015) had a similar result when they found that many tourists were dissatisfied and frustrated with the information they had received elsewhere and therefore decided to visit an information centre to check the quality of that information.

The result of our study is in line with such research; personal information, traditional word-of-mouth, is regarded as trustworthy, both at home in Germany and in Sweden. This result is interesting considering the more sceptical attitudes towards electronic word-of-mouth, i.e. social media, which German tourists have. TICs are used by 50% of our

Figure 39.2 Visitors at the TIC in Vimmerby, Sweden
Source: Vimmerby tourist office 2017.

respondents, and the interviews confirm their importance. Even though tourism actors working in TICs assert that the general need for visiting TICs has changed, visitors continue to value human communication, personal recommendations, and service provided at TICs. Local knowledge is regarded as the main advantage of visiting a TIC, and the presence of German-speaking staff is valued in a highly positive way. In this vein, TICs can be viewed as a trustworthy filter in interpreting the flow of information, triggered by transmedia, on different media platforms as exemplified in Figure 39.2.

Conclusions

From the authors' disciplinary backgrounds, this chapter has contributed with new knowledge regarding how tourists in the digital age combine different media platforms while searching for information. The result of the chapter is interesting in the perspective where transmedia strategies seems to be the norm that organisations follow. As also discussed in this chapter, tourists' information search behaviour is far more complex than anticipated, and the values ascribed to different media channels by tourists play a vital role for their search preferences. If guidebooks or webpages have been preferred, this is unlikely to change even when further channels are offered. It seems in many ways that tourist' old preferences of and values attached to different platforms guide their information search behaviour. Thus, even though there are a number of transmedia platforms and we live in a society saturated with transmedia, some people prefer other ways of accessing information, as discussed in the chapter.

The study shows that contrary to many tourism actors' beliefs, digital tourism information does not substitute analogue channels. In the study, the Internet with its homepages dominates as information channel but analogue ones like guidebooks, TICs, maps, and road signs are also important. Personal meetings and personal services at the destination are regarded as a luxury. Hence, what we can see is a mix of old and new channels rather than the Internet replacing the old ones. The actual choice of media platform is affected by a combination of factors such as accessibility and trustworthiness but also tradition and old preferences. Tourist information is largely accessible both in analogue and in digital form, which points to the fact that the assumed division between analogue and digital information is constructed.

Interestingly enough, social media is not as widely used as expected. Recommendations and advice on social media have low trust, compared to personal meetings. In general, transmedia strategy is a possibility for organisations to spread their tourist information and many tourists are highly interested in both creating, sharing, and accessing tourist-generated content. However, it is important to keep in mind that not all tourists are digitally oriented even though there is an opportunity. We have also identified a counter-trend in form of a group of tourists who want to be digital disconnected while on holiday. The result of the study indicates important managerial implications including the need for organisations to carefully assess their target markets and corresponding information strategies and plan their information strategy and use of different media platforms accordingly.

Acknowledgement

The work has been carried out in co-operation with Regionförbundet Kalmar County, Visit Sweden, Vimmerby Municipality, and Ystad Municipality. It has been financed by the R&D Fund of the Swedish Tourism & Hospitality Industry (BFUF).

Note

1 Parts of this chapter have been reproduced from a previously published report: Zillinger, M., Eskilsson, L., Månsson, M., & Nilsson, J. H. (2018). *What's new in tourist search behaviour? A study of German tourists in Sweden.* Lund, Sweden: Lund University.

References

Bieger, T., & Laesser, C. (2004). Information sources for travel decisions: Toward a source process model. *Journal of Travel Research, 42,* 357–371.

Björk, P., & Kauppinen-Räisänen, H. (2011). The impact of perceived risk on information search: A study of Finnish tourists. *Scandinavian Journal of Hospitality and Tourism, 11*(3), 306–323.

Boulaire, C., & Hervet, G. (2010). Tourism & WEB 2.0: the digital map experience. *International Journal of Management Cases, 12,* 595–607.

Bronner, F., & de Hoog, R. (2016). Travel websites: Changing visits, evaluations and posts. *Annals of Tourism Research, 57*(1), 94–112.

Couldry, N., & Hepp, A. (2013). Conceptualizing mediatization: Contexts, traditions, arguments. *Communication Theory, 23,* 191–202.

de Ascaniis, S., Gretzel, U., & Mistilis, N. (2012). What tourists want to know: An analysis of questions asked at visitor information centres. In: Cauthe 2012: The new golden age of tourism and hospitality; Book 1; Proceedings of the 22nd annual conference. Melbourne, Victoria: La Trobe University, 2012: 16–28. Uploaded at www.researchgate.net/publication/284189833.

Dickinson, J., Ghali, K., Cherrett, T., Speed, C., Davies, N., & Norgate, S., (2014). Tourism and the smartphone app: Capabilities, emerging practice and scope in the travel domain. *Current Issues in Tourism, 17*(1), 84–101.

Dickinson, J., Hibbert, J., & Filimonau, V. (2016). Mobile technology and the tourist experience: (Dis) connection at the campsite. *Tourism Management, 57,* 193–201.

Farias, I. (2011). Tourist maps as diagrams of destination space. *Space and Culture, 14*(4), 398–414.

Fodness, D., & Murray, B. (1999). *A model of tourist information search behaviour: An introduction to theory and research.* Reading, MA: Addison-Wesley.

Freeman, M., & Gambarato, R. (2018). Introduction: Transmedia studies – Where now? In M. Freeman & R. Gambarato (Eds.), *The Routledge Companion to transmedia studies* (pp. 1–12). New York, NY: Routledge.

Grønflaten, Ø. (2009). The tourist information matrix – Differentiating between sources and channels in the assessment of travellers´ information search. *Scandinavian Journal of Hospitality and Tourism, 9*(1), 39–64.

Gursoy, D., & McCleary, K.W. (2004). An integrative model of tourists´ information search behavior. *Annals of Tourism Research, 31*(2), 353–373.

Hepp, A., & Krotz, F. (2014) Mediatized worlds – Understanding everyday mediatization. In A. Hepp & F. Krotz (Eds.), *Mediatized worlds. Culture and society in a media age* (pp. 1–15). Basingstoke, UK: Palgrave Macmillan.

Ho, C. L., Lin, M. H., & Chen, H. M. (2012). Web users' behavioural patterns of tourism information search: From online to offline. *Tourism Management, 33*, 1468–1482.

Ho, C., Lin, Y. C., Yuan, Y. L., & Chen, M. C. (2016). Pre-trip tourism information search by smartphones and use of alternative information channels: A conceptual model. *Cogent Social Sciences, 2*, 1–19.

Jansson, A. (2002). The mediatization of consumption. Towards an analytical framework of image culture. *Journal of Consumer Culture, 2*(1), 5–31.

Jansson, A. (2013). Mediatization and social space: Reconstructing mediatization for the transmedia age. *Communication Theory, 23*, 279–296.

Jansson, A. (2018). *Mediatization and mobile lives: A critical approach.* London, UK: Routledge.

Jansson, A., & Lindell, J. (2015). News media consumption in the transmedia age. Amalgamations, orientations and geo-social structuration. *Journalism Studies, 16*(1), 79–96.

Jenkins, H. (2006). *Convergence culture. Where old and new media collide.* New York and London, UK: New York University Press.

Kah, J. A., & Lee, S. H. (2016). A new approach to travel information and travel behaviour based on cognitive dissonance theory. *Current Issues in Tourism, 19*(4), 373–393.

Lyu, S. O., & Lee, H. (2015). Preferences for tourist information centres in the ubiquitous information environment. *Current Issues in Tourism, 18*(11), 1032–1047.

Månsson, M. (2011). Mediatized tourism. *Annals of Tourism Research, 38*(4), 1634–1652.

Mieli, M. (2017). *The value of travel guidebooks in the digital age.* Master thesis, Department of Service Management and Service Study, Lund University, Sweden.

Mieli, M., & Zillinger, M. (2020). Tourist information channels as consumer choice: The value of travel guidebooks in the digital age. *Scandinavian Journal of Hospitality and Tourism, 20*(1), 28–48.

Peel, V., & Sørensen, A. (2016). *Exploring the use and impact of travel guidebooks.* Bristol, UK: Channel View Publications.

Rosetto, T. (2012). Embodying the map: Tourism practices in Berlin. *Tourist Studies, 12*(1), 28–51.

Snepenger, D., Meged, K., Snelling, M., & Worrall, K. (1990). Information search strategies by destination-naive tourists. *Journal of Travel Research, 29*(1), 13–16.

Tan, W. K., & Chen, T. H. (2012). The usage of online tourist information sources in tourist information search: An exploratory study. *The Service Industries Journal, 32*(3), 451–476.

Vallespin, M., Molinillo, S., & Muñoz-Leiva, F. (2017). Segmentation and explanation of smartphone use for travel planning based on socio-demographic and behavioral variables. *Industrial Management & Data Systems, 117*(3), 605–619.

Yan, L., & Younghee Lee, M. (2015) Are tourists satisfied with the map at hand?. *Current Issues in Tourism, 18*(11), 1048–1058.

Ye, Q., Law, R., Gu, B., & Chen, W. (2011). The influence of user-generated content on traveler behavior: An empirical investigation on the effects of e-word-of-mouth to hotel online bookings. *Computers in Human Behavior, 27*(2), 634–639.

Zillinger, M. (2007). *Guided tourism – The role of guidebooks in German tourist behaviour in Sweden.* Umeå, Sweden: V2007:18.

Zillinger, M., Eskilsson, L., Månsson, M., & Nilsson, J. H. (2018). *What's new in tourist search behaviour? A study of German tourists in Sweden.* Lund, Sweden: Lund University.

40

TOWARDS SUSTAINABLE NAUTICAL TOURISM – EXPLORING TRANSMEDIA STORYTELLING

Fani Galatsopoulou and Clio Kenterelidou

Introduction

Digital media technologies, social media platforms, and virtual travel communities converge, and tourist spaces are augmented with layers of information, comments, reviews, and audio-visual material. Travelogues, personal essays, vlog and blog entries, photo stories, reviews, and recommendations through travel communities created by travellers are known and studied as 'user-generated content' (henceforward UGC). UGC is distributed online through multiple channels and social media networks and allows potential travellers to build knowledge and awareness about tourist spaces while influencing their visiting intention and attitudes toward specific forms of tourism or travel destinations (Minazzi, 2015).

Travellers are digital storytellers. They are both media producers and digital influencers that engage new audiences that help the distribution in multiple channels. For example, Månsson (2011), in her study on mediatized tourism and media convergence, focuses on user-created media products and discusses the active role of tourists in the consumption and production process. Santana, Gil, and Chirino (2018) emphasise the need to study how content is used, produced, and reproduced across the digital media landscape and the need to understand how the tourism industry should deal with media convergence in a dynamic global world.

This study focuses on nautical tourism – a sustainable form of tourism in the age of global climate crisis. This fast-growing niche tourism market may support and contribute to the sustainable development of coastal areas and marine ecosystems, which is one of the priorities of the United Nation's sustainable development agenda 2030. Moreover, it is the focal point of the 'Blue Growth' strategy, launched by the European Union in order to ensure sustainable development of the European seas. This strategy aims to employ the untapped potential of Europe's oceans, seas, and coasts for jobs and growth and is expected to provide innovation, ideas, and opportunities for low impact on the marine ecosystem, the aquaculture, and the coastal environment, contributing to the wealth and well-being of coastal regions (European Commission, 2012b).

Nautical Tourism is also an academic sub-field of travel and tourism studies. In several studies, it appears as a multi-functional activity, and the suggested definitions have some common attributes but different focal points. The only common characteristic between the definitions of nautical tourism is the direct relationship between tourist and the sea (Lam González, León González, & de León Ledesma, 2015).

To identify the tourist's role in nautical tourism and to understand what nautical tourism experiences are and how they are consumed, this study examines media representations of nautical tourism from the perspective of tourists, and the perspective of travel media professionals. Through content analysis of textual travel content, it is examined how nautical tourism is presented within the digital travel media by the professional travel journalists and within the digital travel communities by the travellers themselves, as social media users. The study examines the attributes, similarities, and contradictions, and explores which of the attributes that influence and shape the nautical tourism experience are presented in the travel content as the dominant ones.

The following research questions focus on the onboard and onshore experiences during a travel. The onboard travel experience during a trip on a cruise ship, yacht, boat, or any kind of vessel will be examined as an active or passive experience. The coastal/onshore travel experiences during a trip to a coastal destination will also be examined as passive or active. As coastal travel experiences are considered all coastal, maritime and underwater activities are the primary purpose of a travel on a coastal area (Lam González et al., 2015).

RQ1: Is the onboard and onshore travel experience active or passive?
RQ2: Do the onboard tourists refer to onshore/coastal activities? Do the onshore/coastal tourists refer to onboard activities?
RQ3: What are the dominant attributes of the overall nautical tourism experience?

What is nautical tourism?

To date, there is no official definition of nautical tourism. Scholars use similar terms to define its meaning. The term 'nautical' derives from the ancient Greek word 'naus', meaning the ship, boat, and seamanship (Luković, 2012).

Nautical tourism is sometimes conflated with maritime tourism, the latter predominately refer to water-based activities and cruising (European Commission, 2013). Yachting tourism is also suggested as a component of the broader concept of nautical tourism (Mikulić, Krešić, & Kožić, 2015); however, the focus is on the yacht as a symbol of luxury at sea, which is, according to Luković (2012), more tied to social status than seafaring. In the same vein, leisure boating is considered as a sub-segment of nautical tourism, which has grown significantly locally in recent years. It refers to single-day cruising, short trips on small yachts, motor or sailing crafts (Gon, Osti, & Pechlaner, 2016). Lück (2007) mentions the term marine tourism, focusing on the marine environments which include activities, such as scuba diving and snorkeling, windsurfing, jet skiing, fishing, observing marine wildlife, all beach activities, sea kayaking, visits to fishing villages, marine parks and aquaria, sailing and motor yachting, maritime events and races, the cruise ship industry, and many more (Lück, 2007). Finally, Wild and Dearing (2000) focus on the relationship between tourist, sea, and leisure and define maritime tourism and leisure as 'any maritime based leisure activity engaged in by people in their non-working time involving an element of travel on, below or immediately above water' (Wild & Dearing, 2000).

Most definitions approach nautical tourism as an economic activity and include the operation of landside facilities, manufacturing of equipment, and necessary services (European Commission, 2013). Therefore, coastal and maritime tourism are usually clustered together as subcategories of nautical tourism, as they often demand common infrastructure, facilities, and services and include nautical boating, water-based or beach-based activities (European Commission, 2012a). Thus, it is difficult to define nautical tourism, since the term includes many different forms and activities. These activities have become special forms of tourism and are studied separately. For example, sailing tourism is examined from different academic perspectives. Butowski (2018) reviews studies focusing on the maritime yachting tourism experience, the economic impacts of yachting, the boaters' perceptions, the impacts on the marine environment and the marina management (Butowski, 2018). Surf tourism is another example of water sports activity within adventure tourism in coastal areas, but that recently take the form of surfing boat charters and lodges (Buckley, 2002). Surfing has evolved from a lifestyle and competitive sport to a growing niche of the global tourism industry (Brochado, Stoleriu, & Lupu, 2018). A different example is underwater tourism, which typically includes diving and underwater activities, or just alternative living for non-divers in an underwater structure (Bitterman, 2014).

Luković (2012) analysed and highlighted the complexity of the different aspects and forms of nautical tourism and proposed the following definition for nautical tourism: 'Nautical tourism is a poly-functional tourist activity with a strong maritime component'. This definition has a longer version focusing on the 'poly-functional' element and the 'maritime component'. Based on this definition, and the subcategories mentioned in this section, this study examines a sample of media representations of nautical tourism that circulate online through social media networks and travel communities.

Nautical travel experiences and transmedia user-generated content

The nautical awe has always been a motive to read and write about sea adventures and long maritime trips. The origins of written travel accounts go back thousands of years and scholars hypothesise that this probably began with seafarers' logs (Youngs, 2013). Explorers and navigators used to write about their journeys, sharing information about unknown seas and coastlines. While writing about their experiences and impressions, they inspired other explorers to follow their route.

Today, in the age of Internet with the enormous expansion of digital and social media networks, peer-to-peer applications, platforms and communication technologies, content is published directly by users and can be distributed to multiple channels and reach new audiences. Nautical web series, or sailing vlogs and episodes on YouTube channel, as well as a huge collection of photos on Flickr.com or Instagram about nautical travel experiences inspire travellers and engage them to contribute with comments, reactions, and feedback.

Tourism sites such as Tripadvisor.com provide forums and foster interaction among users in order to exchange information, opinions, and recommendations about travel destinations, products and services (Akehurst, 2009). Some of the forums' topics are *beach vacations, sailing, cruises, and adventure travel*. Communities for bloggers such as Travelblog.org have been created and provide free web space for travellers to publish their travel stories and photos. Apart from the destination-specific categories where plenty of nautical experiences can be found, there are also categories dedicated to *Oceans & Seas, Sailing, Surfing, Diving,* and *Cruising.*

The blogosphere in tourism contains travel blogs as a form of the consumer to consumer (C2C) communication, but also as business to business (B2B), business to consumer (B2C), and government to consumer (G2C) communication (Schmallegger & Carson, 2008). C2C travel blogs, as a form of e-Word of Mouth (WOM) communication, influence travellers search for travel information and decision-making since they are considered as unbiased and trustworthy (Wu & Pearce, 2014). The 'user-generated' content is content by *real people* and fits into the larger discourse on travel (Usha & Divya, 2014).

Nautical tourism and travel journalism

Travel journalism also plays a significant role in mediating tourism. Hanusch (2010) argues that travel journalism is at the intersection between information and entertainment, journalism and advertising (Hanusch, 2010). It is considered lifestyle journalism which exists in a symbiotic relationship with advertising, and its discourse may be influenced by the private travel industry and by government-sponsored tourism departments (Hanusch, 2012). Several travel article categories have been proposed by travel writers (Greenman, 2012; Leffel, 2010; O'Neil, 2006) emphasising on the narrative type, the news type, and the service-oriented and advice type of travel articles. These types of articles are written in different formats, style and tone and can be found in travel magazines, travel sections of newspapers, travel websites, in the blogosphere and the social networks. Their topics vary according to the audience's interests and the travel market.

Nautical tourism is considered to be a broad market with many niche subcategories. According to Greenman's survey about the dominant story types in travel editions, 47% of the travel editors who responded to the survey said that cruising is a significant niche topic to readers (Greenman, 2012). Niches may change over time, but the nautical element still dominates all categories as this is evident in the print and online travel editions. Because of the importance and popularity of this topic in Travel Journalism, 'Cruise Travel' has become a category for the Lowell Thomas Travel Journalism competition.

Studying transmedia storytelling of nautical tourism

This study employs a quantitative approach for the content analysis of travel content, suggested by Stone (2018) which focuses on which attributes are mentioned or presented in the media content and not how it is presented. The final data provides a quantitative count of recounted onboard and onshore experiences and descriptions of specific attributes. The following list of keywords (Table 40.1) was constructed after reviewing the literature about the classification of nautical tourism (Kovačić, Gržetić, & Bosković, 2011; Lam González et al., 2015; Luković, 2012; Moreno & Otamendi, 2017) to define the search of the travel content in online travel magazines, websites, online travel sections of newspapers and blogs.

After the collection of the travel content, the two researchers and authors of the study separately read the travel articles and blog entries and excluded the irrelevant for the study content. The selected content was initially coded according to: (1) the type of article (narrative

Table 40.1 List of keywords used for collecting travel content for the study

Nautical tourism	
Nautical	Diving/snorkeling/underwater
Maritime	Surfing/windsurfing/kitesurfing
Yachting	Watersports/coastal
Sailing	Fishing
Cruising	Lake/river
Boating	Aqua/water

Table 40.2 List of attributes used for coding

Coding attributes
a. Onboard – Onshore
b. Onboard activities – Onshore activities
c. Active – Passive
d. Water activities
e. Adventure experiences
f. Landscape/environment and physical scenery
g. Weather
h. Local people
i. Local culture
j. Onboard services
k. Onshore services

or service/advice), (2) the publication (online magazine/e-zine, online travel section of newspapers, travel site, travel blog, UGC of a travel community), and (3) the genre and type of travel article.

For the last coding category (category c), the typology suggested by O'Neil (2006) was used, and the articles were sorted in destination-specific, personal experience, special interest or subject-specific, and roundup travel articles, in which there is information on different places with a common theme. Another niche category, the 'visitlogue' (Stone, 2018), was added as a new type of travel article, which refers to the itinerary-based travel writing and serves as a guide for exploring a destination by suggesting a visit plan and sometimes an hourly schedule for the whole trip.

The next level of coding was developed according to the attributes, study results, and the framework of (Mikulić et al., 2015; Moscardo, Pearce, Green & O'Leary, 2001); and Whyte, Packer, & Ballantyne (2018) about the nautical travel experiences. In their study, Moscardo et al. (2001) focus on travellers' participation in activities and classify the subjects of their study in eco-coastal, active beach, and passive seaside groups. Mikulić et al. (2015) explore the yachting tourism experience, and their analysis reveals a five-dimensional structure consisting of the core marina services, the basic destination attributes, and the onshore destination experience. They emphasise the results of the onshore destination experience that seems to have a significant impact on the overall travel experience.

Whyte et al. (2018) examine the onboard and onshore attribute factors that shape the overall experience and can influence the decision-making process of booking a cruise vacation. Their study identifies three onboard attribute factors which are related to the onboard environment, social interaction, and recreation, whereas the five onshore attribute factors include all kinds of activities from watersports to entertainment, learning and exploring the local culture, the natural features and infrastructure of the destination, as well as the safety and comfort (Whyte et al., 2018) (Table 40.2).

Sample

A keyword search for online travel publications during the period 2013–2018 revealed an initial sample of 240 travel articles and another sample of 280 blog posts (UGC). During the first reading of the sample, many articles and blog posts were excluded from the study as

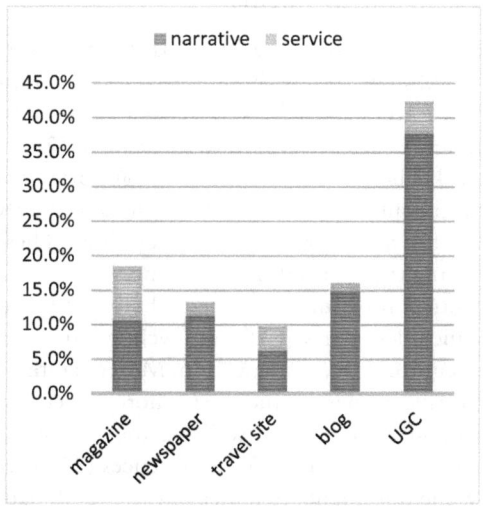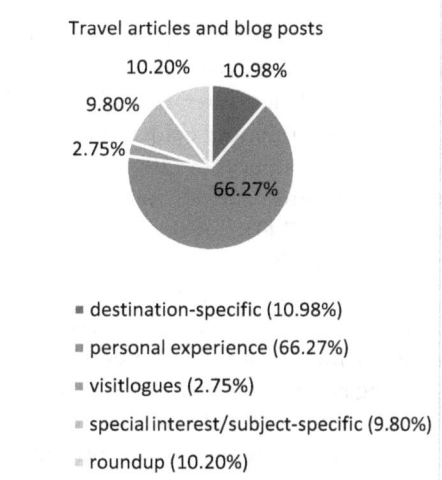

Figure 40.1   The sample of the study

irrelevant, or due to limited text and information. As irrelevant were characterised those that included some of the study's keywords but focused on different themes and topics.

Articles indicated as advertorials, promotional, or paid content were also excluded from the study. The final sample consisted of 147 travel articles and 108 blog posts, divided into 206 narrative and 49 service articles. The sample was further coded into 28 destination articles, 169 personal experience stories, 7 visitloques, 25 special interest or subject-specific articles, and 26 roundup articles (Figure 40.1).

Some narratives (UGC), although they were categorised as personal experience travel pieces, they overlapped the service and advice article category. It was a mixed genre coming from amateur contributors, the travellers, who wanted to tell their story, review the facilities and services provided for their trip, and inspire potential travellers to take this journey. For that purpose, they wrote plenty of advice and service information about the details of their travel, apart from their personal experience.

On the other hand, the professional travel journalists using their personal or the home style of their travel publications offered travel pieces with a literary value that could easily be sorted in specific categories. The online travel editions that were included in the study have large communities of followers and comment contributors and may impact millions of potential travellers. These are *Conde Nast Traveller magazine, Afar magazine, Travel and Leisure Magazine, The Travel Magazine, New York Times, The Guardian, Telegraph, Dailymail, Telegraph, BBC, CNN, National Geographic, Lonely Planet, Roughguides.*

The articles collected from travel blogs written by bloggers who have become influencers and promote travel destinations and touristic products are included in the sample of professional travel journalists. The UGC published by travellers was retrieved from the travelblog.org. The specific platform is in the top rankings of blog communities according to Marine-Roig's (2014) webometrics analysis model, which involves the visibility, the usage, and the content size, as factors for defining ranking of websites (Marine-Roig, 2014).

Findings: nautical tourism in transmedia

The sample consists of 255 travel content pieces. 58% (n=148) of the sample is about cruising or sailing (onboard experiences), while the other 42% (n=107) is about coastal tourism (onshore) that includes water-based and beach-based travel experiences (onshore/coastal). The 88% (n=226) of the onboard and onshore experiences are active, while the 79% (n=23) of the passive experiences (n=29) seem to be onboard. The 113 out of the 148 (76%) stories that describe onboard nautical experiences also write about onshore/coastal activities, but only 5 out of the 107 (4%) of the coastal experiences' articles refer to onboard activities. These activities are daily cruises, diving excursions, fishing trips, or sailing experiences.

The results of the study indicate that the nautical tourism experience whether it is onboard or coastal is described as active since it includes sports, walking, cycling, running, diving, surfing, or other activities or events that require active participation. Moreover, there is a strong connection between the onboard nautical experiences and the onshore and coastal activities, which include sightseeing, side trips and explorations, walking tours, cultural walks, or even extreme and adventurous activities. 76 out of the 255 experiences (29%) are described or referred to as adventurous. Finally, the dominant attributes of the overall nautical tourism experience are, according to the study, the landscape and environment (81%), the culture (77%) and the local people (74%), the weather (67%), and the onboard services for the onboard passengers (65 out of 80–81%). The onshore services seem to be important not only for coastal travellers (52,2%) but also for the onboard passengers (47%).

Conclusions

The purpose of this study was to explore transmedia storytelling and contribute to the conceptualisation and definition of nautical tourism. The analysis of the transmedia representations that was conducted identified the variables involved in the nautical tourism experience and demonstrate the need for re-defining nautical tourism in a sustainable and anthropocentric way.

The most important finding of this study is that the nautical travel experience is referred to as active participation and combines onboard with onshore activities that offer knowledge about the local cultural and natural environment. This conclusion adds to the discussion about sustainable nautical tourism as opposed to cruise ship's 'bubble-based' consumption (Weaver, 2018) or the passive seaside tourists (Moscardo et al., 2001).

The shift to a more sustainable and human-centred approach in nautical tourism and the inclusion of smaller scale practices of alternative tourism that emphasise the local element and the onshore activities are also evident in travel articles about river cruising experiences (Colin, 2017), and other activities, e.g. cycling cruises (Greene, 2018), coastal explorations (Barkham, 2017), and experiences (Bryant, 2013).

Thematic cruises (Bowes, Dunford, Lafferty, Rushby, & Macefield, 2016; Romano, 2018) are the alternative offer of the cruising industry to make their touristic offer more attractive and help passengers to overcome any feeling of boredom or the sense of entrapment (Weaver, 2018).

Another critical element is the adventurous type of activities that accompany nautical tourism. For some travellers, as it is evident in their travel accounts, even the fact that they participate in a sailing trip is for them a brave and adventurous act. Most of the user-generated travel accounts describe past memorable experiences, focusing on their 'adventurous' active participation at coastal, water, or underwater activities and exploration.

Watersports and underwater activities are also defined and categorised as adventurous topics in the travel publications' sections. Articles about cave diving, swimming with turtles, whales and sharks are some of the exciting activities that can be the purpose of the travel or can be just an activity during a sailing trip or coastal exploration. Even cruising is referred according to the destination or the theme as an 'Epic adventure', or 'Expedition Cruise' trying to attract people to an adventurous experience. Hung (2018), in her study of the cruising experience of Chinese travellers, proposed a hierarchical experience model that describes the direct experience as the initial level which includes the various features of cruise travel, the close contact with nature and the exposure on the land. The levitated experience includes the feeling of total freedom and the various activities, whereas, in the ultimate level of cruising experience, travellers find their inner peace, self-worthiness, and happiness of life (Hung, 2018). This model proposes a different perspective to understand the consumption experience in nautical tourism.

What seems to be interesting in this study is that the nautical tourism 'experiencescapes' created by media convergence reveal a new perspective of nautical tourism and the demand for a new approach in the touristic market offer. Media representations of nautical tourism define concrete subcategories with a lot of combined products and tailor-made travel experiences. Local people and culture are always dominant attributes and should be taken into consideration in nautical tourism experience construction and management.

This study was limited to a quantitative approach of content analysis and focused on a broad general theme. It was further limited to the collection of UGC from one travel community as example of transmedia storytelling. This approach can be employed in combination with qualitative content analysis and a different sample for each category of nautical tourism across multiple digital platforms in order to define which nautical experiences are presented and how. Since experience embodies both meanings and feelings (Franjić, Favro, & Perišić, 2012), a qualitative analysis of nautical travel accounts would give valuable information about the quality of products and services in the nautical tourism market, such as thematic cruises, onshore/onboard activities, side trips, etc. and would help for the co-creation of the nautical tourism experience and an integrated human-centred nautical tourism offer.

Transmedia storytelling and how it is used in tourism is another interesting issue for further research. Transmedia tourism content, created by peer-to-peer networks, is affecting the tourism market, as it can influence the tourist's visiting intention and can also result to the design of new touristic products and services. As tourists are involved in the media convergence process and create content in multiple platforms and channels, these transmedia tourism representations are important in the understanding of tourism as a practice and as an act of consumption.

Despite the emerging trend of creating and delivering content through different media channels, little research has been conducted on the use of transmedia storytelling to promote sustainable forms of tourism and raise awareness about environmental topics. In the case of nautical tourism, media convergence and storytelling could help reaching different kinds of audiences and could offer knowledge about climate change and ocean-related issues.

Acknowledgements

This study was part of the research conducted for the EU-funded Project: 'North East Meltemi: The breath of the Archipelago' in the framework of the European Programme 'Nautical Routes for Europe' (EASME/EMFF/2016/1.2.1.12/03/S12.765242).

References

Akehurst, G. (2009). User generated content: The use of blogs for tourism organisations and tourism consumers. *Service Business, 3*, 51–61.

Barkham, P. (2017). *Navigating Norfolk's hidden creeks and salt marshes – In a 1950s whelk boat.* Retrieved from: https://www.theguardian.com/travel/2017/jun/18/norfolk-creeks-salt-marsh-holiday-in-1950s-whelk-boat.

Bitterman, N. (2014). 'Aquatourism': Submerged tourism, a developing area. *Current Issues in Tourism, 17*(9), 772–782.

Bowes, G., Dunford, J., Lafferty, J., Rushby, K., & Macefield, S. (2016). *40 of the world's best cruise holidays.* Retrieved from: https://www.theguardian.com/travel/2016/feb/27/worlds-40-best-cruise-holidays.

Brochado, A., Stoleriu, O., & Lupu, C. (2018). Surf camp experiences. *Journal of Sport and Tourism, 22*(1), 21–41.

Bryant, S. (2013). *The best cruise excursions.* Retrieved from: https://www.telegraph.co.uk/travel/cruises/articles/The-best-cruise-excursions/.

Buckley, R. (2002). Surf tourism and sustainable development in Indo-Pacific Islands. 1. The industry and the islands Author I. *Journal of Sustainable Tourism, 10*(5), 405–424.

Butowski, L. (2018). An integrated AHP and PROMETHEE approach to the evaluation of the attractiveness of European maritime areas for sailing tourism. *Moravian Geographical Reports, 26*(2), 135–148.

Colin, C. (2017). *When you need to escape reality, take a Rhine river cruise.* Retrieved from: https://www.afar.com/magazine/when-you-need-to-escape-reality-take-a-rhine-river-cruise?inspiration=cruise&sub_inspiration=river-cruises.

European Commission. (2012a). *Blue growth.* Retrieved from https://webgate.ec.europa.eu/maritime-forum/system/files/Blue Growth Final Report 13092012.pdf.

European Commission. (2012b). COM(2012) 494- Communication from the commission to the European parliament, the council, the European economic and social committee and the committee of the regions: Blue Growth opportunities for marine and maritime sustainable growth. *Official Journal of the European Communities.* Retrieved from https://ec.europa.eu/maritimeaffairs/publications/blue-growth-opportunities-marine-and-maritime-sustainable-growth_en

European Commission. (2013). *A European strategy for more growth and jobs in coastal and maritime tourism.* Retrieved from: https://ec.europa.eu/maritimeaffairs/sites/maritimeaffairs/files/docs/body/coastal-and-maritime-tourism_en.pdf.

Franjić, R., Favro, S., & Perišić, M. (2012). System concept of experience in nautical tourism. *WIT Transactions on Ecology and the Environment, 166*, 117–128.

Gon, M., Osti, L., & Pechlaner, H. (2016). Leisure boat tourism: residents' attitudes towards nautical tourism development. *Tourism Review, 71*(3), 180–191.

Greene, A. (2018). *A cycling cruise will change the way you experience Europe.* Retrieved from: https://www.afar.com/magazine/a-cycling-cruise-will-change-the-way-you-experience-europe?inspiration=cruise&sub_inspiration=river-cruises.

Greenman, J. (2012). *Introduction to travel journalism.* New York, NY: Peter Lang Publishing.

Hanusch, F. (2010). The dimensions of travel journalism: Exploring new fields for journalism research beyond the news. *Journalism Studies, 11*(1), 68–82.

Hanusch, F. (2012). Travel journalists' attitudes toward public relations: Findings from a representative survey. *Public Relations Review, 38*(1), 69–75.

Hung, K. (2018). Understanding the cruising experience of Chinese travelers through photo-interviewing technique and hierarchical experience model. *Tourism Management, 69*(May), 88–96.

Kovačić, M., Gržetić, Z., & Boskovic, D. (2011). Nautical tourism in fostering the sustainable development: A case study of Croatia's coast and Island. *Tourismos, 6*(1), 221–232.

Lam González, Y. E., León González, C. J., & de León Ledesma, J. (2015). Highlights of consumption and satisfaction in nautical tourism. *Gestión y Ambiente, 18*(1), 129–145.

Leffel, T. (2010). *Travel writing 2.0.* USA: Splinter Press.

Lück, M. (2007). *Nautical tourism: Concepts and issues.* Elsmford, NY: Cognizant Communication.

Luković, T. (2012). Nautical tourism and its function in the economic dDevelopment of Europe. In M. Kasimoglu (ed.), *Visions for global tourism industry – Creating and sustaining competitive strategies* (pp. 399–430). Croatia: IntechOpen.

Månsson, M. (2011). Mediatized tourism. *Annals of Tourism Research, 38*(4), 1634–1652.

Marine-Roig, E. (2014). A webometric analysis of travel blogs and review hosting: The case of Catalonia. *Journal of Travel & Tourism Marketing, 31*(March 2015), 381–396.

Mikulić, J., Krešić, D., & Kožić, I. (2015). Critical factors of the maritime yachting tourism experience: An impact-asymmetry analysis of principal components. *Journal of Travel and Tourism Marketing, 32*(1), S30–S41.

Minazzi, R. (2015). *Social media marketing in tourism and hospitality.* Cham, Switzerland: Springer.

Moreno, M. J., & Otamendi, F. J. (2017). Fostering nautical tourism in the Balearic Islands. *Sustainability (Switzerland), 9*(12), 1–20.

Moscardo, G., Pearce, P., Green, D., & O'leary, J. T. (2001). Understanding coastal and marine tourism demand from three european markets: Implications for the future of ecotourism? *Journal of Sustainable Tourism, 9*(3), 212–227.

O'Neil, L. P. (2006). *Travel writing.* Cincinnati, OH: Writer's Digest Books.

Romano, A. (2018). *This cruise guarantees you'll see the northern lights – or your next cruise is free.* Retrieved from: https://www.travelandleisure.com/cruises/hurtigruten-astronomy-cruise-northern-lights.

Santana, A. A., Gil, S. M., & Chirino, J. B. (2018). The paradox of cultural and media convergence. Segmenting the European tourist market by information sources and motivations, *International Journal of Tourism Research, 20*, 613–625.

Schmallegger, D., & Carson, D. (2008). Blogs in tourism: Changing approaches to information exchange. *Journal of Vacation Marketing, 14*(2), 99–110.

Stone, M. J. (2018). Eat there ! Shop here ! Visit that ! Presenting the city in mass media travel writing. *Current Issues in Tourism, 21*, 998–1013. doi: 10.1080/13683500.2015.1123678

Usha, R., & Divya, C. (2014). Have travelled, will write: User generated contend and new travel journalism. In F. Hanusch, & E. Fürsich (Eds.), *Travel journalism. Exploring procuction, impact and culture* (pp. 116–134). Houndmills, Basingstoke, Hampshire: Palgrave Macmillan

Weaver, A. (2018). Selling bubbles at sea : Pleasurable enclosure or unwanted confinement?, *Tourism Geographies, 21*(5), 785–800.

Whyte, J., Packer, R., & Ballantyne, L. J. (2018). Cruise destination attributes : Measuring the relative importance of the onboard and onshore aspects of cruising and onshore aspects of cruising, *Tourism Recreation Research, 43*(4), 470–482.

Wild, P., & Dearing, J. (2000). Development of and prospects for cruising in Europe. *Maritime Policy and Management, 27*(4), 315–333.

Wu, M. Y., & Pearce, P. L. (2014). Tourism blogging motivations: Why do Chinese tourists create little "Lonely Planets"? *Journal of Travel Research, 55*(4), 537–549.

Youngs, T. (2013). *The Cambridge introduction to travel writing.* Cambridge, UK: Cambridge University Press.

41

THE NEXUS BETWEEN TOURISM HERITAGE ATTRACTION, MEDIA AND FASHION

Kim Williams

Introduction

The purpose of this chapter is to provoke a discourse regarding the nexus between tourism heritage attractions, media and fashion. The chapter examines how transmedia narratives can influence tourism in diverse ways, whilst also providing a framework to understand a systems approach to the syntheses between cultural heritage institutions (CHI), their visitors and the fashion sector. The chapter provides an insight into how tourism is mediatized by means of reinforcement circles which influence visitors' perception and consumption of fashion exhibition held within heritage tourism offerings. This chapter will be addressed from a tourism disciplinary perspective.

A case will be utilised to offer insights. The National Trust of Australia, specifically the Victorian State organisation, will form part of the discourse. The connection between heritage properties and tourism will be examined, and then to what extent mediatization of fashion have influenced attendance numbers and increased potential revenue at CHI properties. The chapter will also provide a discussion concerning the incentives and inducements for CHIs management and curators to forge such partnerships with fashion designers, celebrities, costume creators, and film/television productions.

Tourism heritage attractions

CHIs such as galleries, museums, and heritage properties belong to the category of tourism heritage attractions. These attractions perform a substantial role in the conservation of cultural heritage and also the engagement of visitors with a specific destination (Bąkiewicz, Leask, Barron, & Rakić, 2017; Leask, 2010). Tourism heritage attractions are an important element of the tourism product in numerous countries across the globe (Leask, Fayall, & Garrod, 2002).

The link between heritage and tourism has been a research focus for a number of years (Ashworth, 2000; McKercher, Ho, & du Cros, 2005; Zhang, Fyall, & Zheng, 2015). Tourists in the 21st century desire unusual and thematic encounters to satisfy their thirst for memorable and absorbing visitor and travel experiences (Kempiak, Hollywood, Bolan, & McMahon-Beattie, 2017). It is argued that heritage tourism is able to fulfil this desire, and

CHIs can provide a suitable venue for intellectual and educational encounters. Heritage sites face a complexity of objectives, diverging from conservation and preservation of important cultural legacies, to providing access to visitors wishing to learn and be educated about these legacies and bygone times. There is at times conflict but also synergy between these objectives, for example, 'The approach of heritage organisations is to protect and preserve, while tourism has the overriding aim of becoming a profitable business' (Aas, Ladkin, & Fletcher, 2005, p. 33).

Furthermore, a fine balance is required between cultivating tourism visitation and the preservation of these important sites. Substantial increases in tourist numbers can produce both negative (e.g. exploitation) and positive impacts (e.g. increase revenue) for the site and may influence the choice of significance and the priority given to the differing chief objectives between the numerous stakeholders. Thus, these sites have to incorporate effective forward planning and be carefully managed on a day-to-day operational level (Aas et al., 2005; Bąkiewicz et al., 2017).

CHIs rely on a variety of revenue stream to pay for the maintenance and continued upkeep of these links to the past. The CHIs are custodians of the historical assets as well as a generator of engaging commercial activities. Many of these sites struggle to obtain sufficient funding from government sources. As in the case of the National Trust Australia (Victoria), of the collective total operational revenue generated by the organisation, less than 10% is sourced from the Australian or Victorian government (National Trust, 2018a). Therefore, additional creative and internally motivated revenue streams have become increasingly important. Dynamic pricing, exhibitions, retail and catering, special events and additional entertainment schedules, increased media exposure, and commercial use of the site (hiring the venue as a filming studio) have all become part of the product mix (Leask, Fayall, & Garrod, 2013).

Tourist visitation generates a number of challenges for the management of heritage attractions (Chen & Chen, 2010, 2013). Management of the visitors experience and the resulting satisfaction and possible loyalty for future visitation is a difficult undertaking. Heritage tourism visitors hold preconceived ideas; expectations and understanding, thus the interpretation provided by curators at the site must fulfil this but also be well researched, authentic, and creative in its presentation and not take on a fictional perspective, which may be generated from the association with the media.

Tourism mediatization

The convergence of media products known as 'mediatized tourism' (Månsson, 2011, p. 1635) occurs when tourism is viewed through the lens of multiple media products. Diverse transmedia interactions have the capacity to create a desire for visitors to consume tourism spaces by delivering a perpetual stream of images and information about a particular place (Moores, 2005). There is a convergence of media products interacting with each other to produce a reinforcement circles which may influence the tourist to engage with heritage sites which has intersected with the fashion sector due to mutual benefit (Refer to Figure 41.1).

Fashion mediatization

There has been a continually growing fascination with the film and television productions from the cinematic viewing audiences. Films or television series can generate visitation interest to particular, on and off-site locations. Film-induced tourism has been a topic of

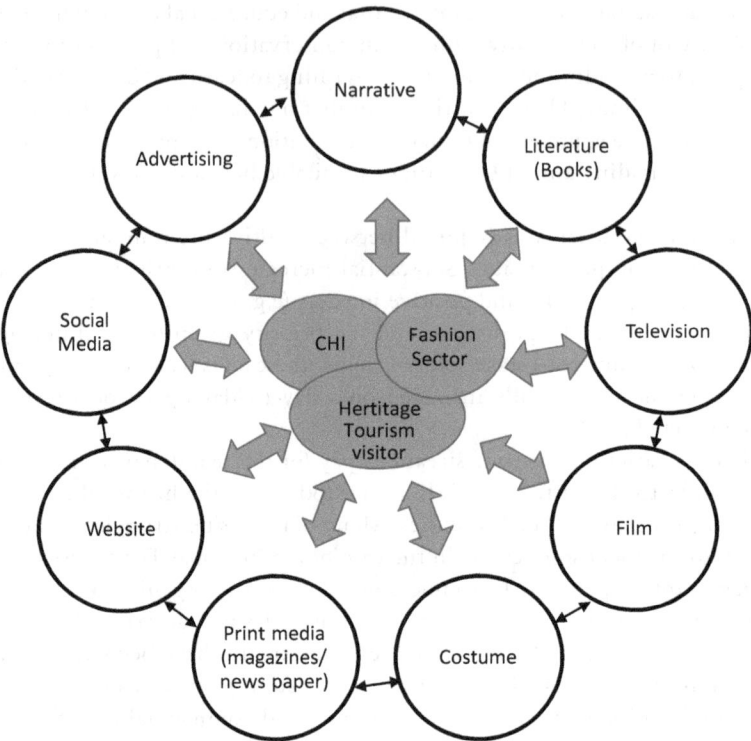

Figure 41.1 Reinforcement circle of transmedia interactions

academic research in contemporary times (Beeton, 2005, 2010; Buchmann, Moore, & Fisher, 2010). The general public can crave assess to interaction with celebrities and filming destinations thus providing a sought-after tourism experience (e.g. Lord of the Rings filmed in New Zealand and tours of Hollywood studios). Characters and locations from fantasies and fables, such, Harry Potter and Narnia, have become popular culture icons who have created fandom consumption which results in visitors developing an attraction to experience and ascertain a greater connection with their chosen icon.

CHIs, especially heritage properties, can be transformed into the backdrop of a film or television series. Costumes and celebrity apparel also have an alluring appeal to the cinematic viewing audience. Fans may clamour for opportunities to witness and immerse themselves in the filming location or be able to scrutinise the intricate detail of the fashion used to portray the time period shown on the screen.

Numerous historical creations for television and cinema can obtain audience popularity by providing a fictitious plot but also a level of historical accuracy. Sargent (1998) explained that costume films provided a particular view of social history which may or may not be completely accurate but nevertheless has the potential to stimulate leisure and tourist activities associated with this genre. The television series watcher can perceive that a dramatisation is an accurate representation of the era being depicted. Fascination can grow and an interest in viewing the costumes can be a motivator to view a selection in situ. Steele (2008) revealed that exhibitions are only able to provide an interpretation and it is the visitor who makes the final decision on the accuracy of what is presented, regardless these exhibitions provoke cultural and historical reflection for the heritage tourism visitor.

Another aspect of this innovative use of television and film costumes is the ability for these items to be valued not only as a prop to the entertainment industry but also as a piece of contemporary art. They hold great merit that is cherished way past the filming process and are able to aid in attracting a wider audience to heritage properties and assist in becoming a funding source for heritage restoration.

Fashion exhibitions

Fashion and costume exhibitions provide a window into the society of the time. These exhibitions may display current or past styles and trends, apparel belonging to celebrities, or offer viewing access to the apparel and accessories utilised in popular cinematic media. Fashion/costume exhibitions contribute to enhancing the spectrum of engagement for tourists and visitors to an urban or regional location. In recent times, there has been a growing partnership between tourism heritage attraction and the fashion industry to provide a novel vehicle for regenerating and enticing new and return visitor engagement.

The trend of exhibiting costumes and fashion is relatively new (Kroening, 2012). Curators are confronted with a variety of important issues to take into consideration when contemplating a fashion inspired exhibition, such as: what assertions are made about the culture and society of the time; how authentic or historically accurate is the display; and what is the order of significance concerning fashion as art, historical interpretation and the construction of commercial entertainment. Palmer (2008) proposed that 'many exhibition reviewer comments focus on the suitability and hierarchical place of fashion in museum settings, rather than the success and merits of the exhibition itself' (p. 123).

Curators have discovered that there is a potential to leverage off the glamour of the fashion exhibition to entice a renewed audience through the doors of art galleries, museums, and heritage homes. This approach has been implemented around the globe by the Victoria and Albert in London, the Musée Galliera in Paris, Solomon R Guggenheim Museum in New York, the Bendigo Art Gallery, the National Gallery of Victoria, and the National Trust of Australia.

By placing fashion within institutional contexts, it has been acknowledged by curators that fashion can be a work of art, and thus worthy of display and to be taken seriously in the art world. 'Some of these fashion exhibitions have attracted up to 500,000 visitors and brought in millions of dollars of sponsorship and considerable publicity' (Steele, 2008, p. 8) providing a paramount incentive to incorporate this genre of installation into the yearly program schedule. Menkes (2000) clarified that along with catwalks and fashion stores, museum sites and heritage properties have become progressively significant exhibition locations for revealing past, present, and future fashions to the public audience.

The National Gallery of Victoria (NGV) introduced textiles in 1895 and fashion in 1948 with the collection currently consisting of more than 8,500 works (National Gallery of Victoria, 2019). The strategy of utilising fashion exhibitions to reinvigorating patronage has been instigated by the NGV. During August to November 2017, the NGV in collaboration with the House of Dior displayed an exhibition which celebrated the 70th anniversary of the fashion house's first couture collection. The exhibition presented an opulent display of more than 140 handcrafted haute couture garments including custom-made Christian Dior Couture wedding dress. This is just one example of the numerous fashion inspired special exhibitions that have been held at the NGV in the last decade. Other fashion inspired exhibitions since 2014 includes but is not limited to the following:

- The Fashion World of Jean Paul Gaultier, From Sidewalk to Catwalk, Oct 2014–Feb 2015;
- Viktor & Rolf, Fashion Artists, Oct 2016–Feb 2017;
- Italian Jewels, Bulgari Style, Sept 2016–Jan 2017;
- Designing Women, Sept 2018–March 2019; and,
- The Krystyna Campbell-Pretty Fashion Gift, March 2019–July 2019.

McNeil (2014) pointed out that 'Australia will see an unprecedented number of exhibitions of sartorial fashion in the next three years' (p. 29). The list above supports the McNeil (2014) prediction of an increase in this genre of exhibition. The recent Krystyna Campbell-Pretty Fashion Gift (2019) showcased more than 150 garments which are now part of the NGV Fashion and Textile collection. Scheduled talks presented over the first weekend of the exhibition attracted a listening audience of over 100 people (author was in attendance). The allure of viewing a fashion/costume exhibition containing a range of apparel connected to iconic designers has the potential to attract visitation from those who may not be ordinarily tempted to visit a CHI general collection, thus assisting in diversifying the customer base.

Cultural heritage institution

In Australia, one of the custodians of a section of the CHIs, heritage properties, is the National Trust, Australia's leading conservation organisation. The National Trust of Victoria (a non–government conservation organisation) is the guardian of a number of both urban and regional properties scattered throughout the State of Victoria. The National Trusts rely heavily on community support generated through membership subscriptions, sponsorship, donations and bequests, property admissions, and retail sales (National Trust of Australia, 2018a). Two of the principle staff involved with the creation of exhibitions and revenue streams are the exhibition curator, Elizabeth Anya-Petrivna and Commercial Manager, Drew Grove. The NTV properties provide a viable space to facilitate historical interpretation, conservation, and preservation, as well as being an excellent venue or site for staging appropriate exhibitions or filming television and cinema productions. In addition to these two properties, there are over 20 other extraordinary heritage sites.

The approach of utilising fashion and costume exhibitions to reinvigorate patronage has been instigated by the NTV in a range of heritage properties across Australia. Review of the National Trust's Strategic Plan 2018–2022 shows there are a number of key objectives which are aligned with this approach. The NTV Strategic Objective Two articulate that:

> We will bring our heritage to life through engaging storytelling and providing memorable visitor experience. To achieve this objective, there is the goal of developing new audiences through major exhibition programs and innovated visitor experience.
>
> *(National Trust of Australia, 2018b, p. 13)*

National Trust of Victoria

The following three cases provide an illustration of the synergy between heritage properties and tourism and then examine to what extent mediatization of fashion have influenced attendance numbers and increased potential revenue at properties managed by the National Trust of Victoria.

Miss Fisher's Murder Mysteries (2012–2015)

The NTV has conducted in partnership with Every Cloud Productions and Marion Bryce (the winner of the 2014 AACTA, award for the Best Costume Design in Television) two special event exhibitions at Rippon Lea House and Gardens, a large urban 19th-century mansion surrounded by seven acres of Victorian pleasure gardens. The exhibitions separately displayed costuming from series two and three of the popular Australian television drama, The *Miss Fisher Murder Mysteries (2012–2015)* was set in Melbourne in the 1920s and was based on a novel of the same name by Kerry Greenwood. The series developed a huge fandom following and a public fascination with the time period (McDowall, 2015a). In addition, the series was partially filmed at a number of NTV properties, including Rippon Lea House and Gardens, Como House, and Labassa in Melbourne. After the considerable success of these two exhibitions in Victoria, the National Trust has embarked on a two-year national tour, travelling to National Trust properties located in other cities across Australia including: Adelaide, Brisbane, Sydney, and Canberra (McDowall, 2015b).

Martin Green, Learning and Interpretation Manager NTV described a number of challenges faced by the NTV. First, the task of creating an engaging three-dimensional exhibition from an action-packed television program. Converting television ideas into a dynamic display space required creative reflection (Green, 2013). The Miss Fisher series pivoted around the exploits of an audacious, superbly fashionable female private detective, Miss Phryne Fisher (Essie Davis) who interacted with the local constabulary, one Detective Inspector Jack Robinson (Nathan Page). The costumes were central to the success of the series. Miss Fisher's fans hold preconceived expectations based on their personal interpretation of the television show. Thus, it was important for costume designer, Marion Boyce, to undertake meticulous research into the twenties period to provide authenticity to that time period. Each fashion item contained elaborate detail and was fastidiously crafted to replicate the lavishness of clothing in this era. More than 30 costumes worn by the characters were on display. Every Cloud Productions' producers Fiona Eagger and Deb Cox said:

> We are thrilled at the enormous success of Miss Fisher's Murder Mysteries costume exhibition hosted by the National Trust which allowed thousands of our show's dedicated admirers to appreciate Phryne Fisher's stunning wardrobe and the work of our amazingly talented costume designer Marion Boyce.
>
> *(McDowall, 2015a)*

To allow the exhibition wardrobe and accessories to be more engaging and accessible to visitors, there was limited barriers between the costumes and the audience. This in itself caused some trepidation for the curator since visitors may try to handle the costumes. Green (2013) explained that the Curator Elizabeth Anya-Petrivna devised an inventive solution. By employing mirrored perspex shapes in diamond patterns around the base of the costume stands, visitors cannot stand too close but feel no sense of a barrier between them and the costumes. In addition to the interpretation panels, many of the costume displays provide sample swatches of the garment fabrics which were designed to be touched, which in itself produced visitor interaction.

The author reflected on the fact that the design of the mansion provided the opportunity to have each room as a specifically themed installation. The Conservatory contained sporting apparel (tennis themed) and the upstairs bathroom displayed intimate lingerie. A number of innovative visitor engagement strategies were employed to encourage the public to linger and also have an experiential tourism encounter. There was an interactive Murder Mystery activity with clues hidden throughout the exhibition (in housewares, on furnishings, in bookcases, and on mirrors and even on the piano). A movement censored talking screen of the character Detective Inspector Robinson spoke to the visitors as they passed through the drawing room. In one of the upstairs bedroom rooms, attendee could complete a drawing template of their own exhibition inspired costume and then pin them up for others to admire and appreciate, building a dynamic visitor centred display via co-production. In addition, an array of different costumes made by our embroidery volunteer Eva Fabian and her team (based on Miss Fisher's 1920s cocoon coat) were available to wear while guests wander through the mansion. An additional avenue to increase monetary contributions for the NTV is the sale of programs, merchandising and memorabilia (including a Miss Fisher inspired jewellery range) for many of exhibitions. (Author's personal observations).

Drew Grove, Commercial Manager of the NTV stated the following about the Miss Fisher exhibition: 'It was our most successful exhibition ever, over 60,000 people came through' (Ross, 2016). This increased patronage assists with revenue and possible future donations but also brought added pressure and aggravation for the local community since the historical property is located in a residential neighbourhood. There is limited parking which is predominantly on the street; however, the mansion is located near to a railway station which takes some of the pressure away from this access and transportation issue. There needs to be a considered balance between visitor access and the focus on conservation and community interests (Bąkiewicz et al., 2017).

Love, Desire and Riches (2014)

After the enormous success of the Miss Fisher exhibition, the NTV embarked on another fashion inspired exhibition, *Love, Desire and Riches (2014)*, which was at that time the biggest exhibition undertaken in the Trust's history. The exhibition spanned 200 years of fashion, reflecting popular culture of Wedding gowns and the proficiency and expertise of those who crafted the significant ensemble for that special day.

The exhibition allowed a CHI the opportunity to present popular culture in a creative and engaging space whilst linking to preconceived expectation of the fantasy of the wedding experience. Anya-Petrivna, curator, took three years to research (Watson, 2014, p. 4) and develop an authentic and creative display of the wedding gown concept. She describes how the Trust's costume collection was a rich source of inspiration:

> Wedding gowns tend to be kept as mementoes. I think they're a really powerful and evocative reminder of the way material culture works in people's lives and how they hold onto things. And as there are a lot of wedding gowns in the Trust's collection it seemed obvious that we should do something with this incredible repository of fashion.
> *(Watson, 2014, p. 4)*

The exhibition included a range of gowns from haute couture, celebrities, princesses, and film and television dresses (McDowall, 2014), all represented from various transmedia components depicted in Figure 41.1. Commercial Manager Drew Grove described how the success of the 2013 Miss Fisher's Costume Exhibition at Rippon Lea gave the Trust courage to think big when using the venue space:

> Having a fashion-driven exhibition in one of our properties was a really good test case to see what we could do; what we were capable of; what the house was capable of. One of our challenges is the fact that our spaces aren't white, rectangular boxes where it is easy to install any type of exhibition. But we've been able to turn this into a positive because our places have got such character and stories, and are aesthetically beautiful.
>
> *(Watson, 2014, p. 4)*

The Dressmaker (2015)

The next fashion instillation produced by the NTV was the costumes from *The Dressmaker (2015)* which was held at Barwon Park mansion, built in 1871 is a lavish 42 room, authentic regional bluestone mansion and stables set in a sweeping rural landscape. Both *The Dressmaker (2015)* and the *Miss Fisher Murder Mysteries (2012–2015)* provided an opportunity for Australian and international viewers to construct a view on what constitutes the national heritage of those times (Orr, 2018; Sargent, 1998). 'Costume shows are a mecca for those interested in historical, social and cultural development about how important costume has been in the past for defining a character and identity on screen' (McDowall, 2015b, para. 13). A costume exhibition allows this audience to gain a first-hand experience of the inner world of the costume department.

The film *The Dressmaker* (2015) depicted a narrative/story of transformation set in the 1950. Myrtle (Tilly) Dunnage a young and accomplished dressmaker returns to her Australian regional home town of Dungatar, 25 years after being accused and exiled due to a violent death of a boy at her school in 1926. The core of the film is the contrast between the dry and arid landscape and the allure of Parisian couture. The costumes took a major role in the movie winning The Australian Academy of Cinema and Television (AACTA) Arts Award for Best Costume Design (2015), and it was the fashion that transformed many of the characters throughout the film and provided them with a rich sense of character.

To continue the success of the film, more than 50 costumes from *The Dressmaker (2015)* were displayed at Barwon Park, then at Rippon Lea Estate. In this case, the NTV formed a partnership with Film Art Media, and once again with designers Marion Bryce and Margot Wilson. Since the film is set in outback Australia in the town of Dungatar in 1950s, it makes Barwon Park mansion an idea exhibition space and venue. Dew Grove said:

> It was important to Producer Sue Maslin and I to have the national premiere of the exhibition in a setting that evokes the fictional town of Dugatar and the juxtaposition between the costumes and the plains of Winchelsea achieves this.
>
> *(Watson, 2016, p. 4)*

Barwon Park mansion provided an ideal venue for the inaugural exhibition for *The Dressmaker (2015)* costumes and was able to attract over 11,000 people in six weeks, in contrast to the usual estimated 10,000 visitors annually (Dittloff, 2016). Janiskee (1996) illuminated that heritage property curators are challenged with the economic interests of finding viable funding sources to finance resource protection of historical properties. Providing crowd pleasing fashion inspired events can assist in gaining new patrons, positive word of mouth promotion and providing increased motivation for return visitation. All these assist in the possibility of greater attendance through the gate currently and into the future.

Drew Grove discussed that as well as being an important source of revenue for the National Trust, exhibitions such as *The Dressmaker (2015)* are crucial for public engagement.

> It's important for us to align ourselves with popular culture and also with history – so with Miss Fisher and the 1920s and The Dressmaker and the 1950s. It gets people to our site, then they fall in love with our site and become advocates for our site.
>
> *(Ross, 2016)*

Conclusions

Distilling the above discussion a framework which depicts a systems approach between CHIs, their visitors and fashion sector has been created (refer to Figure 41.2).

Mediatization has a powerful impact on the overall tourism system. The framework (refer to Figure 41.2) pinpoints key stakeholders and illuminates their individual inputs, processes and outputs. This framework provides a contribution to the understanding of a sympathetic connection between these differing industry sectors. A reinforcement circle of transmedia interactions (refer to Figure 41.1) assists in influencing the heritage tourism visitor to engage with cultural heritage sites whilst also encouraging the fashion sector to collaborate with these stakeholders to secure reciprocal benefits. Tourism mediatization via transmedia components converge on and influence the stakeholders within the heritage tourism attraction system. Each stakeholder delivers inputs, activates process, and produces outputs.

This chapter has contributed to the discourse concerning tourism and fashion mediatization by introducing a framework which illuminates the interconnection of the media, heritage tourism, and fashion offerings (refer to Figure 41.2) which attain a collaboration of mutual benefit. Transmedia interactions have a capability to generate motivation for heritage tourism visitors to engage with particular heritage destinations. The influence of media cannot be under estimated and is abundantly dominant in the way humans now interact with society. Humans are bombarded with the access to instant information, images, and news on a daily basis. In the case of CHIs, this engagement with tourism and fashion mediatization has facilitated desired outcome of substantially increasing revenue and also reinvigorating patronage to CHIs through a more diverse customer base.

Utilising the aesthetically beautiful spaces available in CHIs assists in providing memorable tourism encounters which can provide the visitor viable edutainment. The development of multiple revenue streams will assist in safeguarding social and architectural history. Janiskee (1996) resolved that 'events can greatly increase public awareness of historic house values and enhance the economic, educational and sociocultural benefits associated with the preservation of heritage resources' (p. 412). Thus, it could be inferred that there currently is an important synergy between the future success of visitations to CHIs illustrated in the case of the National Trust of Australia and costume fashion exhibitions.

This chapter has contributed to a richer understanding of the incentives and inducements for tourism heritage attraction, in this case CHIs, to develop meaningful and enduring partnerships with other industry sectors.

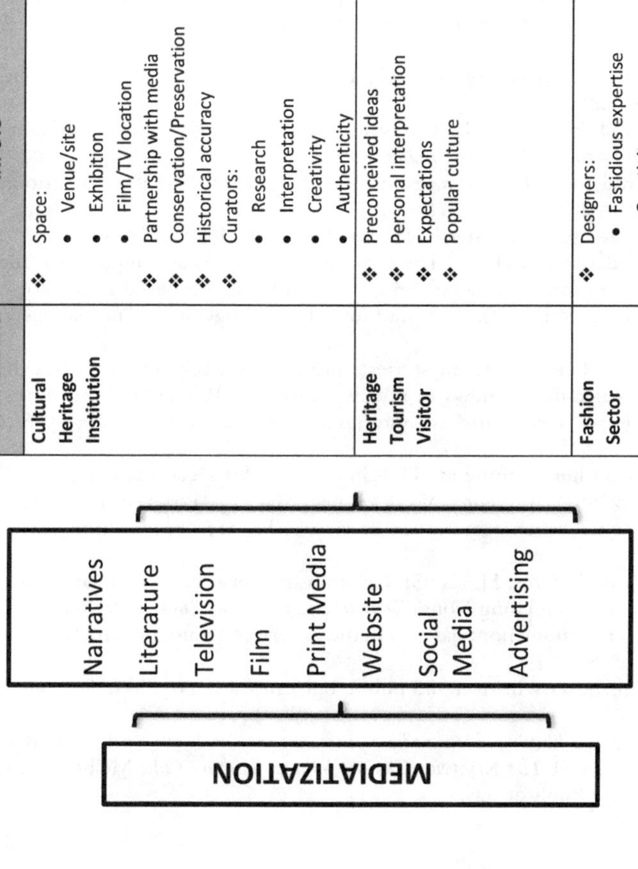

STAKEHOLDERS	INPUTS	PROCESSES	OUTPUTS

Cultural Heritage Institution	❖ Space:
	  • Venue/site
	  • Exhibition
	  • Film/TV location
	❖ Partnership with media
	❖ Conservation/Preservation
	❖ Historical accuracy
	❖ Curators:
	  • Research
	  • Interpretation
	  • Creativity
	  • Authenticity
		❖ Heritage tourism
		❖ Social history
		❖ Stories/narratives
		❖ Cultural awareness
		❖ Themed instillations
		❖ Interactivity
		❖ Dynamic displays
		❖ Learning
		❖ Co-production
		❖ Engagement
		❖ Commercial edutainment
			❖ Revenue stream
			❖ Reinvigorated patronage
			❖ Continued conservation
			❖ Public engagement
			❖ Return visitation
			❖ Memorable experience
			❖ Education
			❖ Aggravated local community
			❖ Diversification of customer base
			❖ Innovative visitor experience

Heritage Tourism Visitor	❖ Preconceived ideas ❖ Personal interpretation ❖ Expectations ❖ Popular culture	❖ Fandom consumption ❖ Enjoyment ❖ Fantasy ❖ Education ❖ Visitor interaction and engagement and accessibility to displayed items	❖ Leisure activity ❖ Modification of expectations ❖ Create a desire to reengage ❖ Validated popular culture ❖ Reflection ❖ Advocate for CHI

Fashion Sector	❖ Designers:
	  • Fastidious expertise
	  • Creativity
	  • Research/authenticity
	❖ Celebrities apparel
	❖ Popular culture icons
	❖ Fashion apparel
	❖ Costumes
		❖ Exposure
		❖ Reuse of costume
		❖ Increased value/merit
			❖ Branding
			❖ Recognition
			❖ Awards
			❖ Art work item

Narratives

Literature

Television

Film

Print Media

Website

Social Media

Advertising

MEDIATIZATION

Figure 41.2 A systems framework depicting the interconnection of the media, heritage tourism and fashion

References

Aas, C., Ladkin, A., & Fletcher, J. (2005). Stakeholder collaboration and heritage management. *Annals of Tourism Research, 32*(1), 28–48.

Ashworth, G. (2000). Heritage tourism and places: A review. *Tourism Recreation Research, 1*(25), 19–29.

Bąkiewicz, J., Leask, A., Barron, P., & Rakić, T. (2017). Management challenges at film-induced tourism heritage attractions. *Tourism Planning & Development, 14*(4), 548–566.

Beeton, S. (2005). *Film-induced tourism.* Clevedon, UK: Channel View.

Beeton, S. (2010). The advances of film tourism. *Tourism and Hospitality Planning and Development, 7*(1), 1–6.

Buchmann, A., Moore, K., & Fisher, D. (2010). Experiencing film tourism. Authenticity and fellowship. *Annals of Tourism Research, 37*(1), 229–248.

Chen, C., & Chen, P. (2010). Experience quality, perceived value, satisfaction and behavioural intentions for heritage tourists. *Tourism Management, 31*(1), 29–35.

Chen, C., & Chen, P. (2013). Another look at heritage tourism experiences. *Annals of Tourism Research, 41*, 215–243.

Dittloff, B. (2016). *The Dressmaker costume exhibition draws bumper crowds to Winchelsea's Barwon Park mansion,* Geelong Advertiser, February 18.

Green, M. (2013) *Miss Fisher's Murder Mysteries: Television, costumes and an App.* Museum Australia's INSITE magazine Sept–Oct: 6. Retrieved from: https://www.nationaltrust.org.au/news/miss-fishers-murder-mysteries-television-costumes-and-an-app (accessed 19 November 2018).

Janiskee, R. L. (1996). Historic houses and special events. *Annals of Tourism Research, 23*(2), 398–414.

Kempiak, J., Hollywood, L., Bolan, P., & McMahon-Beattie, U. (2017). The heritage tourist: An understanding of the visitor experience at heritage attractions. *International Journal of Heritage Studies, 23*(4), 375–392.

Kroening, E. (2012). From runway to Museum: Creating successful exhibitions showing the interrelationship between fashion and art. *Journal of Undergraduate Research at Minnesota State University Mankato, 12*(1), 5.

Leask, A. (2010). Progress in visitor attraction research: Towards more effective management. *Tourism Management, 31*(2), 155–166.

Leask, A., Fayall, A., & Garrod, B. (2002). Heritage visitor attractions: Managing revenue in the new millennium. *International Journal of Heritage Studies, 8*(3), 247–265.

Leask, A, Fyall, A., & Garrod, B. (2013). Managing revenue in Scottish visitor attractions. *Current Issues in Tourism, 16*(3), 240–265.

Månsson, M. (2011). Mediatized tourism. *Annals of Tourism Research, 38*(4), 1634–1652.

McDowall, C. (2014). Love desire & riches – The fashion of weddings at Rippon Lea, the culture concept circle, arts & entertainment news & review, August, 27. Retrieved from: https://www.thecultureconcept.com/love-desire-riches-the-fashion-of-weddings-at-rippon-lea (accessed 19 November 2018).

McDowall, C. (2015a). Miss Fisher's murder mysteries costume show, Rippon Lea 2015, the culture concept circle, arts & entertainment news & review, April, 15. Retrieved from: https://www.thecultureconcept.com/miss-fishers-murder-mysteries-costume-show-rippon-lea-2015 (accessed 19 November 2018).

McDowall, C. (2015b). Miss Fisher costume at NT Rippon Lea: Art Deco Fashion & Fun, the culture concept circle, arts & entertainment news & review, May, 1. Retrieved from: https://www.thecultureconcept.com/miss-fisher-costume-at-nt-rippon-lea-art-deco-fashion-fun (accessed 19 November 2018).

McKercher, B., Ho, P. S. Y., & du Cros, H. (2005). Relationship between tourism and cultural heritage management: Evidence from Hong Kong. *Tourism Management, 26*(4), 539–548.

McNeil, P (2014). The fashion phenomenon: Fashion in the gallery and museum. *Art Monthly Australia, Nov 2014, 275*, 28–35.

Menkes, S. (2000). Museum shows win over the public but can cause conflicts. *International Herald Tribune,* July, 12.

Moores, S. (2005). *Media/theory: Thinking about media and communication.* New York, NY: Routledge.

National Gallery of Victoria (2019) *The Krystyna Campbell-Pretty Fashion Gift.* Melbourne, Australia: National Gallery of Victoria Publications.

National Trust of Australia (Victoria) (2018a) About us. Retrieved from: https://www.nationaltrust.org.au/about-us/ (accessed 19 November 2018).

National Trust of Australia (Victoria) (2018b) Strategic Plan 2018–2022. Retrieved from: https://www.nationaltrust.org.au/wp-content/uploads/2018/03/2017_strategic_plan_DIGITAL.pdf (accessed 19 November 2018).

Orr, A. (2018). Plotting Jane Austen: Heritage sites as fictional worlds in the literary tourist's imagination. *International Journal of Heritage Studies, 24*(3), 243–255.

Palmer, A. (2008). Reviewing fashion exhibitions. *Fashion Theory, 12*(1), 121–126.

Ross, A. (2016). Fashion from the '50s finds fitting home in The Dressmaker exhibition at Rippon Lea House. The Sydney Morning Herald, April 22. Retrieved from: https://www.smh.com.au/entertainment/fashion-from-the-fifties-finds-fitting-home-in-the-dressmaker-exhibition-at-rippon-lea-house-20160422-god2fn.html (accessed 19 November 2018).

Sargent, A. (1998). The Darcy effect: Regional tourism and costume drama. *International Journal of Heritage Studies, 4*(3–4), 177–186.

Steele, V. (2008). Museum quality: The rise of the fashion exhibition, *Fashion Theory, 12*(1), 7–30.

Watson, F. (2014) Love desire & riches, *National Trust of Victoria Vic News* August, page 4–5. Retrieved from: https://www.nationaltrust.org.au/wp-content/uploads/2016/02/NT-Vic-Magazine-2014_3_Aug.pdf (accessed 19 November 2018).

Watson. F. (2016) Finding Dungatar, *National Trust of Victoria Vic News* Feb Issue 5, page 4. Retrieved from: https://www.nationaltrust.org.au/wp-content/uploads/2016/08/NT-Vic-Magazine-2016_1_Feb.pdf (accessed 19 November 2018).

Zhang, C., Fyall, A., & Zheng, Y. (2015) Heritage and tourism conflict within world heritage sites in China: A longitudinal study. *Current Issues in Tourism, 18*(2), 110–136.

42

ONLINE AND ON TOUR

The smartphone effect in transmedia contexts

Susan Carson and Mark Pennings

Introduction

Globally, millions of tourists possess smartphones that provide them with access to extensive computing power on their travels. In the midst of this phenomenon smartphones are, according to Buhalis and Foerste (2015), 'rapidly becoming the remote control of life' (p. 159). This chapter uses qualitative research to analyse this contemporary tourism phenomenon from a communication studies approach that is inclusive of 'the study of the culture with which it is integrated' (Fiske, 2010, p. 2). Whereas tourism studies often address the supply and demand aspects of the industry via a diverse range of social science frameworks, the approach here is to consider the cultural impact of a particular technology as an example of mediatization at work across the tourism industry. For tourists, this 'remote' control extends to planning and booking functions, both before and during travel, as well as to real-time access to online communities, thereby challenging understandings of the word 'remote'. The smartphone accompanies the tourist 'on-location' and enables them to tell their travel 'story' framed by a particular landscape or activity that conveys a type of experiential authenticity that is shared on diverse social media channels. At the same time, tourism businesses are curating content to attract visitors and are monitoring social media channels to assess the efficacy of their content, as well as providing responses to feedback.

Museums and heritage locations offer interesting case studies in relation to tourist use of the smartphone as technological and social interface. Museums and heritage sites were early adopters of mobile technologies including apps that visitors could download onto their own mobile devices to generate self-guided tours (Borda & Bowen, 2017). Many hosts began to use digital media to tell the stories of their collections to engage visitors during and post-visitation. For example, Port Arthur Historic Sites (PAHS) use online stories about convicts as a starting point for visitors to the location. Visitors then create their own story about their visit via platforms such as Instagram or Snapchat so that the images and words that describe their experience are delivered in a transmedia context, across multiple platforms. In this process, as Wozniak, Liebrich, Senn, and Zemp (2016) state, the smartphone is an essential link to the tourist for multiple stakeholders. The smartphones' role in this process is considered in this study to demonstrate some of the challenges of mediatization in the cultural tourism sector.

The museum sites examined here are on the must-see list for Tasmanian tourism. PAHS in Tasmania, Australia, is a premier attraction with a recently completed $A13 million visitor centre. Port Arthur is one of the 11 places that constitute United Nations Educational, Scientific and Cultural Organization (UNESCO) World Heritage-listed Australian Convict Sites. The second, the Museum of Old and New Art, also in Tasmania, is a privately funded gallery and event producer that has won international acclaim for innovation in exhibition practices and its delivery of information and communication technologies (ICTs). The authors conducted interviews with museum professionals and collated visitor opinions about these sites on TripAdvisor during the period 2017–2018. Their intention was to study the impacts of the smartphone and social media on the nature and dynamics of the relationship between providers and users. This was done by recording the perspectives of social media managers in these organisations about the impact of smartphone use on visitor experience and how they intended to use social media to enhance it. The analysis of tourist reviews shared on TripAdvisor was undertaken to assess opinions about the benefits and disadvantages of mobile phone access at Port Arthur and the Museum of Old and New Art in a social media context.

Smartphones and changing patterns of tourism

The information-intense nature of digital delivery and its reliance on ICTs in the tourism industry has been the subject of sustained analysis (Del Vecchio, Mele, Ndou, & Secundo, 2018; Werthner & Klein, 1999). In these discussions, scholars have noted the movement from 'e-tourism', to 'smart-tourism', to developments in augmented reality. For Hunter, Chung, Gretzel, and Koo (2015), 'smart tourism' is a 'social phenomenon arising from the convergence of information technology with the tourism experience' (Hunter et al., 2015, p. 105). The smartphone is central to current discussions of mediatization and smart tourism, most of which refer to Hjaryard's (2008) description of mediatization as a process by which contemporary society is being 'permeated by the media to an extent that [they] may no longer be conceived of as being separate from cultural and other social institutions' (p. 105). Increasingly, the smartphone is being seen as the meeting place for new technologies and associative social behaviours (Wang, Park, & Fesenmaier, 2012). The portability of smartphones plays an important role in smart tourism as described by Hunter et al. (2015), and for Wang et al. (2012) studies involving the mediated gaze illustrate the way that the media constructs the anticipations and motivations to visit places (p. 6). In general, these mechanisms facilitate the trans-mediation of stories about place and experience that flourish in social media networks.

Jansson (2017) has described the effect of these technologies as a movement of digitisation from mass to everyday life where there is a shift from 'stand-alone media fixtures to increasingly integrated and flexible polymedia environments' (pp. 45–46). The complexities of the mediatization of tourism can be seen in the following perspectives. In 2015, Gretzel, Sigala, Xiang, and Koo identified the limitations and disadvantages of the digital interface and ICT dependence, including 'information overload, lack of serendipity that is often essential to meaningful tourism experiences, and an increasing desire to at least escape technology when on vacation' (p. 183). Yet, in 2016, Wozniak et al. wrote about the potential of 'virtual co-creation among a network of multiple tourism stakeholders, with mobile devices such as smartphones as the multi-way link to the tourist' (p. 287). These seemingly contradictory views are supported by the statements from tourists in the research outlined below. Tourists appear to want choice: in some scenarios, they want to use a smartphone as a device through which they can 'co-create' an activity or a visual representation of the experience. This engagement is thought to be empowering and allows tourists to direct their own experience, and, presumably, the digital record of that experience, so that the mediatizing of the tourist experience is pushed further than the conventional photograph or postcard. In other situations, tourists want to free themselves from digital control which they associate with their everyday work life and deliberately step back from a mediatized saturation of their vacation.

Our research considers therefore the nexus of digital adaptability and its social and corporeal co-presence, noting that Larsen, Urry, and Axhausen wrote in 2006: 'tourism involves networking tools such as email, mobile phones, webpages, and access to cars, trains, and planes' (p. 259). Today life without a screen nearby seems unimaginable, as histories of portable devices and apps make clear (Kennedy-Eden & Gretzel, 2012). Many cultural institutions have relied on novel mobile apps and technologies to find new ways to connect with visitors and to encourage repeat visits that help make cultural institutions economically viable. For example, in 2009, the National Gallery in London was one of the first museums to develop an iPhone app, *LoveArt* (Borda & Bowen, 2017; Lagoudi & Sexton, 2010). Museums have also acknowledged new consumer habits of 'wearing' smartphones and using mobile technologies to book travel, accommodation, and ticket entry to cultural sites, and have provided customised travel-related apps to users when at their destination (Kennedy-Eden & Gretzel, 2012; Yoo, Sigala, & Gretzel, 2016, p. 253). Edensor (2018) notes that tourism be 'more broadly considered as multi-sensual in practice and experience' (p. 913). The desire to engage with a range of senses has implications for the development of apps for smartphones. Technology industry observers comment on how the emphasis on 'the visual' in tourism apps (at the expense of sound and touch) has provided a narrow range of experience for tourists who use geo-location apps during self-guided urban tours (Sewell, 2019). Demand for sophisticated apps that include audio, vision, and geo-location means that seamless connectivity to the internet is mandatory. In this environment, non-urban tourism experiences face the challenge of connection and technology support.

Interrogating challenges for visitors and providers

The methodology in this study employed interviews with staff and observations of activities by tourists at both the privately managed Museum of Old and New Art (MONA) and the publicly owned Port Arthur complex at Port Arthur Heritage Site in Tasmania, Australia. The interviews were conducted with these cultural institutions' media managers to ascertain museum perspectives towards the smartphone in particular and social media strategies in general, as well as museum plans for enhancing visitor engagement by improving visitor, smartphone, internet, and Internet of Things (IOT) interfaces. The interviews were supplemented by an analysis of customer reviews on TripAdvisor that referred to smartphones at these sites (211 respondents for MONA and 26 in relation to Port Arthur) to evaluate the effectiveness of existing smartphone offerings and services.

The interviews were also conducted to obtain qualitative information from museum professionals about the benefits and challenges of smartphone use, and secure qualitative data about the use of smartphones in the sites under consideration via content analysis of TripAdvisor reviews. Qualitative content analysis is any 'qualitative data reduction and sense-making effort that takes a volume of qualitative material and attempts to identify core consistencies and meanings' (Patton, 2002, p. 453). Researchers in tourism have relied on the opinions of travellers in a variety of studies. TripAdvisor reviews are 'perceived as highly trustworthy' (Dickinger, 2011), credible and relevant (O'Connor, 2008), up-to-date and engaging (Yoo et al., 2016). Yoo et al. also suggest that TripAdvisor supports both the supply and demand sides of the tourism sector by facilitating transactions as an 'infomediary'.

It specialises in 'big data' and provides 'a technological platform through which content can be created, analysed and distributed to meet the needs of travellers and tourism firms' (Yoo et al., 2016, p. 241). The opinions expressed on the site are also treated as important for market intelligence that informs companies about what is valued by clients and how to improve future services (Sigala, 2012). These clients include the 'urban explorers' that Jansson (2017) writes about as being part of the intensification of everyday lifeworlds 'that are globalized and mediatized' (p. 161).

Digital engagement at Port Arthur and MONA

Social Media Manager for Port Arthur, Jennifer Fitzpatrick, is responsible for user-generated content, engagement creation, advocacy, and partnerships. She acknowledges the benefits and challenges of operating in this mediatized environment and is invested in using this technology to broaden the scope of tourists' experiences:

> The key social media issue is what is picked up and what is left alone on social media; the issue is what is shared by users. The point of having social media channels is to boost the numbers of 'shares'—how many people share the site. The challenge for a historic site is that you have to get people excited about a historic site via the posts, twitter etc or blogs.
>
> *(S.J. Carson, personal communication, 10 October 2018)*

In pursuit of this aim, Port Arthur has a new visitor centre billed as a heritage interpretative site that focuses on the stories of convicts and is designed to develop an emotional connection with visitors. The site covers around 40 hectares of open land and many tourists are as interested in exploring the landscape as they are in convict history. This region in Tasmania, however, was long plagued by poor Wi-Fi connectivity, which was inconvenient for tourists who expected to use their smartphone at all times. Fitzpatrick reports that Optus towers have boosted connectivity in the area and tourists can now book tours or pay for entry tickets from their mobile phone *en route* to the Port Arthur site. Port Arthur has also opened up new market opportunities by using a conventional method of tourism (cruises) coupled with digital strategies in the home country. The result is that visitation rates have also grown quickly since cruise liners began stopping at Port Arthur. In 2018, 22 liners docked at the dedicated pontoon and cruise tourism contributed $4 million to the organisation. Around 80% of the Port Arthur marketing effort is devoted to engaging mainland Chinese tourism with a focus on WeChat and Weibo sites.

Fitzpatrick argues that digital engagement will grow as visitors begin to use voice command technology to book hotels and holiday: 'so much data is determined by the algorithms … Google has unleashed a huge power in their algorithms' (S.J. Carson, personal communication, 10 October 2018). As visitors demand 24-hour access to information, digital strategists such as Fitzpatrick will manage an increasing array of channels. The organisation depends therefore on feedback from digital channels to grow visitation and engagement. For tourists at Port Arthur, Instagram is the premier platform as the 'natural beauty of the site means that a lot of photographs are taken and shared' (S.J. Carson, personal communication, 10 October 2018). Once internet connectivity improved, and smartphone immediacy was guaranteed, the site was able to provide expanded administrative and entertainment options. Yu, Anaya, Li, Lehto, & Wong's (2018) findings that 'the nature of the family vacation may be related to their smartphone usage' (p. 593) and that 'travellers are increasingly dependent on pictures and videos to recollect their experiences' (p. 592) are important to a site such as Port

Arthur. The story of the vacation is shared across multiple platforms via the smartphone in a transmedia convergence. The convergence process encourages a curatorial approach on the behalf of the visitor to recording a vacation. The digital strategists employed by the provider or host in turn need to monitor a range of digital channels to increase visitation or to tailor their offerings for tourist consumption.

MONA, the Museum of Old and New Art, uses a different model to engage with smartphone technology in delivering a MONA-designed digital device, the 'O', to visitors to the Museum. MONA is national and globally famous for its collection of the personal art of the owner, David Walsh. In 2011, the 'O' was introduced as a digital guide to the art collection and contained an internal positioning system to locate visitors in the exhibition space. The 'O' is compatible with social media formats such as YouTube, Twitter, and Facebook. It enables visitors to offer their own opinions about the art and is linked to a MONA website and blog sites. The 'O' device app can be downloaded to a smartphone and can also be used off-site.

The 'O' became internationally famous and was a flagship product that promoted the museum as an innovator in the mediatization of art museum content (Walsh refuses to use conventional wall labels). The development of 'The Pen', for example, at the Cooper Hewitt Museum in New York was informed by the technology and audience engagement seen at MONA (Chan, 2015). Other venues such as The Museum of Human Rights in Manitoba have offered mobile apps for exhibitions that can be downloaded in smartphones, contain still and video images, audio and text, and are linked to the museum's Wi-Fi system. Like the 'O' device, such mobile apps encourage visitors to express their opinions via an interactive 'mood meter', thereby acquiring information about the emotional engagement of audiences with its collection, as well as providing an 'interactive map, online ticketing, and information to help plan the visit' (Borda & Bowen, 2017, p. 14).

Justin Johnson, Manager of Guest Services at MONA, acknowledged that the 'O' device was introduced at a time when the smartphone was rising to prominence. People visiting the site were already familiar with the use of smartphones so easily adjusted to the 'O'. In addition, the growth of social media sites like Facebook coincided with MONA's approach. Johnson has said that with visitors:

> The way the 'O' is used is not too different from how they use their smartphone, and people also use social media in a sophisticated way, so they are very familiar with the 'O' device. They can also click on our website, visit the blog, or they can see You Tube, or they can see twitter.

> *(M. Pennings, personal communication, 6 April 2016)*

The 'O' is a social media app that mediatizes the owner's, artists', and curators' views of the art in the museum as well as allowing the user options that facilitate downloads and sharing information about their experience at MONA online. Everyone who enters the museum is offered a device. Johnson has stated that in future MONA will provide live feedback during which Walsh and others visiting the museum can communicate in real time and thereby generate digital conversations on site.

TripAdvisor posts

TripAdvisor posts during the 2017–2018 period reveal a mixed response to the 'O' and social media options at MONA. A thematic analysis of the TripAdvisor comments showed that visitors offered opinions about the following: (1) Child friendly, (2) Innovation, (3) Quality, or uniqueness of experience offered, (4) Poor experience, (5) Sufficiency of Information, (6) Ease of Use, and (7) General impressions. The responses were evaluated as either 'Positive' (supportive of the digital technology provided) or 'Negative' (critical of the technology provided).

The TripAdvisor reviewers were domestic and international visitors. The highest response rate was in relation to 'sufficiency of information'. MONA has no wall labels for art so visitors are offered the 'O' device, or they can download an app to a smartphone for information about the collection. Some 107 visitors evaluated this technology on TripAdvisor, of which 86 were enthusiastic about the way in which they could access information while 21 found fault with the concept. One visitor stated: 'To help understand the art and displays they provide an iPhone with headphones (free) that you listen to or review things on, which is a good change from the usual cramming around a sign/placard!' (Terran, 2017). Another said:

> You're given a little tablet-type device, which scans the area you're in and lists the artwork with details, commentary and audio if you want it. I liked this because then you don't spend ages standing around reading about the artwork instead of looking at the art itself.
>
> *(Pok, 2018)*

However, around 25% of visitors writing on TripAdvisor expressed frustration with information in the 'O' device. Opinions included,

> the commentary is really bad, no discussion of the piece in a concise few minutes, just long ramblings' (Barnsbury Traveller, 2018) and 'the guide system with the iPhone … does make it easier to find the information on specific pieces in your area but it would be great if it included a map as we kept following signs trying to find exits only to find tucked away dead ends.
>
> *(Gwenall, 2017)*

Some visitors required more warnings about R-rated content. Most visitors appreciated the innovation of the device and the audio with unconventional commentary, but others felt that the technology detracted from the experience and happily discarded the system. So, although 75% of visitors to MONA approved of the 'O', a quarter of visitors were dissatisfied and expressed the desire to 'put aside' what has become everyday technology when viewing artwork, preferring to look at the works *in situ* without relying on the mediated gaze and information contained in the 'O' devices. It was also proposed that the type of information about the artworks was too directorial. These factors diminished the quality of some visitors' museum experience.

The smartphone was also a contested topic for 'ease of use' issues. On TripAdvisor, there was a 50/50 split about how easy this technology was to use. One visitor wrote about the frustration of having to use a screen:

> Utterly disappointed in lack of signage inside museum as well, everyone glued to devices to work out what they were looking at on the wall (we aren't glued to screens enough already, why can't we just look at and absorb the art?)
>
> *(Gay R, 2017)*

Another found that the digital device:

> … means everyone in the museum has their heads down and is focused on the iPhone rather than what's happening around them. People were bumping into one another, they were distracted, and generally, just rude. The requirement of the iPhone creates an atmosphere of independent isolation for the participant with no respect or consciousness for the people around.
>
> *(Webwalla, 2017)*

A significant portion of reviewers (about 30%) did not refer to the 'O' device. These cultural tourists were more attracted to MONA's experience-scape which offers an art museum but also a winery, a gourmet restaurant, bars, and luxury accommodation on-site. These reviewers appeared to be more interested in being physically immersed in a range of cultural activities, rather than seeking fulfilment through MONA's digital technology.

In contrast, visitors to PAHS wanted more engagement with technology. TripAdvisor readers who used audio guides at Port Arthur strongly recommended them to new visitors (Colida2014, 2017; Countrygall35, 2017) and described how this service allowed them to wander slowly through the site and absorb information at their own pace (rather than having the pace determined by the 20-minute tours run by guides). One reviewer stated:

> We also opted to have the iPod tours with us which were also good – the segments were short and informative – good way to get an understanding of why the space you're in was created, how it was used and what it meant to the lives of those that once lived there.
>
> *(Suna R., 2017)*

The few negative assessments of this service commented on the lack of audio in other languages and one called for a greater investment [by the 'National Trust'] on 'rendering text into audio' (Ian B, 2017). There was also critical commentary about inconsistent Wi-Fi reception (and consequent inability to share images) and suggested that the PAHS needed to update its digital engagement:

> Some other big tourist sites I have been you can hire devises (sic) & headphones & go on your own walking tour. This might be a good way of incorporating more stories of the convicts & what happened in certain buildings.
>
> *(Leorke, 2017)*

At the time of writing, there were no apps for tourist engagement at the site. Port Arthur has only recently introduced digital smartphone technology as part of the visitor experience and not all visitors were aware that these were available. The small number of TripAdvisor responses on Port Arthur over the same period of time may indicate that visitors to this site are engaged primarily in 'external' activities, such as wandering through the convict buildings, joining tour groups facilitated by guides, or taking part in the popular night ghost tours. Port Arthur has a well-established tour system that is run by professional guides and it will take time for visitors to become aware of the digital options that are being introduced at Port Arthur. Visitors to MONA, in contrast, are in many respects confronted and sometimes deliberately challenged by such technology. These responses to MONA point to the challenges faced by managers of tourism sites who seek to cater for, or indeed lead the public's familiarity with advanced connectivity when a significant section of reviewers suggested the desire to distance themselves from 'the phone' when undertaking activities such as contemplating art which they wished to do at their own discretion. As Jansson has suggested, 'connective media practices do not replace mobility and face-to-face meetings' and can be 'associated with the erosion of artistic as well as existential forms of authenticity' (2017, p. 161).

Conclusions

The smartphone can provide travellers' information needs at 'any stage of the travel process including anticipatory phase, experiential phase, and reflective phase' (Gretzel et al., cited in Wang et al., 2012, p. 372), but the data gathered from the TripAdvisor sites also indicates that many cultural tourists desire a more nuanced approach to their experience. The evidence suggests that the provision of smartphone access to a museum site does not automatically lead to an improvement in the quality of people's tourism experience. Undoubtedly, smartphones and the ICT context in which they operate will continue to expand the range of services and experiences that appeal to tourists, but not all cultural tourists need to invest in digital technology and its multiple gazes to have a fulfilling experience. Tourists use platforms such as Instagram to plan an itinerary or to pictorially and reflexively place themselves into the 'story' of their journey or experience. In such instances, the dimensions of the story are being structured by and for the technology in order to gain maximum levels of transmedia exposure for the storyteller/s. As Gretzel (2017) argues in relation to 'selfies', 'they are clearly carefully curated representations of the self and are central elements in the travel-related narratives communicated via social media' (p. 124). However, as Yu et al. (2018) state in their research on the family vacation experience: although travellers expect to have real-time access to information 'travelers report perceiving their experience as less adventurous and serendipitous' (p. 592) when a smartphone is used (versus face-to-face).

There is a challenge therefore in balancing the desire for independence and authority *sans* smartphone with the visitor's desire for curating experiences for multi-platform distribution. Whereas many of the users of the 'O' device found that the device did not give them enough flexibility and independence in their tour of the Museum, the visitors to Port Arthur sought greater connectivity. Yu et al.'s (2018) research finding that it is beneficial for travellers to be 'mindful and cognizant of tourism moments that are appreciated first hand and primarily through their own eyes' (p. 593) means that there will be a demand for more nuanced design in smartphone apps (such as offering choice in push notifications) as users become sophisticated curators of their travel experience. With the increased connectivity and speed of the new 5G phones, it seems likely that smartphones, at least in urban locations, will be expected to offer sound, vision, and headphones that allow the tourist to follow an itinerary while remaining in touch with the environment. The digitally connected tourist who distributes material to publishing platforms via a smartphone is an influential contributor to the established processes of mediatization in tourism.

In some circumstances, however, tourists desire a freedom of action where they are willing to combine the benefits of direct experience on-site with digitally directed social media presences as options at their choosing while resisting the compulsory use of digital devices. Jansson (2017) when discussing the future of 'media things' (p. 46) such as smartphones writes that mediatization in general has moved from an emphasis on 'mass media technologies to transmedia technologies' (p. 46). The 5G networks will deliver great speed but the phone will have to provide ever more connectivity so that tourists can plan, book, explore, and co-create via transmedia platforms from a location that also offers disconnected and immersive experiences and meets Edensor's (2018) recommendation that tourism should become 'multi-sensual in practice and experience' (p. 913).

References

Barnsbury Traveller. (2018, January 23). 'Incredibly Disappointing', [online discussion group]. Retrieved from: https://www.tripadvisor.com.au/Attraction_Review-g1783376-d567266-Reviews-Museum_of_Old_and_New_Art_Mona-Berriedale_Glenorchy_Greater_Hobart_Tasmania.html

Borda, A., & Bowen, J. P. (2017). *Smart cities and cultural heritage – A review of developments and future opportunities.* Paper presented at Electronic Visualisation and the Arts (EVA 2017), London, UK, 11 July – 13 July 2017. BCS. doi: 10.14236/ewic/EVA2017.2

Buhalis, D., & Foerste, M. (2015). SoCoMo marketing for travel and tourism: Empowering co-creation of value. *Journal of Destination Marketing & Management, 4*, 151–161.

Chan, S. (2015). Strategies against architecture: Interactive media and transformative technology at the Cooper Hewitt, Smithsonian Design Museum. *Curator. The Museum Journal, 58*(3).

Colida2014. (2017, February 24). 'Holiday', [online discussion group]. Retrieved from: https://www.tripadvisor.com.au/Attraction_Review-g504319-d258126-Reviews-or1580-Port_Arthur_Historic_Site-Port_Arthur_Tasmania.html#REVIEWS

Countrygal135. (2017, January 24). 'A Must!!', [online discussion group]. Retrieved from: https://www.tripadvisor.com.au/Attraction_Review-g504319-d258126-Reviews-or1660-Port_Arthur_Historic_Site-Port_Arthur_Tasmania.html#REVIEWS

Del Vecchio, P., Mele, G., Ndou, V., & Secundo, G. (2018). Creating value from social big data: Implications for smart tourism destinations. *Information Processing & Management, 54(5), 847–860.*

Dickinger, A. (2011). Trustworthiness of online channels in goal directed and exploratory search tasks. *Journal of Travel Research, 50*(4), 378.

Edensor, T. (2018). The more-than-visual experiences of tourism. *Tourism Geographies: An International Journal of Tourism Space, Place and Environment, 20*(5), 913–915.

Fiske, J. (2010). *Introduction to communication studies* (3rd Ed.) London, UK: Routledge.

Gay, R. (2017, July 25). 'Pretentious, overpriced snobby rude staff at entry, confusing no signage', [online discussion group]. Retrieved from: https://www.tripadvisor.com.au/Attraction_Review-g1783376-d567266-Reviews-Museum_of_Old_and_New_Art_Mona-Berriedale_Glenorchy_Greater_Hobart_Tasmania.html

Gretzel, U. (2017). #travelselfie: A netographic study. In S. Carson & M. Pennings (Eds.), *Performing cultural tourism: Communities, tourists and creative practices* (pp. 115–127). London, UK: Routledge.

Gretzel, U., Sigala, M., Xiang, Z., & Koo, C. (2015). Smart tourism: Foundations and developments. *Electronic Markets, 25*, 179–188.

Gwenall. (2017, July 21). 'Allow a few hours, so much to absorb, not really for kids', [online discussion group]. Retrieved from: https://www.tripadvisor.com.au/Attraction_Review-g1783376-d567266-Reviews-Museum_of_Old_and_New_Art_Mona-Berriedale_Glenorchy_Greater_Hobart_Tasmania.html

Hunter, W. C., Chung, N., Gretzel, U., & Koo, C. (2015). Constructivist research in smart tourism. *Asia Pacific Journal of Information Systems, 25*(1), 105–120.

Ian B. (2017, March 24). 'Some Advice' [online discussion group]. Retrieved from: https://www.tripadvisor.com.au/Attraction_Review-g504319-d258126-Reviews-or1490-Port_Arthur_Historic_Site-Port_Arthur_Tasmania.html#REVIEWS

Jansson, A. (2017). *Mediatization and mobile lives.* London, UK: Routledge.

Kennedy-Eden, H., & Gretzel, U. (2012). A taxonomy of mobile applications in tourism. *E-review of Tourism Research, 10*(2), 47–50.

Lagoudi, E., & Sexton, C. (2010). Old masters at your fingertips: The journey of creating a museum app for the iPhone and iTouch. Museums and the Web, Conference paper, Denver, Colorado, USA, April 13–17. Retrieved from: http://www.archimuse.com/mw2010/papers/lagoudi/lagoudi.html.

Larsen, J., Urry, J., & Axhausen, K. (2006). Networks and tourism: Mobile social life. *Annals of Tourism Research, 34*(1), 244–262.

Leorke, J. (2017, April 26). 'Great Day out', [online discussion group]. Retrieved from: https://www.tripadvisor.com.au/Attraction_Review-g504319-d258126-Reviews-or1390-Port_Arthur_Historic_Site-Port_Arthur_Tasmania.html#REVIEWS

O'Connor, P. (2008). User-generated content and travel: A case study on Tripadvisor.Com. In P. O'Connor, W. Höpken, & U. Gretzel (Eds.), *Information and communication technologies in tourism 2008* (pp. 47–58). Vienna, Austria: Springer.

Patton, M. Q. (2002). *Qualitative research & evaluation methods* (3rd Ed.). Thousand Oaks, CA: Sage.

Pok, A. (2018, August 7). 'Absolutely Amazing', [online discussion group]. Retrieved from: https://www.tripadvisor.com.au/Attraction_Review-g1783376-d567266-Reviews-Museum_of_Old_and_New_Art_Mona-Berriedale_Glenorchy_Greater_Hobart_Tasmania.html

Sewell, H. (2019) 'Geo-locative audio storytelling.' Visiting Scholar Presentation, Digital Media Research Centre, Queensland University of Technology, Friday 17 May 2019.

Sigala, M. (2012). Social media and crisis management in tourism: Applications and implications for research. *Information Technology and Tourism, 13*(4), 269–283.

Suna R. (2017, July 4). 'Use the full two days – lots to do and learn!', [online discussion group]. Retrieved from: https://www.tripadvisor.com.au/Attraction_Review-g504319-d258126-Reviews-or1315-Port_Arthur_Historic_Site-Port_Arthur_Tasmania.html#REVIEWS

Terran, N. (2017, September 18,). 'Unique Art.' [online discussion group]. Retrieved from: https://www.tripadvisor.com.au/Attraction_Review-g1783376-d567266-Reviews-or930-Museum_of_Old_and_New_Art-Berriedale_Glenorchy_Greater_Hobart_Tasmania.html.

Wang, D., Park, S., & Fesenmaier, D. R. (2012). The role of smartphones in mediating the touristic experience. *Journal of Travel Research, 51*(4), 371–387.

Webwalla. (2017, January 24). 'Absent of light with a focus on the iPod', [online discussion group]. Retrieved from: https://www.tripadvisor.com.au/Attraction_Review-g1783376-d567266-Reviews-Museum_of_Old_and_New_Art_Mona-Berriedale_Glenorchy_Greater_Hobart_Tasmania.html

Werthner, H. and Klein, S. (1999) Information Technology and Tourism: A Challenging Relationship. Springer-Verlag, Wien.

Wozniak, T., Liebrich, A., Senn, Y., & Zemp, M. (2016). Alpine tourists' willingness to engage in virtual co-creation of experiences. In I. P. Tussyadiah & A. Inversini (Eds.), *Information and Communication Technologies in Tourism 2016* (pp. 284–294). Cham, Switzerland: Springer International Publishing.

Yoo, K. H., Sigala, M., & Gretzel, U. (2016). Exploring TripAdvisor. In R. Egger, I. Gula, & D. Walcher (Eds.), *Open tourism* (pp. 239–255). Berlin and Heidelberg, Germany: Springer.

Yu, X., Anaya, G. J., Li, M., Lehto, X., & Wong, I. A. (2018). The impact of smartphones on the family vacation experience. *Journal of Travel Research, 57*(5), 579–596.

43

SMARTPHONE AS THE INVISIBLE BACKPACK

The impact of smartphone on Chinese backpackers' mobility pattern

Jia Xie

Introduction

Backpacking has gained its popularity in western societies since the 1990s and became an important form of tourism (Cohen, 2003; O'Reilly, 2006). Unlike mass tourists, backpackers prefer travelling on a low budget, contacting with local people as well as other independent travellers, and having a flexible itinerary (Loker-Murphy, & Pearce, 1995). In recent years, researchers notice that the pervasiveness and subsequent normalisation of the Internet and digital technology has a great impact on tourists' behaviour and travelling experience (Kirillova & Wang, 2016; Lamsfus, Wang, Alzua-Sorzabal, & Xiang, 2015; Månsson, 2011; Wang, Xiang, & Fesenmaier, 2014). In the meanwhile, the emergence of 'new global nomads' (Richards, 2015) and 'flashpacker' (Paris, 2012) indicates that more and more backpackers now travel with technological gadgets such as smartphones, tablet computers, and digital cameras. These devices supported by mobile Internet and wireless Internet are employed to record and share travel experience instantaneously. Furthermore, the boundary between 'road' and 'home' is blurred because long-term travellers are allowed to maintain co-presence and intimacy with people both corporeally and virtually proximate to them (Paris, 2010). As a result, the question how does communication technology change the way travellers experience 'place' gains increasing importance in both media studies and geography.

The rapid development of backpacker tourism in China around the millennium is attributed to economic development, a growing number of urban middle class and the Internet (Luo, Huang, & Brown, 2014; Zhang, Morrison, Tucker, & Wu, 2017; Zhu, 2007). Recently, the development of mobile Internet and popularity of smartphone further facilitated self-organised travel and expansion of backpacker community in China (Kristensen, 2013; Luo et al., 2014). However, there is a lack of empirical research focusing on an advanced and sophisticated level regarding the use of smartphone among Chinese backpackers and how it affects their travelling experience. Even less research notice that backpacker culture in China, which is deeply influenced by communication technology and digital culture, has its own characteristics.

This chapter aims to explore the impact of smartphone on Chinese backpackers' mobility pattern. Smartphone with location-based service applications not only provides backpackers the latest travel information but may also change the way tourists linked to the destination and their sense of place. Moreover, with the social media, backpackers become zealous media producers and made their journey into an ongoing drama. Consequently, every aspect of travelling, such as behaviour, experience, host–guest relationship is transferred and transformed by media. With ethnographic methods, this article reflects and captures the concept of transmedia tourism proposed in this section.

Literature review

Since the late 1990s, backpacking has become popular among the Chinese urban middle class, emerging alongside a large number of outdoor clubs that have appeared in major cities (Zhu, 2007). Lacking a history of the Grand Tour or the hippie culture, China's backpacker culture has been largely initiated by the popularity of outdoor activity. In the 1990s, a few outdoor product companies sponsored outdoor activity clubs in order to attract new urban middle-class consumers; this led to the flourishing of relevant clubs in big cities (Zhu, 2007). Meanwhile, Internet communities have also played a crucial role in the emergence and development of backpacker culture in China, as 'self-organised' is regarded as the core of backpacking compared to packaged tour. During different stages of backpackers' self-organised trips, they consistently rely on local news-based BBS and travel-based communities, in order to exchange information and build cooperation (Zhang, 2008). Urban middle classes, as the first generation of backpacker in China, are also the first generation of Internet user in this country.

However, it is worth noting that the profile of Chinese backpackers has changed dramatically over the last five years. Firstly, there is a greater number of young people becoming involved with this activity. Recent surveys (Chen, Bao, & Huang, 2014; Yu, 2012) show that the majority of backpackers are the post-1980s generation, who have grown up with, and responded to, the abrupt social changes incurred by the Reform and Opening-Up Policy (Lian, 2014). More and more student backpackers are found, along with young people who are in the transition of being an adult or just begin their career.

The second development of backpacker tourism is the popularity of budget travel in China. Independent travellers jostle for cheap beds in hostels, hitchhiking as a means to travel, whilst eating street food. The term *qiongyou* (穷游, budget travel) has become particularly fashionable; the term is a combination of '*qiong*' (poor) and '*you*' (travel), promoting an idea that '*qiong yi ke you*' (穷亦可游, anyone can travel – despite being economically poor). The essence of '*qiongyou*' is to spend as little money as possible when travelling. For example, the news reported by *China News Service* (Mo, 2016) indicates how a '*qiongyouer*' (穷游者, budget traveller) spent around seven GBP per day when travelling in Africa. *Qiongyouers* are distinguished from outdoor enthusiasts – the pioneering backpackers in China – although both are recognised as backpackers.

Previous research on Chinese backpackers primarily focused on their demographic and behavioural characteristics and often treated backpackers as a homogeneous group. However, recent surveys have found that backpackers are not as homogenous as they first appear, and the change occurred in China is consistent with the global trend (Cohen, 2003, 2011; Elsrud, 2001; O'Reilly, 2006). In this case, China provides an excellent case to explore the relationship between social, cultural, technological changes occurring in recent years and the change of backpacker tourism under the circumstance.

Methodology

Based on Loker-Murphy and Pearce's definition of backpackers (1995), 30 Chinese backpackers were selected to be interviewed. 20 of them were recruited on-site, when I backpacked in China's Yunnan Province in 2014 and 2015. Other respondents were recruited through snowballing, personal connections, and the social networking sites. All respondents took independent budget travel and had at least one backpacking journey lasting for more than 15 days. Some backpackers had two or three years of travelling experience. Most of them travelled in China, but a few have backpacked in Southeast Asia and Europe. 18 respondents out of 30 were males, and 19 out of 30 were at the age of 25 to 29. Most of them were solo travellers, except four people travelled in a group. The four knew each other through a social networking site and decided to resign in order to take a long trip.

Semi-structured in-depth interview were employed, and topics focused on backpackers' use of smartphone and their online habits. Interview topics also included their travel motivation, experience and backpacking's implication on normal life, in order to locate their travel experiences within the context of the conditions and circumstances of their life, relating it to wider society. Most often, it is the interviewed backpackers who invited me to 'friend' them on Wechat[1]or other social networking sites, as the photos and posts posted on WeChat's Moments[2] or social media platforms always helped them to recall travel experiences. All of the interviews were conducted in Chinese, then I transcribed and translated them into English. Besides in-depth interview, I also conducted participant observation when backpacking. For example, I spoke with people who stayed in the lobby of youth hostel occasionally, and also observed how people used the facilities and interacted with each other in the public space. As a result, photographs and fieldwork notes – collected through participant observation – are used for further analysis. NVivo 10 is employed to organise and analyse all the resources.

Findings

Smartphone as the invisible backpack

During the fieldwork, almost every backpacker, no matter young professionals, low-income factory workers or students, carried at least one smartphone with them. According to the latest report from China Internet Network Information Centre (2016), there are 7.24 hundred million users connected to the Internet through a mobile phone, and this number accounts for 96.3% Internet users. Therefore, it is necessary to further examine the feature of mobile Internet and its role played in backpacker's travel.

In the latest study of backpackers, places without Internet connections were called 'dead zones', and contemporary backpackers took the Internet connectivity for granted (Germann Molz & Paris, 2015). The situation in China is quite similar. A statement from Fong (male, 35), a travel writer, represents the opinion of most backpackers, 'Biological and physiological needs used to be at the bottom of Maslow's hierarchy of needs. Nowadays, it is Wi-Fi. Wi-Fi is like air and water.' As Figure 43.1 shows, the Wi-Fi access notice in Laoxie youth hostel, a popular one, was written in both Chinese and English, emphasising that the hostel provided special VPN, linking to 'blocked' sites such as YouTube, Facebook, and Twitter.[3] The notice was pasted up on the wall in the lobby, to make sure that every customer would notice it immediately.

Ting (male, 40), conducting a global tour independently, always brought three smartphones when travelling. He stated, 'It is always good to be prepared...I don't mean these gadgets make me feel secure, but you know, that's the way you connect to the world'. Bauman has suggested that the mobile phone was a key incarnation of liquid modernity, as it allows individuals living within global insecurity to feel a semblance of security through

Figure 43.1   Wi-Fi access notice
Source: Laoxie, 2015.

the linking in a web of messages (Bauman, 1988; Davis, 2008). In the age of Internet, it is no exaggeration to state that the smartphone becomes the 'invisible' backpack of backpackers.

Smartphone connected to mobile Internet or wireless Internet provides more besides the sense of safety. A few interviewees said that they updated their status every few hours, or every day on social media or WeChat's Moments when travelling; this is precisely why mobile Internet and Wi-Fi are so important. A lot of people's trips were 'live broadcast' to some extent, as they recorded their routes and uploaded photos from time to time. Zang (female, 26, factory worker) said happily that since she had begun to backpack, and that there was an increasing number of people visiting her 'space' on social media. Her friends, who got married and settled down at an early age, were envious of her freedom: 'Most of them said that they admired my courage. Some agreed that one should 'live for myself', like I did. However, they got married and felt they were living for others'. The first thing Zang did every morning was to turn on her smartphone and read people's replies.

Besides building self-identity like Zang, some interviewed backpackers frequently shared their experiences on the Internet with the purpose to promote idea of backpacking. As Yuan (male, 26, student) contended, 'I want people to know backpacking, and sharing my personal experience seems the best way. Friends think my way of travelling is cool. Also, it feels good that they fix their eyes upon me'. Xiao (female, 34, high school teacher), an experienced backpacker, states, 'One girl was afraid of travelling alone because she thought it might be dangerous. After she saw the photos I shared on social media, she changed her mind. Now, she travels alone sometimes'. A few backpackers even developed strategies regarding the use of social media. Ting, the independent global traveller, said, 'You know, people are not interested in your inner feelings. They want to see something funny, something crazy'. Ke (male, 27), a 'shutterbug' and a backpacker, said, 'I rarely shared feelings or personal point of view. I always picked something funny, such as weird food. One reason is I want to protect my privacy. Another reason is that people are looking forward to something funny'. With these strategies, some backpackers attracted lots of followers and fans on social media and became famous. Ke won several photography contests in the last two years as the fans voted for him.

Previous studies on the role of Internet played in backpacker culture emphasised that Internet community contributes to promoting backpacker culture, but this study demonstrates that due to the use of smartphone and mobile Internet, more and more individual backpackers made their travel blog, or social media, into an ongoing drama. A wider range of people understands this way of travelling through the drama. Backpacking is no longer an alternative way of travelling shared by a small interest group met online but gets more attention through numerous individual travellers. Scholars have already noticed that the performativity approach helps to understand travellers' motivation and experience (Cohen & Cohen, 2012; Edensor, 2001), and this study suggests that the use of smartphone has further facilitated tourists' presentation of travel experience. Interviewed backpackers employed the opportunity to convey a personal image, sustain or even enhance social status.

Smartphone and a lower 'price'

As pointed out in the introduction, there have emerged more and more *'qiongyouers'* (budget travellers) in China over recent years. The existing literature on Chinese backpackers focused exclusively on middle-class backpackers (Xiang, 2013; Zhang, 2008; Zhu, 2007), as this group was the main body of pioneering backpackers. However, the emergence of *'qiongyou'* and its popularity demonstrates a more diversified social background of backpackers. In this study, many backpackers were found in their early twenties, with relatively underprivileged backgrounds in terms of family and education. Significantly, the use of smartphone and relevant technologies seems to help narrow the gap between the underprivileged and privileged and contribute to the popularity of budget travel.

The backpacker group mentioned in the section of method illustrates this point well. Beibei (female, 25), Zang (female, 26), Peng (male, 28), and Liao (male, 25) were all young migrant workers who had left their hometowns years earlier to seek opportunity in cities. All of them were born in rural parts of China, and none of them went to college except for Liao. As for Liao, he indicated that his college was 'at the rock bottom' – so it was difficult for him to find a 'not so bad' job in cities. The four can be grouped as the second generation of migrant worker in China. Compared to their parents, these young migrant workers are better educated and have been increasingly exposed to television, mobile phones, as well as the Internet (Ngai & Lu, 2010). As a result, they have arguably been profoundly influenced by city life and globalised cultures.

The four yearned for a long-haul, long-term trip, due to different reasons. Peng wanted to reflect on his previous life and recover from divorce. Beibei, Zang, and Liao attempted to pursue a different way of living as they were tired of the mechanised life and meaningless job. For example, Liao had spent one year at a small workshop, making blueprints for an architecture firm. The work was very simple, and he did the same thing repeatedly each day, with very low payment. However, the living cost in Beijing was very high. In addition, Beibei, Zang, and Peng, all in mid-twenties, were still unmarried, which seems very unusual to their family and friends. Most of their friends married in the early twenties and saved all their earnings for their family and children, as Zang described above. However, the three did not want to live in a similar manner, and they looked forward to taking a long journey which can broaden their horizon.

The travelling plan was postponed due to the lack of Companions. None of them had travelled much before, let alone backpacked. Their friends did not show any interest. Eventually, in 2013, Beibei found a group named 'backpacking to Tibet' on Momo,[4] a location-based meet-ups application, and joined. Zang, Liao, and Peng, who had the same dream, were in the group as well. This is how they found each other and decided to backpack together subsequently. After meeting in the group, they spent several months buying supplies, researching and designing routes. In the summer of 2014, they quit their jobs, met up in the city of Chengdu, and began the trip from there.

It is incorrect to suggest that the four would not backpack without Momo; however, it is fair to state that mobile applications like Momo greatly helped the formation of the group. As for young migrant workers such as Beibei and Zang, it is difficult to imagine another kind of lifestyle without the Internet as 'everyone seems to live in the same way' (quoted from Beibei), and it is even more difficult to put the 'crazy idea' into practice. Most people Beibei met in life married in their early twenties, became parents quickly and settled down. However, Beibei did not want to live her life according to this 'general schedule'. Undertaking a long-term journey may sound common for college students and young professionals, but it sounds 'crazy' for most migrant workers. Previous study states that Internet communities formed by urban middle class contributes to the development of backpacking tourism. This study shows that smartphones may empower the younger generation in China, regardless of social backgrounds, to find like-minded people, form interest-based social groups, and pursue an alternative way of living. Traditionally, it is the middle class, often with a college education, possess high cultural self-confidence and travel more confidently out of their familiar surroundings (Graburn, 1983). Smartphone may help people from an underprivileged background to build cultural self-confidence and further democratise backpacking tourism.

New technology not only reduces the difficulty of finding travel partners but also reduces the travel cost. It is far more convenient for today's backpacker to travel without consulting fellow backpackers or local people, with the support from travel-related websites and mobile applications such as Ctrip and Dianping.com.[5] Particularly, itineraries are able to be more flexible due to the service provided by location-based applications. Interviewed backpackers said they always adjusted their plans instantly, according to the latest deals of budget flights and hostels. Zu (male, 26, student) proudly said that he had successfully snatched a 9.99-euro flight ticket during the journey, and he flight to the destination although did not plan to. Meng (female, 25, self-employed) said she might move several times when staying in a place as long as she found somewhere cheaper online.

As pioneering backpackers have been part of the affluent urban middle classes in China, it was long contended that Chinese backpackers were less concerned about financial budgets when compared with their Western counterparts (Cottrell, 2014). Zhu (2007) also suggested that backpacking in China originated from outdoor activities; therefore, only the middle classes or upper classes could afford the expensive outdoor equipment, training courses, membership fees, and so on. 'Backpacking, as an alternative lifestyle for urban middle classes, is burning up money' (p. 115). However, this study argues that the traditional image of Chinese backpackers has to be changed due to an increasing number of budget travellers (*qiongyouers*). Equipped with a smartphone, people who have a relatively low income such as students, working-class, self-employed, or unemployed, begin to participate in this activity.

Smartphone and its challenge to travel culture

The role of smartphone and networking technologies played in one's trip is deeply ambivalent, although it is more convenient to travel with a smartphone as stated in the above section. First of all, it is worth noting that the classical backpacker culture which emphasises on social interactions may change due to the extensive use of smartphone. For Zhimi, searching via Google Maps was even better than asking local people. He noted:

> The first impression of a place is so important. Imagine I ask a person about the route. However, I get an answer in an indifferent tone. Or sometimes, just a cold 'I don't know'. I became disappointed with the place immediately. Thus, I preferred searching through Google Maps.
>
> *(Zhimi, budget traveller, male, 24)*

Zhimi's idea clearly challenges the belief that backpackers should go native and interact with local people (Loker-Murphy & Pearce, 1995). Ting, the global traveller, providing another example, noted, 'When I was in Brazil, a lot of places did not have Wi-Fi, such as the beach. It was nice sitting at the beach. However, after two or three hours, it became extremely boring. You had nothing to do'. As Germann Molz and Paris (2015) found in their research, since backpackers were able to maintain continual presence and interaction with their personal networks virtually, mobile technologies always tend to disrupt the local travel experience.

Figure 43.2 Table tennis table on terrace
Source: Hump, 2015.

The smartphone and mobile Internet thus may create very individualistic travellers. By avoiding certain kind of unpleasant experience, Zhimi returned to a safety net and refused to take risk. Ting felt bored without the Internet connection. Traditionally, backpackers ate local food, took the local bus, and lived similarly to locals. They tried hard to avoid looking or behaving like tourists who travelled in a group and only went to tourist attractions. In the Internet age, the safety net is no longer provided by travel agencies but by technology. It is a problem if backpackers are overly dependent on the smartphone and even trapped by it, because the meaning of backpacking journey may be changed significantly in the case. The lobby of youth hostel was designed as public space for backpackers getting together. However, more and more people preferred to play with smartphone individually even when they were sitting in the lobby. The manager of Hump, which is a famous youth hostel in Kunming, complained in interview (Figure 43.2).

Smartphones enable people to capture the transient feeling and moment, as well as remain in contact with their social network continuously. As stated above, many backpackers convey their reliance on the smartphone as it helps to build personal identity and enhance social status. However, not all of the backpackers who participated in this study liked sharing their experiences frequently through social media. There are increasing reflections on the use of smartphone in this study. Ning (female, 27), a long-term traveller, pointed out that if one shared too much on the Internet, she/he might be too concerned about others' comments, rather than ones' inner feelings. The experience of Shen (male, 25) provides another example. When in college, he backpacked a lot during school holidays. Shen said that he was addicted to social media when he launched his backpacking trip for the first time. However, he quickly noticed that a lot of his time was spent on writing, editing, and deleting posts, rather than travelling. He contended:

> It seems that I was getting sick. I updated a lot every day, then deleted all of the information on social media suddenly one day. Probably I was afraid of revealing too much, or perhaps I was upset that day. Then I started to use microblogging again, and then deleted the account. It was a vicious circle. Now, I control myself. I keep away from social media and the Internet.
>
> *(Shen, budget traveller)*

Backpackers like Ning and Shen suggested that the feeling of achievement did not come from others' comments but from one's own heart. As Ning and Shen imply, the desire for others' attention through social media may go against the statement to 'travel for myself', as both of them suggest their purpose of the journey is to reflect on oneself. Therefore, the reliance on smartphone and mobile Internet conflicts with backpackers' intentions to immerse themselves in the local community and to reflect on themselves. Cohen and Taylor have argued, 'the electronic media have thoroughly undermined our distinctive sense of place' (1992, p. 9). Whenever you want to escape, your device connects you back. As a result, the experience of 'being alone' is constantly challenged by the new technological environment.

It is also noteworthy that more and more researchers begin to notice the impact of 'fear of missing out', or FOMO, resulted from compulsive checking for status updates and messages via smartphone (Dossey, 2014; Hetz, Dawson, & Cullen, 2015; Hodkinson, 2016). FOMO is a social construct that people are concerned that they are missing out on experiences that others are having (Hetz et al., 2015). It may further cause social anxiety because people regret they make a wrong decision on how to spend time as they imagine there is another better choice. In the case of travelling, constantly logging in social media not only aims to post new travel photos but also aims to check updates from others. Although backpackers did not express their FOMO directly in this study, it is necessary to further examine whether there is extensive use of smartphone related to FOMO and if the performance of happiness on social media authentic or contrived.

Conclusions

The central theme of this book is to explore the interrelationship between media and tourism by introducing theoretical concepts such as convergence (Jenkins, 2008). The story of Chinese backpackers told in this chapter shows that media can both refer to the content (text and photo shared online) and the device (smartphone). Therefore, media convergence not only signifies the blurred boundary between media texts and tourism but also indicates how tourists' behaviour and experience are shaped by the media they used. As media device like smartphone connects people, objects, information, place in a new way through high speed mobile Internet, tourism is existed as transmedia in nature. There are three main findings:

First, smartphone is perceived as the invisible backpack by Chinese backpackers as it not only provides the sense of security but also serves as a tool to build personal identity and sustain social status. Smartphone, supported by mobile Internet and wireless Internet, enables backpackers to record and share travel experience instantaneously.

Second, backpackers are empowered by the advanced technology as the use of smartphones and mobile Internet makes independent travel more convenient for them, and backpackers with underprivileged background are able to travel individually more easily.

Third, the new technological environment may create very individualistic travellers who rarely interact with fellow travellers or the locals. Also, the continual connection with friends, family, and colleagues via smartphone frequently distracts backpackers from exploring the destination and from exploring the inner self. As a result, the classic travel culture of backpacker is changed due to media convergence.

Notes

1 It is a Chinese multi-purpose messaging, social media, and mobile payment app developed by Tencent.
2 'Moments' is WeChat's brand name for its social feed of friends updates, and it allows users to post images, text, comments and to share music and articles.
3 As the Chinese government has never admitted that they blocked certain websites, they cannot demonstrate the VPN service which helps getting access to foreign websites is illegal. Otherwise, the demonstrations of the government would be self-contradictory. In China, one could pay extra money to get the VPN service, as the owner of the youth hostel did.
4 It is a Chinese social networking platform that is used for all manner of meetups. The Chinese name for the platform is 陌陌.
5 Both of Ctrip and Dianping.com are famous travel websites in China, like Booking and TripAdvisor.

References

Bauman, Z. (1988). *Freedom*. Milton Keynes, UK: Open University Press.
Chen, G., Bao, J., & Huang, S. (2014). Developing a scale to measure backpackers' personal development. *Journal of Travel Research, 53*(4), 522–536.
CNNIC. (2016) *Statistical Report on Internet Development in China*. China Internet Network Information Centre (CNNIC), Beijing, China.

Cohen, E. (2003). Backpacking: Diversity and change. *Journal of Tourism and Cultural Change, 1*(2), 95–110.

Cohen, E., & Cohen, S. A. (2012). Current sociological theories and issues in tourism. *Annals of Tourism Research, 39*(4), 2177–2202.

Cohen, S. A. (2011). Lifestyle travellers. *Annals of Tourism Research, 38*(4), 1535–1555.

Cohen, S., & Taylor, L. (1992). *Escape attempts* (2nd Ed.). London, UK: Routledge.

Cottrell, C. (2014). Young Chinese backpackers hit the road. Retrieved November 10, 2016, from: https://www.theguardian.com/travel/2014/oct/11/young-chinese-backpackers-hit-the-road.

Davis, M. (2008). *Freedom and consumerism: A critique of Zygmunt Bauman's sociology.* Farnham, Surrey: Ashgate Publishing Limited.

Dossey, L. (2014). FOMO, digital dementia, and our dangerous experiment. *Explore, 10*(2), 69–73.

Edensor, T. (2001). Performing tourism, staging tourism: (Re)producing tourist space and practice. *Tourist Studies, 1*(1), 59–81.

Elsrud, T. (2001). Risk creation in traveling. *Annals of Tourism Research, 28*(3), 597–617.

Germann Molz, J., & Paris, C. M. (2015). The social affordances of flashpacking: Exploring the mobility nexus of travel and communication. *Mobilities, 10*(2), 173–192.

Graburn, N. (1983). The anthropology of tourism. *Annals of Tourism Research, 10*(1), 9–33.

Hetz, P. R., Dawson, C. L., & Cullen, T. A. (2015). Social media use and the fear of missing out (FoMO) while studying abroad. *Journal of Research on Technology in Education, 47*(4), 259–272.

Hodkinson, C. (2016). 'Fear of Missing Out' (FOMO) marketing appeals: A conceptual model. *Journal of Marketing Communications, 25*(1), 1–24.

Jenkins, H. (2008). *Convergence culture: Where old and new media collide.* New York: New York University Press.

Kirillova, K., & Wang, D. (2016). Smartphone (dis)connectedness and vacation recovery. *Annals of Tourism Research, 61*, 157–169.

Kristensen, A. E. (2013). Travel and social media in China: From transit hubs to stardom. *Tourism Planning & Development, 10*(2), 169–177.

Lamsfus, C., Wang, D., Alzua-Sorzabal, A., & Xiang, Z. (2015). Going mobile: Defining context for on-the-go travelers. *Journal of Travel Research, 54*(6), 691–701.

Lian, H. (2014). The post-1980s generation in China: Exploring its theoretical underpinning. *Journal of Youth Studies, 17*(7), 965–981.

Loker-Murphy, L., & Pearce, P. L. (1995). Young budget travelers: Backpackers in Australia. *Annals of Tourism Research, 22*(4), 819–843.

Luo, X., Huang, S., (Sam) & Brown, G. (2014). Backpacking in China: A netnographic analysis of donkey friends' travel behaviour. *Journal of China Tourism Research, 11*(1),67–84.

Månsson, M. (2011). Mediatized tourism. *Annals of Tourism Research, 38*(4), 1634–1652.

Mo, H. (2016). Budget backpacking in vogue among young Chinese travellers. Retrieved November 10, 2016, from: http://www.ecns.cn/cns-wire/2016/03-09/202190.shtml.

Ngai, P., & Lu, H. (2010). Unfinished proletarianization: Self, anger, and class action among the second generation of peasant-workers in present-day China. *Modern China, 36*(5), 493–519.

O'Reilly, C. C. (2006). From drifter to gap year tourist. *Annals of Tourism Research, 33*(4), 998–1017.

Paris, C. M. (2010). *Understanding the virtualization of the backpacker culture and the emergence of the flashpacker: A mixed-method approach.* Tempe, Arizona State University.

Paris, C. M. (2012). Flashpackers: An emerging sub-culture? *Annals of Tourism Research, 39*(2), 1094–1115.

Richards, G. (2015). The new global nomads: Youth travel in a globalizing world. *Tourism Recreation Research, 40*(3), 340–352.

Wang, D., Xiang, Z., & Fesenmaier, D. R. (2014). Adapting to the mobile world: A model of smartphone use. *Annals of Tourism Research, 48*, 11–26.

Xiang, Y. (2013). The characteristics of independent Chinese outbound tourists. *Tourism Planning & Development, 10*(2), 134–148.

Yu, Z. (2012). [Research on backpackers' self-realization through tourist experiences]成己之路:背包旅游者旅游体验研究. Dalian, China: Dongbei University of Finance.

Zhang, J., Morrison, A. M., Tucker, H., & Wu, B. (2017). Am I a backpacker? Factors indicating the social identity of Chinese backpackers. *Journal of Travel Research, 57*(4), 425–439.

Zhang, N. (2008). *Donkey friends: Travel, voluntary associations and the new public sphere in contemporary urban China.* Pennsylvania: University of Pittsburgh.

Zhu, X. (2007). [Backpacker tourism: Theoretical and empirical study based on China]背包旅游：基于中国案例的理论和实证研究. East China Normal University, Shanghai.

AFTERWORD

44

AFTERWORD – PARTICIPATORY PLACEMAKERS

Socio-spatial orderings along the nexus of tourism and media

Szilvia Gyimóthy

Introduction

On a sunny September evening 2019, I was taking selfies with the lead editor of this Companion at one of the most frequently shared locations in Visby on digital media. This small Swedish town is renowned for its UNESCO listed, fortified city centre and Hanseatic connections across the Baltic Sea. Die-hard fans of Astrid Lindgren would also recall Visby as the home of Pippi Longstocking, the most beloved character of her children's stories. Despite the abundance of both historical and popcultural points of interest, the most 'instagrammable' local attraction in town is of far more recent origin – and it was the exact selfie-sharing location of our sunset stroll. This is *# I Love Visby*, a meta-referential art installation, which was erected on the 28th June 2019 with the aim to facilitate a viral spread of local pride on social media. The artwork was crowdfunded by residents and businesses of Gotland (Visby Centrum, 2019), and it was envisaged that digitally active visitors would act as part-time marketers, assisting the local Destination Management Organization to direct attention, awareness, and eventually visitor flows towards this quaint Baltic town. As such, it is a topical illustration of converging tourism and media consumption and a fitting point of departure for the afterword of this Companion (Figure 44.1).

Brandscapes and hashtag geographies

The art installation is a minimalistic sculpture, consisting of a two-meter tall hashtag sign and five letters (*# I Love Visby)* on a portable wooden pedestal measuring 3 × 6 meters. Its lacks in sightseeing qualities and references to Gotland's cultural history are compensated by its sight-creating capabilities. Without the hashtag and the verb, it could be taken for a simple signpost marking a geographical locality from a distance. But the placement of the artwork in the heart (rather than the edge) of Visby is a visual provocation begging for attention and action. Its shiny chrome and rusty palette stand in sharp contrast with the rest of the medieval buildings, and changes the affective vibrancy of the sleepy historical town. It is an aesthetically framed, physical manifestation of the destination as a *brandscape –* and resonates with similar, purpose-built pop-artworks in international tourism destinations. The meta-referential connection to the letter sculptures of HOLLYWOOD and

Figure 44.1 Participatory placemakers caught in the act of putting Visby on the map
Source: Szilvia Gyimóthy.

IAMSTERDAM suggests that the initiators aspired to create a photogenic 'brand touch-point' (Botschen, Promberger, & Bernhart, 2017; Davis & Dunn, 2002) in the pursuit of international tourist appeal. From this perspective, # I Love Visby could be mistaken for an empty signifier; being nothing more than a purely representational display or brand label.

But # *I love Visby* is more than just a fashionable epithet, which boldly imitates well known, iconic installations in global destinations. I would contend that it is a powerful prop of placemaking practices and meaning-making, an intentionally designed device to assist visitors in documenting, sharing, and geo-tagging unique experiences. By 'marking the spot' for taking pictures; the artwork serves as an instructive and manipulative stage for hegemonic touristic performances. It incites visitors to engage in the conventional documentary practice of taking photos at points of interests, but also to conduct playful performative experiments, and to personalize visual accounts of their Gotland visit. Consumer culture scholars would interpret these acts as conspicuous and staged expressions of identity positioning, which also affect the creation and circulation of place meanings. The view on tourism as a media-induced, culturally embedded phenomenon existing in the interplay between places and stories is well established (Goodman, 2007; Tzanelli, 2018), but how can we build on and move beyond those research perspectives?

From consumer culture manifests to shaping of places

I would argue that a solely discursive approach allows little room to engage with concrete, micro-level practices at a deeper level. It may also impede more fine-grained explorations of how apparently insignificant manifests of mediatized tourism lead to long-term socio-spatial transformations. Thus, instead of asking what the interactive letter sculpture *means* on a networked global marketplace, let us reflect upon what it *does* to the space framed by Visby's town square. Soon after the artwork's inauguration during the Swedish political summit *Almedalsveckan*, passers-by would engage with it in trivial ways, using the pedestal for an improvised repose, or as place to wait, meet, or eat. It opened op a new platform of socializing and mooring in public space. Occasionally, it would stimulate creative choreographies and new, elevated positions to take pictures for social media. Some visitors apparently went so far that the curators of the artwork felt necessary to discipline inappropriate behaviour and put a sign on to prohibit climbing on the letters. Seen from this perspective, *# I Love Visby* is a tactical, material intervention that affects the spatial order and pedestrian flow: it multiplies individuals' opportunities to use, engage, dwell, and view the cityscape. It emphasizes the emergent *communicative textures* characterizing the spaces of tourism. A textural approach (Adams & Jansson, 2012; Jansson, 2007) allows capturing patch-worked entanglements of representational, performative, and imaginary spatial dimensions in which diverse mobility patterns create 'scraps of interwoven communicative threads' (ibid., p. 308). As the production site of the new destination brandscape, the artwork is both shaped by and shaping tourism performances, and ideologically challenges the notion of what may be considered a 'proper' attraction or a 'correct' way of sightseeing. This banal illustration excellently showcases the socio-spatial (re/dis)ordering power of entwined tourism and media practices. Tourists are part-time, participant placemakers, who intersect public planning and strategic communicative endeavours through unruly appropriations of destinations.

The acknowledgement of tourists' active role in not just replicating but constructing the visual discourse of places is aptly elaborated by Garner (2020) in the chapter 'I came, I saw, I selfied', which stresses the growing supremacy of the instagramming public as location creators. But what is the nitty-gritty of intertwined popular culture imaginaries and individual's shared stories? Several contributions in this Companion go beyond descriptive or critical readings of mediatized tourism and work towards a robust conceptualization of participatory placemaking. I will in the remainder use illustrations from these chapters to sketch out three distinct processes of participatory placemaking: Place-wrecking, Place-assembling, and Place-enhancing. In the concluding remarks, I will reflect on the development potential of these concepts.

Place-wrecking: the reproduction of exclusion and disquiet

This pioneering Companion takes up the challenge to explore the socio-spatial transformations of participatory placemaking along different subsets of media tourism and establishes an impressive interdisciplinary powerhouse with novel conceptual perspectives on the nexus of tourism and media. Inspired by disciplines of cultural studies, media aesthetics, experience economy, spatial anthropology, and the sociology of tourism, the authors depart from the macro-level reading of tourism as a social and discursive ordering process (Franklin, 2004). Several chapters reiterate the view on tourism driven by hierarchies and producing asymmetries, by telling stories of *objectification* (during host-guest encounters between an Ethiopian tribe and wannabee anthropologist-youtubers, by Sintobin and Tonnaer, 2020), *contested imaginaries of heritage sites* (McAdam's (2020) study of Irish monastic heritage history being replaced by franchised fiction), or the *uneven distribution of benefits* from commodified place narratives in the South Pacific (favouring more recent colonialist narratives to native ones, by Seaton and Beeton, 2020). These chapters revisit problematic power dynamics tilting in favour of resourceful, external actors in tourist encounters and content production. Both the cases from Global South and Global North contexts (providing rich analyses of the amateur documentarist footages of the Mursi, the commodification of Skellig Michael as a Star Wars fanspace and Hanging Rock's aboriginal sense of place sinking into oblivion) demonstrate that seemingly innocent, co-constructed, and user-generated placemaking is not necessarily democratic, equitable, or inclusive for all parts.

The staging of tourist performances never happens on neutral grounds but always involves a symbolic takeover – *stage-taking* – that may marginalize local people, bypass historical recollections, or overlook indigenous narratives. Place-wrecking emphasizes participant placemaking along the notions of a predatory (or at best, ignorant) gaze and confrontational dyads, which raises deep concerns about privileged outsider positions above locals (in which the former's story is being told, while the latter's voices and spatial rights will be silenced).

However, the battlefield metaphor is analytically problematic, as it is often based on crude dichotomies that are not necessarily valid in a de-differentiated tourism context. First, the notion of a sedentary population intact from external influences is, in many cases, an illusion. As the boundaries between hosts-guests, locals-outsiders, and residents-visitors have become increasingly blurred in a mobile world, the quest to locate the 'rightful' representatives and 'original' narratives of a place will be increasingly complex (Broegaard, Larsen, & Larsen, 2019). Global migration and tourism flows have always contributed with new stories and hybrid cultural manifests; hence, the conservation (or radical particularization) of an authentic place DNA is not possible. Second, subaltern communities are neither passive, nor defenceless in a mediatized world – as we could follow locals' dissent against unflattering representations of Melania Trump's hometown in American media (Turnšek, Trdina, & Pavlakovič, 2020). Participatory media provides new platforms to let subversive voices to be heard. As Sintobin and Tonnaer's documentation of crowdshaming of unscrupulous stage-taking shows, social media audiences are self-regulative and quick to respond to perceived injustice. Apart from drawing awareness to place-wrecking disruptions, future research could equally explore new communicative and texturation practices arising from such conflicts. To some extent, the confrontational lens may hinder the understanding of participatory place-making practices, where collaborative and community-enhancing processes are in focus.

Place-assembling: the production of convergence platforms

Let us turn our attention to the contributions that deal with instances of social ordering processes that are collaborative, participatory, and inclusive. While place-wrecking uniquely focused on dialectic performances turning the spaces of tourism into ideological battlefields, *place-assembling* refers to the construction of platforms which manifest identity, belonging, and community. Contributions illustrating place-assembling in this Companion are inspired by fandom studies as well as community studies, which are preoccupied with both the social dynamics of collectives and the spaces of community convergence. Enthusiastic fans, consumer tribes, and other ephemeral or lasting communities have a tendency to create their own space to meet, exchange ideas, and engage with their particular pursuit or interests, should that be football, cosplay, or cult TV-series. These convergence spaces may be physical events and gatherings (Graakjær & Grøn, 2020) but also spaces that are purposefully created by these groups. For instance, a finely crafted account of the Olsen Gang Fan Club (Mortensen, 2020) reveals how a precious site of popular cultural memory [lieux de mémoire] arose, when the fan club initiated the preservation of the Yellow Mansion, an iconic railway building from the series for the benefit of future generations. Hence, we are witnessing a new wave of heritagization processes, which is driven by bottom-up participatory cultures, independent of authoritative cultural institutions.

Place-assembling may also take place in virtual and imaginary domains, such as the case of the fan group of *Der Bergdoktor* (tv-series from Austria). Trdina and her colleagues (2020) observe that feelings of embarrassment and intimidation by highbrow peers made Slovenian followers of *Der Bergdoktor* to keep their interest to themselves and only expose it in intimate settings with other fans. Apparent distinctions in popular cultural taste have not only produced enduring hierarchies among cultural snobs and the uncomfortable aficionados of TV-melodramas but also triggered the assembly of clandestine spaces of consumption. Tourism (in this case, the organized fan trips to Bergdoktor-locations in Tirol) may provide a temporary protective shelter, where the 'underdogs' of popular culture can evade amused or condemning comments and live out their fascination with the bucolic Alpine stories.

Place-enhancing: affective and sensuous intensification

Mobility, interactivity as well as cultural convergence have rendered places into market commodities, so that place meanings are constantly rearticulated and recirculated by participatory placemaking. The previous sections addressed relevant discussions on the social dynamics (structures, positions, and hierarchies) of participatory placemaking, where the ordering of physical or imaginary spaces reflects actual or shifting power balance within or between actor groups. Researchers' interest for space was primarily expressed in a proprietary sense, demarcating it either as 'ours' (i.e. the residents' or the community's space) or 'someone else's' (i.e. the native Other's space, the claimed/occupied or commodified space). This preoccupation raises issues of place rights and governance, which can, for instance, determine the boundaries of legislative or symbolic control, each being topical for social science enquiries of contemporary media tourism. However, the study of participatory placemaking may be further enriched with experiential, non-representational and phenomenological perspectives.

Recently, tourism scholars have started to consider people's sensuous and corporeal engagement with the places visited (Tzanelli, 2018), as well as with the haptic (Waade & Jørgensen, 2010) or auditory (Bolderman, 2018) aspects of 'location creation'. Surprisingly, atmospheric dimensions received relatively limited attention in this Companion. For instance, even when exploring immersive environments, such as theme parks (Mittermeier, 2020), the focus stays on crossmediation and meaning circulation processes, rather than on the sensory and affective intensification of places. A fascinating exception is Graakjær and Grøn's (2020) chapter, who explains how the soundscape of televised football soundscapes creates a very special atmosphere of 'liveness'. Their analysis not only confirms the importance of visual and sonic dimensions during in-stadium experiences, but also that mediatized football events can (to some extent) reproduce sensations of oneness with the cheering masses. Rhythmical crowd chanting evokes a sense of 'being there', also for spectators following the match on screens at a distance. Paradoxically, experiencing football matches through telepresence may become more real, more immersive and thrilling than attending the game in person, as home environments or sports bars are free from negative ambient conditions or security restrictions of contemporary stadiums.

The topic of place-enhancement and affective and sensuous intensification may open up empirically rich and conceptually uncharted research opportunities for media tourism scholars. For instance, more studies are required to entirely understand texturation (Jansson, 2007) and processes through which tourist performances are creating new destination textures (Gyimóthy, 2018). With the rise of overtourism and congestion in urban destination contexts, there is a need to nuance and problematize the effect of crowds. While ample literature deals with positive tourism and the design of memorable experiences (Filep, Laing, & Csikszentmihalyi, 2016), we know very little about mediatized anxiety, apathy, stress, rage, and boredom – and how do these produce intensities in spaces already dense with sensory impressions. Finally, we need in-depth explorations on the sensory aspects of telepresence as well as into the characterization of mediatized and manipulated affect in different contexts. How do collective moods and sentiments travel through digital platforms?

Concluding remarks

These lines of enquiry may not only enrich interdisciplinary scholarship but also provide valuable insights for practitioners. We are living in precarious times, where global mobility and binge consumerism put local communities, resources, and markets under immense pressure. Therefore, there is a need to insource and engage with participant-placemakers in tactical and strategic ways to improve (rather than deteriorate) livehoods and atmospheres. This requires a shift where tourists' and other temporary citizens' privilege of passivity is extended with a request to contribute to the places visited in more resourceful and imaginative ways.

References

Adams, P. C., & Jansson, A. (2012). Communication geography: A bridge between disciplines. *Communication Theory, 22*(3), 299–318.

Bolderman, L. (2018). *Musical Topophilia: A critical analysis of contemporary music tourism.* Retrieved from: https://core.ac.uk/download/pdf/154410665.pdf.

Botschen, G., Promberger, K., & Bernhart, J. (2017). Brand-driven identity development of places. *Journal of Place Management and Development, 10(2), 152–172.*

Broegaard, R. B., Larsen, K. T., & Larsen, L. H. (2019) Translocal communities and their implications for place branding. In C. Cassinger, A. Lucarelli, & S. Gyimóthy (Eds.) *The Nordic wave in place branding: Poetics, practices, politics (pp. 109–123). Cheltenham, UK: Edward Elgar.*

Davis, S. M., & Dunn, M. (2002). *Building the brand-driven business: Operationalize your brand to drive profitable growth.* San Francisco, CA: Jossey-Bass.

Filep, S., Laing, J., & Csikszentmihalyi, M. (2016). *Positive tourism.* Abingdon, UK: Routledge.

Franklin, A. (2004). Tourism as an ordering: Towards a new ontology of tourism. *Tourist Studies, 4(3), 277–301.*

Garner, A. O. (2020). I came, I saw, I selfied: Travelling in the age of Instagram. In M. Månsson, C. Cassinger, L. Eskilsson, & A. Buchmann (Eds.), *The Routledge Companion of media and tourism* (Chapter 33). Abingdon, UK: Routledge.

Goodman, E. (2007). *Destination services: Tourist media and networked places.* UC Berkeley: School of Information. Retrieved from: https://escholarship.org/uc/item/0919c6sv.

Graakjær, N. J., & Grøn, R. (2020). Football tourism and the sounds of televised matches. In M. Månsson, C. Cassinger, L. Eskilsson, & A. Buchmann (Eds.), *The Routledge Companion of media and tourism* (Chapter 10). Abingdon, UK: Routledge.

Gyimóthy, S. (2018). Transformations in destination texture: Curry and Bollywood romance in the Swiss Alps. *Tourist Studies, 18*(3), 292–314.

Jansson, A. (2007). Texture. *European Journal of Cultural Studies, 10*(2), 185–202.

McAdam, A. (2020). Star gazing: The nexus and disparity between the media, tourism and cultural heritage in Ireland. In M. Månsson, C. Cassinger, L. Eskilsson, & A. Buchmann (Eds.), *The Routledge Companion of media and tourism* (Chapter 27). Abingdon, UK: Routledge.

Mittermeier, S. (2020). Theme parks – Where media and tourism converge. In M. Månsson, C. Cassinger, L. Eskilsson, & A. Buchmann (Eds.), *The Routledge Companion of media and tourism* (Chapter 4). Abingdon, UK: Routledge.

Mortensen, C. H. (2020). Commemorating popular media heritage: From shrines of fandom to sites of memory. In M. Månsson, C. Cassinger, L. Eskilsson, & A. Buchmann (Eds.), *The Routledge Companion of media and tourism* (Chapter 28). Abingdon, UK: Routledge.

Seaton, P., & Beeton, S. (2020). Rewriting history, revitalizing heritage: Heritage-based contents tourism in the Asia-Pacific region. In M. Månsson, C. Cassinger, L. Eskilsson, & A. Buchmann (Eds.). *The Routledge Companion of media and tourism* (Chapter 21). Abingdon, UK: Routledge.

Sintobin, T., & Tonnaer, A. (2020). Tourists' filmic representations on YouTube: A case study analysis of two mediatised visits to the Mursi in Ethiopia. In M. Månsson, C. Cassinger, L. Eskilsson, & A. Buchmann (Eds.), *The Routledge Companion of media and tourism* (Chapter 26). Abingdon, UK: Routledge.

Trdina, A., Pavlakovič, B., & Turnšek, M. (2020). Cultural intimacy of fans/travellers: Popular culture and the politics of classification. In M. Månsson, C. Cassinger, L. Eskilsson, & A. Buchmann (Eds.), *The Routledge Companion of media and tourism* (Chapter 35). Abingdon, UK: Routledge.

Turnšek, M., Trdina, A., & Pavlakovič, B. (2020). What do Melania Trump tourism and Dracula tourism have in common? 'Othering' in the Western media discourse. In M. Månsson, C. Cassinger, L. Eskilsson, & A. Buchmann (Eds.), *The Routledge Companion of media and tourism* (Chapter 6). Abingdon, UK: Routledge.

Tzanelli, R. (2018). *Cinematic tourist mobilities and the plight of development: On atmospheres, affects, and environments. Abingdon, UK: Routledge.*

Visby Centrum (2019). #ILOVEVISBY. Retrieved from: http://visbycentrum.se/ilovevisby.

Waade, A. M., & Jørgensen, U. A. (2010). Haptic routes and digestive destinations in cooking series: Images of food and place in Keith Floyd and The Hairy Bikers in relation to art history. *Journal of Tourism and Cultural Change, 8(1–2), 84–100.*

INDEX